T5-CQZ-617

MARINE ECOLOGY AND OIL POLLUTION

The Proceedings of an Institute of Petroleum/Field Studies Council meeting, 'Marine Ecology and Oil Pollution', held at Aviemore, Scotland, 21–23 April 1975

The office of the Oil Pollution Research Unit, Orielton Field Centre, Pembroke, Wales.

MARINE ECOLOGY AND OIL POLLUTION

Edited by

JENIFER M. BAKER

Field Studies Council

A HALSTED PRESS BOOK

JOHN WILEY & SONS
NEW YORK—TORONTO

PUBLISHED IN THE U.S.A. AND CANADA BY
HALSTED PRESS
A DIVISION OF JOHN WILEY & SONS, INC., NEW YORK

Library of Congress Catalog Card Number: 75–41514

ISBN: 0 470–04541–8

WITH 168 ILLUSTRATIONS AND 72 TABLES

Printed in Great Britain by Galliard (Printers) Ltd Great Yarmouth

Contents

SAND AND MUD FAUNA AND THE EFFECTS OF OIL POLLUTION AND CLEANSING

REFINERY EFFLUENTS AND THEIR ECOLOGICAL EFFECTS

TOXICITY TESTING OF OILS, EFFLUENTS AND DISPERSANTS

OFFSHORE MONITORING AND WORLD OIL SPILLAGES

Opening Address

J. D. DEWHURST
(*President, Institute of Petroleum*)

May I welcome you to this Conference and thank you all for coming, particularly those who have come from great distances. It might interest you to know that about a third of us come from the oil industry, including many experts in the field; about a quarter of us come from various conservation groups; about 15% of us come from abroad from some eight countries, and I extend a particular welcome to all of you. We have about ten from universities, another ten from Government departments—not to forget one member of the Royal Navy. Quite a good cross-section, I think, to provide us with an interesting few days.

Now I am here to talk about environmental responsibility, but I do not think that this is really something I should be talking about because I think the oil industry as such is perfectly responsible about these matters already, without discussing it any further. Indeed, it has already written the Eleventh Commandment—'Thou shalt not pollute'.

What we are really here to discuss is the fact that there are certain problems related to the whole of this subject. One of them is, of course, that a conservationist probably lurks in everyone's breast, but because of human nature it affects different people in different ways. The man who does not care very much about whether there is oil on the beaches is the man who will blow his top about the fact that there may be soot on his yacht or that his propeller is jammed with plastic. Consequently there has been a proliferation of associations and organisations dealing with a number of aspects of the problem and a conference such as this enables us all to get together to see the other person's point of view and perspective.

The next point is that what we are really here for as much as anything is to get more facts; I do not think anyone who has even glanced through the printed papers can feel they are short of those. The facts lead to the setting of standards and the standards lead to codes and the codes can even lead to legislation. So it is extremely important that we get these facts straight at the beginning.

So, ladies and gentlemen, may I again welcome you all here. I will just read out the real purpose of this which is printed in our programmes: that the Institute of Petroleum and the Field Studies Council are jointly sponsoring this Conference which is dealing as much as anything with the work of the Field Studies Council's Oil Pollution Research Unit at Orielton, and I think anyone who has read all the papers must be extremely impressed at the work that they have done.

Environmental Responsibility

1
Environmental Responsibility: Industry

H. JAGGER
(*Esso Petroleum Co. Ltd.*)

The very fact that this meeting on the work of the Orielton Oil Pollution Research Unit is taking place under the auspices of the Institute of Petroleum is evidence of both interest in and concern for the environment by people engaged in the oil industry. Since the Institute is also heavily dependent on support from member companies both in terms of financial contributions and company time, it is evidence of oil company support for such interest. Furthermore, a significant part of the work to be reported during this Conference has been funded under an Institute research project financed jointly by several oil companies.

British industry got off to a bad start in the early days of our coal-based industrial revolution. There are still many monuments to this in the form of slag heaps, back-to-back housing and derelict areas. Smoke emissions especially from domestic consumers were essentially uncontrolled until the Public Health Act of 1936 and the Clean Air Act of 1956. But there is evidence of concern long before those dates. Increasing population and the need to improve river quality resulted in a Royal Commission making recommendations for the quality of sewage discharges in 1912—standards which are not unreasonable in many situations today. The fact that the report of the Alkali Inspector for 1974 will be the 111th Annual Report also demonstrates some long-standing national concern for air pollution prevention.

The oil industry in Britain is relatively new. As recently as 1938 oil consumption was only 9 million tons per annum, of which less than one-quarter was refined here. By 1973 consumption had increased to 105 million tons and refinery capacity exceeded 140 million tons. There have clearly been adequate opportunities for random developments with accompanying widespread environmental damage. In fact, the combination of planning controls, environmental legislation and a responsible approach by industry has resulted in a situation which in large measure has minimised the environmental damage despite the very rapid growth.

The first responsibility of any industry is to meet the legal requirements placed on it. To the extent that industry can be concerned in setting those requirements, the easier compliance will be in most cases—not because the objectives are changed but because the objectives are met in a more economical manner. The oil industry has therefore concerned itself with establishing, together with Governments and other regulatory bodies, desirable and attainable environmental standards. It has also sought to establish the best way of achieving those standards at least cost to the industry—and therefore to the community.

Industry has frequently taken initiatives to reduce environmental effects ahead of legislation and to establish criteria before either public opinion or regulations have required such standards. Action taken by the UK oil industry through the Oil Companies Materials Association (OCMA), in drawing up standards for the limitation of noise in plant and equipment, preceded legislation on this subject and, incidentally, these standards were in advance of any other studies on this subject and have been widely adopted in other countries.

In the oil pollution field, the oil industry has frequently taken initiatives both to reduce the incidence of pollution and to minimise the damaging effects. The voluntary limitation of oil discharges through 'load on top' procedures for sea-going tankers and the active steps taken by oil companies to have these procedures adopted as part of the IMCO Convention amendments are but one example. The adoption of elaborate spill prevention procedures during the loading and discharge of vessels at marine terminals and the setting-up of co-operative schemes such as the Milford Haven oil spill clean-up organisation are further examples of oil company concern.

More recently the development of offshore oil exploration in Scottish waters has highlighted the many developments in oil spill prevention and other safeguards to reduce the chances of disaster from these operations. These range from sleeve exploders to minimise damage to fish during seismic survey work, to elaborate shut-off devices fitted to producing wells and submarine pipelines.

In refineries and oil storage installations, care is taken to minimise oil discharges in liquid effluents including all drainage water which may become contaminated by oil. Atmospheric emissions are frequently controlled beyond the requirements of the Alkali Inspectorate, and while refineries are large fuel consumers, much of the fuel requirement is met by burning very low sulphur fuel gas. The sulphur content of diesel fuels and heating oils for domestic and commercial users is reduced by hydro-desulphurisation so that emissions from user sources are not excessive even in urban areas.

I have mentioned a few of the steps taken by the oil industry to reduce environmental pollution, but what of monitoring itself?

Initially, oil companies were concerned only with the testing of their effluents to ensure that these complied with the requirements placed on them by the responsible river or harbour authority. Such 'testing' frequently consisted of periodical examination of the receiving water to ensure that no visible oil film was formed. Many river authorities preferred the 'no visible oil' limit since it simplified the task of the river inspector as compared with the sampling and testing required to confirm compliance with some numerically expressed criterion.

As refinery capacity expanded and such operations as catalytic cracking and petrochemical processes increased the risk of phenolic and other pollutants, greater awareness developed. The desirability of having baseline data on sulphur dioxide in the Fawley area was recognised by my company before construction on the new refinery started in 1949. Monitoring by ground-level concentration measurement and by observation of possible vegetation damage has continued.

On the liquid effluent/oil spill side, although volumes of liquid effluent had been reduced by an order of magnitude with the change by 1960 from water cooling to air cooling for new process equipment, the development of Milford Haven as an oil port and refinery centre has been accompanied by marine monitoring on a scale not previously undertaken in the UK for a refinery location. The careful control and reporting of all oil spills and clean-up operations by the Milford Haven Conservancy has facilitated the correlation of spill frequency with ecological consequences.

This monitoring work has had financial assistance from some small funds available in the Institute of Petroleum in addition to more substantial support from several oil companies through an Institute-coordinated research project. The monitoring techniques developed at Orielton in connection with their monitoring in the Haven have been applied to some other oil-refining areas in close co-operation with, and in some cases with financial support from, the refining companies.

At least one oil company has gone so far as to establish a measure of experimental monitoring capability at all its refineries. While this is more than most companies would claim, the level of public awareness and of legislative control across Europe is such that each refinery must have within it, or readily available to it, people with the ability to recognise sensitive environmental areas and recommend steps to minimise disturbance to that environment. Liquid effluent control, odour assessment, and pollution measurement and noise monitoring are the stock-in-trade of such people. Monitoring in the sense of observing or measuring the effects as opposed to the level of pollution is less commonly carried out by industry—except to the extent that effects such as vegetation damage either from atmospheric pollution or from chronic oil pollution are used as indicators of unacceptable pollution levels.

In addition to monitoring the levels of pollution from its own operations, the oil industry has in recent years financed many university and similar studies and has itself been very active in studies relating to the quality of its products and the effects of the increasing use of those products. Examples of the former include studentships at universities to study the fate and effect of oil in estuaries—work which is co-ordinated with the MAFF studies of UK coastal waters. There have been extensive studies of the oil levels in Lake Maracaibo, off the coast of California and along major tanker routes of the oceans. These have confirmed that even in areas where oil has been produced offshore for up to 50 years—and despite operating procedures which may have been less carefully controlled than under present conditions of environmental awareness—there is no significant accumulation of oil. Petroleum oil contents of sea water are of the same order of magnitude as those of non-petroleum oils—each around 5–10 ppb.

Of less interest in the context of this Conference, the work on atmospheric monitoring has centred largely around sulphur dioxide and the more emotive pollutant lead. Major oil companies have carried out surveys of SO_2 and other pollutants in major cities to supplement National Survey data in attempts to assess the effects of and justifications for changing fuel specifications or fuel use recommendations. Equally, major gasoline manufacturers have been heavily involved in measuring automotive emissions—carbon monoxide, hydrocarbons and nitrogen oxides—and in methods for reducing these emissions. An IP research project has investigated human uptakes of lead from gasoline engine exhausts.

The Institute of Petroleum's atmospheric conservation committee has recently initiated a detailed investigation of the long-range transportation of sulphur dioxide under a research project financed by a group of oil companies.

There is no available tabulation of UK industry-sponsored environmental activities comparable to that of Government and Government-supported schemes listed in the report on monitoring by the Central Unit on Environmental Pollution published in 1974. I have attempted in this brief survey to indicate that the oil industry is very much concerned in both environmental conservation and in monitoring to establish the extent to which pollutants resulting from its operations and from the use of its products persist in the environment. In so doing it seeks to be better equipped to participate constructively in the establishment of meaningful environmental standards and to carry out its own operations with a minimum of disturbance to the natural environment, whilst still meeting the massive demands that are made on it to provide fuels and other products. At the same time, the industry is concerned to avoid unnecessary restrictions, to utilise the natural absorption capacity of the environment and certainly to prevent situations in which the removal of some last trace of a pollutant is achieved only by the use of equipment and materials the manufacture of which causes more pollution than they eventually remove.

2
Environmental Responsibility: Education and Research

C. A. SINKER
(Director, Field Studies Council)

I wish to start, as the Director of the Field Studies Council who are co-sponsors of this Conference, by expressing our collective gratitude to the Institute of Petroleum and its staff for the major role that they have played in planning and organising the Conference. Without their help we could not have done this ourselves, and it gives us a very much valued forum for presenting the work that our staff have been doing and opening it up for discussion and critical review by other interested people.

After the reception yesterday evening, the dinner and our discussions subsequently, I began to sense in the atmosphere of the Conference a collective interest in environmental problems which totally altered anything that I might have said before, and so I start now from a new viewpoint.

It seems to me that we do have a very large measure of common ground, of common concern for the environment. I would doubt whether there is anyone here who is not deeply concerned about his environment, the things that we are doing to it, and the ways in which we can mitigate the more harmful effects. Any points of contention between us are sectional; they are limited to particular areas. Any points of misunderstanding are largely a matter of technical and semantic obstacles rather than fundamental differences of view, and so in many ways we can avoid the less fruitful preliminaries to dialogue and get straight down to facts and claims and to the discussion of problems on this common ground.

But as well as these advantages, there is a very considerable danger in having this common ground, and that is the danger of complacency. Henry Jagger has given us a masterly summary of the efforts made by the oil industry over the past few decades to moderate the particular effects of their activities on the environment; these achievements, the counter-claims from other industries, the statements from Government departments and from research institutions on all that they have done, could lead us to suppose that very little further was needed—that the proper attitudes are there, that the money is available, that the right things are being done.

We have heard a lot of facts already. We shall hear many more, and so I am deliberately bending the course of the Conference away from facts, because I want to break through to you at a different level of communication if I can: I shall stay within my title 'Environmental Responsibility: Education and Research', but I shall leave the facts and the claims for others more competent to deal with. Whether I succeed in reaching you, and whether I am justified in trying, will be for you to judge afterwards.

I had intended to speak in some detail about the origins and the work of the Field Studies Council, the organisation which I have served throughout my working career. I was going to tell you about its growth in the mind of a remarkable idealist, Francis Butler, a London science teacher, later an Inspector of Schools, who believed passionately in the benefits to deprived children of an opportunity to get out into the countryside and to learn about it, to gain knowledge and solace from something which contrasted in its richness with the privation of their home setting. His belief and his powers of persuasion over one or two influential people in the academic world during the early years of the war led to the foundation of the Council for the Promotion of Field Studies in 1943, and to the opening of its first residential centre at Flatford Mill in 1946. The organisation grew fast and very successfully to four Centres by the end of the 1940s, and to ten by the early 1970s; we cover a very wide range, not only of subject-matter in the environment, but also of ages and backgrounds of customers: about 70% of our intake are from senior forms of secondary schools coming on short residential courses of a week's duration to learn more about the environmental aspects of such subjects as geography, geology and biology. At the same time we take in teachers, university students and large numbers of adult amateur naturalists who wish to learn more about some particular aspect of the regions in which our Centres are located.

In more recent years we have expanded into ecological research because this enriches our teaching, and we have begun to provide courses for professions other than our own to help them to understand and use the environment better and more wisely. This new role includes courses over the last two years, and again this year, for middle and senior management in industry, introducing some of the special problems of the environment and the way the professional ecologist thinks about them and studies them. I could have described a most remarkable integrated study of the hydrology of the stream catchment of a freshwater lake in South Devon at our Slapton Ley Field Centre, together with input and output of major nutrients, turnover of the biological systems in the freshwater lake, behaviour and population levels of fish, fish parasites, otter, feral mink and a large number of other organisms which are part of this complex system; few other organisations would have the flexibility to mount such a wide-ranging (even if geographically small-scale) research programme. I could also have introduced you to the Oil Pollution Research Unit at Orielton, a research arm of the Council which has already established an international reputation. All these things I was going to tell you about in some depth, but I shall not.

The Field Studies Council occupies in some ways a unique position in the educational and research sphere in this country. It is outside the state education system, but now provides an essential service for it. We can make modest claims to having accumulated a certain amount of experience and expertise over the 30 years of our active existence, but what is 30 years? Environmental concern is not a new post-World War II phenomenon. We can find it as far back as recorded history goes and probably beyond, and we can find with that concern, with that urge to protect what is good in the environment, the built-in conflict with man's need to use, to manage, to exploit—a deliberately loaded term—his own environment.

There is an ancient Sumerian epic, predating Homer by at least a thousand years. In it Gilgamesh, hero-king of the city of Uruk, determines to kill the evil giant Humbaba, whose name means 'hugeness' and who is guardian of the vast forest of the Cedars of Lebanon. Gilgamesh takes with him his companion Enkidu, a child of the wilderness, who leads him to the forest gate:

> 'Great cedars towered against the mountain, and in their lovely shade lay comfort; mountain and valley were green with undergrowth.'

They seek out Humbaba, and together they slay him. Gilgamesh then proceeds to cut down every cedar from Mount Hermon to the Euphrates.

Millennia later Sophocles, in one of his greatest plays, has the Old Men of Colonus address the blind exile Oedipus as follows:

Εὐίππου, ξένε, τᾶσδε χώ-
ρας ἵκου τὰ κράτιστα γᾶς ἔπαυλα,
τὸν ἀργῆτα Κολωνόν, ἔνθ'
ἁ λίγεια μινύρεται
θαμίζουσα μάλιστ' ἀη-
δὼν χλωραῖς ὑπὸ βάσσαις,

> 'Stranger, this is good horse country you come to, white Colonus, the finest home on earth, where the pure-noted nightingale abounds, filling every green glade with fluted song . . ..'

and so on. Later again, in 16th-century France, the poet Ronsard says to his idle love who is lying late abed:

> 'Sus debout, allons voir l'herbelette perleuse'
> ('Get up! Come and look at the dew on the grass!')

Gerard Manley Hopkins asks in South Wales:

> 'Margaret, are you grieving over Goldengrove unleaving?'

In America, Robert Frost writes:

> 'The woods are lovely, dark and deep . . .'

These are personal environments ranging from the almost inconceivable landscapes of the prehistoric Middle East, through the bleached rocks of classical Greece and the pearly lawns of a French château, to the layered Carolinian forest of the eastern United States. They span a world in climate and landscape, they span the whole of history in their social context, and yet there runs through them all this dual feeling of concern for one's own environment and of the need to share that concern.

Great literature throughout the ages, when it is not obsessed with the tribulations of the individual and his self-concern—and sometimes even when it is—is full of the environment, the surroundings of self. So what is environment? This is really something that I think we must take far more time than we do to examine, to discuss, to explore in our own minds. What on earth do we mean by this clumsy word 'environment', or by '*the* environment'? Is it the agglomeration of facts with which we are dealing at this Conference: all the things we can see, count, measure in our surroundings? Is it our own much more subjective vision of the physical and social world in which we as individuals live?

One thing we can say quite certainly about 'environment' in purely logical and semantic terms is that without a living individual—an organism, a man, a self—there can be no environment, because the environment is the effective setting of that self. Abstract the individual and the environment is meaningless, so there is certainly a personal and subjective element in it. But there are many conservationists who would not put man, or even themselves as individuals, in that focal position. They would put the whale, or the Teesdale Violet, or whatever else happens to be their particular interest and speciality at the middle point. It is not then man's environment: it is the environment of some other organism, or of some group of organisms with which man shares it, that concerns them; so this extends in yet another dimension our concept of the environment and its range of variation.

To whose environment, then, should we give priority? Whose environment are we concerned about? For whose environment are we taking responsibility? And whose responsibility should these environments be?

A few years ago I might have been tempted to answer by suggesting that it was the environment of the expert, the person who knew about it, that we should interpret, and that the experts should be left to look after it; by the experts I suppose I would have meant the ecologists and the geographers—those detached, high-minded, superhuman, objective academics who know it so much better than anyone else. But I have met oilmen and other industrialists who actually know far more about the environment in practical ways than the great majority of the academic environmentalists whom it has been my privilege to work with: an oilman brought up on a farm in Yorkshire, who can herd cows, walk behind a horse-drawn plough team, even milk mares, slaughter pigs, dress geese! How many ecologists can claim that range of environmental experience, that capacity to support themselves on a direct relationship between man and land? I know an industrialist who is writing an ecological monograph on his own chosen and well-loved landscape, and who knows far more about the geology of the Highlands than most ecologists do.

I have also met a research biologist who happened to be the first person to introduce me to the wonders of the scanning electron microscope. It was a devastating experience for me, peering through this little porthole into a green world where he had displayed some lake bottom sediments. They included the eroded frustules of freshwater diatoms mixed up with decaying plant debris, and they lay there in that gloomy submarine light like window-frames from Atlantis in the poem by Edgar Allan Poe. I expressed my pleasure at this vision opening up a new world on a scale that I had never seen before, and my guide commented: 'I'm fed up with everyone telling me how beautiful it is. I want to get on with my job. It is not my business to see it in aesthetic terms.'

Surely the answers to questions about 'whose environment and whose responsibility' must include an element of compromise? After all, we are dealing with everyone's environment—and not only mankind's at that. We are dealing with an environment in which most of us share the benefits of a material, mechanical civilisation as well as the risks. An eminent physiologist, in the bad old days when Salvarsan and certain other Hg compounds were the only known remedy for venereal diseases, is said to have warned his students that 'One night with Venus may mean a lifetime with Mercury'.

We all owe our presence here to 'one night with Venus'. We all spend a lifetime with mercury, for there are heavy metals, there are other so-called pollutants in the environment at low, natural levels which man is not responsible for. There are meres in the Midlands which were 'eutrophicated' long before farmers came to put fertilisers on their fields, long before sewage systems were set up to release high levels of phosphate and nitrate into their waters. We live in a world which is naturally 'polluted', if that means anything at all; we are surrounded by mixtures of substances which may be harmful at levels so low that we can barely control them, or which may not be harmful at levels far above what present legislation will accept. 'Zero emission' policies are going to get us nowhere. Compromise between the practicable and the desired ideal is obviously the only course we can reasonably pursue.

Yes, it is everyone's environment we are concerned about and responsible for. It is everyone's responsibility, too; quite obviously those who have the dedication, the energy and the experience, those who can contribute extra bits of factual knowledge, further interpretations and tested hypotheses, are going to carry more weight politically and elsewhere. Decisions about priorities are far too important for the individual environmental scientist, whether he works in a university laboratory, a Government establishment, or an industrial complex, to take on his own. It is neither for the industrialist nor for the ecologist to decide whether water supply is more or less important than the unique plant communities of Upper Teesdale. It is neither for the engineer nor for the architectural historian to rule whether the risk to Salisbury Cathedral is more serious than the advantages of storing natural gas in a convenient geological structure underneath it. These must be public decisions based on the evidence and the arguments put forward by the experts from whatever quarter they come, and ultimately it will be the politicians, democratically representing the public, who determine these decisions; but we all take a part in electing them, and those of us who work in and are concerned for the environment can take a part in influencing them.

Research workers in and teachers of environmental sciences have a particular and special role to play in the provision and transmission of facts, interpretations and the balancing of judgements. When the scientist is also a teacher (and many of them are), there can be a fruitful feedback of first-hand information into what he communicates to his students from his own experience; questions from his students will generate fresh questions in his own mind and these in turn will lead him to look anew at his chosen bit of the environment. The power and the responsibility of this teacher-scientist call for a peculiar combination of qualities and attitudes: three, to my mind, are of paramount importance.

The first is *commitment*—be it simply the loyal enthusiasm of the Old Men of Colonus for their homeland, a feeling that this is the best part of the world and is worth looking after; or a well-argued dialectic, a skilfully woven and tailored garment of beliefs which our ideal teacher-scientist can wrap around his mind and his soul, fitting comfortably at every point and meeting all his observations. More often it will be something between these two, a faith of some kind, a belief in what he is doing and in the value of what he is studying.

The second attribute is *honesty*—intellectual honesty, moral honesty—which may conflict with the first. The agricultural ecologist who happens to like hedges and to think they are a beautiful part of his surroundings, that they contribute to the quality of the rural scene, may deliberately seek data and interpretations of those data to support his contention that hedges should remain a part of the landscape. It is all too easy for the ecologist to use his specialised knowledge and his technical jargon (just as the industrialist or the engineer can do) to attempt to pull the wool over the eyes of his opponents, and of the public who are laymen in these matters. It is all too easy for the defender of the hedge to point out that the hedgerow, as a diverse pocket of food-plants and microclimates, supports a wide range of insect pests whose life histories are spread over the year, and can therefore sustain a good population of wide-spectrum predators—predators who are lurking there in the hedge ready to pounce on any specialised pest of the monocultural crop in the neighbouring field; from their well-stocked stronghold in the hedge they can guard the crop against seasonal infestation, but if you remove the hedgerow the predators are no longer there to exercise this biological control. The argument is an elegant one, it can very easily be extended, it can very easily be put over—ignoring the fact that all the evidence to date suggests that this mechanism is highly effective for approximately and not much more than one and a half times the height of the hedge into the field. You would need some remarkably high hedges to protect the fields of Lincolnshire!

On another example, it can be argued that exhaust emissions from Concorde will punch holes in the tropospheric 'window' by the combination of uncombusted hydrocarbons with the ozone, and so let in more harmful ultra-violet radiation, and let out more of the infra-red and heat, thus possibly chilling the climatic circulation of the earth to a level where the polar ice caps build up and the seas are lowered drastically; it can be claimed, alternatively, that Concorde, by leaving particulate precipitates in the atmosphere as nuclei for condensation, will blanket the earth with a more effective 'greenhouse' cover which will keep the heat rays bouncing to and fro inside, and raise the ambient temperature to levels where the polar ice caps melt altogether. Civilisation frozen or drowned—it is easy to sell one or other of these points of view as an alarmist, 'doomwatch' line, but it is not honest. We do not have the evidence to pursue any particular postulated conclusion at the moment to the exclusion of other possible arguments.

The third attribute is a *desire to communicate*. All of us here feel this. All of us see in our environment things we desperately want to show and to discuss with other people. Driving up here through the Western Highlands and seeing primroses growing in situations where I am unfamiliar with them in England, I very much want to know if they are mainly confined to the glacial boulder clays or to disturbed soils, soils where leaching has not removed the unsaturated bases on the clay particles which primroses seem to prefer. Do they need recently cleared woodland cover? These are ecological questions of a very simple kind which I would like to encourage people to think about.

Certain kinds of midges—the bane of Scotland later in the year—sometimes swarm over the top of a small bush or a pointed object or even a man standing still, in a little cloud, and sometimes, on other days, they swarm under the end of a branch. What is it that determines this change of behaviour? Are they different species? Is it due to different environmental conditions on these different occasions? Are they in some way communicating with one another in a feedback mechanism to maintain the swarm, or are they responding individually and totally selfishly to some common ambient condition of air movement, humidity, temperature, CO_2 concentration, or whatever it may be?

Look at the River Spey outside here, working its way laboriously across a broad strath of glacial meltwater deposits far greater than it could possibly have laid down with its present discharge and volume. This shrunken river has so far failed to develop a system of regular meanders whose mean wavelength shows the proper mathematical relationship to its bank-full bedwidth (somewhere between 8 and 12 times). Is it then another beautiful example of an underfit river of the Osage type, first described in the Upper Mississippi basin? Does the Spey have a series of pools and riffles (the anglers would know) spaced with a rhythmic sequence which corresponds to the still-absent meander pattern? Here is another fascinating environmental question which I would passionately like to find an answer to, which I would like to discuss with and to communicate to people.

My much more expert colleagues in the Oil Pollution Research Unit, and the other contributors who have kindly agreed to speak at this Conference, will shortly bring us back on to the factual plane. I, speaking purely for myself, am very proud to direct and work for a small, off-beat, charitable organisation which is genuinely dedicated to the objective of its slogan: to guide people 'towards a better understanding of our environment'.

Our environment is Everyman's, as should be its better understanding. John Donne, writing of human mortality, used an apt environmental metaphor: 'Everyman is a piece of the continent, part of the main. If a clod be washed away by the sea, Europe is the less as well as if a promontory were, as well as if a manor of thy friends or thine own were. Any man's death diminishes me because I am involved in mankind; and therefore never send to know for whom the bell tolls; it tolls for thee.'

3
The Field Studies Council Oil Pollution Research Unit

JENIFER M. BAKER

(*Head, Oil Pollution Research Unit, Orielton Field Centre, Pembroke, Wales*)

In January 1967 a tanker called the *Chryssi P. Goulandris* came into Milford Haven with a hole in her side and, while unloading, spilled over 250 tons of crude oil. The Warden of Orielton Field Centre at this time, Eric Cowell, realised that little was then known about the biological effects of oil or of the different cleaning methods, so he set about finding grants to start a small research unit. About two months later the *Torrey Canyon* ran aground, the need for research became more widely obvious and the necessary funds became more easily obtainable.

The Oil Pollution Research Unit started at Orielton Field Centre at the end of 1967 with grants from the World Wildlife Fund and the Institute of Petroleum Jubilee Fund. These grants totalled £9000, which supported for three years a zoologist (Dr Crapp), a botanist (myself), a technician (Mr Alexander), a Land Rover and various bits and pieces of equipment. There were two main projects, one on rocky shores and the other on saltmarshes, and results were published at a symposium held at the end of 1970. Close contacts were established with the University College at Swansea, Dr Nelson Smith of the Zoology Department there having done surveys of the rocky shores of Milford Haven before industrialisation. His results provided valuable baseline data. Contacts were also established with local oil companies and with the Milford Haven Conservancy Board, and we are grateful for their continuing co-operation.

After the first three-year programme, Sheila Ottway (now Sheila van Gelder-Ottway) joined the Unit as its one research assistant and valiantly produced a thesis entitled 'Some effects of oil pollution on the life of rocky shores' before leaving at the end of 1971.

Further grants from the Natural Environment Research Council, the Institute of Petroleum and an oil companies group during 1971 meant an expansion, with Brian Dicks, David Levell, John Addy and more recently Susan Hainsworth joining the Unit. We also now have a part-time secretary, first Janet Dicks and now Patricia Elsmore, and have been able to improve the Research Unit facilities.

I hope you will gain a good idea of what we have been doing for the past four years from the papers to be presented at this Conference. Briefly, monitoring of the rocky shores and saltmarshes of Milford Haven has continued. Field surveys and preliminary experimental work concerning refinery effluents have been carried out. Progress has been made in the hitherto unknown areas of the sand and mud of Milford Haven both in the intertidal zone and sublittorally. Benthic surveys have been carried out in the North Sea and similar work is in progress in the Celtic Sea. Toxicity testing of oils and dispersants has continued.

As well as these research projects, we have been pleased to discuss our work with the many visitors who come from all parts of the world to see what goes on in Milford Haven and we have helped with some of the Field Studies Council's courses, particularly the environment courses for industry.

An interesting experimental educational project was a primary schoolchildren's survey of Britain's beaches, carried out through the Advisory Centre for Education. There is no doubt that both schoolchildren and students are very interested in pollution studies.

Recent grants help ensure that our work can continue and we are grateful to the Natural Environment Research Council, the oil companies group, the EEC Environment programme and the Leverhulme Trust for their support. We are now supplementing the grants by undertaking some part-time contract work.

The Research Unit is part of the Field Studies Council, an organisation which is concerned with education and research through field work. The Research Unit emphasises field work in pollution studies and this takes various forms. There are some 'one-off' surveys following oil spills, or in refinery effluent discharge areas; there are baseline surveys and continued monitoring schemes; and there are field experiments in which we pollute patches of the intertidal zone with known oils or dispersants following detailed recording of flora and fauna.

We are concerned about accuracy of field methods. It is obviously important to know the precision of different methods in order to know if changes observed from year to year are real or not. But precise methods are very time-consuming, and we sometimes have to deal with large areas in a short time. So we have been using two main types of field method. First, for field experiments and some field surveys, fully quantitative data have been obtained, and I should like to say briefly what this entails. For the saltmarsh experiments it meant sticking thousands of random pins into the saltmarsh vegetation and recording what each touched. For the work on lugworms it has meant counting thousands and thousands of worm casts inside quadrats. For the sublittoral work it has meant taking usually ten grab samples per sampling area, sieving the mud, and extracting and identifying thousands of worms and other animals. All this means a very great deal of work. For example, the grab samples taken during one week's boat work in Milford Haven may take up to a year to sort out. The second main type of method we use is the semi-quantitative abundance scale. This is usually used on the rocky shore but can be used for saltmarsh vegetation or on sublittoral rock by divers. Basically, each species is given an abundance category, scales in common use having five, seven or ten different categories ranging from rare to very abundant. This quicker approach is often used for surveying whole estuaries and is useful for comparing different shores, or a set of shores from year to year.

Though the method is relatively crude, it has shown for example a number of changes in Milford Haven and these will be described in further papers. However, I think that the development and improvement of such practical monitoring methods must be a research priority.

Laboratory work is related as closely as possible to field work, and in fact usually arises from field observations. The toxicity of different oils and dispersants may be compared in toxicity ranking tests using a convenient test organism, or the relative susceptibilities of different species to one pollutant may be determined. The effect of oils and dispersants on the behaviour of various species is important as it is being increasingly realised that finding the concentration of a compound which kills 50% of the test animal population in the laboratory does not usually tell you much about what is happening in real life, and we hope that the papers to come will demonstrate the importance of considering behavioural responses, diurnal rhythms and young stages.

This Conference provides a good opportunity for the presentation of the Research Unit's work and the discussion of different points of view. I should like to thank Mr Dewhurst, Mr Jagger, Mr Sinker, Captain Dudley, Mr Roberts, Mr Wardley Smith and Dr Holdgate for their valuable contributions, Mr Gillies, Mr Mackay, Dr Shelton and Dr Cormack for chairing sessions, the members of the organising committee, and Jill de Wardener and Colin Maynard who have put in a vast amount of work behind the scenes. And lastly, I should like to thank my colleagues for weathering so many financial, nautical and experimental problems.

Discussion

Lake Maracaibo

In reply to a question from **Dr A. G. Bourne** (Hunting Technical Services Ltd), **Mr Jagger** referred to a detailed study carried out on Lake Maracaibo by the Battelle Institute. Reports are now publicly available. Both the producing and non-producing areas were surveyed, the former occupying something like 25–30% of the lake. Production has been taking place for 50 years. Comparisons were made between pollutants in the producing areas and in the more remote areas of the lake, and hydrocarbon levels of water from both areas were of the same order of magnitude. Fish and other marine life samples were taken and compared. Oil contents of fish tissues were compared with those of fish caught in well-established American fishing areas such as Galveston Bay and were found to be similar. **Mr P. H. Monaghan** (Exxon Production Research Company), who had followed the course of this work as Battelle carried it out, said that the hydrocarbon content of the fishes did vary somewhat both with the species and with the location from which they were taken in the lake, but the extractable hydrocarbon levels were related more to the kind of fish than to the position where any particular specimen was taken. He added that one of the most prominent features of Lake Maracaibo is floating blue-green algae which cover large portions of the lake. Samples of algae taken next to producing platforms and samples taken 15–40 miles away from any producing platform had exactly the same assemblage of hydrocarbons—a few normal alkanes and unsaturated hydrocarbons that are characteristic of algae in general. The hydrocarbon levels in the water were essentially uniform over the entire lake. Total extractables as measured by the carbon tetrachloride extraction technique ranged from 0·2 to 0·4 ppm. Considerably more details were available.

Dr Bourne commented that the distance from the platform was not really relevant. It is the assimilation rate and capacity of the organisms and the concentration in the water which are important. If the concentrations of hydrocarbons are higher nearer the platform, then the assimilation will be quicker but in time those organisms further away will have reached their maximum capacity for assimilation and any sampling will then provide the same levels of hydrocarbons over a wide area. Put another way, provided the hydrocarbons are available saturation will be reached by an organism, but it will take longer in regions of low hydrocarbon concentration. In reply, **Mr Monaghan** said that with one major food fish, Curvina (*Cynoscion maracaiboensis*), hydrocarbons were extracted from the tissue in the order of 1–3 ppm, from 7–8 locations throughout the lake. The hydrocarbon levels in another species, Bocachico (*Prochilodus r. reticulatus*), which has much more natural oil in it, were in the order of 38 ppm and from the gas chromatograph traces it was not possible to identify any predominance of petroleum hydrocarbons. There were a lot of unidentified little peaks and an unresolved envelope on the GC trace. The hydrocarbon levels in the same kind of fish taken from Lake Maracaibo and from other lakes up in the mountains were within the range of variation of that same species collected within the lake itself. Battelle therefore could detect no clear evidence of petroleum hydrocarbon build-up in the tissues of the fish.

Dr R. G. J. Shelton (Department of Agriculture and Fisheries for Scotland) asked if the Battelle Institute study included analyses of fish livers, and **Mr Monaghan** replied that the choice of material to analyse was the muscle, the material which is normally eaten by man, because the question basically concerned a possible build-up which would be dangerous to the people that are consuming the fish. There were no analyses made on livers or any of the organs that are normally discarded. In reply to a question from **Mr C. B. Duggan** (Department of Agriculture and Fisheries, Ireland), **Mr Monaghan** said that the flavour of fish in Lake Maracaibo was not affected by oil.

UKOOA Contingency Plans

Mr J. Moorhouse (Rio Tinto Zinc Corporation) said that everyone probably felt that the most complete steps should be taken to minimise the possibility of a blowout in the North Sea, but if it should come to the worst, then the oil industry was ready through the UK Offshore Operators' Association to tackle oil spills. A lot of money had been invested in equipment at four main bases along the east coasts of Scotland and England. The oil industry, through its Clean Seas Committee, would work closely with the Government pollution control officers. **Mr J. A. A. Johnston** (Dee and Don River Purification Board) asked:

1. If the UKOOA contingency plans had been tested, either on a pollution incident or as an exercise.
2. If sufficient dispersant was available, bearing in mind that a North Sea blowout could take three months to contain and that the emphasis of the UKOOA plans is on dispersant techniques.
3. If a high seas recovery system would not be a more efficient approach.

He added that it had been admitted that there is no high seas recovery system capable of operation during a North Sea storm. The oil industry is working at the forefront of technology in order to discover North Sea oil, so surely, knowing the difficulties, it could have designed a recovery system over the last eight years.

In reply, **Mr Jagger** said that:

1. There has been no pollution incident to test the UKOOA plans. There have, however, been 'paper exercises' to check the plans, and major revisions, including stockpile relocations, have been made as North Sea operations have developed.
2. UKOOA has adopted dispersant concentrates to maximise both storage and vessel carrying capacity. Sufficient stocks are carried to meet foreseeable treating requirements until back-up stocks from manufacturers could be available.
3. Very considerable effort has been expended on the development of high seas oil recovery systems both in this country and in the US. It is unlikely that such a system could be used—even if it were available—in anything but favourable North Sea conditions.

The main emphasis has therefore been to take all possible steps to minimise the risk of a major spillage, but should one occur under bad sea conditions, natural dispersion would probably ensure that no oil reached either coast. Should oil persist towards coastal waters, then dispersants would be employed.

Other Discussion

This was concerned with impact statements, education and policy-making. **Mr Moorhouse** thought that in the future there would be a need to formalise the system of environmental impact statements, though some companies have already prepared such statements without being required to do so by legislation or by regulations. There was a need to state quite clearly what the pollution levels were expected to be, but account should also be taken of jobs and the social environment so that in the event of a public inquiry, or when a planning decision has to be made, there is the fullest appreciation of what will be entailed. Mr Moorhouse also said that he subscribed to the view that extensive monitoring of pollution should be undertaken and serious consideration should be given to the effects outside the industrial plants concerned. **Mr B. L. Fuller** (Mid-Glamorgan Fire Service), accepting the commitment of senior management in the oil industry to environmental control, asked what commitment the industry had to educating its operating staff in lower management levels where profit maximisation is the motivating force and not pollution control. **Mr Jagger** replied that at present and forecast oil prices, spill prevention and profit maximisation must surely be synonymous. The oil industry included employee instruction and education on oil loss and environmental protection in its training programmes and in day-to-day management. In his experience, a small investment in employee awareness programmes was more effective in avoiding pollution incidents than larger investments in equipment which can never be fully effective without employee commitment at all levels.

Dr M. W. Holdgate (Institute of Terrestrial Ecology) said that he was concerned about the need for the scientific information gathered to be relevant to the policy decisions that have to be taken. The gap between research and policy-making is a serious one today, emphasised by the delusion (which even recent Government actions have done something to foster) that the policy-maker who lacks information can commonly afford to commission research taking three years, and await its outcome. Generally the policy-maker needs the best evaluation he can get, in something like three weeks.

To balance the conservation of the environment and the most productive industrial development it is necessary to gather good data, use them to evolve standards, appraise the likely environmental impact of a new development, and design monitoring systems that give a good feedback. For this an efficient administrative as well as scientific system is needed.

Dr Holdgate pointed out that the OPRU is small and is not the only research body in the field, and continued by asking several questions. How should its work be co-ordinated with other groups? Is the existing system adequate? How should the results of all this research be drawn together in 'state of the art' evaluations? Is there a role for OPRU as an information-gathering and assessing group? Such evaluations are what the policy-maker needs—but if the communications gap is to be bridged, the evaluation machinery must be responsive to this need, and provide the statement in the format and on the time-scale required.

How should environmental impact predictions be co-ordinated? Is there a role for OPRU—or an OPRU-like body—working out the methodology that should be applied to oil-related impact? Should not the industry and central and local Government be discussing this? Should OPRU or a comparable body also be involved in developing monitoring methods—and if so, how should this work be linked to policy-makers in Government and industry, on whom operational responsibility will rest? And finally, what should be the relationship between a research and evaluation group like OPRU and the Government machinery concerned with co-ordinating action against pollution—at a central level the Central Unit on Environmental Pollution or locally by groups like the Sullom Voe Environmental Advisory Group?

Effect of Oil in Milford Haven

4

The Incidence and Treatment of Oil Pollution in Oil Ports

CAPTAIN G. DUDLEY
(Milford Haven Conservancy Board)

INTRODUCTION

This paper does not pretend to contain any particular flashes of fundamentally new thinking; it is in the nature of a summary and updating of previous papers I have presented elsewhere. Since it deals with pollution within oil ports it may be at variance with other papers dealing with oil pollution in general, since what is suitable at sea is not always applicable within a port.

Statistics and other references are taken from our records over 14 years, which are subject to the usual limitations. We do not maintain that we have recorded every single spill but, on the other hand, some of our records undoubtedly include a single pollution reported in several different places which may have been recorded as more than one spill. Any quantities recorded are based on visual assessment which is necessarily only an approximation.

In the best of all possible worlds there is no reason why oil pollution should ever occur during loading or discharging of a tanker, which is merely a matter of connecting up pipes on a terminal to the manifolds aboard a ship by means of flexible or solid pipes. Unfortunately, this statement assumes that the equipment on the terminal and aboard the ship is of perfect design, has been perfectly maintained and is perfectly operated. It also assumes that the human beings involved are infallible and that no mechanical failure of plant can occur.

Regretfully, it must be admitted that all these prerequisites cannot be guaranteed at all times. Furthermore, external forces, such as adverse weather conditions, the passage of other ships and collisions, fires and explosions, mean that some oil pollution in oil ports is inevitable.

The amount of pollution and thus its overall effect on the environment can, however, be greatly reduced if loading and discharging procedures are carefully controlled and if planning ensures that plant and labour are readily available to deal with pollutions as quickly as possible.

The reduction of avoidable spills will only occur where there is determination by every organisation, from the head man downwards, to achieve this. Willing and open co-operation and continuous effort and vigilance are necessary to ensure that the reduction is permanent and continuous.

OIL POLLUTIONS IN OIL PORTS

All oil port pollutions are likely to fall into four categories:

(a) *Fairly frequent minor pollutions involving a few gallons.* These are caused by minor overflows of cargo tanks, tank cleaning operations, malfunctions of sea valves, carelessness during the connecting and disconnecting of hoses, and sometimes by breaking the rules, such as pumping bilges. They may occur anywhere within a port and are not restricted to operations at terminals.

(b) *Infrequent medium pollutions involving up to 5 tons of oil.* These pollutions will frequently be the result of damage or mechanical failure. They are only likely to occur during loading or discharging operations in the vicinity of terminals.

(c) *Extremely rare serious pollutions.* Very occasionally a pollution of 100 or 500 tons will occur within a port. Whilst there have been avoidable instances of this type of pollution occurring owing to human error, there may also be a very occasional spill of this size due to damage, grounding, etc. They may, therefore, occur anywhere within a port, for instance on the edges of channels and anchorages or at terminals.

(d) *Catastrophic pollutions.* For reasons which appear later in this paper, pollutions of the *Torrey Canyon* type are very unlikely to occur inside a port. However, it has to be admitted that in any port handling large numbers of tankers, a collision or grounding could occur, from which a pollution of more than 500 tons might result.

With so many oil ports now being constructed it is important that, whilst accepting that some oil pollution is inevitable, we should not be guilty of exaggerating the problem. In Tables I and II details are given of the pollutions which have occurred in one major oil port—Milford Haven. Whilst these figures are not necessarily representative of what will happen in other oil ports, they do give some idea of the relative size of the problem. These statistics are sub-divided into different categories to those used above, but in summary they demonstrate:

(a) Over 35 000 tankers have been handled in 14 years, involving perhaps nearly 100 000 movements, and 704 pollutions have occurred (1·95 spills per hundred ships).

(b) Of these 704 pollutions, 612 have been under 160 gal, and of the remaining 92 not more than 20 have been above 5 tons (1 per 2000 ships handled).

(c) The frequency of pollutions is decreasing, both absolutely and, more significantly, relatively to the number of ships handled.

(d) Over 450 million tons of oil have been handled. Excluding six major spills involving damage or grounding, about 250 tons have been spilt (about half a ton per million tons handled).

(e) Of the six major spills which have occurred, two amounted to approximately 100 tons, three to between 200 and 500 tons, and one of about 2300 tons.

(f) Of the five major spills from ships, three were caused by grounding, one from a fire and explosion during discharging and one from hull damage suffered before the vessel entered the port. The remaining major spill was caused by failure of equipment in a refinery ashore.

TABLE I

Milford Haven Oil Spill Statistics: Tankers and Oil Terminals

	1961	*1962*	*1963*	*1964*	*1965*	*1966*	*1967*	*1968*	*1969*	*1970*	*1971*	*1972*	*1973*	*1974*	*Total*
Total no. of pollutions	45	33	28	34	83	72	50	52	58	55	49	56	48	43	706
Severity:															
Class 1 (<80 gal)	25	18	18	19	44	38	27	27	36	35	36	36	29	30	418
Class 2 (80–160 gal)	9	4	5	13	21	21	16	24	17	17	10	17	15	7	196
Class 3 (>160 gal)	11	10	5	2	18	13	6	—	5	3	2	3	3	6	87
Class 4 (>20 tons)	—	1	—	—	—	—	1	1	—	—	1	—	1	—	5
No. of ships	1066	1192	1236	1392	1985	2378	2680	2669	3266	3359	3490	3465	3886	4200	36 264
Tonnage of cargo (million tons)	9·9	11·5	13·0	17·7	24·9	28·9	28·2	30·0	39·9	41·2	43·2	45·7	53·1	59·2	446·4
No. of spills/100 ships	4·2	2·8	2·3	2·4	4·2	3·0	1·9	1·9	1·8	1·6	1·4	1·6	1·2	1·0	1·95
No. of spills/million tons cargo	4·5	2·9	2·1	1·9	3·3	2·5	1·8	1·7	1·4	1·3	1·1	1·5	0·9	0·7	1·58

Notes:
(*a*) 41 further pollutions have occurred from fishing vessels, tugs, non-oil company industrial outfalls, etc.
(*b*) 1 Class 4 pollution occurred in 1960 before full statistics were kept.

TABLE II

Milford Haven Oil Pollution Statistics: Causes of Pollutions

	1963	*1964*	*1965*	*1966*	*1967*	*1968*	*1969*	*1970*	*1971*	*1972*	*1973*	*1974*	*Totals*
Tankers in passage and at moorings	5	—	—	2	1	2	3	3	2	2	3	2	25
Tankers:													
Loading	1	8	11	17	10	9	11	15	16	14	5	9	126
Discharging	5	3	9	13	17	10	22	13	8	5	2	10	117
Ballasting	1	2	6	3	3	4	—	4	6	2	2	1	34
Deballasting	2	3	3	—	1	—	4	3	1	2	1	2	22
Hull defects	—	—	—	—	—	—	—	—	—	11	11	5	27
			New category										
Tankers:													
Bunkering	1	3	13	8	4	9	9	8	4	6	5	2	72
Jetties	8	8	24	15	5	12	6	6	11	8	11	9	123
Tankers:													
Miscellaneous	5	7	17	14	9	6	3	3	1	6	6	3	80
Total no. of spills	28	34	83	72	50	52	58	55	49	56	46	43	626

(g) Approximately 80% of the pollutions came from vessels and 20% from shore installations.

It might be convenient, at this stage, to make several important points which experience in Milford Haven has demonstrated, which may not always be appreciated:

(a) Major spills should be very small in number and it should be possible to deal with them without any long-term effect on flora and fauna or on the enjoyment of the general public. For instance:

(i) Fisheries can co-exist with oil. In Milford herrings still spawn, salmon use the rivers, shellfish are still gathered and lobsters are stored.
(ii) The tourist industry can live alongside a major oil port. The recreational use of Milford Haven for yachting, water ski-ing, skin diving, swimming and general holidaymaking has increased over the last few years; indeed, the tankers themselves are a tourist attraction for some.
(iii) There is no evidence to show that major bird sanctuaries on the doorstep of the largest oil port in the United Kingdom have been affected by the oil trade.

(b) If an adequate anti-pollution service is available it should be possible, in estuarial ports, to ensure that only a small number of pollutions reach the shore (in the case of Milford Haven approximately 15%, or seven per annum).

(c) A conflict of interests is likely to arise between biologists and the tourist industry if shore pollution occurs. The tourist will want it cleaned up as quickly as possible, whilst biologists may well prefer the oil to be left untreated. The probable solution is to clean amenity beaches but not to clean inaccessible foreshores as a general rule.

(d) Tankers can run aground without any pollution whatsoever occurring. Even in Milford Haven, which has the disadvantage of a rocky bottom, this has occurred on at least four occasions and, in one case in particular, a 150 000 ton vessel was aground throughout a whole tide without any pollution resulting.

(e) Collisions should be extremely rare. Milford has experienced about 100 000 tanker movements and probably over a million movements of tugs, service craft and others, without a single collision.

(f) Even if a major disaster occurs, such as a tanker exploding whilst discharging, it does not mean that all the oil within the ship will escape. Our experience suggests that within the relatively sheltered waters of a harbour serious damage can be done to a tanker and only a very small proportion of the cargo will escape. This is due to the fact that tankers are heavily subdivided and each tank is naturally oil- and water-tight and, in the case of bottom damage, which is the most likely during grounding, the only oil which can escape is that laying above the waterline in the tank or tanks affected.

(g) The *Torrey Canyon* scale of disaster should not occur within a port, since the natural shelter provided will almost certainly mean that oil within the undamaged parts of the ship can be transferred to other tankers before further damage occurs.

(h) Types of oil carried in tankers vary considerably and not all pollutions require treatment, nor will they cause much damage. The most serious pollution in Milford in terms of volume of cargo lost, resulted in some 2000 tons plus of gasoline escaping from a vessel which went aground. No treatment was considered necessary and, indeed, owing to the volatile nature of the cargo it would have been dangerous to put men and craft into the area. Furthermore, the resulting damage was restricted to a one mile strip of the foreshore which is recovering naturally.

(i) Not insignificant chronic oil pollution occurs in *all ports* from sources unconnected with the oil industry, such as fishing vessel bilges, land drainage systems and industrial outfalls.

(j) Regular carefully controlled use of chemicals does not necessarily result in biological damage. (In justifying the use of chemicals I would stress the importance of correct application in minimum necessary quantities.)

CAUSES OF OIL POLLUTION

(a) *Design faults in terminals.* This is a rare source of pollution, particularly in the case of terminals operated by major oil companies. It is a cause which is always capable of remedy and certainly continuous research is necessary to improve the situation further.

(b) *Design faults in tankers.* Although all tankers look much alike, closer examination shows that there are great differences in type and equipment and the recent increase in size of ships has inevitably led to new designs being incorporated, all of which have not proved to be entirely satisfactory. We believe that the design of tankers could be improved to reduce pollution and this is referred to in greater detail below.

(c) *Breakages and mechanical failure.* Major fractures are a rare source of pollution, but mechanical failure of valves is a relatively frequent cause of minor pollutions. This applies particularly to ships' overside valves used in connection with ballasting operations.

(d) *Incorrect operating procedure.* In the case of major oil company terminals this is unlikely to be a cause of more than a very small proportion of pollutions, but in the case of tankers it is a very frequent cause and it is vitally important, therefore, that shore personnel should ensure that the correct operating procedures are fully understood and enforced. The importance of good communications and liaison from ship to shore during loading and discharging cannot be over-emphasised.

(e) *Tiredness.* In many ships one officer is responsible for the entire loading or discharging operation and this can result in him obtaining little or no sleep for 24 hours. This will inevitably result in occasional errors of judgement and, in my view, a system which requires such hours of work can only be described as a bad operating procedure.

(f) *Human error.* Human error is, and always will be, the most important cause of oil pollution. Sometimes an excuse can be found, such as language difficulties and possibly even ignorance, but it is undoubtedly true that sheer carelessness is a major contributory factor.

DESIGN FEATURES OF TANKERS

The design of tankers could be improved so as to eliminate a considerable number of small pollutions and reduce the number of larger spills which occur in ports. Unfortunately, some of the steps which would reduce the number of pollutions may not be wholly acceptable, since they would affect other factors, such as stability. However, I believe that the following suggestions are worthy of serious thought:

(a) Cargo and bunker overflows on to the deck form a substantial proportion of all pollutions and many of these could be prevented from entering the sea if deeper scupper bars were provided at the ship's side. It is not practical to suggest any fixed height, since ships will vary and objections will be voiced to this proposal on the grounds that large retaining bars could result in a large free surface of trapped water on deck which would impair the stability of the ship at sea. In my view, this problem could be overcome by careful planning and the provision of adequate openings which would only be closed whilst in port and whilst actual loading or discharging was taking place.

(b) As a further means of preventing overflows from deck, tankers could be provided with facilities which would ensure that oil is automatically drained off the after end of the deck into slop tanks.

(c) An improved design of scupper plug is required. Innumerable small pollutions occur through leaks in scupper plugs and surely it is not beyond the wit of man to design an efficient plug.

(d) High-level warning devices could be fitted to all tanks so that audible or visual warning is given when tanks are nearing their full capacity. Some very experienced operators believe that there would be disadvantages to this system, since it might provide a false sense of security and a resulting reduction in efficient watchkeeping, with the result that pollutions might be even more likely owing to failure of automatic equipment.

(e) Pollutions from overboard discharges and inlets, which are a frequent source of small pollutions, could be eliminated by fitting a positive blanking facility in the ship's design to be compulsorily used in ports. For a number of reasons this may be unacceptable in main sea valves, but a big improvement could be achieved if double block and bleed systems were incorporated in all overboard discharges, so that oil could be drained from the lines before the outboard valves are opened. Warning devices could also be fitted which would indicate the presence of oil in overboard ballast lines between double valves. (It is interesting to note that a Department of Trade 'M' Notice, published in November 1974, supports the desirability of double valve separation and detection devices between the valves.)

(f) Some of the pollutions which occur from manifold leakages and fractures could be overcome if manifold designs and sizes were standardised internationally and, more importantly, if the manifolds and valves were constructed of steel and not cast iron. Furthermore, pollution from these sources could be reduced if ship manifolds were fitted with a pipe leading back into a cargo tank to enable effective draining of ship to shore lines before they are disconnected.

(g) Many modern ships are fitted with butterfly and ball valves as standard, but experience has suggested that they are not always satisfactory for overboard discharge and sea suction lines, since leakages can occur through distortion of rubber seatings, the failure of the seal between rubber and metal and the malalignment of keyways and stops in the valve control gear.

DESIGN FEATURES OF TERMINALS

I have no experience of single-point moorings or other types of offshore discharging facilities and would therefore not wish to comment upon them. However, since some pollutions and failures will inevitably occur on the more traditional type of oil berth, it might be of value if I listed some of the design features which we consider to be desirable:

(a) Whilst the spills which occur on to jetty decking during disconnection of hoses may, in themselves, be very small in quantity, any heavy accumulation of rainwater could easily cause flooding of the berthing-head and, if no precautions are taken, the small amount of oil present will float on the surface of this rainwater and eventually flow over the side and into the harbour. It is also desirable that, should any heavy pollution occur, it is capable of being retained and pumped away. My Board, before giving statutory approval to plans, ensures that:

(i) The deck itself is dished, with a drain at a low point in, or near, the centre of the berth deck.
(ii) A continuous water-tight lip is built round the extreme edge of each berthing-head to assist in the containment of liquids.
(iii) The berthing-head is capable of retaining rainfall at the rate of $\frac{1}{2}$ in per hour for one hour.
(iv) The jetty drains are led to an individual slop tank underneath, or in the immediate vicinity of, each berthing-head, and two pumps are provided with high-level alarms and fitted with automatic operation. Each pump to be capable of handling 1 in of rain per hour and separate sources of power provided to each pump.
(v) Vent systems from slop tanks are constructed so that any liquid coming from them falls on to the berthing-head.

(b) Emergency valves are placed on every line leading from the berthing-head to the shore, preferably at both ends. This is necessary to ensure that a fractured line will not result in all the contents, right back to the refinery or storage tank, escaping into the harbour.

(c) Automatic remote controls at the berth are provided for shutting off loading pumps in the event of an emergency.

(d) Adequate road access is provided to every berthing-head to ensure that emergency services and additional pumping equipment can reach the hose-handling area at all times.

TREATMENT OF OIL POLLUTIONS

There can be no single type of organisation which is applicable to all ports and there is certainly no single type of equipment best suited to dealing with all types and sizes of pollutions. In all ports there is certainly an absolute necessity to plan ahead and to ensure that there is a rapid response on all occasions, since oil spreads very quickly.

The type of equipment required to deal with pollution will depend upon the severity of the pollution and the type of oil involved. The best solution is obviously the physical removal of all oil from the surface of the sea, but complete removal is probably impossible and, I suggest, impracticable and unnecessary in the case of minor pollutions by most crude oils and all lighter products.

Booms have their uses, particularly in non-tidal docks or harbours, but there are a number of factors which severely restrict their use in estuarial ports with significant tidal currents, or which are subject to severe weather conditions. To be effective in anything other than relatively calm water they need to be large and are therefore both expensive and difficult to handle. Furthermore, in tidal conditions of over about one knot they will not stop the spread of oil. The successful use of a boom depends upon the provision of efficient means of removal of the oil contained within the boom and this, in turn, depends upon efficient skimming devices and adequate reception facilities.

Booms can, however, be of great value in non-tidal harbours or in the open sea, where they can be used to corral the oil and make treatment much easier. Indeed, in some particularly fortunate ports it may be possible to provide booms as a permanent feature around all vessels working cargo. Furthermore, even in tidal harbours a boom can be used to reduce the rate of flow of oil and to alter its course of flow somewhat and thereby protect particularly vulnerable sites, such as areas of scientific importance or industrial intakes such as those used by power stations.

I suggest, therefore, that in our present state of knowledge, the best methods of dealing with oil pollutions are:

(a) *Minor pollutions in the vicinity of tankers and terminals.* Dispersal by chemicals in all cases other than heavy fuel oil, which should be scooped up from the surface, possibly with the use of a small boom to assist collection. The equipment required for this type of pollution must be small and extremely manœuvrable and easily handled, since it will have to operate between piles and under mooring ropes.

(b) *Pollutions in enclosed docks or non-tidal harbours.* Collection of the oil by means of booms, followed by physical removal. This is possible since speed is no longer so important and the oil contained within the boom can be moved to a position where skimmers and collection craft can operate.

(c) *Moderate or heavy pollutions in tidal harbours.* Physical removal of areas of heavy pollution and dispersal of areas of lighter pollution. Booms may well be used to control and concentrate pollution to assist physical removal. The use of 'Herder' chemicals, as a pre-treatment of amenity beaches, may be considered to reduce deposit of oil on sandy bays.

(d) *Pollution of the shoreline.* It is a matter of judgement whether it is worthwhile to clean pollution from the shore, since all cleaning methods cause ecological damage which might even exceed the damage caused by the oil. Circumstances in each individual port should be taken into consideration, but probably rocky foreshore and areas not used for amenity purposes should be left untreated, except in the most severe pollutions. Amenity areas should normally be cleaned by physical removal of the oil, rather than treatment by chemicals.

I suggest that no port can be expected to provide equipment, etc., for dealing with the 'once in a lifetime' catastrophic spill. It should, however, ensure that plans are prepared to mobilise the national, or indeed international, resources available. In Britain, wholehearted and rapid assistance can be expected from the Department of Trade, who will provide experts and equipment. Indeed, if their equipment is stored near the port area, it could be used, at minimum cost, to supplement port equipment in all serious pollutions.

USE OF DISPERSANT CHEMICALS

Although modern chemicals are very much less toxic than the chemicals used on *Torrey Canyon*, even the best of them are toxic to some degree. It is essential, therefore, that they should be used with care and in the minimum possible quantities.

I suggest that, in ports where pollution is likely to occur fairly regularly, a craft should be provided capable of carrying reasonable quantities of chemicals in bulk tanks with efficient spraying equipment and efficient means of agitation. The advent of water-dilutable chemicals, for use on the sea only, now coming on the market, may well make agitation much less important and, quite apart from the enormous gain which will be achieved logistically by these chemicals, since a given quantity will do perhaps ten times as much work, the fact that there is no necessity to agitate will make the design of the craft much simpler and their use much easier in confined spots around piles and vessels.

If it is accepted that chemicals of any sort will be used in any significant quantities, I suggest that bulk storage and bulk handling should be employed, since the use of drums is extremely wasteful of labour and, perhaps, dangerous under certain circumstances aboard small craft.

Whilst one specialised craft may well be sufficient to deal with the frequent minor pollutions, arrangements should be made whereby additional harbour craft or tugs can be used for spraying purposes and this can be achieved relatively easily by the provision of portable pumping, spraying and agitation equipment, together with collapsible pillow tanks, to contain the chemicals.

PROBLEMS OF PHYSICAL REMOVAL

Until recently, physical removal of oil in anything but flat calm, and preferably non-tidal conditions, was impracticable.

Plant is, however, now available which allows us to look again at this ideal solution. Certainly in open sea conditions it is practicable, but in harbours there are still considerable difficulties, due to the difficulty of deploying the equipment in restricted waters. This applies particularly to the attendant booms.

The difficulty in the past has always been one of separation of the oil from the recovered oil/water mixture, since equipment was dependent upon gravity separators which are slow in operation and very cumbersome. The breakthrough which has recently occurred is that relatively lightweight floating plant is now available which can achieve a high degree of separation at source. The first of these new equipments was, however, relatively bulky and was not suitable for harbour use, but a smaller version is now being produced which will be of considerable assistance in ports. Its use will, however, probably be restricted to more significant pollutions, with chemicals continuing to be used on the much more frequent very light pollution.

The main problems connected with physical removal are:

(a) the need to provide relatively large mobile reception facilities at short notice. Barges may not provide sufficient capacity and coastal tankers are unlikely to be immediately available; and
(b) the problem of disposal of the resultant waste oil and oil-contaminated flotsam.

In summary, therefore, physical removal is now fully practicable in major open sea pollutions, such as *Torrey Canyon* or *Pacific Glory*, but may only be applicable in ports in the cases of severe pollution when reception vessels are readily available.

THE NEED FOR PLANNING

I suggest that the need for pre-planning is obvious. It is probably only possible to produce an effective plan if it is an agreed, voluntary, document rather than a legally binding one. Our experience has suggested that any plan should cover the following points:

(a) *Organisation and responsibility*. The main policy-making body should be recruited at the highest level. The head man in each organisation involved is thereby personally committed. This body is responsible for policy and finance. As a second tier, to control day-to-day operation, an operating sub-committee is required which will meet frequently and should consist of middle-management technical personnel.

(b) *Financial responsibility*. Acceptance of responsibility must not be allowed to slow up anti-pollution measures. This means that financial responsibility must be firmly laid down in the plan and an official designated to attribute responsibility in cases of doubt. In our case, the Harbourmaster has that responsibility. In general, the bulk of the financial responsibility should fall on the oil companies and ship owners and not on the Harbour Authority, particularly if it does not, itself, provide facilities for handling oil.

(c) *Prosecution*. Whilst prosecution will not, in itself, stop oil pollution, it may well act as a deterrent. The responsible authority for instituting proceedings will be the Port Authority, and this will involve it in a detailed examination of each and every pollution, the collection of evidence and statements and the serving of summonses. If a foreign-flag vessel has to be prosecuted it is desirable to have a system whereby the summons can be served before the vessel sails, since that ship, together with that Master, may not return to the jurisdiction of British courts.

(d) Each and every pollution, whether a case for prosecution or not, should be considered in detail by the operating sub-committee, who should satisfy themselves that the oil company involved is taking reasonable steps to stop a recurrence and the lessons learnt should be passed on to all other operators. (To be effective, this 'soul-searching exercise' must be conducted on a confidential basis.)

(e) Detailed check lists in many languages should be provided, setting out the loading and discharging procedures and anti-pollution measures required.

(f) All oil should be dealt with on the surface of the water, where possible, and this entails planning a rapid response throughout the 24 hours.

(g) A constant look-out should be provided, both on ships and on the waters of the port, to ensure that every pollution is quickly discovered. This does not necessarily mean the setting-up of special staff and special equipment, since the patrol work can be carried out by craft employed on other duties, such as tugs, pilot boats, etc.

(h) Detailed measures should be set out to deal with damaged vessels arriving in the port, since oil ports must accept that they are the logical places to take damaged tankers, if overall pollution is to be reduced. Basically, this means that damaged vessels should be examined before they are permitted to enter, anti-pollution equipment should be laid on in advance and methods of discharge should be approved by the Port Authority.

(i) The various conservation interests should be fully consulted in the preparation of the plan, so that guidelines may be laid down concerning, for instance, areas where beach cleaning should be restricted or closely supervised.

(j) A detailed system of beach patrolling should be prepared to ensure complete coverage in the event of heavy pollution. In this respect the public and local authorities may be co-opted.

(k) Adequate equipment should be permanently provided to deal with all minor spills rapidly and additional equipment should be provided so that harbour service craft can be rapidly equipped as anti-pollution vessels in the event of more serious pollutions.

(l) If chemicals are to be used, a central bulk store should be provided in a position in which all craft likely to be employed can load direct.

(m) If physical removal is to be employed, places and methods of dealing with the recovered oil and oiled flotsam should be planned in advance.

(n) The plan should contain a directory of all those likely to be involved in the event of pollution occurring, and thought should be given to methods of obtaining assistance from outside the area, including the provision of barges and coastal tankers for lightening purposes.

CONCLUSIONS

The conclusions which I hope you would draw from this paper are:

(a) Small accidental pollutions are inevitable in oil ports, but can be satisfactorily dealt with.

(b) Collisions should be extremely rare in any port, particularly modern estuarial ports.
(c) Grounded and damaged vessels do not necessarily cause pollution, and even if pollution occurs it will probably only involve a small proportion of the cargo aboard the vessel.
(d) Damaged oil tankers must be allowed to enter oil ports in order to reduce overall pollution, even if this results in relatively severe pollution of the oil port itself, but always subject to predetermined conditions aimed to reduce this to a minimum.
(e) Oil can most easily and satisfactorily be dealt with on the surface of the sea, and therefore any system which is evolved must be one which ensures a rapid response. The importance of regular look-outs, particularly at night, cannot be exaggerated.
(f) Whilst physical removal of oil pollution is always to be recommended, except in minor spills, chemicals can often be used in estuarial ports without damage to the environment.
(g) Chronic shore pollution is more damaging than a single severe pollution. It is therefore more important for any plan to concentrate on keeping numerous small pollutions from the shore, rather than concentrating on the rare very heavy pollution.
(h) Good intelligence is essential and, in the case of the more severe pollutions, this will entail a rapid reconnaissance of all beaches. This requires a lot of manpower and the general public and, in particular, the conservationists should be encouraged to help.
(i) Any shore cleaning operations require skilled supervision if they are not to do more damage than the oil itself.
(j) In the case of severe pollution it is always wise to ensure that representatives of the owners of the vessel and, if possible, their insurers are on the scene quickly to deal with consequential matters.
(k) Any plan should be carefully drawn up and given wide circulation. A better plan will be obtained if it is not a legally enforceable document. The plan should be exercised regularly and should be kept constantly under review by a technical sub-committee.
(l) Good communications are essential and these should be firmly under the control of the Port Authority. This will involve the provision of numerous portable radio telephones.

Finally, I would make two points:

(a) Serious conservationists have a very important role to play. Firstly, by the pressure which they bring to bear they can ensure that all reasonable steps are taken to reduce the effects of pollution, and secondly, they can be of the utmost value in helping to deal with severe pollution when it occurs. I suggest that the worst enemies of the serious conservationists are the publicity-seeking minority who are frequently guilty of gross exaggeration and thereby tend to reduce the credibility of their own serious brethren.

(b) Experience has taught us that it is a very short step in the public mind from an organisation accepting, as a moral duty, responsibility for cleaning up oil pollution when not legally compelled to do so, to that organisation being held directly responsible for the oil pollution itself. I would suggest that anyone accepting such a responsibility, and producing such a plan, should do his best to make it clear that he is not accepting responsibility for oil pollution itself and I would further suggest that those who attempt to impose this responsibility are doing a disservice since the only result can be a reluctance on the part of organisations to carry out such work voluntarily.

5

Biological Monitoring—Principles, Methods and Difficulties

JENIFER M. BAKER

(Oil Pollution Research Unit, Orielton Field Centre, Pembroke, Wales)

SUMMARY

Biological monitoring is defined as the use of living organisms to determine the presence, amounts, changes in and effects of physical, chemical and biotic factors in the environment.

Different approaches and methods for monitoring estuarine and coastal areas are covered: these include a discussion of different types of indicator organism; and a description of the shore transect technique used in several monitoring programmes. An outline monitoring programme for a typical estuary is given.

INTRODUCTION

The subject of biological monitoring is receiving increasing attention and it is common for new industrial developments, particularly those of the oil industry, to include monitoring schemes. Such schemes have also been used at older installations in conjunction with improvement programmes (for an example see Dicks, 1975a). The emphasis of most existing monitoring programmes, and therefore of this paper, is, for various reasons, on the intertidal zone. Off-shore areas are, however, increasingly important and this subject is discussed by Dicks (1975c).

The increasing numbers of people and organisations concerned with monitoring (industry, County Councils, environmental groups, consultant groups, schools and universities) may find it difficult to find information on the subject because publications are few in number and diffused through the scientific literature. A need identified by the Conference on Pollution Criteria for Estuaries (1973) was a manual of methods for study of the estuarine environment, and this paper, together with the cross references to other papers in this conference, can be regarded as a preliminary stage in such a project.

THE OBJECTIVES AND SCOPE OF BIOLOGICAL MONITORING

Biological monitoring is the use of living organisms to determine the presence, amounts, changes in and effects of physical, chemical, and biotic factors in the environment. Data collected in monitoring programmes are used to provide information upon which standards can be set; to give warning of changes in, for example, estuarine and coastal ecosystems, as a basis for making policy decisions for action to control pollution, and to justify action and expenditure (Conference on Pollution Criteria for Estuaries, 1973).

As changes in conditions are being followed, biological monitoring necessarily involves repeated surveys. The first of a series of surveys is the baseline survey, and this is often more detailed than subsequent work. In an area which is biologically virtually unknown there is an obvious need for a thorough first survey before sensible decisions on methods, species, and sites for monitoring can be made. The latter should include suitable control areas in order to separate natural changes from other changes. For example, when monitoring the effects of a refinery effluent, areas at different distances along the pollution gradient would be studied, the furthest from the discharge being regarded as controls. It is usual in such cases to find some effects which can be correlated with the gradient, and others, caused perhaps by climatic or other natural changes, which cannot.

Biological monitoring may use selected indicator organisms, or groups of organisms, and some different approaches are described in more detail below.

INDICATOR ORGANISMS

The term 'indicator organism' can mean many different things, and is often used in a confused way. A broad view of indicator organisms covers the following types (modified from Jenkins (1971) and Goodman (1974)).

(a) *Sentinels*. Sensitive organisms introduced into the environment as early warning devices (e.g. canaries in coal mines) or to measure pollutant effects and area affected, e.g. limpet, winkle and *Spartina* transplants into effluent pollution gradients (see Baker, 1975).

(b) *Detectors*. Naturally occurring species that show a measurable response to pollutants or environmental changes, such as death, change in growth or reproduction, or change in behaviour. For example, limpets drop off rocks under the influence of oil or detergents (see Dicks, 1975b).

(c) *Exploiters*. Organisms whose presence indicates the probability of pollution. They are often abundant in polluted areas due to lack of competition from eliminated species. For example, *Enteromorpha* following oil spills, or in sewage contaminated areas.

(d) *Accumulators*. Organisms that take up and accumulate chemicals in measurable quantities. For example, accumulation of heavy metals in Bristol Channel molluscs (Crothers, 1973) or accumulation of hydrocarbons in mussels in the lagoon of Venice (Fossato, 1975).

(e) *Bioassay organisms*. Selected organisms used as laboratory reagents to detect the presence and/or concentration of toxic pollutants, or to rank pollutants in order of toxicity. For example, the brown shrimp is used by the Ministry of Agriculture, Fisheries and Food at Burnham on Crouch for testing the toxicity of oil spill dispersants (Wilson, 1974).

Different classes of indicator are suitable for different classes of pollutant and different purposes. For example, the accumulator type of indicator ideally has a neutral or indifferent response to a pollutant, but accumulates it as it increases in the environment. Conversely, a sentinel, detector or exploiter should be sensitive to the presence of the pollutant, reacting in a negative or positive way. It is important to bear in mind that an ecosystem is a process in time, and the status of indicator organisms may change as a result of succession. For example, the annual saltmarsh plant *Salicornia* is a detector in that it is particularly susceptible to oil and dispersant pollution, however, it is also eliminated as the saltmarsh succession proceeds.

OTHER APPROACHES

For many marine systems, it has so far not been possible to identify suitable indicator organisms. For others, as indicated above, there are promising species such as limpets, or the possibility of using a limpet/seaweed or other ratio. However, we do not yet know enough about these organisms to confidently arrange marine monitoring programmes using only a few indicator species. The following approaches, therefore, tend to utilise several or most of the common species in any particular community.

Mapping. Mapping over a large area, perhaps combined with aerial and other photography, is suitable for areas such as saltmarshes. It is particularly useful if the abundance of the different species rather than straightforward presence or absence is mapped, and this can be done using estimates on an abundance scale, density measurements and so on. An area to be mapped has to be worked over systematically, often using a grid system, and the method is therefore only useful for certain types of terrain. An example of a mapping method used to record changes in saltmarsh vegetation is given by Dicks (1975a). Existing maps of *Salicornia* and *Zostera* in Nigg and Udale Bays, Cromarty Firth (Natural Environment Research Council) could be used as a partial baseline should a monitoring scheme be operated in this area.

On precipitous rocky shores, mapping is not practical over a large area, but is sometimes used on a small scale (e.g. a 1 × 1 m or 5 × 5 m quadrat) to follow seasonal and yearly fluctuations.

Transects. The transect approach is most commonly used on rocky shores and forms the basis of the original Milford Haven monitoring programme, the Bantry Bay programme (Crapp, 1973) and a survey of Sullom Voe (Addy *et al*., 1973). The methods used for these three programmes are based on those of Crapp (1970, 1971, 1973) and are outlined below.

The transects are made from mean low water of spring tides (M.L.W.S.) to the top of the supra-littoral zone. Levelling is carried out with a cross staff instrument devised by Dr A. Nelson-Smith, in which a spirit level is mounted on a vertical leg and surmounted by an angled mirror. So long as the horizontal distances are fairly short a line of stations can be established at regular vertical intervals by sighting along the horizontal spirit level. In any area the mean range of spring tides in metres is determined, and the vertical leg is set to a length one tenth of this (for Milford Haven this is 0·6 m, for Bantry Bay 0·3 m and for Sullom Voe 0·2 m).

A list of about sixty common littoral species is used, and at each station of each transect the abundance of these species is assessed in terms of the abundance scales devised by Crisp and Southward (1958) and expanded by Ballantine (1961), Moyse and Nelson-Smith (1963) and Crapp (1970).

The abundance categories are:

Ex (Extremely abundant)
S (Superabundant)
A (Abundant)
C (Common)
F (Frequent)
O (Occasional)
R (Rare)

The first two of these categories were late additions used by Crapp (1973) and originally suggested by Dr J. R. Lewis.

Abundance scales for common groups of organisms mentioned in subsequent papers are given below in Table I. Part of a standard recording sheet, filled in for a Milford Haven transect, is shown in Table II. It is common to represent the abundance data in the form of kite histograms (Fig. 1) and these are used in some of the subsequent papers.

Crapp (1973) recommends that abundance values should be assessed over a transect width of about 10 m and this is what has been done in Milford Haven.

The advantages of this method are that it provides semi-quantitative information relatively quickly and easily even on very rough and broken shores. Mapping in detail, or large numbers of quadrat counts, are not feasible when perhaps 50 miles of shore has to be monitored, although limited detailed studies of this kind can be fitted into the reference framework provided by the transect surveys described.

The method may be criticised for not being fully quantitative and objective, because recorders are required to estimate rather than count everything. In order to investigate differences between workers, the Watch House Point transect in Milford Haven was surveyed within a few days by three people who had no reference to each other's record sheets. The people were two Oil Pollution Research Unit staff and a university undergraduate with a few weeks' experience. The results for the common species, shown in Fig. 2, indicate that differences in one abundance category between workers were common, and that some estimates differed more. However, taking into account the results from all the stations for all the species shown, the correspondence is good.

TABLE I

Abundance Scales for Some Common Littoral Organisms

Lichens and other encrusting species:	
Ex.	More than 80% cover
S	50–80% cover
A	20–50% cover
C	1–20% cover
F	Large scattered patches
O	Widely scattered patches, all small
R	Only one or two patches
Seaweeds (algae):	
Ex.	More than 90% cover
S	60–90% cover
A	30–60% cover
C	5–30% cover
F	Less than 5% cover, zone still apparent
O	Scattered plants, zone indistinct
R	Only one or two plants
Barnacles:	
Ex.	More than $5/cm^2$
S	$3–5/cm^2$
A	$1–3/cm^2$
C	$10–100/dm^2$
F	$1–10/dm^2$, never more than 10 cm apart
O	$1–100/m^2$, few within 10 cm of each other
R	Less than $1/m^2$
Limpets and winkles:	
Ex.	More than $200/m^2$
S	$100–200/m^2$
A	$50–100/m^2$
C	$10–50/m^2$
F	$1–10/m^2$
O	$1–10/dam^2$
R	Less than $1/dam^2$
Top-shells and dog-whelks:	
Ex.	More than $100/m^2$
S	$50–100/m^2$
A	$10–50/m^2$
C	$1–10/m^2$, locally sometimes more
F	Less than $1/m^2$, locally sometimes more
O	Always less than $1/m^2$
R	Less than $1/dam^2$
Mussels:	
Ex.	More than 80% cover
S	50–80% cover
A	20–50% cover
C	Large patches, but less than 20% cover
F	Many scattered individuals and small patches
O	Scattered individuals, no patches
R	Less than $1/m^2$

TABLE II

Part of a Standard Recording Sheet for the Milford Haven Monitoring Programme

Site Watch House Point *Date* 22.7.74
Height of lowest station above Chart Datum 0·9 m

Station number	1	2	3	4	5	6	7	8	9	10	11	12	13	14	15	16	17	18	19	20	21	22
Species																						
Green Algae																						
Enteromorpha spp.	O	F	C	A	C	C	C	F	F	R	—	—	—	—	—	—	—	—	—	—	—	—
Ulva lactuca	—	—	—	—	—	—	—	—	—	—	—	—	—	—	—	—	—	—	—	—	—	—
Cladophora spp.	R	R	R	—	—	—	—	R	—	—	—	—	—	—	—	—	—	—	—	—	—	—
Brown Algae																						
Alaria esculenta	—	—	—	—	—	—	—	—	—	—	—	—	—	—	—	—	—	—	—	—	—	—
Laminaria digitata	C	—	—	—	—	—	—	—	—	—	—	—	—	—	—	—	—	—	—	—	—	—
Fucus serratus	C	C	R	—	—	—	—	—	—	—	—	—	—	—	—	—	—	—	—	—	—	—
Fucus vesiculosus	—	—	—	—	—	—	—	—	—	—	—	—	—	—	—	—	—	—	—	—	—	—
F.v. var. linearis	—	—	F	F	F	C	C	F	—	—	—	—	—	—	—	—	—	—	—	—	—	—
Ascophyllum nodosum	—	—	—	—	—	—	—	—	—	—	—	—	—	—	—	—	—	—	—	—	—	—
Fucus spiralis	—	—	—	—	—	—	—	—	—	—	—	—	—	—	—	—	—	—	—	—	—	—
Pelvetia canaliculata	—	—	—	—	—	—	—	—	R	O	R	—	—	—	—	—	—	—	—	—	—	—
Red Algae																						
Corallina officinalis	F	C	F	O	—	—	R	—	—	—	—	—	—	—	—	—	—	—	—	—	—	—
Gigartina stellata	C	C	F	R	—	—	R	—	—	—	—	—	—	—	—	—	—	—	—	—	—	—
Chondrus crispus	—	—	—	—	—	—	—	—	—	—	—	—	—	—	—	—	—	—	—	—	—	—
Laurencia pinnatifida	O	F	F	O	—	R	R	R	—	—	—	—	—	—	—	—	—	—	—	—	—	—
Lithothamnion spp.	S	C	O	F	R	F	C	O	—	—	—	—	—	—	—	—	—	—	—	—	—	—
Lomentaria articulata	F	F	F	—	—	—	—	—	—	—	—	—	—	—	—	—	—	—	—	—	—	—
Rhodymenia palmata	C	F	F	R	—	—	R	—	—	—	—	—	—	—	—	—	—	—	—	—	—	—
Porphyra umbilicalis	R	R	F	F	C	C	C	F	F	O	—	—	—	—	—	—	—	—	—	—	—	—
Catenella repens	—	—	—	—	—	—	—	—	—	—	—	—	—	—	—	—	—	—	—	—	—	—
Lichens																						
Verrucaria mucosa	C	C	O	R	R	O	R	R	F	—	—	—	—	—	—	—	—	—	—	—	—	—
Verrucaria maura	—	—	—	—	—	—	—	—	F	C	S	C	F	O	O	C	C	C	O	R	R	R

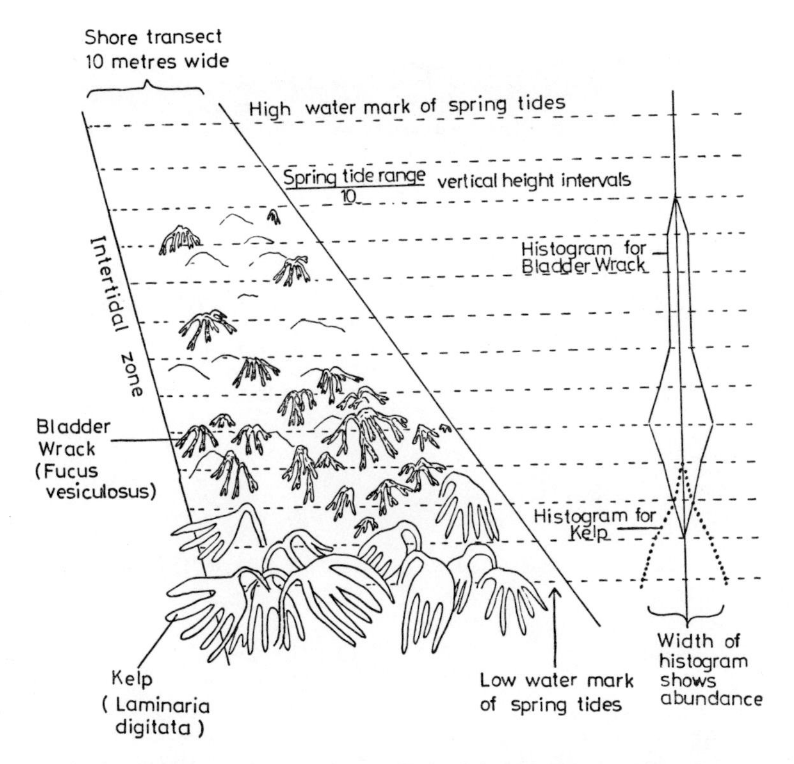

FIG. 1. Representation of part of a rocky shore transect showing, for two seaweed examples, how abundance is expressed as histograms.

Another transect method which is being used for some programmes involves stretching a line down the shore and recording everything it touches. This is an improvement on the previously described method in terms of objectivity and quantification, but because the distribution of species on shores is usually patchy, some important species may not be recorded at all, and some may assume a false importance in the records. The belt transect/abundance scale method gives a much better idea of the biology of the shore and makes possible useful comparisons between shores. A line transect can only be compared with itself from year to year.

Settlement plates. The settlement plate technique involves the fixing of uniform substrates (e.g. 20 cm asbestos squares) onto fixtures such as jetties and buoys. Growth of organisms is recorded over a period of time, and this is helpful in showing pollution effects uncomplicated by local variations in type of substrate. It is perhaps the most practical method of assessing effects on at least some of the plankton species.

For example, if barnacle larvae in the plankton are locally reduced in numbers, or if their behaviour is altered, this is reflected in the settlement of barnacle spat.

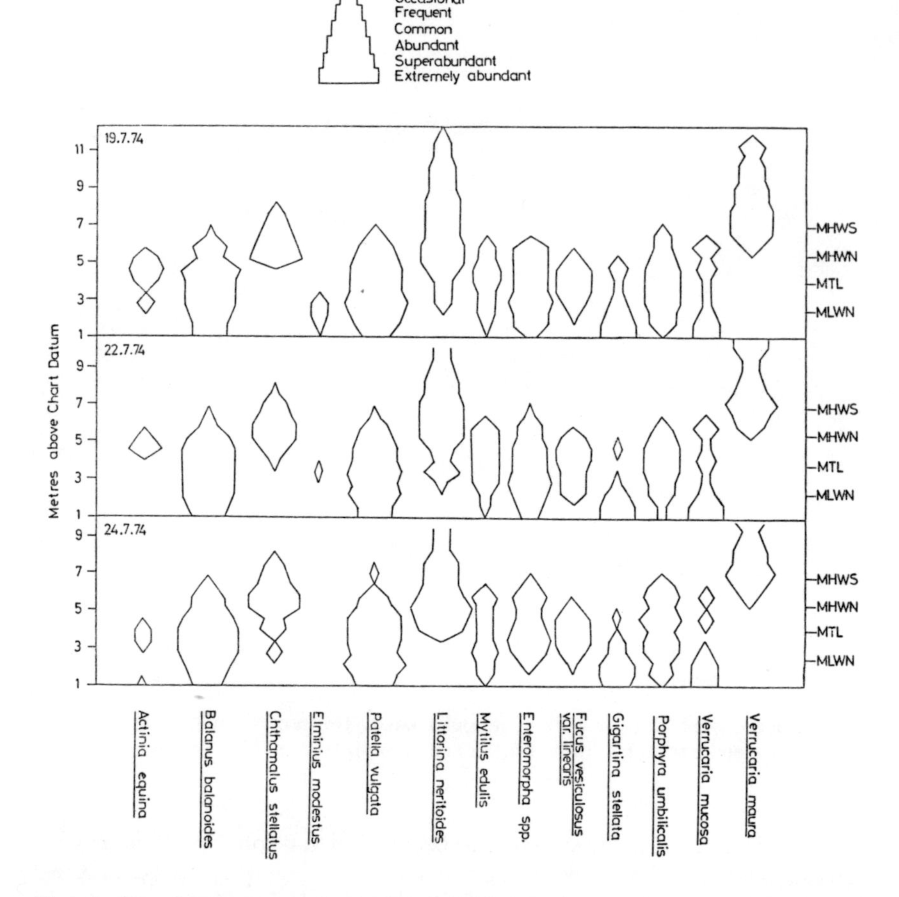

FIG. 2. Watch House Point transect, Milford Haven. Results obtained by three different workers.

Diversity indices. Diversity indices are described by Dicks (1975c). A variety of different diversity indices can be used, but in general a number is calculated which reflects the number of species and individuals in a community and how evenly the species are spread through that community. The diversity index, in other words, shows the richness of a community, and this is believed to be reduced as a result of pollution. It is important to bear in mind that diversity can also change as a result of natural causes and is also a function of the biological age of the ecosystem.

It might be useful in a monitoring scheme to compare the diversity at a particular site from year to year to detect trends of increase or decrease of diversity, but inter-site comparison is meaningless for detecting pollution effects or gradients unless the sites are physically comparable.

PROBLEMS

Some problems associated with monitoring of estuarine and coastal systems are:

(a) Large areas may be involved giving rise to data collecting problems. In some cases it is useful to develop simple methods which unspecialised workers can carry out, at least in part.

An experimental survey of Britain's shores was carried out by schoolchildren during the summer of 1974 (ACE/FSC 1974). Several hundred children, mainly in the age group 6–14, assessed the abundance of groups of organisms (limpets, barnacles, green seaweeds, etc.) and obvious pollutants (oil and tar, garbage, etc.) on a simple four point scale (absent, rare, frequent, abundant). Most reports received were sober and objective and many contained well-executed maps and other information. Some reports provide useful if not comprehensive baseline data for particular localities. Plotting the biological data on a nationwide basis produced very crude distribution maps of little use for monitoring but nevertheless of some interest. For example, it is clear from the maps that there are many more limpets in the west and north (as one would expect) and that starfish were stranded along several miles of East Anglian coast during the summer of 1974. The oil pollution map (Fig. 3) corresponds well with R.S.P.B. records of oiled birds on beaches (see, for example, Bibby and Bourne, 1972) and shows promise as a monitoring exercise.

A tanker accident in the eastern Straits of Magellan (Baker *et al.*, 1975) brought to light a need for biological monitoring in this large area which is itself a small oilfield with numerous shore flares, hitherto uncontrolled discharging of oily ballast water, and (proposed) dumping of drilling muds. The problem here is how the two marine biologists who work in the area can monitor hundreds of miles of shoreline and sea. Two possibilities discussed are:

(1) Growth records of mussels from selected sites, with non-specialist staff to do the actual mussel measuring (the mussel *Mytilus edulis chilensis* is one of the most common and widespread inter-tidal organisms in the eastern Straits).

(2) the use of settlement plates fixed to buoys, with the navy installing and collecting them.

(b) Data collection may have to be a very long term proposition in order to show trends of increase or decrease against a background of large natural seasonal and annual fluctuations.

(c) Monitoring schemes are usually started too late to gain any idea of the yearly biological fluctuations before industrial development and there are few possibilities of 'retrospective monitoring'. Some organisms may record changes up to about ten years (for example, changes in growth pattern of the long-lived seaweed *Ascophyllum nodosum*) but would need to be investigated in much greater detail.

It may sometimes be necessary to estimate what was on an unsurveyed

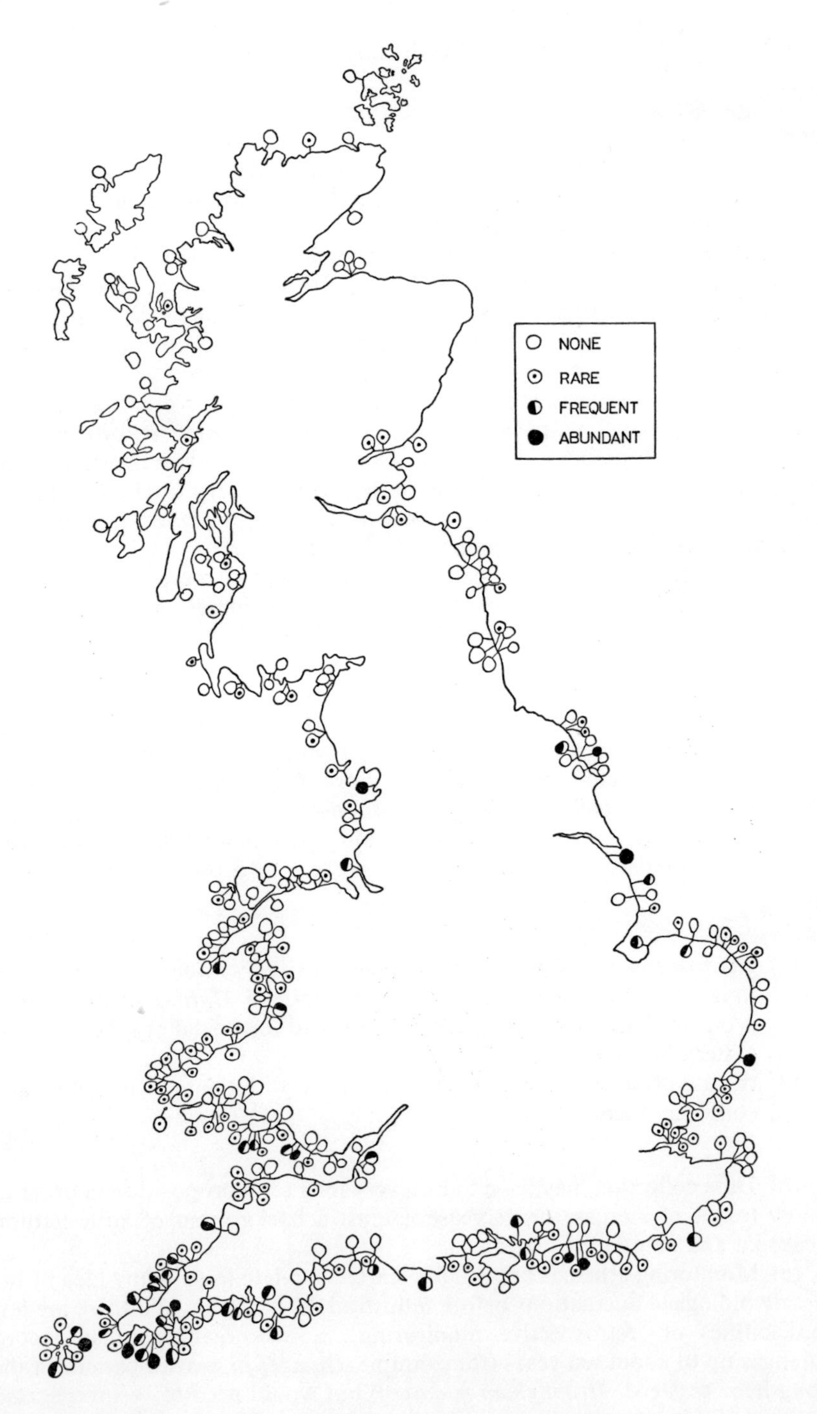

FIG. 3. Distribution of oil and tar on the British coastline, as recorded by schoolchildren during the summer of 1974.

shore before environmental damage occurred and in some cases this may be done from limited data collected after the damage (see, for example, the paper by Cowell *et al.*, (1974) which describes how the general biology of a rocky shore (which is related to exposure) can be deduced from the maximum height of the lichen *Verrucaria maura* (which is also related to exposure).

(d) Monitoring measures trends but does not explain them. Even good correlation between pollutant and biological data does not prove a cause and effect relationship.

PLANNING A MONITORING SCHEME

A typical inlet scheduled for industrial development will contain sand and mud flats with saltmarshes above the high neap tide mark, stony shores and bedrock shores near the estuary mouth. The intertidal flats and marshes may be important bird feeding grounds and there are likely to be small fisheries and shellfisheries.

Described below is a monitoring scheme for such an inlet in the event of a refinery development with associated terminal and tank farm. It is by no means the last word in monitoring scheme proposals, but is a compromise between what is ideal and what is practical. Factors which have to be borne in mind are time, expense, and the desirability of involving local people whenever possible.

Essentials

Rocky Shore Biology

Rocky shores may be affected by oil slicks and dispersants used for cleaning. In the generalised inlet described above, the refinery effluent would also probably be discharged in a rocky shore area near or outside the mouth of the inlet where dispersion is good. Rocky shore transects should be established at different distances from installations, including some distant 'control' transects.

Sand and Mud Flat Biology

With sand and mud flat areas there is a greater danger of long term accumulation of oil, especially if dispersants carry it into the sediments. It is only possible to collect accurate data quickly with species which can be counted or estimated without digging and sieving. Assuming the presence of the lugworm *Arenicola marina*, the most convenient monitoring method is cast counting (Levell, 1975) to give worm densities at different points over the flats.

Saltmarsh Biology

Here again there is a danger of long-term accumulation of oil. Depending on the size and slope of the marsh, mapping of important species could be used, or a series of marsh transects using frequency or cover measurements for the species instead of abundance estimates (Baker, 1971).

Bird Populations
Counts of important resident and migrant species should be made at different times of year. This aspect of monitoring can best be carried out by local ornithologists.

Hydrocarbon Analysis
Samples of sediment and selected inter-tidal and sub-littoral organisms should be analysed for hydrocarbons to find if there are any areas of accumulation.

Suitable organisms include fish and shellfish of commercial importance, especially as such populations are very difficult to monitor in other ways.

Desirables

Sub-littoral Biology
Sub-littoral monitoring of abundance and distribution of species is expensive and very time-consuming. It is, however, desirable when possible, not least because it is a subject about which we know very little. It is important not to confuse ignorance of effects with no effects. Local diving clubs may provide a great deal of help, provided straightforward methods are devised, suitable for use by non-biologists.

Settlement Plates
These may be useful in, for example, an effluent discharge area, for partly defining planktonic effects.

Abundance of Oil Deposits on the Shore
This is included as a project which could be carried out by local people, perhaps schools, who would be in a better position to inspect miles of shore than biologists visiting selected transect sites.

ACKNOWLEDGEMENTS

I am grateful to Catharine Howe and Brian Dicks for assistance with field work and to Janet Dicks and Marion Knight for assisting with the children's survey which was originated by Geoffrey Young of the Advisory Centre for Education.

REFERENCES

ACE/FSC (1974) Shore Watch pack. Advisory Centre for Education, Cambridge.

Addy, J., Baker, J. M., Dicks, B. and Levell, D. (1973). The inter-tidal biology of Sullom Voe, Shetland and proposals for a biological monitoring scheme, O.P.R.U. Annual Report, 1973.

Baker, J. M. (1971). The effects of oil pollution and cleaning on the ecology of salt marshes, Ph.D. thesis, University of Wales.

Baker, J. M. (1975). Experimental investigation of refinery effluents (this volume).

Baker, J. M., Campodonico, I., Guzman, L., Texera, J., Texera, W., Venegas, C. and Sanhueza, A. (1975). An oil spill in the Straits of Magellan (this volume).

Ballantine, W. J. (1961). A biologically-defined exposure scale for the comparative description of rocky shores, *Field Studies*, **1** (3), 1–19.

Bibby, C. and Bourne, W. R. P. (1972). Trouble on oiled waters, *Birds*, **4** (6), 160.

Conference on Pollution Criteria for Estuaries, Southampton University, July, 1973, Conference statement, 4 pp. (unpublished ms.).

Cowell, E. B., Syratt, W. J., Saubrekka, H. and Sage, B. L. (1974). Some observations on the vertical distribution of the supra-littoral lichen *Verrucaria maura* (Wahlend ex Ach) in the region of Lindås, Bergen, Norway, 11 pp., ms.

Crapp, G. B. (1970). The biological effects of marine oil pollution and shore cleansing, Ph.D. thesis, University of Wales.

Crapp, G. B. (1971). Monitoring the rocky shore, in *The Ecological Effects of Oil Pollution on Littoral Communities* (ed. E. B. Cowell), pp. 102–13, Institute of Petroleum, London.

Crapp, G. B. (1973). The distribution and abundance of animals and plants on the rocky shores of Bantry Bay, Irish Fisheries Investigations, Series B, No. 9, 35 pp., Stationery Office, Dublin.

Crisp, D. J. and Southward, A. J. (1958). The distribution of inter-tidal organisms along the coasts of the English Channel, *J. Mar. Biol. Ass. U.K.*, **37**, 157–208.

Crothers, J. (1973). Three assessments of the heavy metal pollution in the Bristol Channel, *Field Work*, 2nd series, March, 1973, 90–102 (Field Studies Council duplicated bulletin).

Dicks, B. (1975a). The effects of refinery effluent: the case history of a saltmarsh (this volume).

Dicks, B. (1975b). The importance of behavioural responses in toxicity testing and ecological prediction (this volume).

Dicks, B. (1975c). Offshore biological monitoring (this volume).

Fossato, V. U. (1975). Elimination of hydrocarbons by mussels, *Marine Pollution Bulletin*, **6** (1), 7–10.

Goodman, G. T. (1974). How do chemical substances affect the environment? *Proc. R. Soc. Lond.*, B, **185**, 127–48.

Jenkins, D. W. (1971). Biological monitoring of the global chemical environment, 54 pp. (unpublished ms.).

Levell, D. (1975). The effects of Kuwait crude oil and the dispersant BP 1100X on the lugworm *Arenicola marina* L (this volume).

Moyse, J. and Nelson-Smith, A. (1963). Zonation of animals and plants on rocky shores around Dale, Pembrokeshire, *Field Studies*, **1** (5), 1–31.

Natural Environment Research Council (Nature Conservancy Council). A prospectus for nature conservation within the Moray Firth. 81 pp. + maps.

Wilson, K. W. (1974). Toxicity testing for ranking oils and oil dispersants, in *Ecological Aspects of Toxicity Testing of Oils and Dispersants* (ed. L. R. Beynon and E. B. Cowell), pp. 11–22, Institute of Petroleum, London.

6
Ecological Changes in Milford Haven During its History as an Oil Port

JENIFER M. BAKER

(*Oil Pollution Research Unit, Orielton Field Centre, Pembroke, Wales*)

SUMMARY

The current ecological state of Milford Haven, after fifteen years of industrialisation, is described. There are still no overall changes which can be attributed to the oil industry, though a varying pattern of short-term localised effects persists. The most severe of such effects over about one mile of shore are a result of the Dona Marika *petrol spillage of 1973. Localised chronic effects in Little Wick Bay are associated with a refinery effluent, and the death of a patch of* Spartina *(cord-grass) marsh at Martinshaven may also be a result of chronic pollution.*

Attention is drawn to the problem of unidentified slicks outside the mouth of the Haven.

INTRODUCTION

In the past, a number of papers dealing with various aspects of Milford Haven ecology have been published and it was concluded at a previous symposium (Cowell, 1971) that ten years of industrialisation (1960–70) had not caused overall ecological changes. The only striking general change (concerning barnacle populations) could be attributed to a climatic deterioration. Localised changes were observed round one of the refinery effluents and on a few rocky shores and salt marshes following oil spills and, in some cases, cleaning.

This paper refers to and updates earlier work. It was felt that an account of the current ecological state of Milford Haven, after fifteen years of industrialisation, would be useful to those concerned with oil developments elsewhere, though straightforward extrapolation of results are often not possible (Dicks, 1975).

ROCKY SHORES

Moyse and Nelson-Smith (1963), and Nelson-Smith (1964, 1967) described the distribution of littoral plants and animals in Milford Haven, basing their studies on a series of rocky shore transects. These results, dating from the earliest stages of the industrialisation of Milford Haven, form a baseline for all subsequent rocky shore surveys. The transects were re-surveyed by Crapp during 1968–70 and results are discussed in full in his thesis (Crapp, 1970). He concluded that no general impoverishment of the rocky shore fauna and flora had occurred during Milford Haven's first decade as an oil port. There had been, however, a marked change in the barnacle populations, attributable to climatic deterioration, a decline in the abundance of the topshell *Monodonta lineata* possibly attributable to climatic deterioration, and an inexplicable decline in the numbers of *Littorina saxatilis tenebrosa*. Localised effects attributable to the oil industry were observed in the case of Little Wick Bay near a refinery effluent and on Hazelbeach following oiling and dispersant treatment.

Using Crapp's methods which are summarised in an earlier paper (Baker, 1975a), a further survey of some of the Milford Haven transects was carried out during 1973 and 1974. These included areas of particular interest, namely, shores affected by the *Thuntank* fuel oil in 1971, the *Dona Marika* petrol in 1973, and the refinery effluent in Little Wick Bay (Fig. 1).

The *Thuntank 6*—West Angle Bay

This incident is described by Ottway (1971). Light fuel oil affected rocky shores mainly in the West Angle Bay area during March 1971. Cleaning involved careful use of BP 1100X and some pumping from rock pools.

The animals affected were gammarid crustacea, which were trapped in the surface oil film of rock pools; gastropod molluscs (mostly *Littorina littorea*, *Littorina saxatilis* and *Gibbula umbilicalis*) which detached and fell into crevices and pools; and limpets (*Patella vulgata*) which detached. Loose and detached limpets were mostly in the smaller size range. Two months after the spill, few traces of oil could be found and surveys by students from Orielton Field Centre indicated that there was little change in the shore biology. The West Angle Bay and Angle Point transects were re-surveyed in 1974 and these results are compared with earlier data in Fig. 2. The major change in abundance between 1968 and 1974 concerns the green seaweed *Enteromorpha* which has increased both at West Angle Bay and Angle Point. *Littorina saxatilis* at Angle Point shows a change in distribution, being further down the shore in the later survey. This is possibly a seasonal effect as the 1974

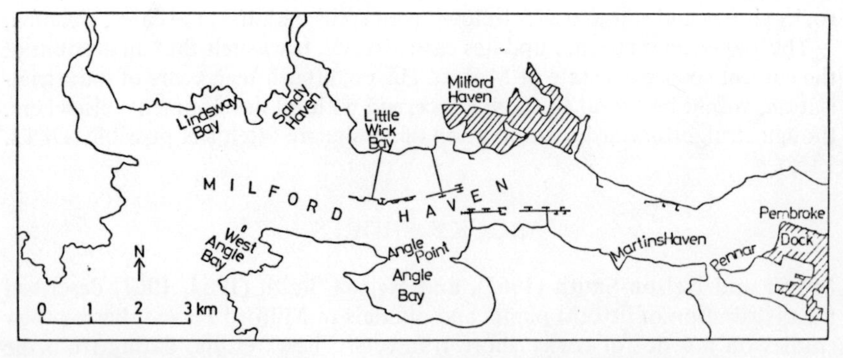

FIG. 1. Milford Haven, showing sites mentioned in the text.

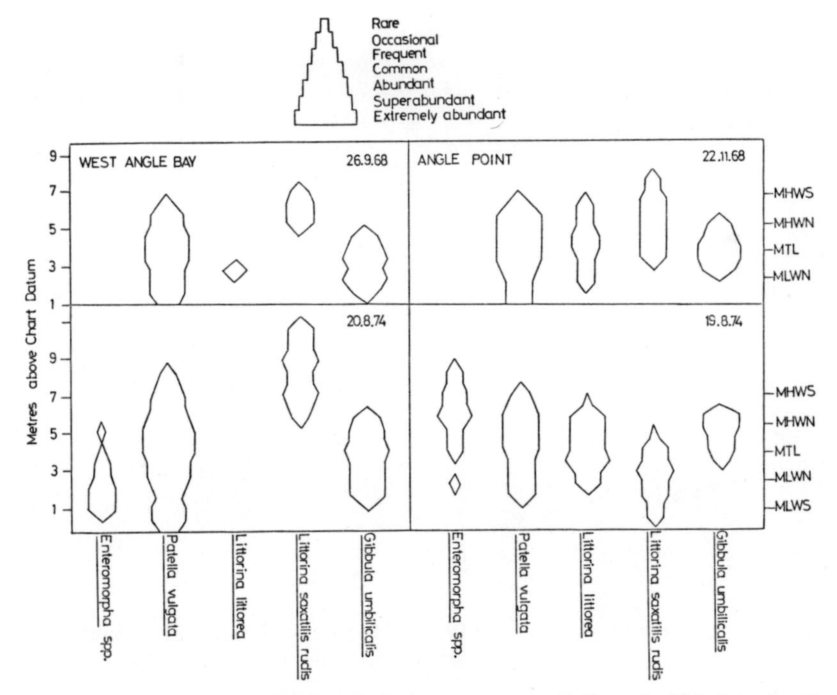

FIG. 2. West Angle Bay and Angle Point transects: 1968 and 1974 data for five species.

survey was in very hot weather in mid-August, whereas the 1968 survey was at the end of November.

Pennar

This shore was affected by light Iranian crude oil weathered for only two to three hours, on 22 November 1972. It is a sheltered, weedy shore dominated by *Ascophyllum nodosum* and was heavily oiled between about 3 m above Chart Datum and high spring tide mark.

Unfortunately no previous data were available for this site; however, a transect survey (Fig. 3) carried out on 27 November 1972 showed a fairly normal result for this type of shore. The winkle *Littorina littoralis* was observed feeding on oily *Ascophyllum*, specimens brought back to the laboratory had guts full of oil droplets and passed oily faeces, but remained active for five weeks in the laboratory before being returned to the shore.

Limpets were not badly affected—they are present at Pennar in relatively small numbers and may have been partly protected by the *Ascophyllum*; in addition, the oil reached the shore at a time when the limpets were probably inactive.

The *Dona Marika*—Lindsway Bay

A full report of the *Dona Marika* incident of August 1973 is given by Blackman *et al.*, (1973). Between 2000 and 3000 tons of petrol were spilled over two periods of about 48 and 36 hours, into a small area of enclosed bays and

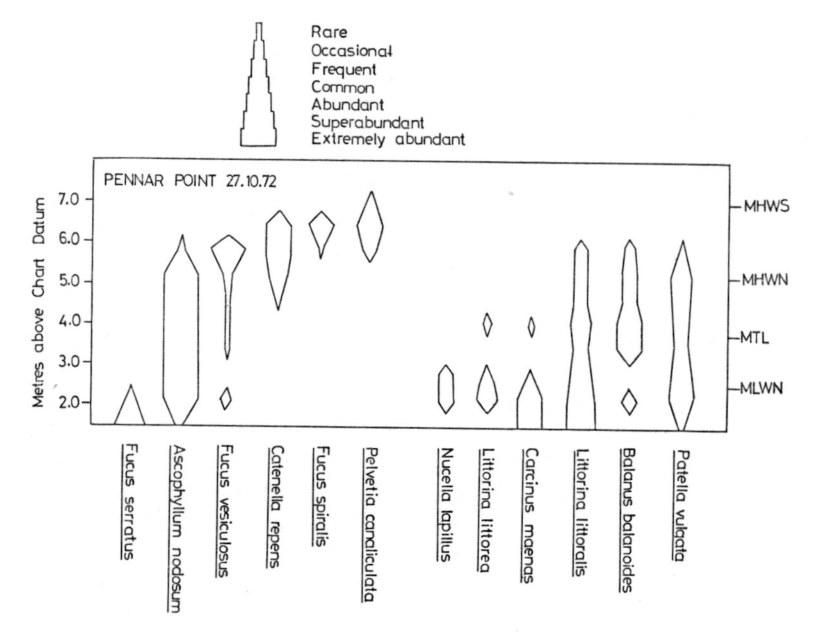

FIG. 3. Transect survey data from Pennar, five days after a spill of light Iranian crude oil.

during high winds and heavy seas. Under these conditions, the petrol did not behave as a non-persistent oil. Leakage was so rapid that the surf apparently formed water-in-petrol emulsions before much evaporation could take place.

The shore between Little Castle Head and Watch House Point (Fig. 4) was examined two days after the accident. Over much of this area, there was evidence of limpet (*Patella* spp.) detachment, and retraction of winkles

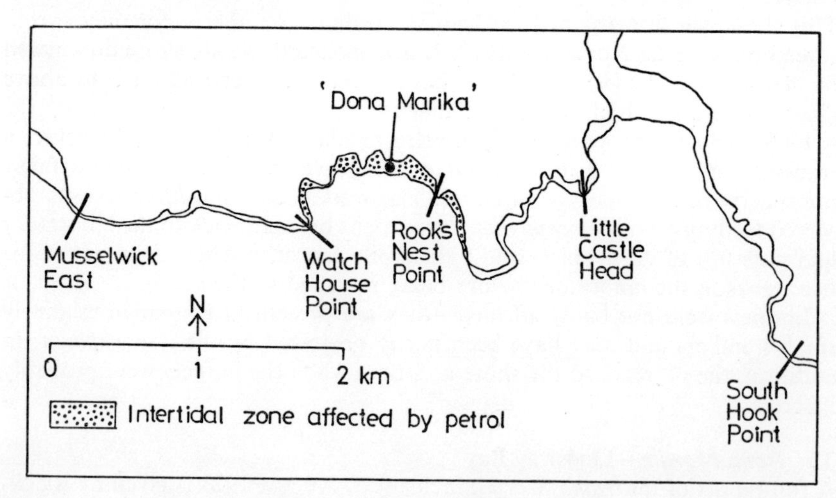

FIG. 4. Part of the north shore of Milford Haven showing area affected by the *Dona Marika* petrol and positions of transects.

(*Littorina* spp.), top shells (*Gibbula* spp.), and dog whelks (*Nucella lapillus*) into their shells. Mussels (*Mytilus edulis*) near the ship were gaping. Effects were severe between Watch House Point and Great Castle Head, and minor between Great Castle Head and Little Castle Head.

Detached *Patella*, and retracted *Littorina*, *Gibbula* and *Nucella* were taken back to the laboratory, washed, and kept in clean aerated seawater. Within a day at least 50% of each species had apparently recovered. Thus the initial effect of the petrol was narcotic rather than lethal.

Five of the transects (for positions, see Fig. 4) which form part of the Milford Haven monitoring programme, were re-surveyed. Some of the data from three surveys are shown in Figs. 5 and 6.

In 1973, shortly after the spill, the Watch House Point and Rook's Nest Point transects showed a large reduction in limpet numbers from abundant (i.e. over 50 per m^2) to occasional (i.e. 1–10 per dam^2). The other transects show changes between abundant and common, i.e. between more than 50 per m^2 and 10–50 per m^2. Changes of this magnitude between surveys are common because of natural yearly fluctuations, seasonal variations and different observers.

Following limpet detachment, there was abundant growth of the green seaweed *Enteromorpha*, followed by growth of fucoid seaweeds. Limpet numbers were back to normal the following year (1974) but nearly all limpets were at this time very small as recolonisation was from larval stages in the plankton. Thus the grazing activity was insufficient to have cleared the abundant growth of *Enteromorpha* during 1974.

Quite apart from *Enteromorpha* growth following limpet detachment, there appears to be an overall increase in *Enteromorpha* on all transects between the 1969 and 1974 surveys. This could be a seasonal difference (a February–March 1969 survey compared with later dates in 1974), but even so it is rather large. A similar effect was observed at West Angle Bay, where it is unlikely to be purely seasonal. The question of whether *Enteromorpha* is really increasing in Milford Haven as a whole will have to be elucidated by a complete re-survey of the monitoring transects. Numbers of gastropod molluscs such as the dog whelk (*Nucella lapillus*) had recovered by 1974, but in this case recolonisation was largely by adult animals. This sequence of events is similar to that following oiling and cleaning on Hazelbeach (Crapp, 1970, 1971) where numbers of winkles and topshells decreased and then recovered. Crapp deduced that animals which retracted under the influence of pollutants were washed into the sub-littoral zone and eventually crawled back into the littoral.

It is interesting to compare the effects of oils or dispersants on sheltered rocky shores (such as Pennar) with exposed rocky shores (such as Lindsway Bay). An unpolluted sheltered shore naturally has fewer limpets and large masses of weed such as *Ascophyllum*, so in some respects resembles a polluted exposed shore. In other words, oils or dispersants have a more profound effect upon the ecology of exposed rocky shores, in particular through limpet detachment (Dicks, 1973) and subsequent chain reactions.

Little Wick Bay

Re-surveys of transects in this area of effluent discharge are described in the

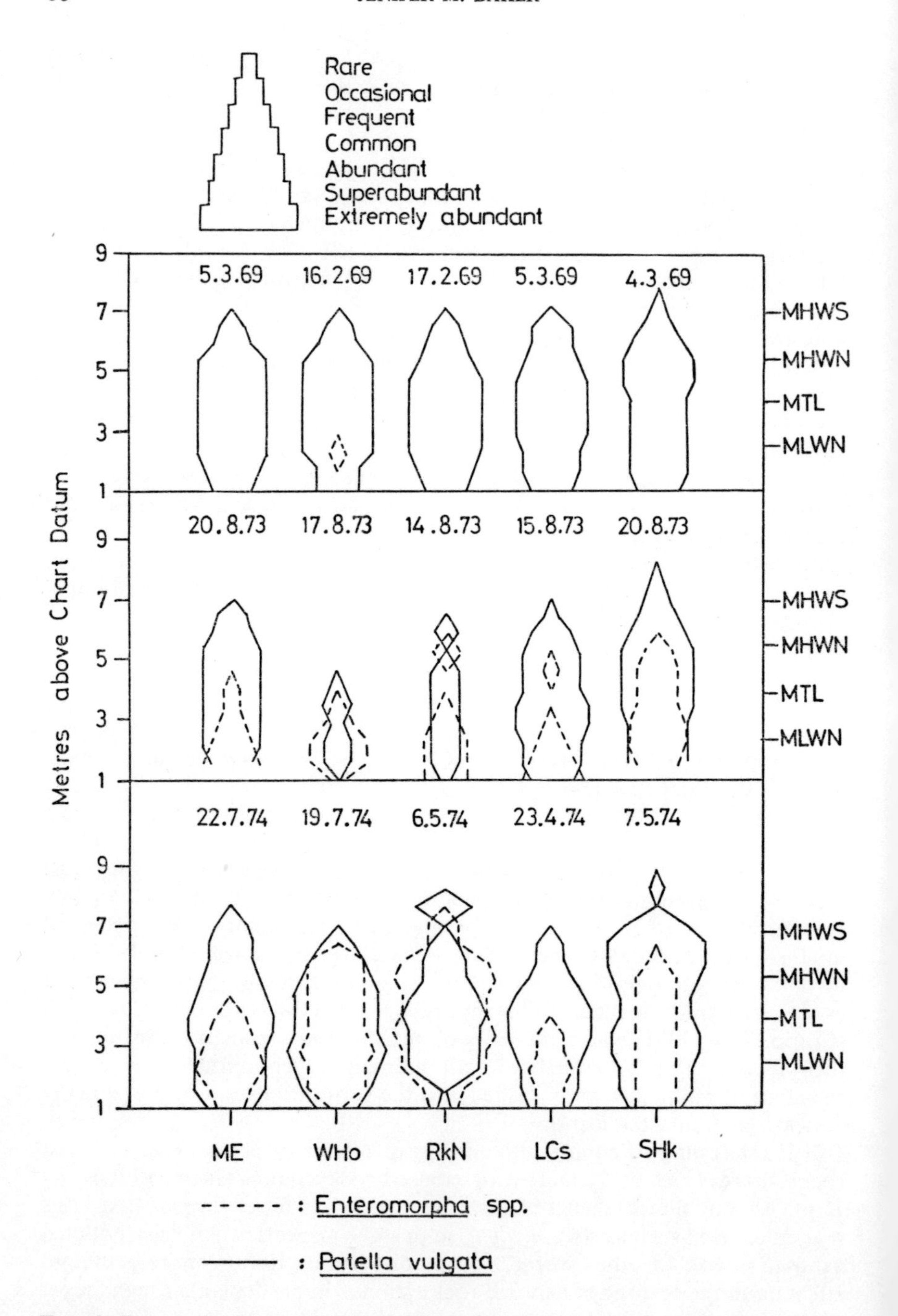

FIG. 5. Pre- and post-*Dona Marika* data for the green seaweed *Enteromorpha* and the limpet *Patella vulgata*.

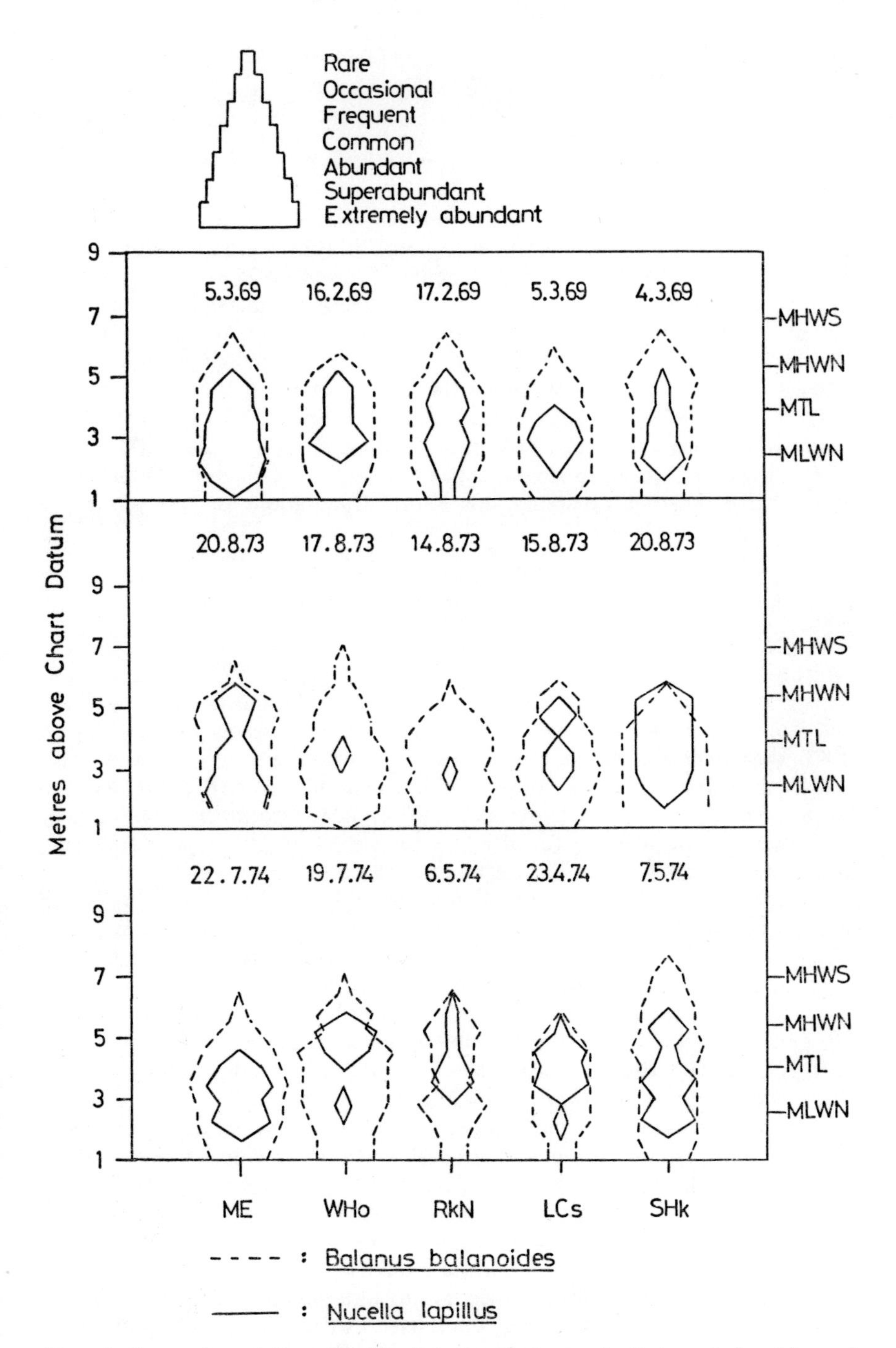

FIG. 6. Pre- and post-*Dona Marika* data for the barnacle *Balanus balanoides* and the dog whelk *Nucella lapillus*.

paper by Baker (1975b). There is little change from the situation in 1970 (Crapp, 1970, 1971).

SAND AND MUD SHORES

Following the *Thuntank 6* accident, oil was stranded on the sand at West Angle Bay and no lugworm casts were observed (Ottway, 1971). This subject has been covered by the field experiments of Levell (1975).

Thuntank oil also reached Sandyhaven, where numbers of cockles were found on the surface, some dead, others still alive but weak and gaping. As no traces of oil could be seen on these cockles, Ottway concluded that dispersants could be responsible for the high mortalities.

Oil spills have occasionally occurred in the Angle Bay area and the semi-commercial cockle fisheries there have declined from a peak of 100 tons in 1971; however, as cockle stocks appear to be low throughout West Wales this decline must be attributed to natural fluctuations or to over-fishing rather than oil pollution.

SALTMARSHES

Martinshaven

This marsh, comparatively near oil company installations (Fig. 1), has been oiled on a number of occasions. Baker (1971a, b) describes incidents up to 1971; subsequently small patches of thin oil film have been seen at the seaward end of the marsh on two occasions and a heavier film on part of the *Spartina anglica* (cord-grass) area was observed at the end of December 1974. Refinery effluent has been discharged down the main creek for short periods during the winters of 1972 and 1973 (see Baker, 1975b).

Oil dating from the *Chryssi P. Goulandris* spillage of January 1967 is still present in mud at the seaward end of the marsh—it is present down as far as 50 cm where it was carried by dispersants. Degradation is presumably slow because the mud is anaerobic.

A 5 cm thick patch of heavy fuel oil stranded in the middle of the marsh at the end of February 1969 is still visible in places. Dispersants were not used in this case, and the oil has remained on the surface. Recolonisation, by *Triglochin maritima* and other species, is not yet complete.

The most interesting recent change in the saltmarsh vegetation is the death of a patch of *Spartina* at the seaward end of the marsh (Fig. 7). Signs of this were noticed during 1973 and the dead patch became well defined during 1974. Death has not been obviously associated with any one oiling incident or the effluent discharges, so the cause is debatable. It could be a chronic pollution effect or it could be a natural die-back. There is some evidence that *Spartina anglica* can eventually succumb to waterlogged reducing conditions (which the plant may help to bring upon itself by trapping fine silt). Goodman and Williams (1961) decided that *Spartina* 'die-back' was caused by a toxic reduced ion, possibly sulphide, in the substrate. The Martinshaven effect does not, however, entirely fit this picture because some of the dead plants are

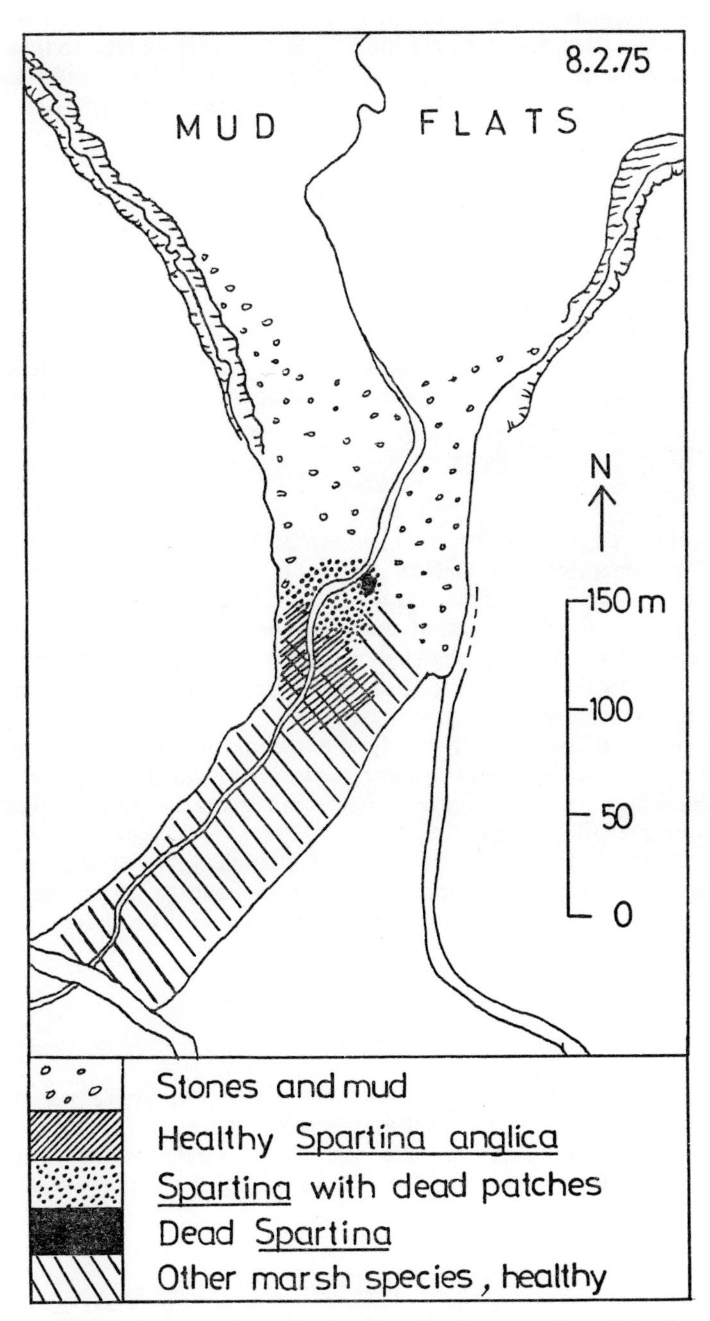

FIG. 7. Martinshaven saltmarsh showing features mentioned in the text.

in relatively well-drained stony mud and in other parts of the marsh *Spartina* is flourishing in more badly drained conditions. The dead patch is, on the other hand, in the position where small oil films land from time to time, where the short effluent discharges fanned out, and where oil is still, in places, present in the mud. The evidence suggests, therefore, that it is a chronic pollution effect.

SUBLITTORAL FAUNA

Unfortunately, no baseline data are available from pre- or early industrial times. A survey of the current situation is described by Addy (1975). It is also worth noting an observation by Peter Hunnam (Dale Fort Field Centre) who dived in Lindsway Bay following the *Dona Marika* accident. He reported narcotised bivalves and large numbers of dead sea-urchins (*Echinocardium*).

PLANKTON

The work of Dias (1960) on the plankton of Milford Haven was used as a basis for comparison by Gabriel (1974). His samples were taken during 1970–71. He concluded that phytoplankton abundance was similar to that found by Dias, though he found small changes in the seasonal distribution, and smaller fluctuations in the size of the standing crop. Only one diatom species has disappeared while thirteen new diatoms and four new dinoflagellates have appeared.

Dias recorded a far greater variety of zooplankton. However, many of the species which have apparently disappeared were only taken from Dale Point—the greater majority of species recorded from regular monthly samples inside the Haven are still present.

Gabriel concluded that on the basis of a single year's study, and in the absence of detailed hydrographic data, it is difficult to assign the changes which appear to have occurred to any specific cause. The changes are relatively small and do not constitute a deterioration.

FISH

Very little detailed information is available. There is no obvious oil industry effect on movements of sea trout and salmon, and a small isolated herring fishery survives at Llangwm. After the *Dona Marika* accident, mackerel tasting of petrol were reported from Sandyhaven, but reports of dead fish were not verified.

BIRDS

Oiled birds, particularly mute swans, are reported in the Haven occasionally—usually the numbers involved are from one to five, though more were affected by the *Chryssi P. Goulandris* spill of 1967. Some birds become oiled outside the Haven, e.g. 309 scoters, oiled in the summer of 1973 by an unidentified slick, came ashore on the south Dyfed (Pembrokeshire and Carmarthenshire) coast. 224 of them were dead and small numbers of other species were also affected. After the *Dona Marika* accident up to 100 dead gulls were reported from St Brides Bay (north of Milford Haven). It was suggested that these were killed by eating petrol-contaminated limpets, but no dead gulls were found in the area where they were actually feeding.

MAMMALS

There have been no reports of adverse effects within the Haven, but 52 seals (including 25 pups) were oiled in the Skomer area during September and October 1974. Survival was good, but growth of pups, both oiled and unoiled, was below average. This was another case of unidentified slicks, apparently of several different types of oil.

CONCLUSIONS

On rocky shores there is some evidence of an unexplained increase of *Enteromorpha* over the areas of Milford Haven resurveyed during 1974. This will have to be further investigated through complete re-survey. Oil spillages, and in particular the *Dona Marika* petrol spillage, caused short-term localised changes, mainly involving gastropod molluscs and the limpet/seaweed ratio. About one mile of coastline was badly affected by the *Dona Marika* in 1973, but recolonisation by young limpets occurred during 1974.

Little change has occurred in the effluent discharge area of Little Wick Bay, so the species and area affected appear to have reached some sort of equilibrium.

Concerning salt marshes, a patch of *Spartina anglica* at Martinshaven has died, possibly as a result of chronic pollution by oil films and two periods of effluent discharge.

It is still true to say that there are no overall ecological changes in Milford Haven which can be attributed to the oil industry, and this is a reflection of an efficient harbour administration, the co-operation of the oil companies, and a well-organised clean-up system. It is, however, clear from effects noted above that there is no room for complacency.

Unidentified (and untreated) slicks off the Dyfed coast are a cause for concern, two bad incidents being the oiling of scoters in 1973 and seals in 1974.

ACKNOWLEDGEMENTS

I am grateful to Brian Dicks, Marion Knight and Susan Reynard for assistance with field work, to Roger Bray for information on seals, and to David Saunders for information on scoters.

REFERENCES

Addy, J. (1975). The sublittoral fauna of Milford Haven (this volume).
Baker, J. M. (1971a). The effects of oil pollution and cleaning on the ecology of salt marshes, Ph.D. thesis, University of Wales.
Baker, J. M. (1971b). In: *The ecological effects of oil pollution on littoral communities*, ed. E. B. Cowell, Institute of Petroleum, London, pp. 17–18.
Baker, J. M. (1975a). Biological monitoring—principles, methods and difficulties (this volume).
Baker, J. M. (1975b). Investigation of refinery effluent effects through field surveys (this volume).
Blackman, R. A. A., Baker, J. M., Jelly, J. and Reynard, S. (1973). The *Dona Marika* oil spill, *Marine Pollution Bulletin*, **4** (12), 181–2.
Cowell, E. B. (ed.) (1971). *The ecological effects of oil pollution on littoral communities*, Institute of Petroleum, London.
Crapp, G. B. (1970). The biological effects of marine oil pollution and shore cleansing, Ph.D. thesis, University of Wales.
Crapp, G. B. (1971). In: *The ecological effects of oil pollution on littoral communities*, ed. E. B. Cowell, Institute of Petroleum, London.
Dias, N. S. (1960). The plankton of Milford Haven, M.Sc. thesis, University of Wales.
Dicks, B. (1973). Some effects of Kuwait crude oil on the limpet, *Patella vulgata*, *Environmental Pollution*, **5**, 219–29.
Dicks, B. (1975). The applicability of the Milford Haven experience for new oil terminals (this volume).
Gabriel, P. L. (1974). The plankton of Milford Haven: a report on the plankton of the estuary following ten years of industrialisation, Duplicated report, University College of Swansea.
Goodman, P. J. and Williams, W. T. (1961). Investigations into 'die-back' in *Spartina townsendii* agg. III, *J. Ecol.*, **49**, 391–8.
Levell, D. (1975). The effects of Kuwait crude oil and the dispersant BP 1100X on the lugworm *Arenicola marina* L. (this volume).
Moyse, J. and Nelson-Smith, A. (1963). Zonation of animals and plants on rocky shores around Dale, Pembrokeshire, *Field Studies*, **1** (5), 1–31.
Nelson-Smith, A. (1964). Some aspects of the marine ecology of Milford Haven, Pembrokeshire, Ph.D. thesis, University of Wales.
Nelson-Smith, A. (1967). Marine biology of Milford Haven: the distribution of littoral plants and animals, *Field Studies*, **2**, 435–77.
Ottway, S. (1971). The *Thuntank 6* spill, *OPRU Annual Report*, 1971.

7

The Applicability of the Milford Haven Experience for New Oil Terminals

BRIAN DICKS

(*Oil Pollution Research Unit, Orielton Field Centre, Pembroke, Wales*)

SUMMARY

Over the last decade, Milford Haven has expanded rapidly as an oil port. Biological surveys during this period have shown that the oil industry has caused little overall change in the life on the shores of the Haven, though localised damage has been caused by occasional large spillages or in the immediate vicinity of refinery effluent outfalls.

As little damage has been done to the marine fauna and flora of Milford Haven, there is a tendency to assume that similar development elsewhere will have little ecological effect. This is not necessarily the case, as certain special characteristics of Milford Haven have aided in keeping it clean. To illustrate some of these special characteristics, and some of the differences between Milford Haven and other development sites, a brief comparison is made of Milford Haven with two other potential development areas: Sullom Voe in the Shetlands, and the Cromarty Firth in the east coast of Scotland.

It is concluded that a number of coincidental factors have contributed toward keeping Milford Haven clean. If these factors are duplicated at new sites, or fully understood and taken into account, then a wide range of experience, both in terms of port management and the effects of oil industry development on shore life, become applicable to new development areas. If this is not the case, extrapolation of the Milford Haven experience to new oil terminals can only be regarded as dangerous.

INTRODUCTION

Since the early 1960s, the oil industry in Milford Haven has expanded rapidly. Recent figures (Dudley, 1975) show that 4000+ tankers handled in excess of 60 million tons of oil and refined products in 1974. Even with this high level of shipping activity, plus the presence of four oil refineries, the largest oil-fired electricity generating station in Europe, and an oil tank farm supplying a refinery elsewhere, biological damage to the marine environment seems small. Extensive biological work since 1960 (summarised in Crapp, 1970, and Baker, 1975a) on the shores suggests that the biological damage which has occurred has been localised to areas where large spillage incidents have occurred, e.g. the *Dona Marika* spill (Baker, 1975a) or in the immediate vicinity of refinery effluent outfalls (Baker, 1973, 1975b).

As a result of the high level of oil industry activity combined with little extensive ecological damage in Milford Haven, there is a tendency to assume that similar levels of activity elsewhere will also have little ecological effect. This is not necessarily the case, as a number of factors influencing the impact of pollution on Milford Haven may not apply elsewhere. Some of these factors are discussed herein, and to illustrate differences between Milford Haven and other sites, a comparison with two other superficially similar oil development areas is made.

FACTORS AFFECTING THE BIOLOGICAL IMPACT OF OIL POLLUTION IN COASTAL DEVELOPMENT AREAS

Three categories are distinguishable:

(1) those factors dependent on management of the port or coastal area, and the efficiency with which pollution hazards are dealt;
(2) those factors dependent on the hydrographical, geographical and biological make-up of the area;
(3) others.

Port management and pollution policy

The influence that port management can have on pollution has been dealt with by the Milford Haven Harbourmaster, Captain G. Dudley, in this volume ('The incidence and treatment of oil pollution in oil ports'). It is evident that Milford Haven has a port authority aware of the inevitability of at least some pollution, with its attendant problems, and as a result has developed an efficient and successful policy to deal with pollution promptly as and when it occurs. There is a considerable amount of co-operation in this anti-pollution effort by the oil companies. These have undoubtedly been major factors in maintaining Milford Haven in a relatively clean condition. However, this is a factor dependent on the human element, and is bound to vary from site to site.

Physical and biological factors

When assessing the possible impact of pollution on the marine ecosystem, there are three important variables which depend on the geographical, hydrographical and biological make-up of the area which should be considered. They are:

(i) the rate at which pollutants are dispersed and diluted (either naturally or after treatment by man);
(ii) the type of shores likely to be affected;
(iii) the occurrence of rare species, SSSI's (Sites of Special Scientific Interest), and birds (migrants and residents) feeding on the shore.

Dispersion and Dilution of Pollutants

The importance of adequate dispersion of oil and other pollutants is twofold. First, the greater the dispersion and dilution, the less time the pollutant can affect any one area, and the progressively lower its concentration. Secondly adequate dispersion will allow biodegradation to take place most efficiently, and avoid the overloading of these natural systems which would result in build-up of pollutants. Whilst dispersion of oil can be speeded up by chemical dispersants (if recovery of the oil is not possible), it is pointless doing this in areas where this dispersed mixture is not adequately dispersed by natural water movements.

The rate at which pollutants will be dispersed and diluted in inshore areas depends on several factors, which include currents and tidal movements, prevailing winds, wave action and the run-off of freshwater from the land catchment area. Wind direction is important, in that it will determine the likelihood of oil slicks being driven ashore into small bays or inaccessible areas. Currents and tidal movements will also affect slick movements as well as the mixing of coastal water with water further offshore. The greater the currents and tidal flows, the higher the chance of dispersion of effluents and oil, and the less chance of oil being ponded in an area for any length of time. Ponding is potentially a great danger, as build-up of pollutants to a critical level may occur almost unnoticed in areas where water movements are restricted or non-existent. The rate at which pollutants are dispersed to the sea may also be influenced by freshwater run-off, a large river producing greater dispersion and dilution than a small stream. Wave action assumes importance in shore pollution incidents when it, in combination with tidal movements, will determine the speed with which oil is removed from the shore, either naturally or after treatment.

In areas where dispersion and dilution may be partly or severely restricted, input of pollutants should be reduced to levels with which the natural breakdown systems can cope. If this is not possible, serious consideration should be given to alternative areas.

Shore Type

It is convenient here to divide shore types roughly to two major divisions—hard substrates (rocky and stony shores) and soft substrates (sandy and muddy shores, including salt marshes), as certain general comments can be made about each type.

Hard substrates. Many of the organisms of rocky shores are resistant to all but the most severe oil pollution, for example many of the fucoid seaweeds, or can use their natural protective mechanisms to withstand short spells of pollution stress, e.g. *Littorina littorea* (Hargrave and Newcombe, 1973), *Littorina saxatilis* (Parsons, 1972). However, some organisms can be very sensitive, e.g. *Patella vulgata* (Dicks, 1973) and *Balanus balanoides* (Crapp 1970; Dicks, 1975). There is obviously a wide range of sensitivity and response. These shores are also relatively easy to clean by spraying with dispersant, although this usually results in greater ecological damage. Once cleaned, however, the shores are usually suitable for immediate recolonisation.

Soft substrates. Whilst sandy shores do occur in areas of heavy wave action, these shores are very often poor in flora and fauna. The majority of muddy and sandy shores occur in areas of reduced water movement and little wave action, which may mean the presence of the problem of limited dispersion discussed in the last section. Some soft shore organisms have been shown to be very sensitive to pollution stress in certain conditions, e.g. *Arenicola marina* (Levell, 1975), *Nereis diversicolor* (Baker, 1975b and 1975c and Dicks, 1975), saltmarsh plants to repeated oiling (Baker, 1971) and chronic pollution (Dicks, 1975). Others have been shown to thrive under a pollution stress to which they are resistant but which removes other organisms, allowing them to grow well, e.g. *Capitella* (Reish, 1965).

Whilst there is undoubtedly as wide a range of sensitivity to pollution stress in the biota of these shores as on rocky shores, a further danger occurs to soft shore organisms, and this is the danger of incorporation of toxic substances into the sediments. Oils can penetrate the sediments of the shore (Levell, 1975) and penetration can be deep if the oil is treated with dispersant. Once trapped in the sediment, oil may remain for considerable periods of time (Baker, 1975a) as dispersion and evaporation are slowed or prevented and degradation may not continue due to a lack of oxygen. Once sediments contain oil they may prevent or slow down recolonisation by inhabitant species (Levell, 1975). Oils and other contaminants may also be trapped in sediments by adsorption onto suspended silt particles, which then sink, or in the bodies of dead organisms. Whilst most of such oil will biodegrade as it lies on the substrate surface where plenty of oxygen is available, if sedimentation rates are high, or oil abundant, biodegradation may not be able to keep pace with deposition.

Further difficulties occur when trying to prevent oil from reaching muddy shores or when cleaning sandy and muddy shores. The shallowness of water over a mud flat on a rising tide will make the area inaccessible to boats cleaning up the oil or handling booms, and softness of the substrate may prevent access from the land. Hence oil slicks taken over a mud flat by the tide may be completely inaccessible.

Whilst it is evidently best to keep oil off the shore where possible, being easiest to deal with at sea, oil produces greater problems over soft substrates than hard if it arrives on the shore. This is almost inevitable in an oil port or around oil terminals where arrival on the shore can be very soon after spillage if strong winds or currents move the oil.

A further difference between hard and soft shores is that soft shores are often feeding or roosting areas for many resident and migrant birds. Hence oil affecting soft shores may also affect bird populations. This is considered in the next section.

Rare species, SSSI's and bird populations

Rare species and SSSI's can be dealt with together, for it is usual that any sites where rare species occur are listed as SSSI's. The danger of pollution stress at such sites is obvious, as it may result in loss of rare species or destruction of particularly unusual or interesting habitats. Where possible, such sites should be given extra protection or, preferably avoided altogether, as their loss is an extra loss to the diversity of our environment.

The effects of oil on, and dangers of oils to birds have been dealt with in great detail by several authors (Bourne, 1968; Cramp *et al.*, 1974; IMCO, 1973). Birds are susceptible at two levels, first by the direct action of oil on their plumage, and secondly by action of pollution on their feeding grounds, in particular on the soft shores, the feeding grounds of many waders and migrants. Pollution stress on both the feeding grounds and the birds is particularly important in Britain where many migrants visit our shores. If these are affected in Britain, environments over a wide area of the globe are being affected. Any oil developments in areas of abundant residents or migrant bird feeding grounds immediately place these animals at risk due to the inevitability of at least some pollution.

Other Factors

Most of the factors in this section are not directly related to the biology of the area, but result in biological effects if development of the oil industry occurs. These include socio-economic pressure and political pressure for industrial developments to take place, and the development of amenity areas, where pressure to clean up oil pollution from whatever source may be intense and result in considerably greater damage than the original pollution. Further factors include the importance of commercial fisheries, and aesthetic factors such as destruction of scenic beauty by development. These factors, whilst of obvious importance in the development areas, are outside the scope of this paper.

A COMPARISON OF MILFORD HAVEN, SULLOM VOE AND THE CROMARTY FIRTH

A brief comparison is made here of Milford Haven, which has already been extensively developed by the oil industry, with two sites of proposed oil industry development: Sullom Voe in the Shetland Islands and Cromarty Firth on the east coast of Scotland. Comparison is made on points discussed above:

geographical location;
tidal and other water movements;
throughput of freshwater;
shore type;
rare species and SSSI's;
bird populations.

Comment on the previous section is not possible in the context of these new development areas, and detailed comment is not appropriate here.

Geographical Location

The location of the three areas is shown in Fig. 1, and their relative sizes and aspects in Fig. 2. All three sites are very different in aspect, though superficially similar in being long inlets in the coastline. The inlet of Milford Haven

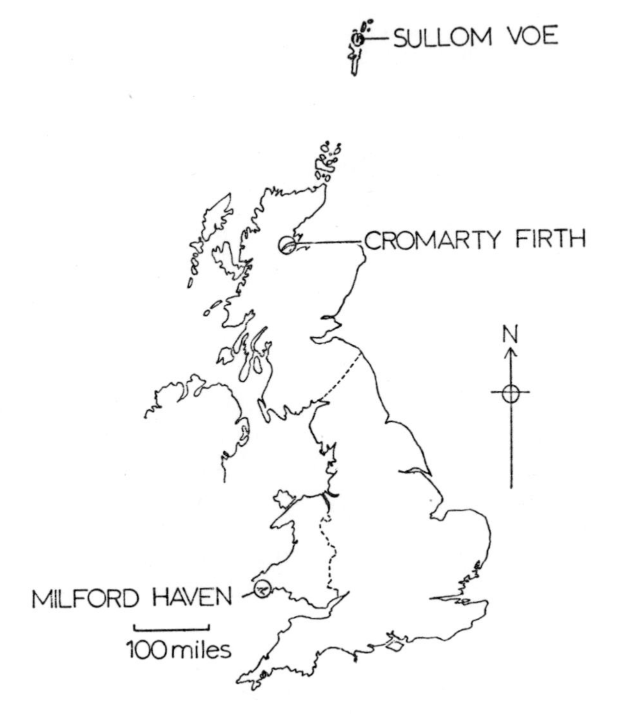

FIG. 1. The respective positions of Sullom Voe, the Cromarty Firth and Milford Haven in the British Isles.

runs from east to west, of Sullom Voe from south to north and of the Cromarty Firth from west to east. As the prevailing wind direction in Britain is from the south west, Sullom Voe and Cromarty Firth are relatively more sheltered than Milford, though the bend in Milford Haven inside the mouth protects the inner Haven from heavy wave action (Nelson-Smith, 1965). However, the winds in Britain are extremely variable, and can blow strongly from any direction.

Tidal and Other Water Movements

Whilst the water volumes of Milford Haven have been calculated elsewhere (Nelson-Smith, 1965), little published data is available on either Sullom Voe or the Cromarty Firth. A crude method is used here to estimate the low tide and tidal volumes involving calculation of cross-sectional areas of profiles at various distances along the three inlets and multiplying by distance for each profile. Volumes calculated were the empty volume (low spring tide volume) and full volume (high spring tide). This gives some idea of the rate of water change, and (assuming that the tidal volume flows evenly through the mouth each six hours) a rough idea of current speeds in the mouths of the three inlets produced by the tides. The same method was used for all three sites, and the results for Milford Haven compared with those of Nelson-Smith (1965). Whilst Nelson-Smith regarded the mouth of Milford Haven as lying at Thorn Island, the mouth is regarded here to be between

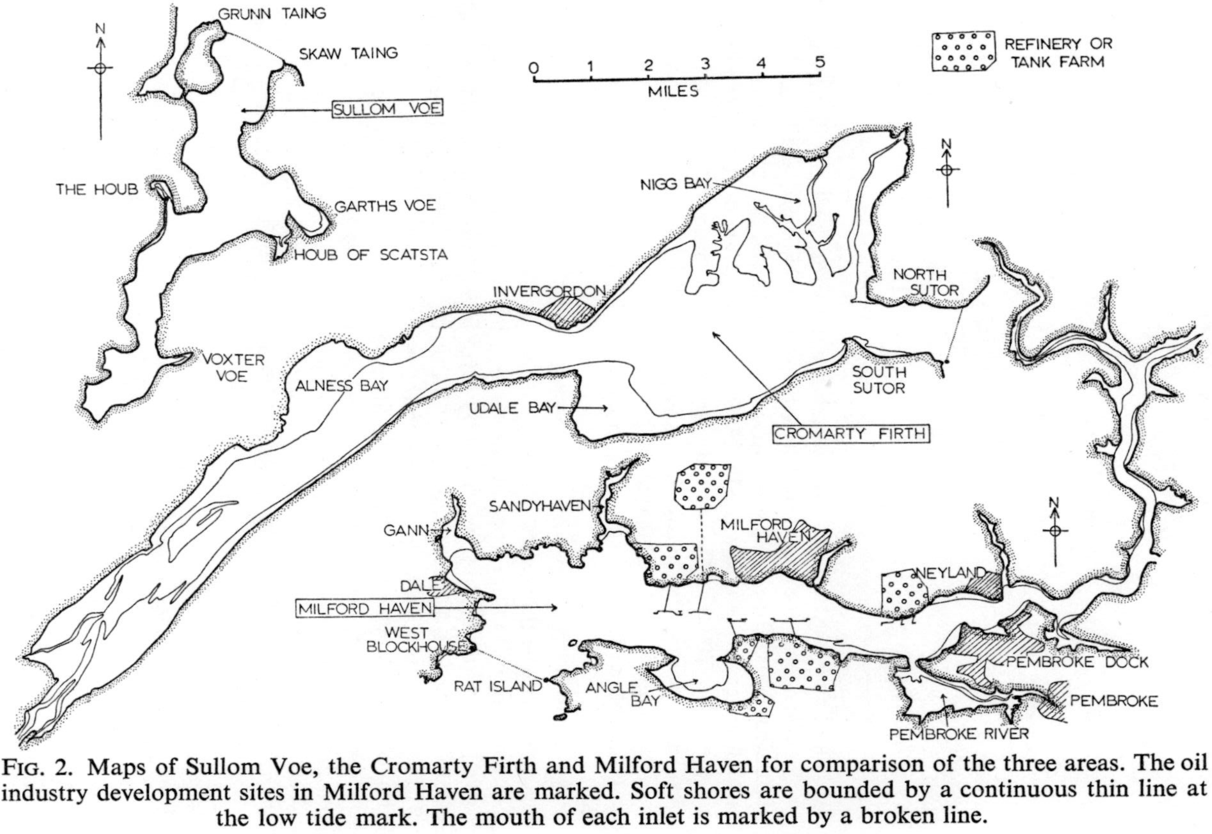

FIG. 2. Maps of Sullom Voe, the Cromarty Firth and Milford Haven for comparison of the three areas. The oil industry development sites in Milford Haven are marked. Soft shores are bounded by a continuous thin line at the low tide mark. The mouth of each inlet is marked by a broken line.

TABLE I

Volumes of Sea Water (m^3) in Sullom Voe, Cromarty Firth and Milford Haven at Low Spring Tide and High Spring Tide. Also Shown is the Range of Spring Tides in the Three Areas

	Volumes (m^3)		
	Cromarty Firth	*Sullom Voe*	*Milford Haven*
Volume at low water, spring tide	405 000 000	325 000 000	285 000 000
Volume at high water, spring tide	683 000 000	362 000 000	598 000 000
Tidal volume	278 000 000	37 000 000	313 000 000
Spring tidal range	3–4 m	2 m	8 m

Rat Island and West Blockhouse, which accounts for the higher figures given here, a considerable extra water area being included inside Milford Haven (about 200 000 000 m^3 at high water). The volume found by Nelson-Smith for high water spring tide was 345 000 000m^3 and that here as 498 000 000m^3, but the extra 200 000 000 included by the extra piece of Milford Haven considered here brings the two results very close. Likewise, the empty volume of Nelson-Smith (154 000 000m^3) when adding the extra 85 000 000m^3 for the tidal volume of the extra area makes the results similar (285 000 000m^3 here).

The results for the three sites are tabulated in Table I along with the spring tide range for the three sites.

The mouth of Sullom Voe was taken as a line between Grunn Taing and Skaw Taing whilst that of the Cromarty Firth from Sutor Stacks on the south side to the eastern edge of the North Sutor on the north side (see Fig. 2).

Using these tidal volumes as passing through the mouth area at low spring tides, rough calculation of tidal velocities were made. These velocities will be mean velocities, and obviously very variable. The mean velocities are tabuated in Table II.

The figure given as mean velocity (spring tides) for the mouth of Milford Haven (Admiralty Charts 3274 and 3275) is 36 cm/s, which agrees very closely

TABLE II

Mean Tidal Velocities (Spring Tides) Through the Mouth of Milford Haven, Cromarty Firth and Sullom Voe

	Milford Haven	*Cromarty Firth*	*Sullom Voe*
miles/h	0·8	1·44	0·1
cm/s	37	65	4·6
approximate width of mouth (m)	3000	1350	1800

to that calculated here. It should be stressed, however, that these calculations are very crude and are only useful to give an idea of mean volumes, tidal flow, and tidal velocities, and their relative proportions.

Tidal water movements at spring tides in Milford Haven are obviously extensive, with the tidal exchange being greater than the low-water volume. Tidal velocity in the mouth (0·8 miles/h) is such that booms would probably be of little use for keeping oil slicks stationary (speeds in excess of 1 knot render booms useless for retaining oil (Dudley, 1975). In the Cromarty Firth, the tidal volume is about two thirds of the low tide volume and this, in association with the relative narrowness of the mouth, has been calculated to produce tidal flows of around 1·44 miles/h.

Though the flushing of the Firth is extensive, the high water velocities through the mouth will impart great mobility to any slicks in this area, and render their containment within stationary booms impossible. Once away from the mouth, flows may well be lower, but any slicks brought in through the mouth by the currents may well end up in Nigg Bay or Udale Bay before much can be done.

In Sullom Voe, tidal flushing is very small, about one tenth of the low water volume. Therefore, tidal currents will be very slow (about 0·1 miles/h). This may make the area one where build-up of oil is possible if natural degradation rates are exceeded. This may apply particularly to the inner end of the Voe, the Voe being long and narrow (Fig. 2) and little water exchange being probable in this area.

Throughput of Freshwater

As a rough method of comparing the amounts of freshwater entering the three inlets, the catchment area of the rivers supplying each inlet was marked on a map and calculated in square miles. The greater the amount of fresh water entering the area, the greater the flushing of potential pollutants out to the sea. The catchment areas are marked on Fig. 3 and the approximate areas in square miles are given in Table III.

Though the catchment areas of rivers supplying Milford Haven and the Cromarty Firth appear extensive, the actual volume of water is probably small compared to flushing by tide. For Milford Haven, Nelson-Smith (1965) has shown that the catchment area of the combined Cleddau rivers (200+ square miles) produces on average about 260 million gallons per day of freshwater, compared to a spring tidal volume change of 40 000 million gallons per tide or about 80 000 million gallons per day. The freshwater is, in this case, insignificant in comparison. This is probably also true of the

TABLE III

The Approximate Catchment Areas of Cromarty Firth, Milford Haven and Sullom Voe in Square Miles

	Cromarty Firth	*Milford Haven*	*Sullom Voe*
Catchment area	780	320	19

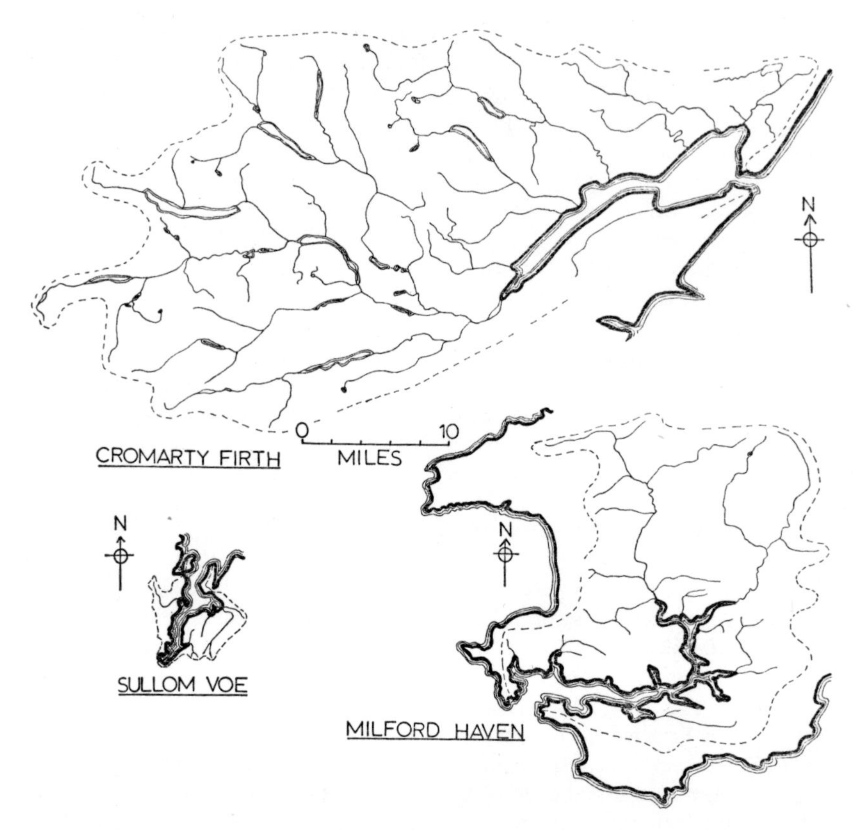

FIG. 3. The catchment areas (bounded by a broken line) of Sullom Voe, the Cromarty Firth and Milford Haven. The approximate catchment areas are 780 square miles (Cromarty Firth), 320 square miles (Milford Haven) and 19 square miles (Sullom Voe).

Cromarty Firth with a tidal exchange of around 278 million m^3 per tide (about 5500 million gallons). Though freshwater volumes are probably small compared to tidal volumes, the freshwater may play a significant part in flushing the inner ends of these two areas.

Sullom Voe is entirely different from both other sites, having a relatively very small catchment area, and no major river input. There is, therefore, little possibility that freshwater input could contribute to the water exchange in the Voe, which is relatively small, particularly at the innermost end.

Shore Type

The shores of Milford Haven are predominantly rocky especially in the area of refinery developments (Fig. 2) although four large soft substrate areas occur; the mouth of the Pembroke River, Angle Bay, Gann Saltings and Sandy Haven. Sullom Voe is also predominantly lined by rocky or stony shores, with small areas of sand or saltmarsh in The Houb, Voxter Voe, The Houb of Scatsta and Garths Voe. The Cromarty Firth, however, is almost entirely lined by soft shores (although these are occasionally stony) with extensive mudflats in Nigg Bay, Udale Bay, Alness Bay and in the whole of the head of the Firth. Therefore, by the arguments outlined in the previous section, the Cromarty Firth is at the greatest risk of oil pollution damage to the shores of the three sites, and will present more difficulties regarding the treatment and removal of oil slicks due to shallowness of water over shore areas on a rising tide.

Rare Species and SSSI's

It is not my purpose here to go into details of all rare species or SSSI's at the three sites. This information can be obtained from the Nature Conservancy Council and brief comment has already been made as to the risk of pollution of such areas. However, as an example of such a site in one of the three areas, the eel-grass (*Zostera*) beds in the Cromarty Firth are worth considering. Their importance is two-fold, for not only do they constitute one of a very small number of such areas in Britain, but they also form an important feeding area for some migrant birds (wigeon, teal, greylag geese—from 'A prospectus for Nature Conservancy within the Moray Firth'). The growth of the eel grass on the mud surface renders it easily accessible to pollutants, although I have not been able to find any information as to pollution effects on these plants.

Birds

Once again, full details on the bird life of the three sites for detailed comparison can be obtained elsewhere (Nature Conservancy Council, Royal Society for the Protection of Birds, British Trust for Ornithology, Tulloch and Hunter (1970), Cramp *et al.* (1974)), but a brief comparison of the three sites in a more general way is useful here. Within both Milford Haven and Sullom Voe there are no large bird interests, although both areas are near to bird sites of national and international importance (Skomer, Skokholm and Grassholm Islands around the mouth of Milford Haven; Yell, Unst and Fetlar near the mouth of Sullom Voe). However, the abundant bird life of the Cromarty Firth lies within the Firth (Nigg Bay, Alness Bay and Udale Bay) and is therefore presumably at greater risk if developments go ahead than at either of the other sites which have the important bird areas outside. It may only be good fortune that serious spillage has not yet affected the bird islands near Milford Haven, but it is probably significant that all major spillages have so far been inside the Haven.

CONCLUSIONS

This paper cannot and does not pretend to be a comprehensive coverage of all factors which make comparison of the Milford Haven experience with other sites possible or otherwise, but the factors chosen, compared at just three sites, show all three to be different in some ways and similar in others. The relatively small impact of oil developments on Milford Haven is obviously due to a combination of a large number of coincidental factors, which include good port management, extensive tidal flushing, only a relatively small number of muddy shores and small numbers of the very susceptible birds within the port. For this reason, and the often considerable differences between sites, the Milford Haven experience only becomes valuable to new developments if all the factors which have resulted in this port remaining relatively clean are fully understood. If this is the case, then Milford Haven offers a wide range of managerial and biological information of use in predicting effects of development elswhere. If these factors are not taken into account, extrapolation of what has happened in Milford Haven to new sites can only be regarded as dangerous.

Two further points should be made. First, whilst a considerable amount of information is available on the effects of oil pollution on certain shores, much less is known of effects on sub-littoral regions or on some soft shore types. Milford Haven is also unusual in having a lot of biological data available through most of its developmental phase. More biological work needs to be carried out on little known shore types and in association with new developments, for this will provide information useful in the future, and, in the case of new developments, provide an early warning of any pollution effects. Secondly, advances in design of terminal and refinery installations and better knowledge of biological effects of pollution mean that a better start is at least possible in new areas.

REFERENCES

Baker, J. M. (1971). The effects of oil pollution on saltmarsh vegetation, Ph.D. thesis, University College of Wales, Swansea.

Baker, J. M. (1973). Biological effects of refinery effluents, Proceedings of EPA/API/USCG World Conference on the prevention and control of oil pollution.

Baker, J. M. (1975a). Ecological changes in Milford Haven during its history as an oil port (this volume).

Baker, J. M. (1975b). Investigation of refinery effluent effects through field surveys (this volume).

Baker, J. M. (1975c). Experimental investigation of refinery effluent (this volume).

Bourne, W. R. P. (1968). Oil pollution and bird populations, *Field studies*, **2**, 99–121.

Cramp, S., Bourne, W. R. P. and Saunders, D. (1974). *The Seabirds of Britain and Ireland*, Collins, London, 287 pp.

Crapp, G. (1970). The biological effects of marine oil pollution and shore cleaning, Ph.D. thesis, University College of Wales, Swansea.

Dicks, B. (1973). Some effects of Kuwait crude oil on the limpet, *Patella vulgata*, *Environ. Pollut.*, **5**, 219–29.

Dicks, B. (1975). The importance of behavioural responses to toxicity testing and ecological prediction (this volume).

Dudley, G. (1975). The incidence and treatment of oil pollution in oil ports (this volume).

Hargrave, B. T. and Newcome, C. P. (1973). Crawling and respiration as indices of sub-lethal effects of oil and dispersant on an intertidal snail, *Littorina littorea*, *J. Fish. Res. Board, Canada*, **30**, 1789–92.

I.M.C.O. (1973). The environmental and financial consequences of oil pollution from ships, Report of study No. VI (U.K.), Appendix 3.

Levell, D. (1975). The effect of Kuwait crude oil and the dispersant BP 1100X on the lugworm, *Arenicola marina* L. (this volume).

Nelson-Smith, A. (1965). Marine biology of Milford Haven: the physical environment, *Field studies*, **2**, 155–88.

Parsons, R. (1972). Some sub-lethal effects of refinery effluent upon the winkle *Littorina saxatilis*, Annual Report 1972, Oil Pollution Research Unit, Field Studies Council.

Reish, D. J. (1965). The effect of oil refinery wastes on marine animals in Los Angeles harbour, California, Symposium sur les pollutions marines pars les micro-organismes et les produits petroliers, p. 355.

Tulloch, R. and Hunter, F. (1970). *A Guide to Shetland Birds*, Shetland Times Publication.

Discussion

The Incidence and Treatment of Oil Pollution in Oil Ports

Dr A. M. Jones (Dundee University) asked Captain Dudley for comments upon the relative dangers of pollution from loading and from discharge operations by tankers in ports.

In reply, **Captain Dudley** said that statistics for Milford Haven show that on a percentage basis pollutions occur more frequently from vessels discharging than from vessels loading. However, it was important that two points should be made:

1. The number of vessels discharging represents about 5% of the total ships handled, and therefore there is a distinct possibility of distortion of statistics occurring owing to the relatively small numbers involved.
2. The scale of pollutions occurring from vessels unloading is usually very small, a typical example being very minor pollutions occurring at the completion of discharging when ballasting starts.

Captain Dudley did not think it would be wise to draw any conclusion from Milford Haven experience on the relative dangers involved in loading and discharging.

Mr J. Wardley Smith (TOVALOP) asked Captain Dudley:

1. What was done to prevent pilots from hazarding ships and so causing pollution?
2. Did prosecuting offenders in the local court help prevent pollution?
3. Could design, for example of the shape of valve-operating wheels, help to avoid error?
4. Who was responsible for shore clean-up?

Captain Dudley replied that:

1. Groundings and collisions are probably far and away the most likely cause of really serious pollution and it was therefore vital that adequate navigation aids should be provided to assist those handling ships and reduce the possibility to a minimum. He thought Mr Wardley Smith was correct in implying that adequate navigation aids are a significant contribution to reducing pollution. In Milford Haven they take three forms:

(i) Very extensive communication networks to ensure that all vessels moving in the Haven are in constant contact with the port control and the oil terminals and can therefore be programmed in such a manner that dangerous close-quarter situations do not exist.

(ii) The entire area of the Haven and the entrance to approximately 25 miles to seaward is covered by a harbour surveillance radar system which provides the Port Authority marine officers with 24-hour cover of the area. This again helps to programme vessels safely and to maintain a watch on the positions of all ships, including those at anchor, so that warnings can be issued if vessels are standing into danger.

(iii) Very extensive navigation lighting is provided, including high-intensity transit lights capable of being used in poor visibility in daylight as well as at night.

2. He believed that if a port takes a firm line from the very beginning that all those who pollute will be prosecuted unless they have a valid legal defence, that port will quickly get a reputation among the world's tanker fleets as being a place where special care must be taken. He therefore believed that prosecutions for oil pollution do reduce the amount of pollution which occurs and he did not think that any exceptions should be made in the cases where the pollution is small, since the culpability may be equal to that of the most severe pollution.

3. There have been examples of pollution occurring through open valves where those on board the ship believed that the valve was closed, possibly by reason of a broken spindle between the operating wheel and the guard itself. So the answer to the question was yes, a more accurate system of ensuring that valves are closed would be a help.

4. If a reasonable anti-oil pollution service is available and the harbour is sufficiently large, it should be possible to deal with all small pollutions before they reach the shore. Therefore, significant shore pollution is likely to be the result of large spills, the source of which can be established without difficulty. In these cases the responsibility for cleaning up the spill ashore is a matter for the culprit. In Milford Haven the oil terminal operator involved usually accepts responsibility for this work and, where necessary, bills the offender. However, local authorities do have a responsibility for cleaning up shore pollution and do carry out this work in Dyfed outside the Haven, frequently utilising the expert knowledge and equipment available to them from the Port Authority or the oil companies within the Haven.

Professor A. R. Halliwell (Heriot-Watt University) agreed with all except one of Captain Dudley's conclusions. He had some reservations about the conclusion that damaged oil tankers must be allowed to enter oil ports in order to reduce overall pollution. The inherent assumption made was that a spill outside the oil port is more damaging to the 'environment' than a smaller spill within the oil port and this may not always be the case, although there seems to be some evidence that this is true for the port of Milford Haven. The rate of recovery from the spill of the oil port area compared with that for the outside area needs to be considered: this rate of recovery depends among other things upon the physical water movement (e.g. flushing time) and the environment itself (e.g. mussel beds, large bird colonies), and these factors may be very different for different ports. Perhaps a more reasonable conclusion is that there should be agreed and recognised havens where damaged tankers can shelter and the pollution be dealt with in a satisfactory manner: of course many (but not necessarily all) oil ports will be suitable for this purpose.

Captain Dudley replied that his point was that the less oil that escapes on to the surface of the sea, the better. If a damaged ship is left at sea there is always the possibility that very large quantities of oil will escape, probably fairly close to the port which the ship is attempting to enter; whereas if that damaged ship is taken into a well-equipped oil port it should be possible to remove significant quantities of oil which would otherwise have escaped. He admitted that this would not apply in all circumstances and it would perhaps have been wiser to suggest that 'in many cases oil ports must be prepared to accept damaged tankers in order to reduce overall pollution'. In general, he adhered to his view that an oil port is the best place to take a damaged tanker and it is usually better to remove the oil from a damaged tanker than to allow it to escape into the ocean.

Monitoring

Dr I. C. White (Ministry of Agriculture, Fisheries and Food) did not entirely agree with Dr Baker's definition of biological monitoring and favoured the stricter definition that requires any exercise to be repeated regularly and to involve the measurement of the level of a variable for reasons related to the assessment or control of a pollutant. He agreed that organisms can be used valuably in a great variety of ways to get information on the effects of pollutants and suggested that chemical monitoring can also be valuable. As well as the obvious public health aspects concerned with the protection of the consumer of marine organisms, he suggested that the repeated and regular sampling of organisms, sediment and water for chemical analysis may provide early warning of a change in the quality of that area. This information would permit the design of specific investigations into the effects of such a change. He thought that Dr Baker's comments about the value of mounting chemical monitoring stemmed from the fact that monitoring was often required before adequate background information was available on such aspects as dose/response relationships and the fate of various substances in the marine environment.

Points concerning selection of sites, organisms and sampling frequency were raised by **Dr A. G. Bourne** (Hunting Technical Services Ltd), **Dr M. W. Holdgate** (Institute of Terrestrial Ecology) and **Dr Jones**. **Dr Bourne** commented that as changes observed may be seasonal or caused by factors other than pollutants, trends over a long period of time may also be due to other factors and have nothing to do with pollution. He asked how sites and organisms were selected and thought that it would be more to the point to know the level of a pollutant that will impair the metabolism of an organism. It is this which governs, in the long run, the efficiency of an ecosystem and it seemed pointless to wait until the indicator organism dropped off its substrate. Had not the metabolic efficiency of the organism been impaired a long time before death occurred? Were there not methods of detecting and measuring pollutant levels which could indicate impairment of metabolism? He thought that it was essential to work towards such methods, otherwise it was not really possible to understand the role of pollutants in the environment.

Dr Holdgate said he was concerned about the frequency of sampling in time and space under different circumstances. Would Dr Baker agree that one should normally begin with some system of objective sampling to define the variability of the environmental situation whose changes one is seeking to detect and use this to stratify subsequent sampling—and hence to ensure that the sites examined in detail were adequately representative of the main environmental categories? His second concern was with the time frequency of observations. This surely also depended on some initial characterisation of the rate of change to be expected in the situation and a judgement about the amount of change to be detected with the methods being used.

Dr Jones supported the concept of site selection as being a vital part of planning a monitoring programme. Sites should be selected which are representative of the principal habitat types but should be chosen to ensure a fairly uniform distribution rather than using random points, particularly on rocky shores. In Scapa Flow a number of sites were found suitable for quantitative analyses even when using criteria such as the requirement of a fairly smooth bedrock area, the fairly uniform distribution of species over an area some 10 m wide, and the absence of human interference. Much of the littoral zone of Scapa Flow is composed of boulders which are totally unsuitable for transect studies. His second point referred to the vital role played by population studies in monitoring programmes. By studies of growth, year-classes, recruitment, mortality and condition, sub-lethal effects of pollutants may be detected, thus enabling a consideration of the environmental problems before they cause drastic community changes.

Replying to points raised, **Dr Baker** said she agreed in principle with initial objective sampling and characterisation of rate of change but had to contend with considerable practical difficulties. Random sampling was particularly difficult on precipitous rocky coasts. Access to sites by road, footpath or boat was an important consideration. An effort was made to sample a wide variety of site types; for example the Milford Haven rocky shore monitoring scheme covered the whole range of exposure grades represented within the Haven. Sampling frequency was again determined by practical considerations, namely time and manpower available, and ideally she would like more frequent observations at least at some sites in order to characterise seasonal changes. She thought that some judgement about rates of change to be expected following pollution incidents could be made from field experimental results which were available. A large number of organisms were included in the Milford Haven monitoring scheme and she thought there was now scope for elimination of some of these species from the scheme, but they were included initially because at the time no one knew which would prove to be the best species for monitoring purposes. Dr Baker agreed that methods of detecting pollutant levels which could impair metabolism would be useful—growth studies might be relevant here—but behavioural responses such as the limpet 'drop-off' appeared to be all-or-nothing responses following certain levels of certain types of pollutant.

Mr K. Hiscock (Coastal Surveillance Unit, Menai Bridge) said it had been suggested that the results of different workers at the same site might be at variance by as much as plus or minus two points on a five- or seven-point abundance scale. Was this acceptable in a scheme which aims to follow changes taking place on rocky shores in Milford Haven? If the error is unacceptable, did Dr Baker consider it desirable to change the long-established techniques, which are comparable with past data, to methods which are more accurate? In reply, **Dr Baker** said this was currently under consideration. There seemed to be little prospect of getting away from abundance scales when large numbers of sites had to be considered, and although the method was crude it did pick up the larger changes. However, she thought additional objective quantitative data, for example on limpet or barnacle recruitment, should be collected from selected sites.

The purposes of environmental monitoring were felt by **Dr R. H. Cook** (Environmental Protection Service, Canada) to be:

1. To determine background or baseline environmental criteria in an area where new industrial developments are anticipated so that the impact of this activity on the environment can be assessed quantitatively.
2. To carry out biological investigations in the general vicinity of existing pollution sources in order to quantify the impact of wastes on the environment and by so doing to evaluate the effectiveness of the pollution control standards applied to that industry.

With the high priority and requirement for the above monitoring activities, he wondered how much monitoring effort should be devoted to the assessment of single spill occurrences such as the one reported on the grounding of the *Dona Marika*. **Dr Baker** thought that the limited effort which had been directed to this particular spillage was justified; it was unusual to have a large gasoline spill and several things had been learnt about the behaviour and effects of this pollutant. Moreover, the investigations formed part of the overall Milford Haven monitoring programme. She did not think that effort should necessarily be made to follow up effects of all isolated spills.

Mr A. D. McIntyre (Department of Agriculture and Fisheries for Scotland) illustrated the problem of distinguishing between natural events and pollution effects by two examples. In long-term studies of an intertidal and shallow water sand ecosystem at the field station in Loch Ewe, one of the dominant species of that ecosystem, the bivalve mollusc *Tellina tenuis*, did not have a fully successful recruitment to one particular bay after 1963. There is no pollution in the area, and the failure to recruit can probably be attributed to a number of natural causes which are being examined. However, if during the mid-1960s a source of pollution had been introduced into the bay, this *Tellina* case history might well have been quoted as an example of a pollution effect. Again, during the past 15 years or so, a survey of soft-bottom benthos round the whole Scottish coast has been made. In the deep water at the head of Sullom Voe, Shetland, the sediment appeared anoxic and macrobenthos was virtually absent. If this observation had not been made until some time after oil had been flowing to Shetland, it might have been difficult to refute the suggestion that it was an effect of oil.

Retention and Accumulation of Hydrocarbons

The need for better comparability of quantified data was pointed out by **Dr H.-J. J. Marcinowski** (Stichting CONCAWE). Examples of approved criteria were the critical organs and the residence time in the body (biological half-life time, BHLT). In addition, information on accumulation and depletion mechanisms is useful. All such data give useful references to the biological pathway of the pollutant. **M. C. R. Gatellier** (Institut Français du Pétrole) said that factors as high as 200 for accumulation of hydrocarbons had been published. In his opinion, such accumulations could proceed in two ways, either instantaneous take-up followed by a large depression when the water was clean (e.g. the case of oysters or clams accidentally polluted by oils), or accumulation of the pollutant, whatever the level in the water, without an appreciable depressive effect (this might happen with mercury). Only the latter type will be useful for monitoring a past pollution. In the case of hydrocarbons, was there evidence of such a permanent pollution? **Dr White** pointed out that there is some disagreement on the retention of hydrocarbons by marine organisms, especially for aromatic compounds. An appraisal of the available information, backed up by experiments conducted with the Ministry of Agriculture, Fisheries and Food on the uptake and loss of labelled benzpyrene by fish, showed that the half-life of these compounds is short—a matter of a few days—and that, at least in the case of fish, most activity is concentrated in parts of the fish not normally consumed (e.g. gall bladder and guts). **Mr McIntyre** described a relevant experiment carried out jointly with Torry Research Station at the field station at Loch Ewe. Cod were fed 1 mg crude oil per day for six months, and kept for a further six months on an oil-free diet. Analyses of *n*-alkanes showed no significant change in the amount or pattern in the flesh. In the liver, on the other hand, there was a marked build-up. After six months on an oil-free diet the level of alkanes had fallen but was still higher than in the control and the oil-induced profile was retained. **Dr White** added that different effects were noted in two species of fish used—cod and plaice. Other work had been done using radioactive benzpyrene in fish; the accumulation concentrations in the liver were very high, but loss was rapid.

Scope and Standardisation of Monitoring

Dr D. Scarratt (Department of the Environment, Canada) thought that standardisation was not necessary in fine detail and possibly not desirable or attainable, but studies must be comprehensive, accurate, non-destructive, repeatable, stand rigorous analysis and stand up in court. Monitoring can be biological, physical, chemical, economic or sociological, but the comprehensiveness depends upon the priorities the monitoring is designed to meet. These may include academic curiosity, the detection of changes to permit feedback to the operation, the support of legal claims, the prediction of effects of new installations or the aiding of site selection. **Dr Jones** drew attention to the apparent consideration of chemical and biological monitoring as alternative methods by many of the participants. He saw these as complementary methods which both require baseline studies and suggested that biological monitoring should then be used until apparent effects on the biota are observed. At that point, chemical techniques should be reintroduced to establish whether pollutants may be responsible. For selected pollutants, e.g. heavy metals in commercial species, routine chemical monitoring is vital at all stages of a monitoring programme.

Mr K. Hay (American Petroleum Institute) commented that efforts to perfect the science of marine monitoring would be a great contribution, both to programmes being developed in North America and to many other nations. There was a need to develop standard biological procedures for such monitoring and hopefully employ them on an international basis. It would be useful, 25 years from now, to have comparative biological monitoring data on an international basis. He was sure there would be a lot of argument as to why this cannot be achieved, but saw no reason for not aspiring to some degree of standardisation in this area. **Dr Holdgate** thought it was most important that there should be intercomparability between results, and this would partly be achieved by understanding the methods used and quantifying the relative reliability of the results. He did not believe in standardisation in the narrowest sense, of saying that there are only two or three accepted ways of studying a particular facet of the environment. He believed strongly in standardisation in the sense that it was the responsibility of any scientist, before he quotes a result, to attach to it information about how it was obtained and the likely error. Much intercomparison of methodology was necessary.

Further discussion on this subject took place later in the Conference and is reported in the final discussion on monitoring.

Applicability of the Milford Haven Experience

Captain Dudley said that oil pollution occurs outside Milford Haven but little comes from within the Haven. Of two serious pollutions outside the Haven, one came from a tanker which broke up 3000 miles away, the other was of a Venezuelan crude not handled for some years in Milford Haven. New ports can be expected to be relatively well run—pollution from major oil company sources will always be seriously dealt with—but the same cannot be said for some single-ship flag of convenience shipowners. He suggested that as part of any base survey in a new oil port, careful photographs be taken of any oil pollution occurring before the port handles oil. This will be vital to show what is true: that oil pollution occurs in all ports and on all shores used by vessels or passed by vessels of all types—particularly, he suspected, by small fishing vessels in the case of Scottish ports.

Mr C. B. Duggan (Department of Agriculture and Fisheries, Ireland) referred to an atlas drawn up by the Irish Department of Agriculture and Fisheries for the use of oil pollution officers. This showed the location of commercial fisheries and shellfish. He agreed with Dr Dicks' statement on the need for a good Harbour Authority. The lack of such an authority had been felt in Bantry, where there had been two major spillages, and there was now a promise to introduce a Harbour Authority when the necessary legislation had been passed.

Mr A. Currie (Nature Conservancy Council) pointed out that there are differences in political approach to the uses of firths. Depending upon aims and intentions, people may visit Milford Haven or read the literature, and reach different conclusions. Conservationists are as guilty as planners or developers in selectively interpreting data. Since care should be taken about using Milford Haven information as a basis for the forecasting of the effects of oil-related development in Cromarty Firth, a selection of dissimilarities should be listed:

	Milford Haven (Inner)	*Cromarty Firth*
Sand/mud flats	*c.* 800 acres	*c.* 8800 acres
Zostera	Very little	*c.* 4000 acres
Birds	Good winter wader populations	Wintering wildfowl and wader populations, each normally in excess of 10 000. Wildfowl are internationally significant
Spartina	Abundantly present	Very small quantity, introduced

Dependence of birds upon the food resources of the available intertidal mud and sand flats must be stressed in the case of the Cromarty Firth. The huge area supports an important *Zostera* resource, and a considerable invertebrate fauna (the latter has been assessed by Miss S. Anderson, and a paper will be published).

A comparison in terms of management can be made. Milford Haven is clean because of the vigorous policy of Captain Dudley in controlling the port. The Cromarty Firth is a clean firth at present, and can only remain so if the newly appointed Harbourmaster is equally vigilant.

In the Cromarty Firth the close proximity of a proposed National Nature Reserve to the planned oil refinery constitutes a special matter for consideration and conservation. In assessing the impact of oil-related developments, we must bear in mind UK international commitments to protect wildfowl and wildfowl habitats.

Sand and Mud Fauna and the Effects of Oil Pollution and Cleansing

8

Preliminary Investigations of the Sublittoral Macrofauna of Milford Haven

JOHN M. ADDY

(*Oil Pollution Research Unit, Orielton Field Centre, Pembroke, Wales*)

SUMMARY

Twenty-four sample stations on a variety of substrates were sampled in May 1974 in Milford Haven. 114 taxa were isolated, most of which were identified to species level.

A preliminary particle size analysis of sediments at each station has been completed, enabling the approximate characterisation of sediment type at each station in terms of median particle size, phi quartile deviation and phi quartile skewness.

A simple association analysis failed to indicate more than one grouping of stations. Little evidence has been found for the existence of classical benthic 'communities' in Milford Haven. The data show a high degree of overlap in the distribution of different species.

Proposals for a monitoring scheme for Milford Haven macrobenthos are discussed. It is recommended that at least ten samples per site be taken as a matter of routine. The areas most suitable for sampling regularly in an intensive way are thought to be Angle Bay, Dale Shelf and Dale Roads.

INTRODUCTION

The littoral fauna of rocky shores and saltmarshes of Milford Haven is well documented both qualitatively (Dale Fort Marine Fauna) and quantitatively (Crapp, 1970; Baker, 1971). However, virtually no quantitative sampling had been done on the soft shores and sub-littoral deposits in the Haven. This work is intended to widen the range of the Oil Pollution Research Unit's monitoring programme by examining the soft sub-littoral deposits of Milford Haven.

The initial biological survey, as the first stage in a monitoring scheme for the Milford Haven benthos, was necessary primarily to identify communities within the Haven, and the spatial and numerical distribution of these communities in relation to environmental influences, particularly the nature of the sediment which supports them. Once such information is obtained, a suitable programme can be designed.

THE PHYSICAL ENVIRONMENT OF MILFORD HAVEN

The type of bottom deposits throughout the Haven are shown on the Admiralty Charts (Nos. 3274 and 3275). Though imprecise, they indicate a great variety of bottom types ranging from exposed bedrock through pebbles and sand to mud and clays. Nelson-Smith (1965) describes the physical oceanographic and geological characteristics of Milford Haven, which is a ria cut estuary, more properly regarded as an inlet of the sea, at least below Pembroke Ferry.

The depth profile of the estuary is variable, with the main channel flanked by extensive areas of relatively shallow shelves in a number of places, both on its north and south sides (see Fig. 1). At the entrance to the Haven, the bottom is largely exposed bedrock or coarse block or pebble deposits, though in the several bays in this region, which are very exposed to the prevailing south-west winds and sea there are areas of fine to medium sand interspersed with projecting ribs of exposed bedrock. Off Thorn Island, the channel bends to the east, bounded on its northern side by Dale Shelf and Dale Roads. Dale Roads are relatively shallow with a bottom of sand mixed with shells which is continuous with the rather deeper area of Dale Shelf. South-east of Stack Rock, which marks the eastward limit of Dale Shelf Anchorage, the deep-water channel is narrow with a less extensive sandy shelf to its south which is continuous with the large shallow sandy basin of Angle Bay. Beyond Popton Point, the southern flank of the channel is bounded largely by rocky shores and tidal mud flats at the base of low cliffs extending to West Pennar.

BIOLOGICAL SAMPLING

Sampling Stations

A grid of sampling stations was laid out in the wide western end of the Haven. In the more restricted waters of the central and eastern regions, the stations were arranged to form a regularly spaced chain as far as Pennar Beacon—the eastern limit of this survey. In Angle Bay a transect line of closely spaced sampling stations was used, in order to connect up with, and overlap by one station, a projected intertidal transect across this very uniform shallow basin. (See Fig. 28 for particle size analysis of Angle Bay transect station 15.)

Having established the locations of sampling sites, a number of the stations were not sampled due to the unsuitability of sediments for grab sampling.

Position Fixing

After anchoring at each station, the boat was allowed to adjust its position under the effect of wind and tide before fixing its precise position. This was achieved by selecting where possible at least two lines of sight on fixed landmarks, though navigation buoys were used when this was not possible. (See Appendix II for precise sampling station locations.)

Sampling Methods

As the use of divers to take such a large number of samples was impracticable, a Day grab designed to sample 0·1 m^2 of the sea bed to a depth of about 10 cm was used.

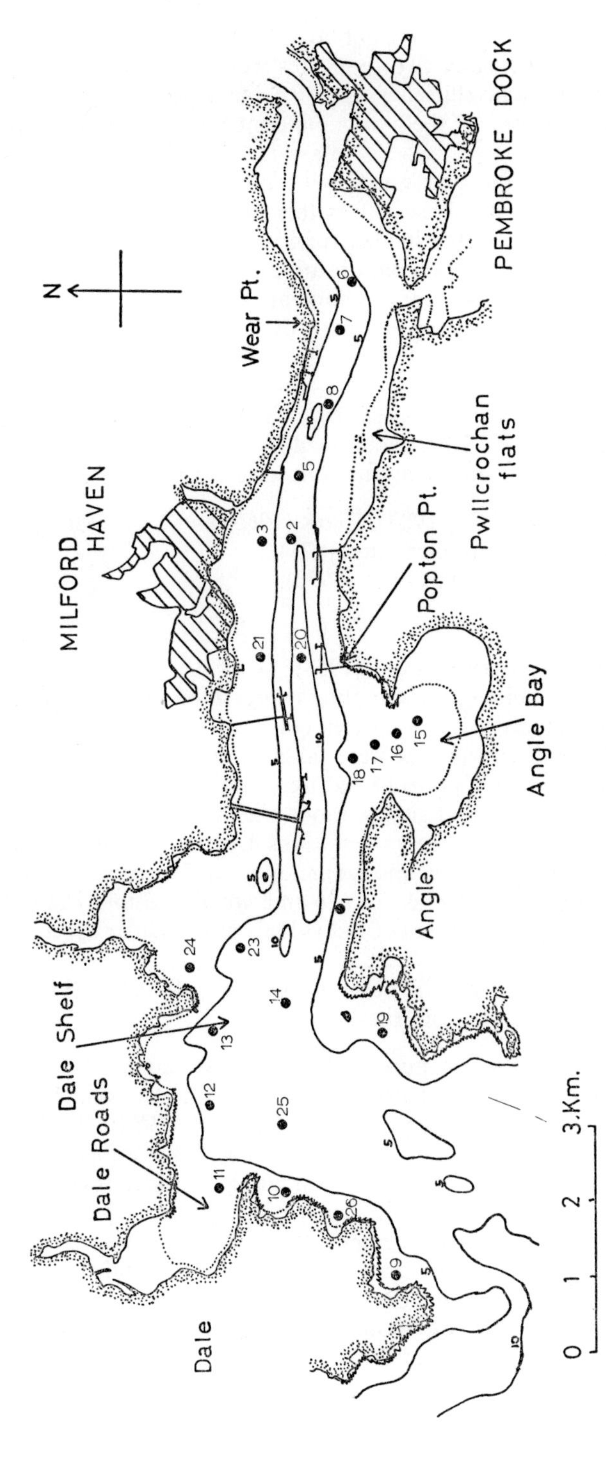

FIG. 1. The outline of Milford Haven showing the general bathymetric character indicated by the mean low tide contour and the 5 and 10 fathom depth contours. Place names mentioned in the text are also shown. Sample stations are indicated by black circles.

At each station five replicate samples were taken, but at eight of the 24 sampling locations a further five replicates were taken, in order to test the sampling sensitivity and efficiency and provide a more accurate indication of the population density of the more abundant species at these eight selected sites.

The sediment from the grab was sieved on board the vessel using a 1 mm mesh sieve and sea water hose. After this initial screening, everything retained by the sieve was stored in 10% formalin in sea water and stained with Eosin to facilitate sorting in the laboratory.

In the laboratory, all remaining sediment of less than 1 mm was washed through the sieve and the samples sorted by hand.

Treatment of Biological Data

A total of 114 taxa were isolated from 160 grab samples taken in this survey, most of which have been identified to species level. Species identification was largely carried out using the following works:

Polychaeta: Fauvel (1923, 1927), Clark (1960) and Day (1967).
Mollusca: Tebble (1966), Graham (1971).
Sipuncula: Stephen (1960), Gibbs (1973).
Crustacea: Allen (1967).

Also useful was Barrett and Yonge (1958).

The data from the biological sampling is presented in Appendix I as total numbers of individuals of each species collected in the five grabs taken at each station. Where ten samples were taken, it was thought appropriate to treat the first and last five grabs as two separate groups, in order to permit comparison between different sites without bias due to the increased number of samples. Thus station 1 becomes stations 1.1 and 1.2 for the purpose of comparison. The distribution and abundance of 24 species or groups of similar species are shown in Figs. 3 to 26 and, where possible, the histograms are divided to show the numbers of individuals of particular species collected in successive grabs.

Significant species in terms of distribution were selected for illustration in this way by using a number of criteria:

(i) Occurrence at 50% or more of the sample stations.
(ii) Significant population density at any particular station. 'Significance' was determined as suggested by Feder *et al.* (1973). Ranking of the percentages each species contributed to the total number of individuals at each station was followed by summing in descending order until a cut-off was reached at 50%. The species contributing to the first 50% were regarded as having significant densities at that particular station.
(iii) More subjectively, a number of species were included because their distributions appeared to be clearly defined without being restricted to one or two stations, though their densities were found not to be 'significant' by the test above.

Nelson-Smith (1965) divides Milford Haven proper into two ecological zones. The limit of south-west swell marks the outer margin of the zonal

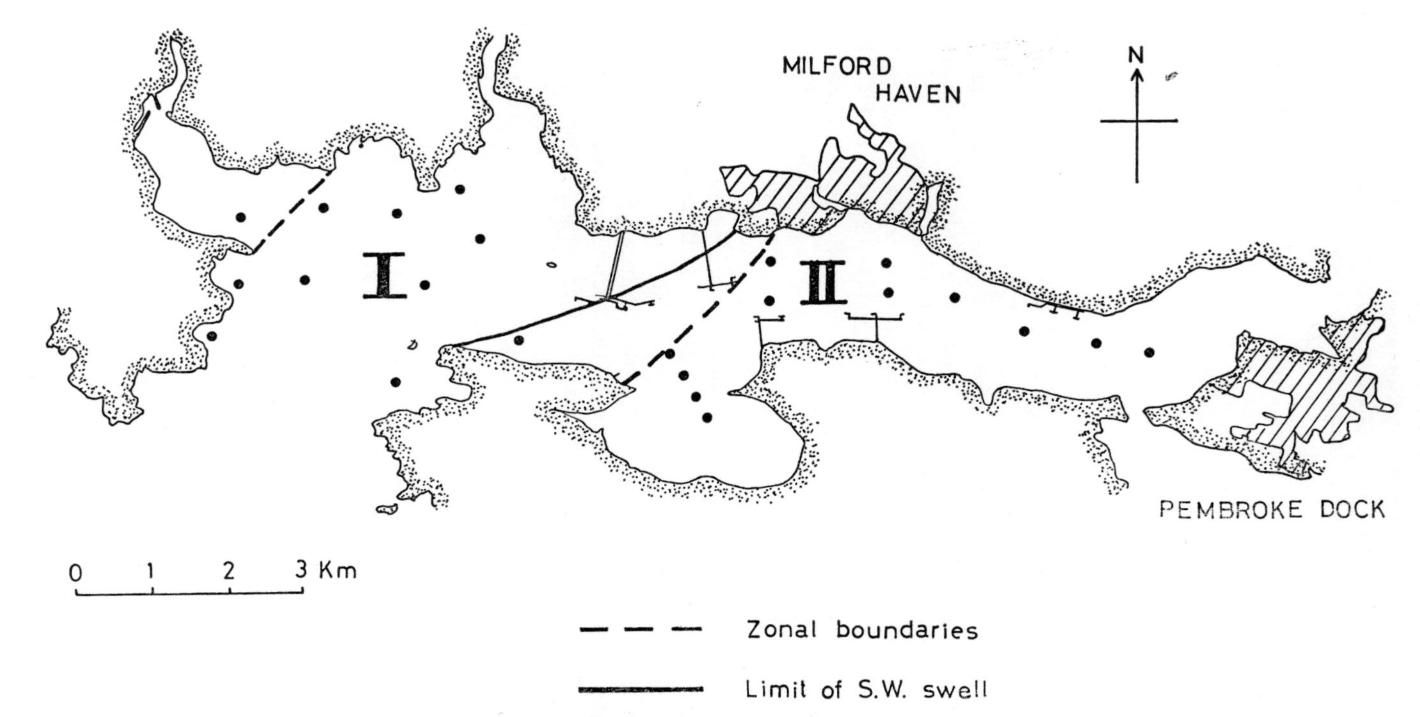

FIG. 2. The outline of Milford Haven showing the limit of south-west swell (after Nelson-Smith, 1965), and the extent of oceanic zone I and depositionary zone II referred to in the text.

boundary (see Fig. 2). In his outer, oceanic zone, salinities are always greater than 32‰. The clean beaches and bottom deposits are probably a reflection of the exposure of the region to strong Atlantic swell. In zone II, characterised by mixed bottoms, are included Angle Bay and Pwllcrochan mud flats, both indicative of a depositionary environment.

Broadly speaking, the distributions of the sub-littoral species shown in Figs. 3 to 26 can be divided into three categories:

(i) Ubiquitous.
(ii) Restricted to zone I.
(iii) Restricted to zone II.

(i) Ubiquitous

Of the 114 taxa isolated in this survey, five were distributed throughout the area examined. However, as Figs. 3 to 7 show, the densities of these ubiquitous species varied widely at different sites. Polychaetes, such as *Notomastus latericeus*, *Nephtys* spp., *Lumbrineris* spp. and *Cirriformia tentaculata*, and the bivalve *Abra alba* were found at almost all sample stations, indicating their ability to exist on a wide variety of substrates.

Notomastus latericeus (Fig. 3) is at its greatest densities at stations 11, 12, 13 and 25, in the medium sands and sandy pebble deposits of Dale Roads and Dale Shelf. At station 18, where *N. latericeus* thrives, the depositionary environment favours a finer sediment, but is of a sand and shelly texture similar to that of Dale Shelf.

Lumbrineris spp. (Fig. 4) are similarly distributed to *N. latericeus*, and with the exception of station 11 are present in very similar densities.

Nephtys hombergi (Fig. 5) are most abundant at those stations where *N. latericeus* and *Lumbrineris* are present only in low numbers, notably in the shallowest three stations of the Angle Bay transect. Clark and Haderlie (1960) have carried out an analysis of the sediment particle size of the substratum frequented by the different *Nephtys* species around the south western coasts of Britain, and found that *N. hombergi* achieved its highest densities in muddy substrata. This preference effectively isolates *N. hombergi* from *N. cirrosa* which is characteristic of clean coarse sand. (See Clay, 1967, for a review of the literature on *N. hombergi.*) *Nephtys cirrosa* was only found at station 24, at which station *N. hombergi* was absent.

The organism with the highest abundance at any sample station was *Cirriformia tentaculata* (Fig. 6), which, at station 8, achieves densities greater than 1200 per m^2, many of which are large specimens. At stations 5, 7 and 8 the sediment is very poorly sorted (see Table I) and *C. tentaculata* was observed in the undisturbed grab sample to be concentrated mainly against stones or under shells.

Abra alba (Fig. 7), though ubiquitous, is only abundant at stations 10 and 11, and does not favour the finer muds of the upper areas of the survey.

(ii) Restricted to Zone I

Nelson-Smith (1965), in his distinction between zone I and zone II, included Dale Roads in the oceanic region, although in terms of the benthic fauna it might be more properly grouped with the depositionary areas in the Haven, and hence be treated as a part of zone II (Fig. 2).

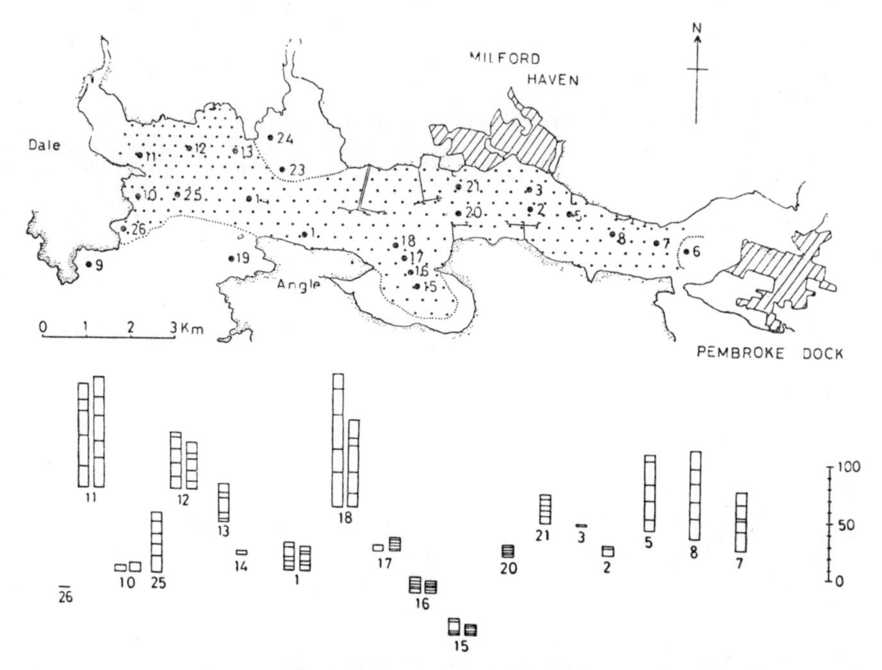

FIG. 3. The outline of Milford Haven, showing the grid of sampling stations and the distribution of *Notomastus latericeus*. Divisions on the histograms represent the numbers of animals found in each grab.

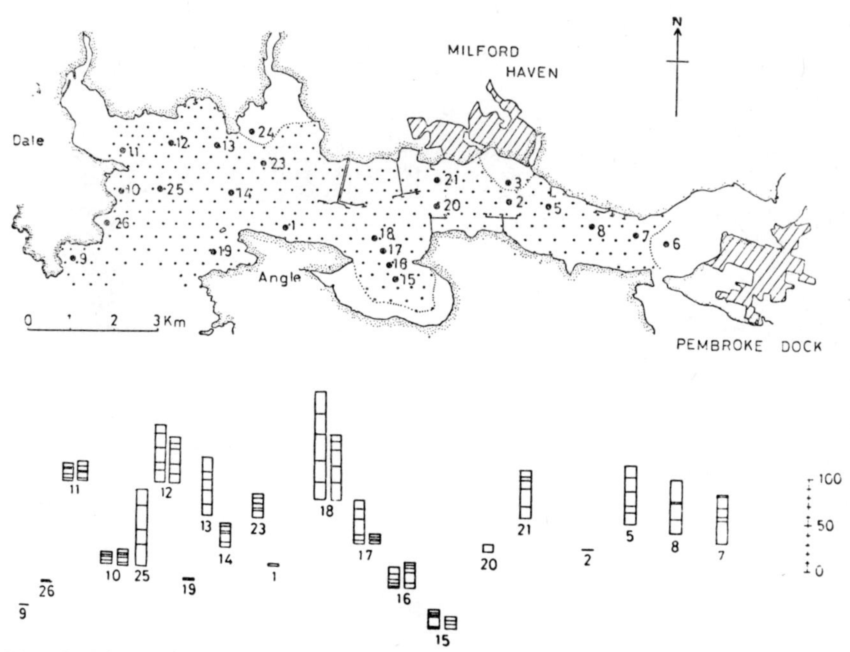

FIG. 4. The outline of Milford Haven, showing the grid of sampling stations and the distribution of *Lumbrineris* spp. Divisions on the histograms represent the numbers of animals found in each grab.

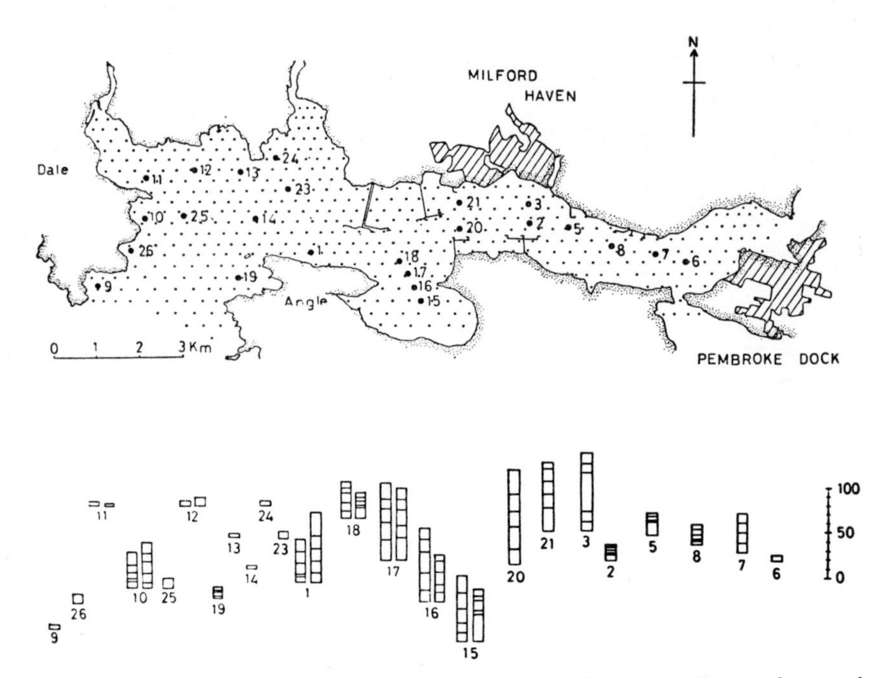

FIG. 5. The outline of Milford Haven, showing the grid of sampling stations and the distribution of *Nephtys* spp. Divisions on the histograms represent the numbers of animals found in each grab.

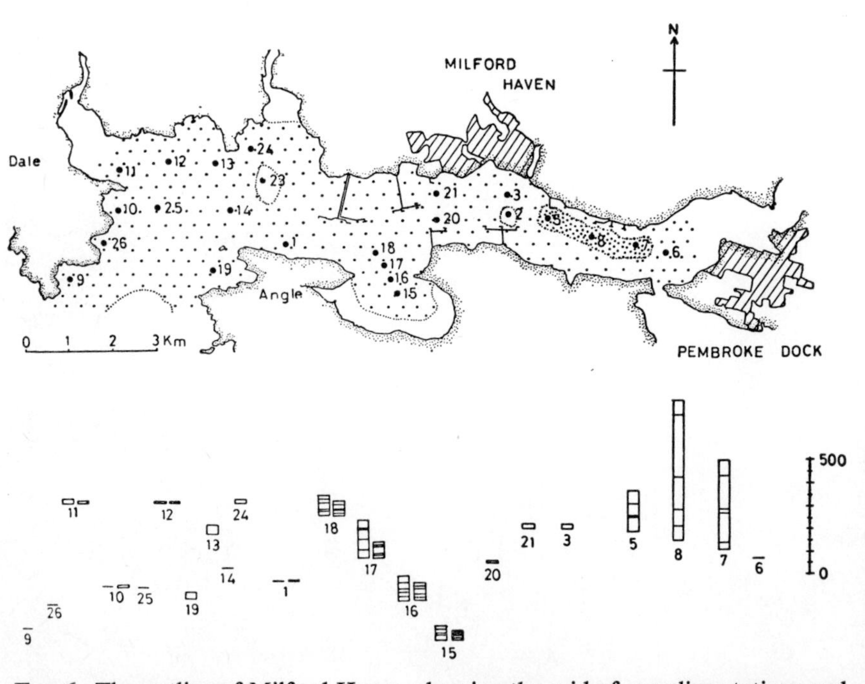

FIG. 6. The outline of Milford Haven, showing the grid of sampling stations and the distribution of *Cirriformia tentaculata*. Divisions on the histograms represent the numbers of animals found in each grab.

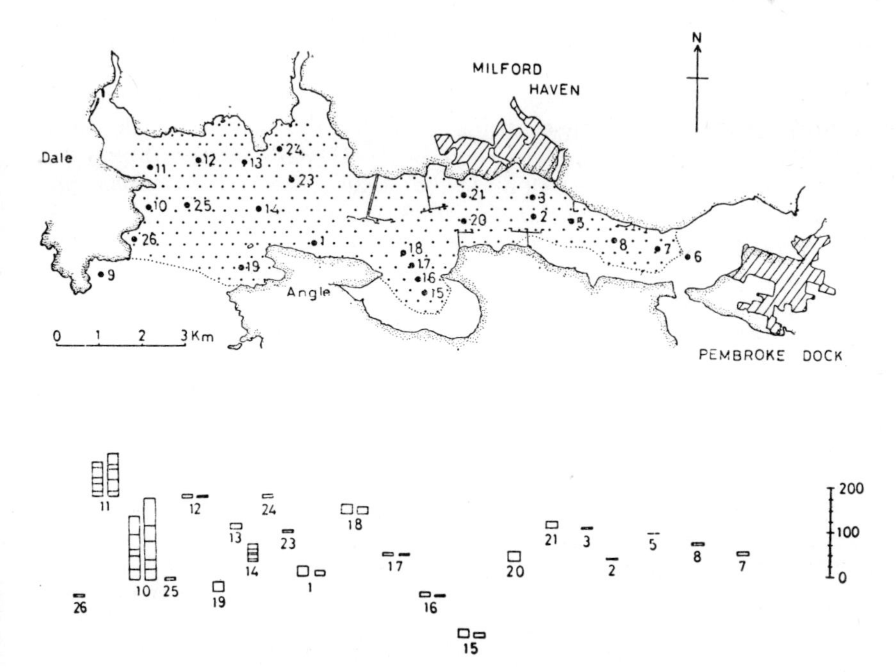

FIG. 7. The outline of Milford Haven, showing the grid of sampling stations and the distribution of *Abra alba*. Divisions on the histograms represent the numbers of animals found in each grab.

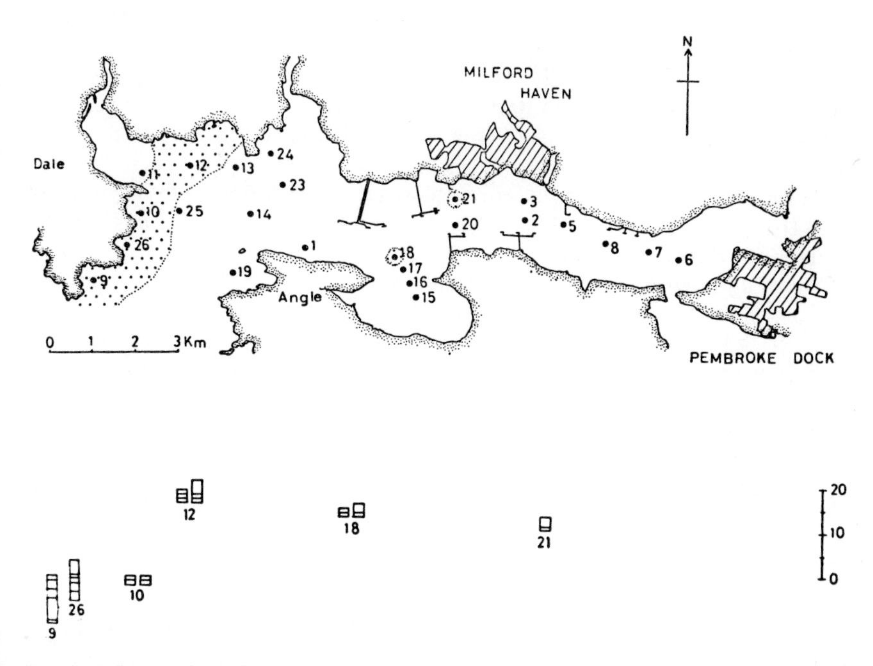

FIG. 8. The outline of Milford Haven, showing the grid of sampling stations and the distribution of *Glycera convoluta*. Divisions on the histograms represent the numbers of animals found in each grab.

Glycera convoluta (Fig. 8) is distributed in a band along the western margin of the mouth of Milford Haven, being absent from station 11 in Dale Roads and only occasionally found elsewhere in the Haven (stations 18 and 21).

Most striking in their restriction to the oceanic zone are *Magelona* spp., *Owenia fusiformis* and *Nucula nucleus* (Figs. 9, 10 and 11). These species are present on the 'cleaner' sediments only, though *N. nucleus* is absent from the coarse pebbles and gravels in the centre of the mouth.

It is possible that the *Nucula* sp. recorded within zone II at stations 5, 7 and 8 is either *N. turgida* or *N. tenuis*, which prefer rather muddy bottoms (Tebble, 1966), and not *N. nucleus*. However, neither *N. turgida* nor *N. tenuis* has been recorded in Milford Haven (Dale Fort Marine Fauna, 1966).

The polychaete, *Ophelia limacina,* and bivalve *Dosinia lupinus* are restricted to stations 10 and 11 (Figs. 12 and 13). *Golfingia procera* and *Amphiura filiformis* (Figs. 14 and 15) have similar distributions, being concentrated around stations 10, 12 and 25. At station 10 the bivalve *Mysella bidentata* (Fig. 16) is also present in significant densities.

Nematonereis unicornis (Fig. 17) occurs in Dale Roads and Dale Shelf and, avoiding the coarser deposits, is found at its highest density off Angle Bay at station 18. Usually found together with *Lumbrineris, N. unicornis* is not, however, found in the shallower sample stations of the Angle Bay transect.

(*iii*) *Species Restricted to Zone II*

For the purposes of this discussion, Dale Roads will be considered as part of zone II, although it is separated from the eastern part of Milford Haven by the oceanic zone (Fig. 2).

Only *Scoloplos armiger* (Fig. 18) was found uniquely in zone II, this polychaete being more typical of intertidal regions and found in greatest numbers in Angle Bay. *Lanice conchilega* (Fig. 19) is also confined mainly to Angle Bay.

A number of species are present in association with the sandy bottoms of zone II (both east and west of zone I), though *Clymene oerstedii* (Fig. 20) extends out into Dale Shelf.

Thyasira flexuosa is restricted more definitely to zone II (Fig. 21), though its eastern limit is reached at Newton Noyes. Where the substrate becomes muddier, *Melinna palmata* (Fig. 22) has a similar distribution to *C. oerstedii*, but it is present beyond Newton Noyes, although in very low numbers.

Four of the species illustrated here do not show strict preference for either zone, notably *Sthenelais boa* (Fig. 23) which seems to prefer poorly sorted deposits with large stones and mud together. The distribution of *Leptosynapta* sp. 55 (Fig. 24) is very patchy, though its population density is significant at station 6.

Pomatoceros triqueter (Fig. 25) is widely distributed around the shores of Milford Haven, and empty tubes were found at most stations wherever stones were exposed on the surface. Live specimens, however, were found in two main regions: Dale Shelf, and between Wear Point and Newton Noyes.

Tellina fabula (Fig. 26) is very abundant off West Angle Bay, where it is present in densities of about 200 per m^2 and is also present in Angle Bay, below low water.

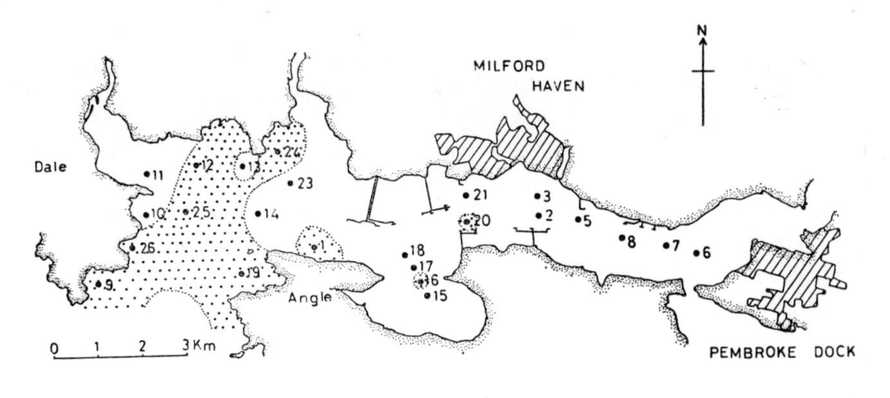

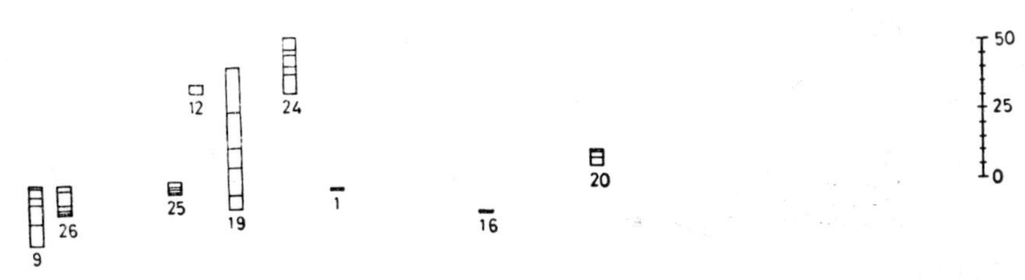

FIG. 9. The outline of Milford Haven, showing the grid of sampling stations and the distribution of *Magelona* spp. Divisions on the histograms represent the numbers of animals found in each grab.

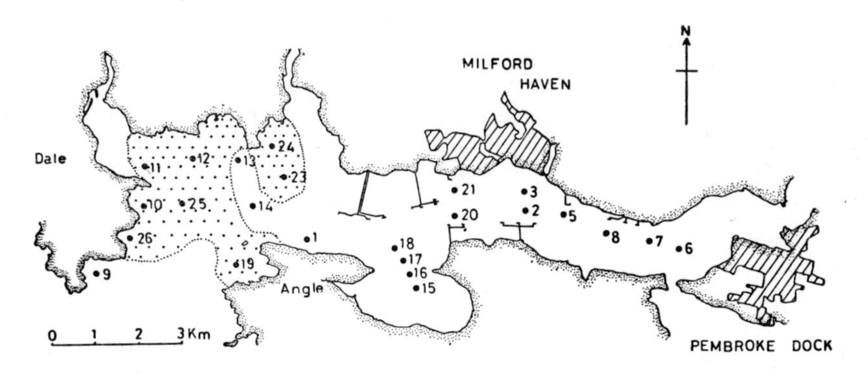

FIG. 10. The outline of Milford Haven, showing the grid of sampling stations and the distribution of *Ophelia limacina*. Divisions on the histograms represent the numbers of animals found in each grab.

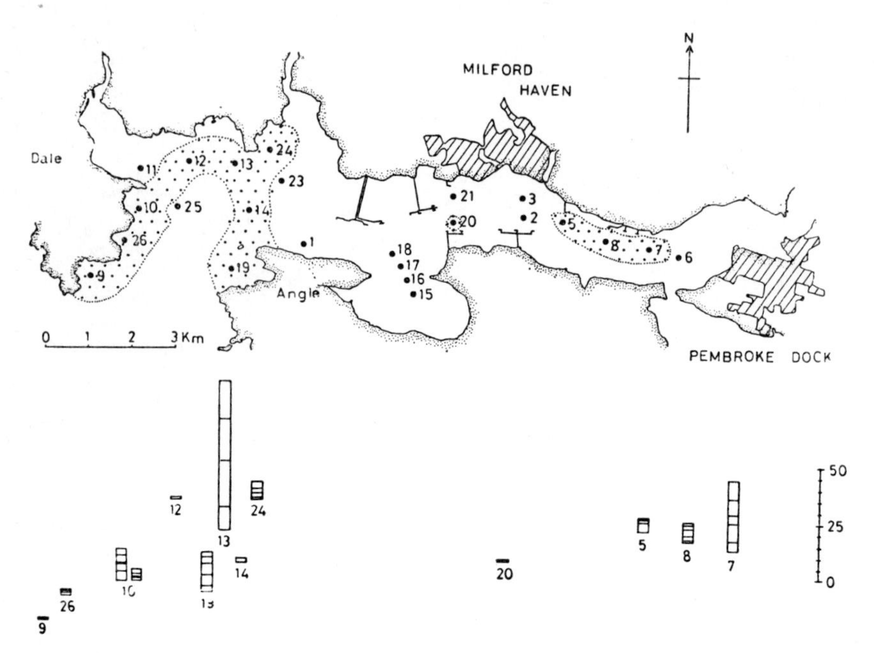

FIG. 11. The outline of Milford Haven, showing the grid of sampling stations and the distribution of *Nucula nucleus*. Divisions on the histograms represent the numbers of animals found in each grab.

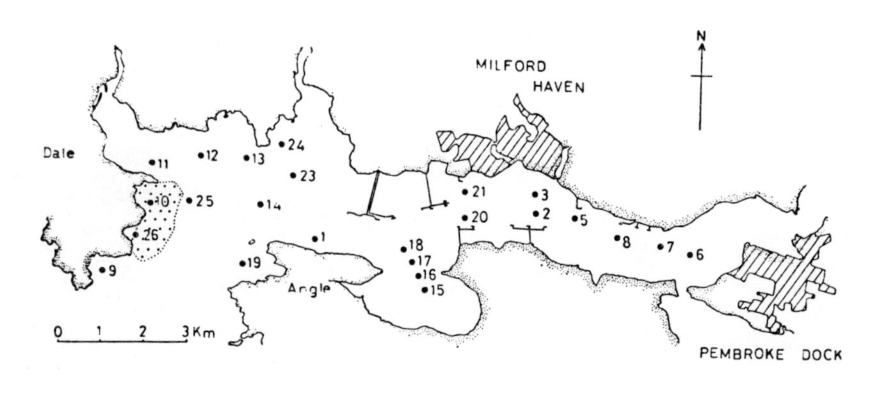

FIG. 12. The outline of Milford Haven, showing the grid of sampling stations and the distribution of *Ammotrypane aulogaster*. Divisions on the histograms represent the numbers of animals found in each grab.

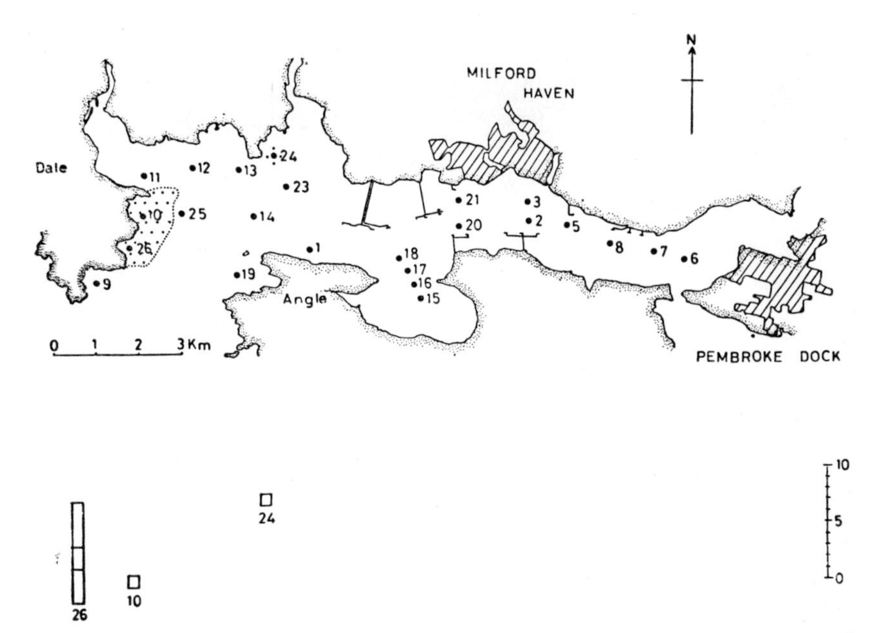

FIG. 13. The outline of Milford Haven, showing the grid of sampling stations and the distribution of *Dosinia lupinus*. Divisions on the histograms represent the numbers of animals found in each grab.

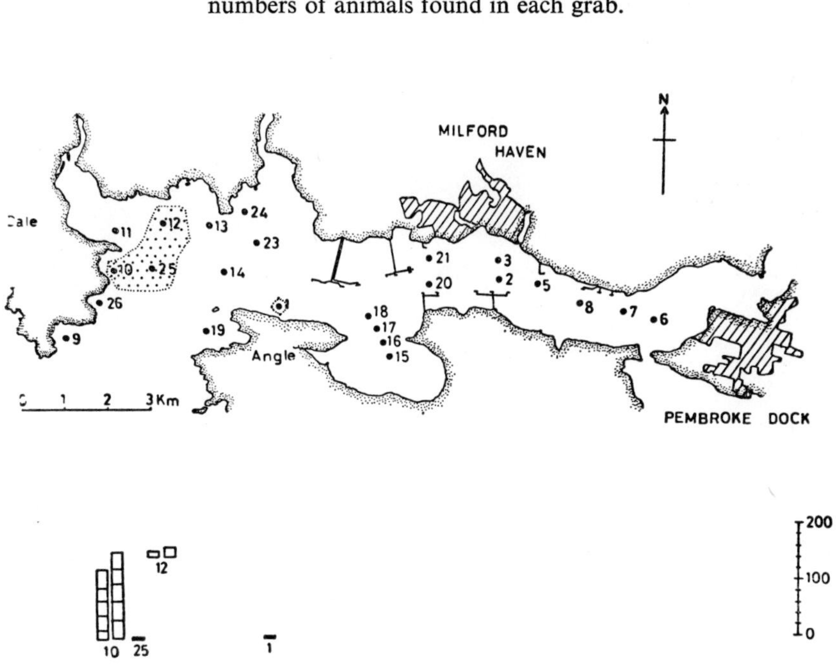

FIG. 14. The outline of Milford Haven, showing the grid of sampling stations and the distribution of *Golfingia procera*. Divisions on the histograms represent the numbers of animals found in each grab.

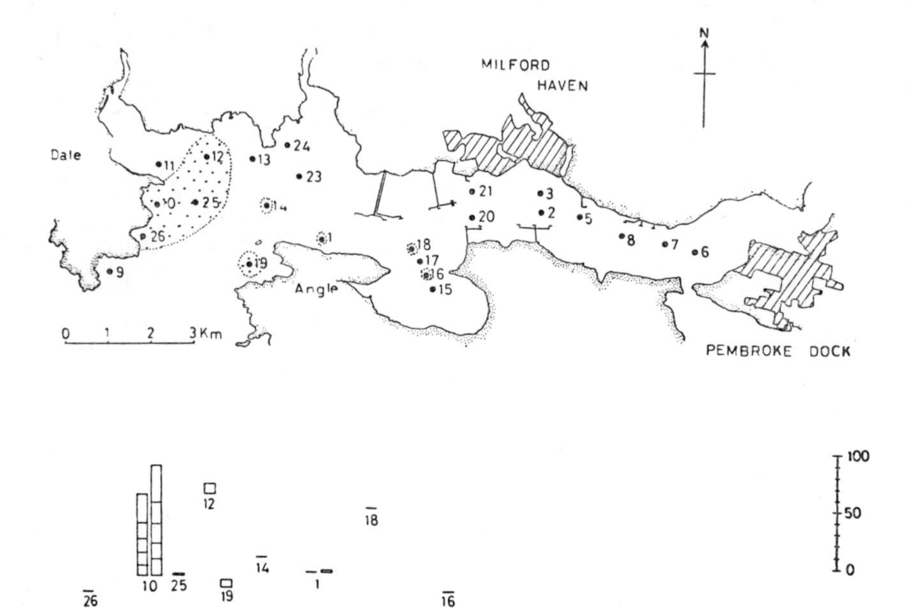

FIG. 15. The outline of Milford Haven, showing the grid of sampling stations and the distribution of *Amphiura filiformis*. Divisions on the histograms represent the numbers of animals found in each grab.

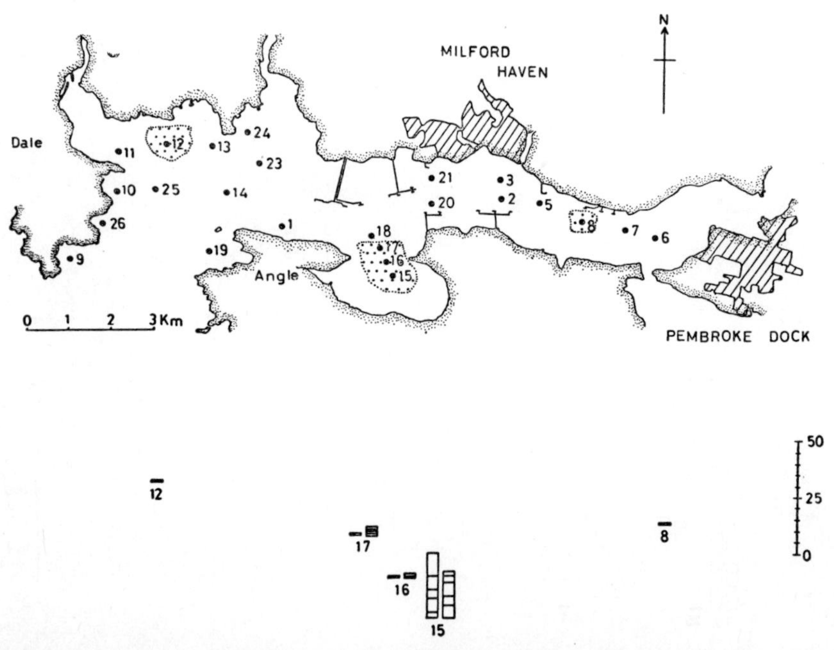

FIG. 16. The outline of Milford Haven, showing the grid of sampling stations and the distribution of *Mysella bidentata*. Divisions on the histograms represent the numbers of animals found in each grab.

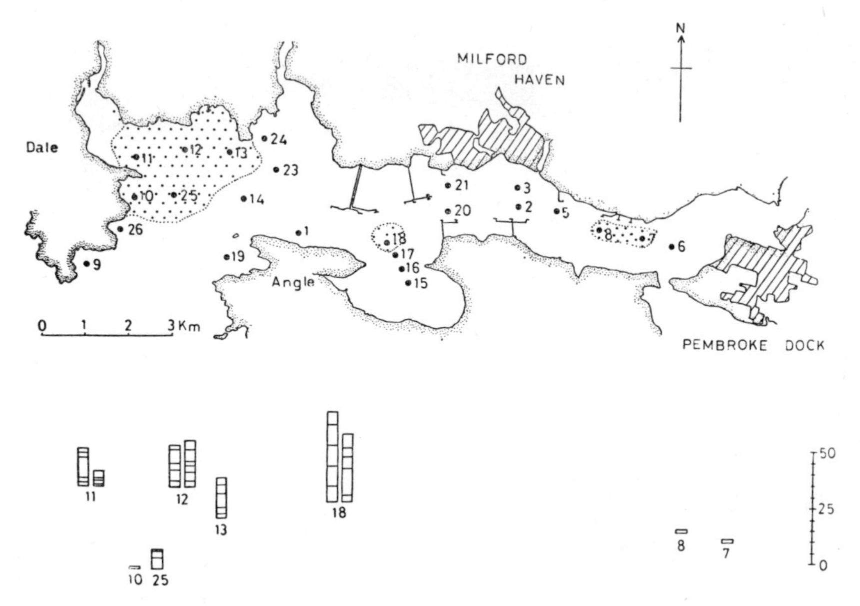

FIG. 17. The outline of Milford Haven, showing the grid of sampling stations and the distribution of *Nematonereis unicornis*. Divisions on the histograms represent the numbers of animals found in each grab.

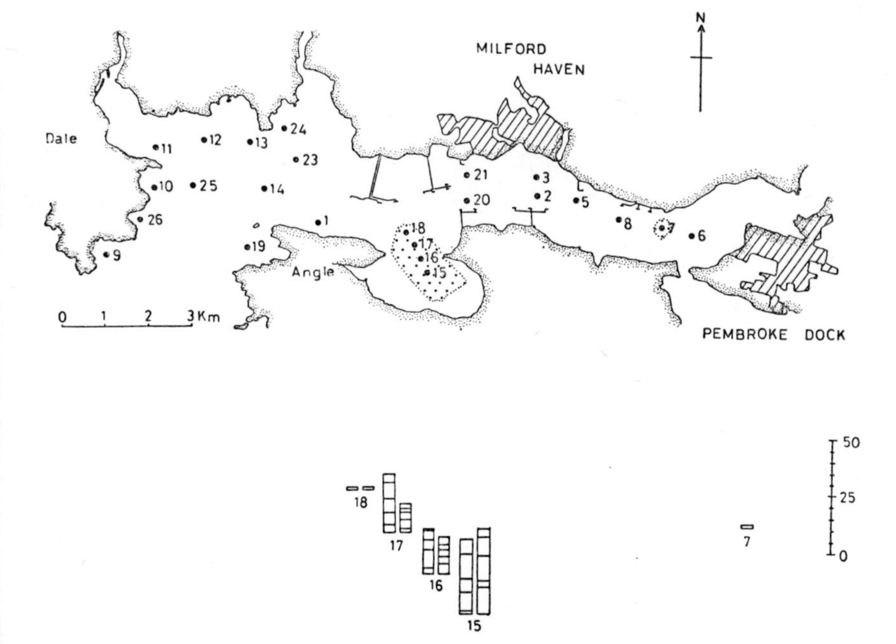

FIG. 18. The outline of Milford Haven, showing the grid of sampling stations and the distribution of *Scoloplos armiger*. Divisions on the histograms represent the numbers of animals found in each grab.

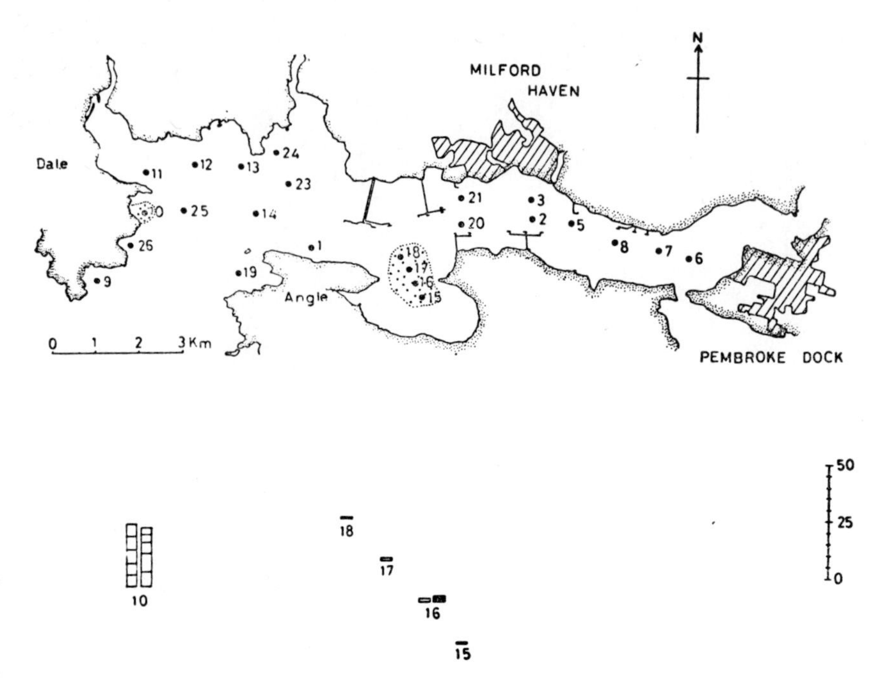

FIG. 19. The outline of Milford Haven, showing the grid of sampling stations and the distribution of *Lanice conchilega*. Divisions on the histogams represent the numbers of animals found in each grab.

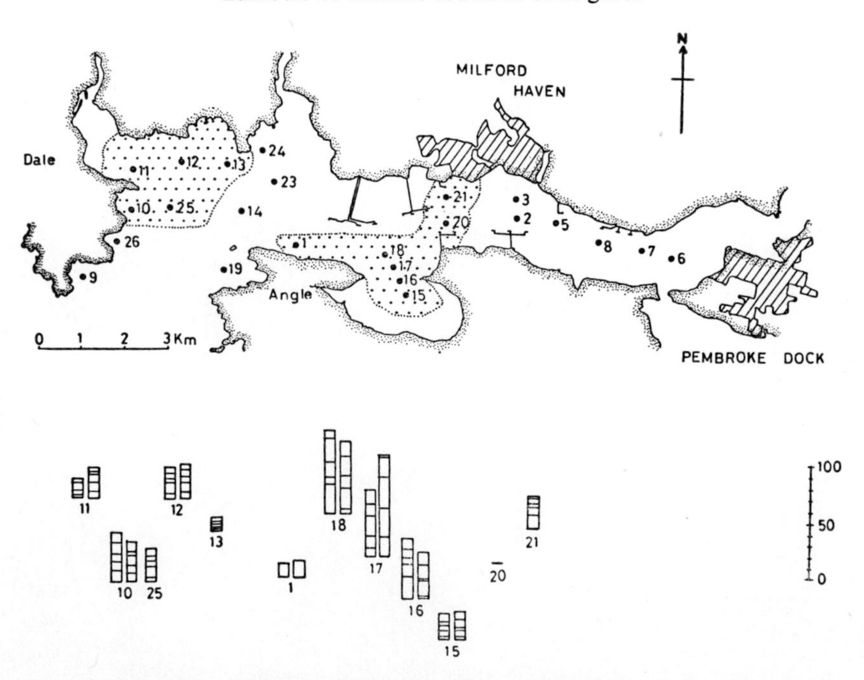

FIG. 20. The outline of Milford Haven, showing the grid of sampling stations and the distribution of *Clymene oerstedii*. Divisions on the histograms represent the numbers of animals found in each grab.

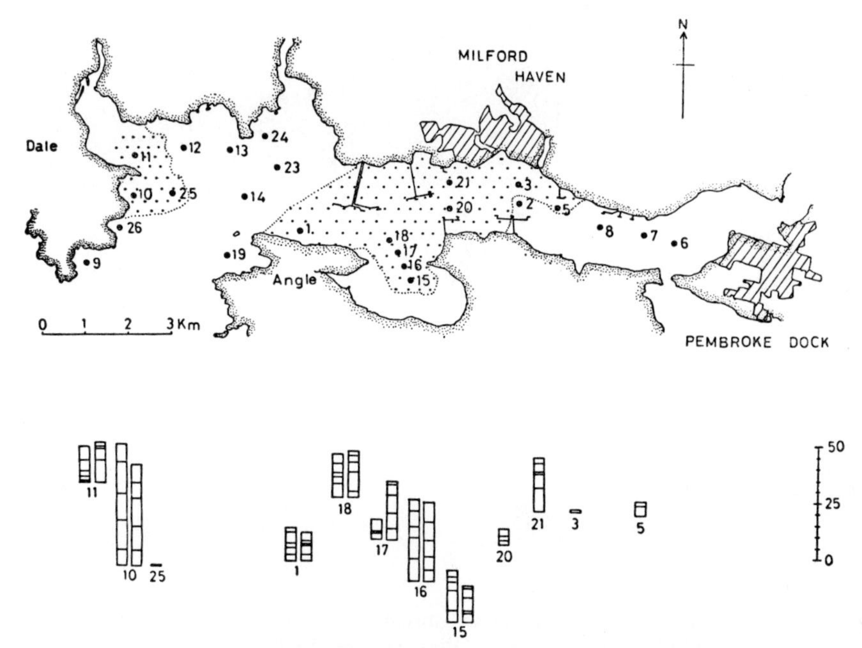

FIG. 21. The outline of Milford Haven, showing the grid of sampling stations and the distribution of *Thyasira flexuosa*. Divisions on the histograms represent the numbers of animals found in each grab.

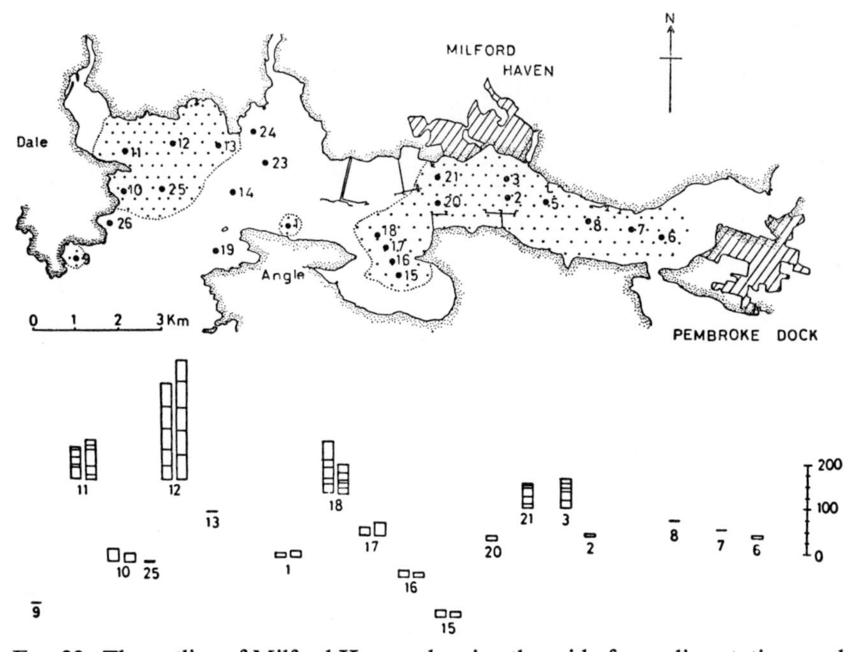

FIG. 22. The outline of Milford Haven, showing the grid of sampling stations and the distribution of *Melinna palmata*. Divisions on the histograms represent the numbers of animals found in each grab.

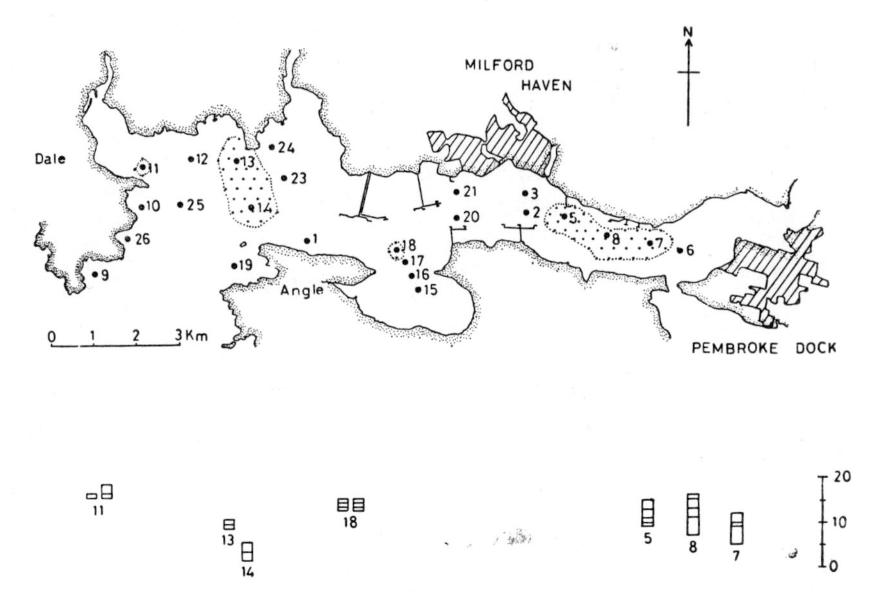

FIG. 23. The outline of Milford Haven, showing the grid of sampling stations and the distribution of *Sthenelais boa*. Divisions on the histograms represent the numbers of animals found in each grab.

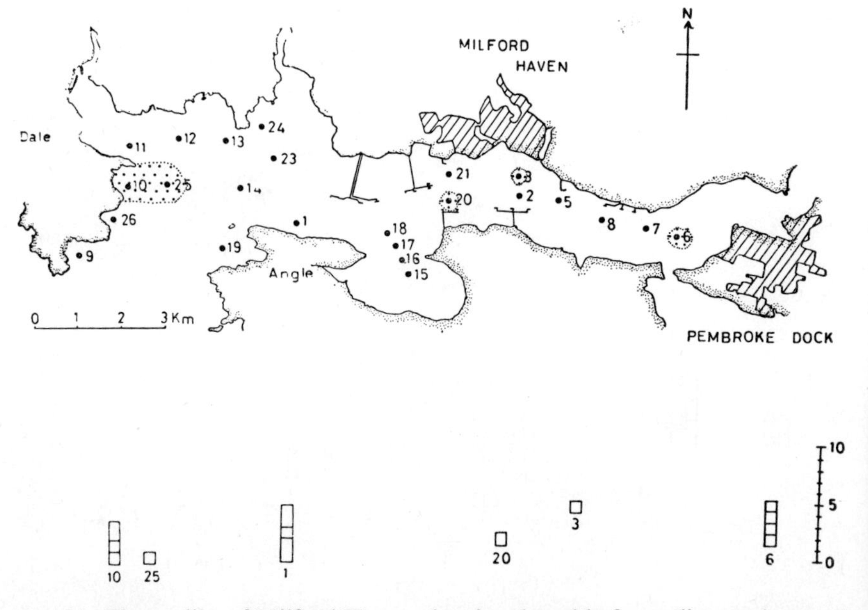

FIG. 24. The outline of Milford Haven, showing the grid of sampling stations and the distribution of *Leptosynapta* sp. 55. Divisions on the histograms represent the numbers of animals found in each grab.

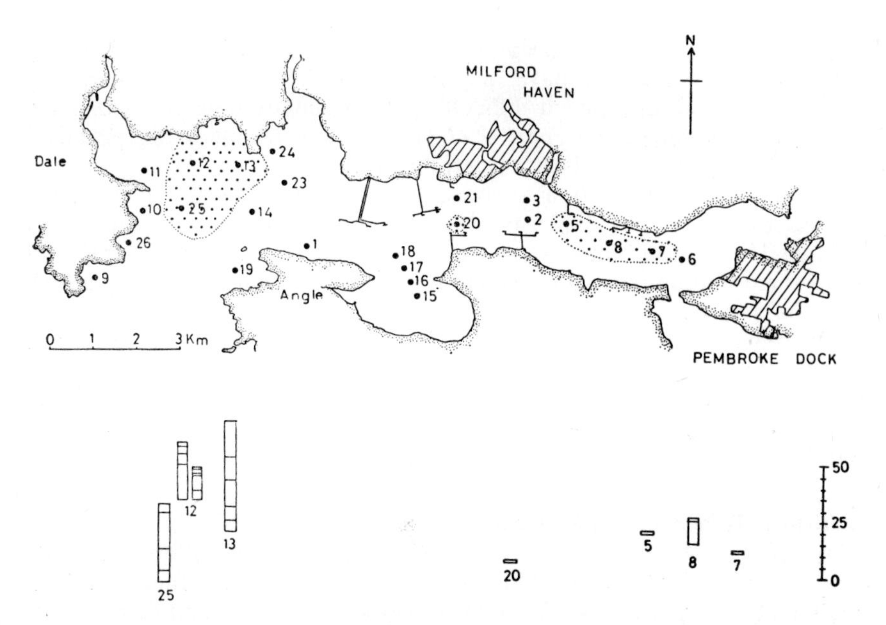

FIG. 25. The outline of Milford Haven, showing the grid of sampling stations and the distribution of *Pomatoceros triqueter*. Divisions on the histograms represent the numbers of animals found in each grab.

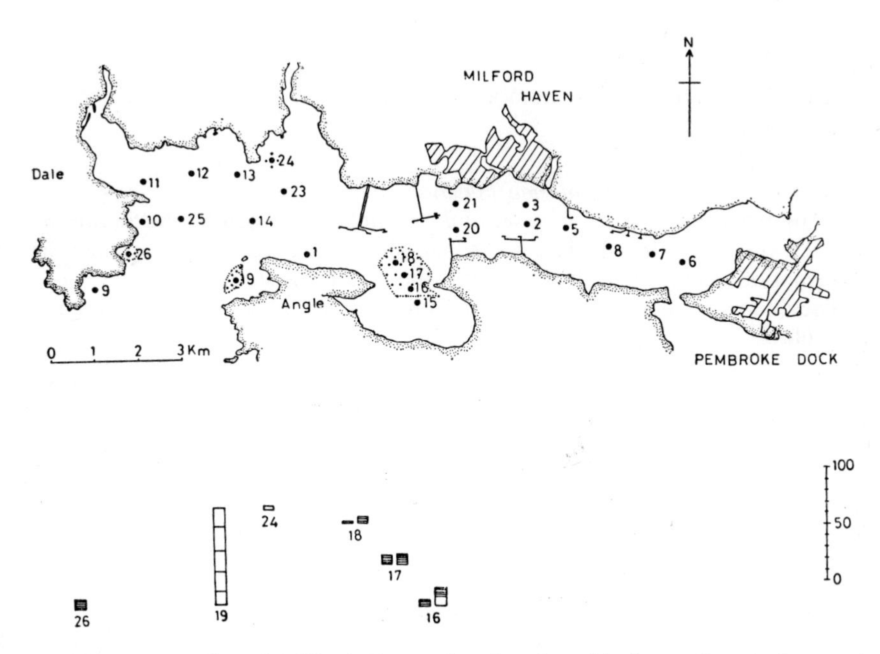

FIG. 26. The outline of Milford Haven, showing the grid of sampling stations and the distribution of *Tellina fabula*. Divisions on the histograms represent the numbers of animals found in each grab.

The Petersen–Thorson concept of benthic communities has been shown by a number of workers to have only limited applicability in semi-enclosed areas. Gage (1972a) found it difficult to draw definite boundaries between discrete communities, and Feder *et al.* (1973) also encountered the problems of overlapping species in Port Valdez, Alaska.

A preliminary association analysis, using a trellis diagram constructed for Sørenson's index of similarity, was only of limited success. It placed only stations 15, 16 and 17 into a clearly defined group, and with the exception of these three stations, spatial proximity of stations was not reflected by a high index of affinity.

Milford Haven has a wide variety of substrates in a relatively small area. The benthic fauna possibly consists of a series of overlapping zones marking the boundaries between potential communities which never attain the status of offshore communities because of restricted areas and other limiting physical factors.

Statistical Treatment of Biological Data

Two measures of species diversity were calculated for each station, Gleason's index of diversity and the Shannon–Wiener information function. The equitability component proposed by Lloyd and Ghelardi and described in Southwood (1966) is also calculated for each sample station.

(*i*) *Gleason's Index of Diversity* (Southwood, 1966)

$$D = \frac{S - 1}{\log_e N}$$

where D = diversity, S = number of species and N = total number of individuals.

(*ii*) *Shannon–Wiener Information Function*

The 'information content' of a set of data can be expressed by the Shannon–Wiener function ($H(S)$). The more complex a community, the greater its information content and, it is thought, the greater its stability.

The Shannon–Wiener information function can be calculated from the formula:

$$H(S) = -\sum_{i=1}^{s} p_i \log_2 p_i$$

where s = total number of species and p_i = the observed proportion of individuals which belong to the ith species; or, using logarithms to the base 10, we can use the modified equation (Southwood, 1966):

$$H(S) = c\left\{\log_{10} N - \frac{1}{N}\sum_{r=1}^{s} n_r \log_{10} n_r\right\}$$

where n_r = the number found in the rth species, N = the total number of individuals, and c = 3·3219 (the constant needed to convert the base of logarithms from 10 to 2).

The Shannon–Wiener information function has been used by Buchanan and Warwick (1974), who also calculated the equitability component ε based on the use of MacArthur's 'broken-stick' model as a yardstick of maximum equitability attainable:

$$\varepsilon = S'/S$$

where S = the actual number of species and S' = the number of hypothetical species characteristic of MacArthur's model for a Shannon–Wiener function of $M(S')$. S' is extracted from a table of $M(S')$ and S' by entering the survey data derived $H(S)$ in the $M(S')$ column and reading off the value of S' (Southwood, 1966).

PARTICLE SIZE ANALYSIS OF SEDIMENT SAMPLES

At each station, a small sub-sample was taken from the undisturbed contents of each grab, before sieving, for analysis of particle size distribution. These samples were oven dried to constant weight at 100°C and dry sieved in stacked sieves on a sieve shaker for 30 min. The weight retained on each sieve and in the bottom pan was recorded.

To measure the degree and efficiency of sorting exhibited by each deposit sample, the phi quartile deviation (QDϕ) and the phi quartile skewness (Skqϕ) were calculated using the methods of Morgans (1956).

Phi quartile deviation (QDϕ) is given by the formula:

$$\mathrm{QD}\phi = \frac{\mathrm{Q}_3\phi - \mathrm{Q}_1\phi}{2}$$

where $\mathrm{Q}_3\phi = \phi$ value corresponding to the third quartile and $\mathrm{Q}_1\phi = \phi$ value corresponding to the first quartile (see Fig. 27), and is a measure of the degree of sorting of the sample between the first and third quartile.

Phi quartile skewness (Skqϕ) is given by:

$$\mathrm{Skq}\phi = \frac{\mathrm{Q}_3\phi + \mathrm{Q}_1\phi - 2\mathrm{Md}\phi}{2}$$

where $\mathrm{Md}\phi = \phi$ value of median particle size, and is a measure of the efficiency of sorting between the first and third quartiles. Thus positive values indicate a greater sorting efficiency in the first quartile, and zero value for Skqϕ represents equally efficient sorting in the second and third quartile, and negative values indicate most efficient sorting in the third quartile. (See Fig. 27 for a well-sorted substrate with little skewness, and Fig. 29 for a poorly sorted substrate.)

The median particle size, Wentworth classification, phi quartile deviation and phi quartile skewness are presented in Table I.

Morgans (1956) and Buchanan and Kain (1971) recommend wet sieving to achieve an initial splitting of the silt clay sample from the bulk sample, and Morgans points out that the dry sieving techniques widely used are not

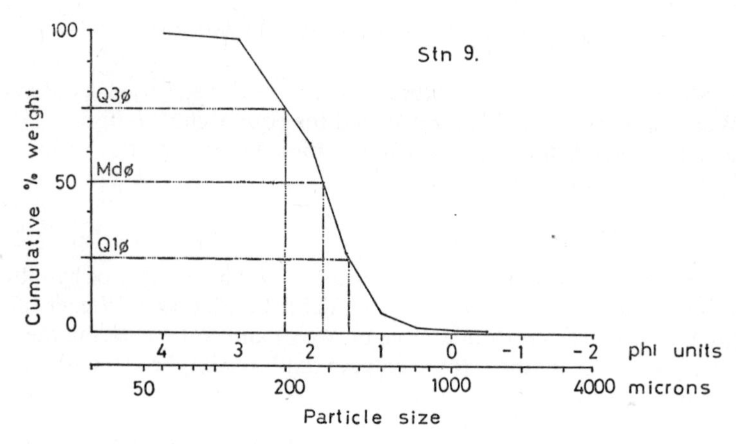

FIG. 27. The graph of cumulative weight percentage against particle size for station 9, with the first and third quartiles shown.

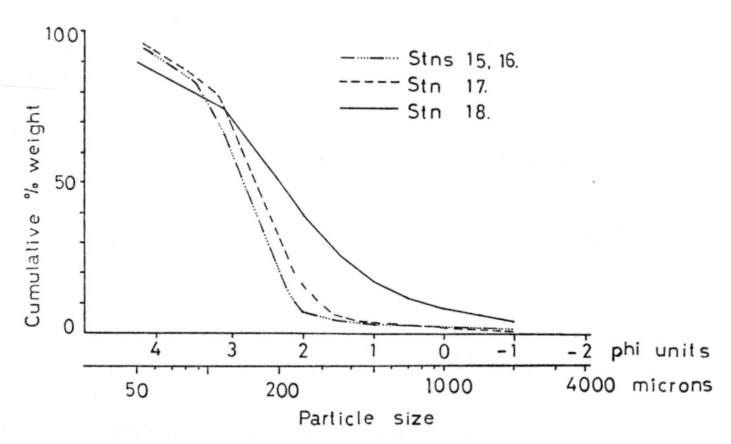

FIG. 28. The graphs of cumulative weight percentage against particle size for stations 15, 16, 17 and 18.

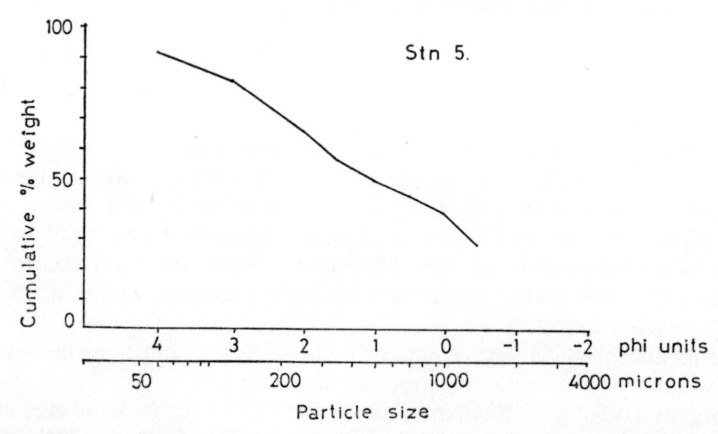

FIG. 29. The graph of cumulative weight percentage against particle size for station 5.

TABLE I

Particle Size Analysis Data for Each Sampling Station

Station	*Wentworth grade*	*Med. particle size* Md (μm)	Md ϕ	QD ϕ	Skq ϕ
1	Fine sand	160	2·6	0·6	0·2
2	Medium sand	275	1·9	0·5	0·1
3	Fine sand	195	2·3	1·2	0·3
5	Coarse sand	500	1·0	2·0	1·7
6	Very fine sand	~ 112	~3·2	—	—
7	Pebble	~1600	~ −0·7	—	—
8	Very coarse sand	~1150	~ −0·2	—	—
9	Medium sand	280	1·8	0·4	0·1
10	Medium sand	435	1·2	2·3	−1·0
11	Pebble	>1400	< −0·5	—	—
12	Fine sand	190	2·5	0·5	0·1
13	Pebble	>1400	< −0·5	—	—
14	Medium sand	370	1·4	1·5	−0·6
15	Fine sand	135	2·8	0·4	0·1
16	Fine sand	135	2·8	0·4	0·1
17	Fine sand	160	2·7	0·4	−0·03
18	Fine sand	210	2.3	0·9	0·1
19	Fine sand	180	2·5	0·4	0·03
20	Medium sand	265	1·9	1·4	0·6
21	Coarse sand	870	0·2	—	—
23	Medium sand	420	1·3	1·5	−0·8
24	Fine sand	185	2·4	0·8	0·1
25	Medium sand	450	1·2	1·2	−0·7
26	Fine sand	225	2·2	0·5	0·03

strictly relevant in terms of the biological significance of particle size. However, in this survey, time did not permit precise analysis of physical parameters, so the data presented here, which are derived from dry-sieving without initial splitting of the sample, can only be used in a general way to define areas of similar substrate. The particle size distribution of the substrate is unlikely to be changed except by gross pollution, so the precise relationship between fauna and substrate, which would require more elaborate analysis of the deposits, need not be investigated in a monitoring programme.

EFFICIENCY OF SAMPLING

Operation of the Day Grab

The grab worked effectively on a wide range of deposits, though the depth of penetration varied from about 15 cm on the fine silt at stations 2 and 6 to 5 or 6 cm on the hard sand of Angle Bay. The weights attached to the grab were varied in an attempt to overcome this problem, though even with 2 cwt of lead attached, the jaws had not closed before the grab started its ascent, so that on the hardest substrates only the surface 5 cm or so was sampled.

Another problem was encountered on the shelly sands and gravels of stations 7, 11 and 13, at which up to 50% of the grabs taken had a large shell or stone between the jaws resulting in a partial loss of the sample. Whenever this occurred, the sample was discarded and another grab sample taken. This problem could probably be overcome by the use of a Shipek grab at these sites in future surveys.

Number of Samples Taken at Each Station

Benthic grab-sampling usually employs instruments designed to sample an area 0·2 m^2 or 0·1 m^2, to a depth of 10–15 cm. McIntyre (1971) notes that the use of a large grab to save ship-time can have disadvantages. The additional information gained by taking more replicate samples with a smaller grab to cover the same total area is lost with the larger grab. The size of vessel available, in this case a 60 ft motor fishing vessel, precluded the use of a larger grab.

A compromise must be reached between sensitivity of sampling at each sample site and time and effort available for the whole survey. Various numbers of samples have been used by different authors. Feder *et al.* (1973) used only three replicate samples of 0·1 m^2 in Port Valdez, and Dicks (1975) used ten replicate samples in the North Sea. In this survey, five samples were taken routinely at each site, but at eight selected sites ten replicate samples were taken. These sites were chosen for two reasons:

(i) A large number of individuals were present.
(ii) The suitability of the substrate for grab-sampling suggested it as a useful area to be included in a grab-sampling monitoring programme.

Using data from these eight sites, graphs of cumulative percentage of individuals (C%I) and cumulative percentage of new species (C%S) were plotted for successive samples in the order they were taken at sea (Figs. 30–33). The values of C%S range from 65% to 85% for the first five replicate samples, which agrees with similar tests made by other authors (Feder *et al.* 1973; Dicks, 1975).

The suitability of cumulative species percentage analysis as a measure of the proportion of the population samples has been discussed (Longhurst, 1959; Lie, 1968; Holme, 1969). Feder *et al.* (1973) recommend the use of newly recruited individuals at each grab, by which they showed the sampling of abundant species to be adequate in the first two or three grabs.

However, in a monitoring programme designed to detect changes due to abiotic influences, it is also important to sample as completely as possible. Low densities and scattered distribution of a species do not necessarily indicate its lack of biological importance. The densities of some prey organisms might, for instance, be depressed by continual predation. Thorson (1966) reviews the factors influencing the establishment of marine benthic communities. Friedrich (1969) discusses the predator/prey relationships of marine benthos.

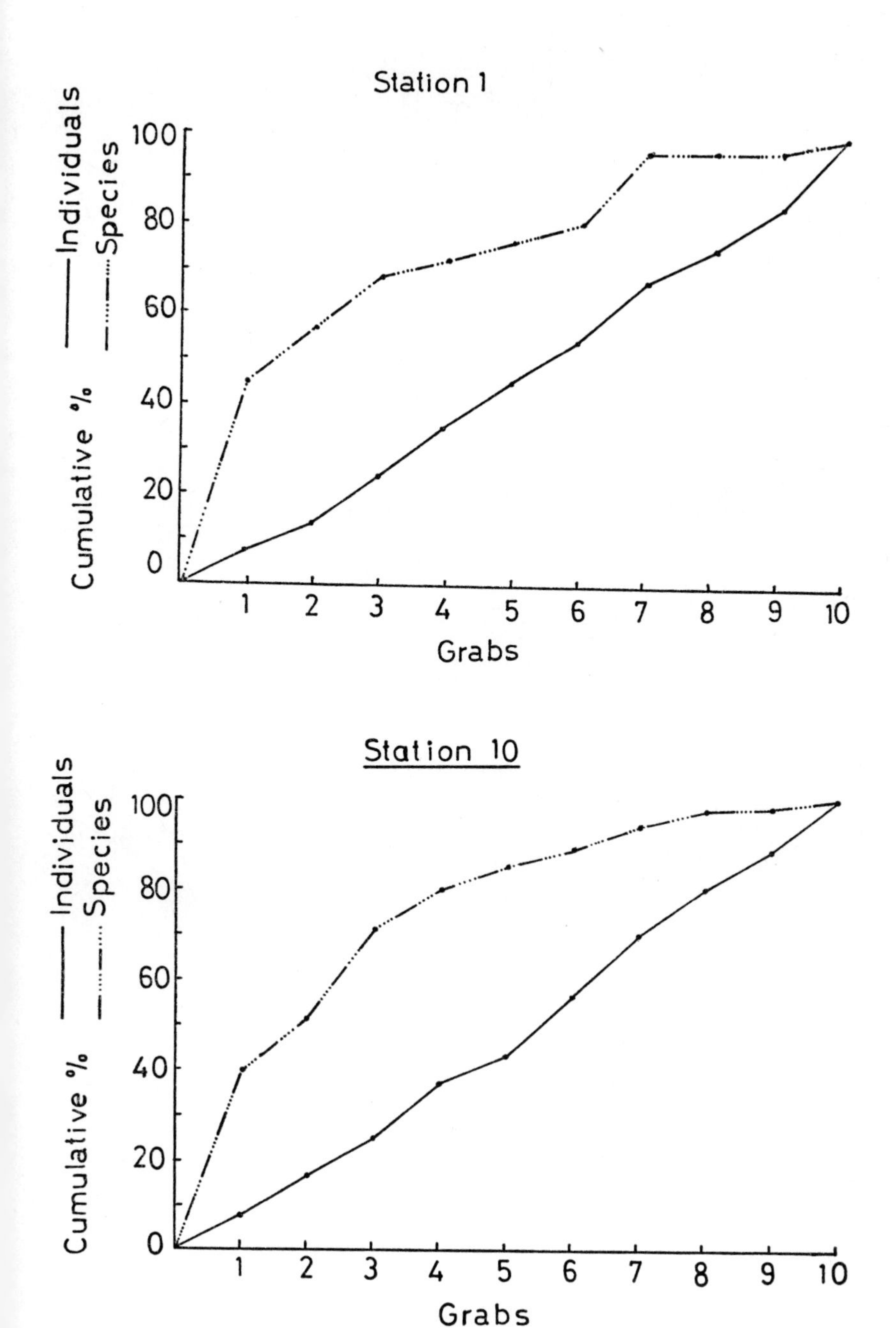

FIG. 30. Graphs of cumulative percentage of new species and cumulative percentage of individuals occurring with ten successive replicate grabs at sampling stations 1 and 10.

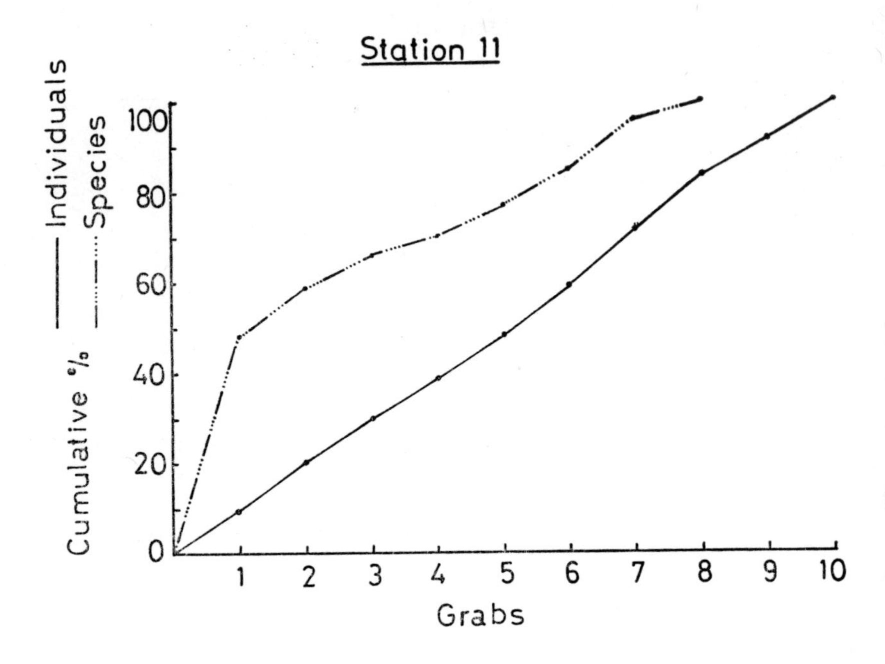

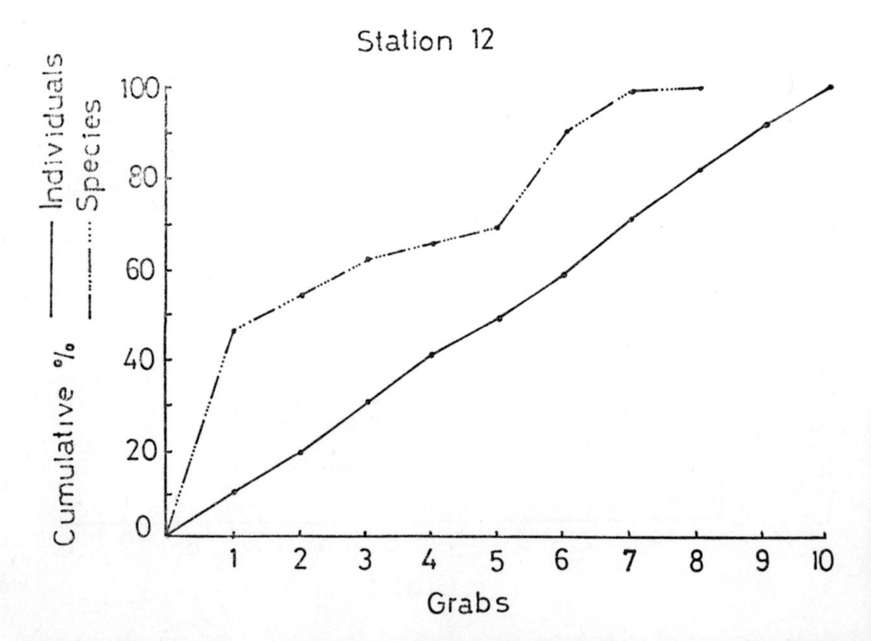

FIG. 31. Graphs of cumulative percentage of new species and cumulative percentage of individuals occurring with ten successive replicate grabs at sampling stations 11 and 12.

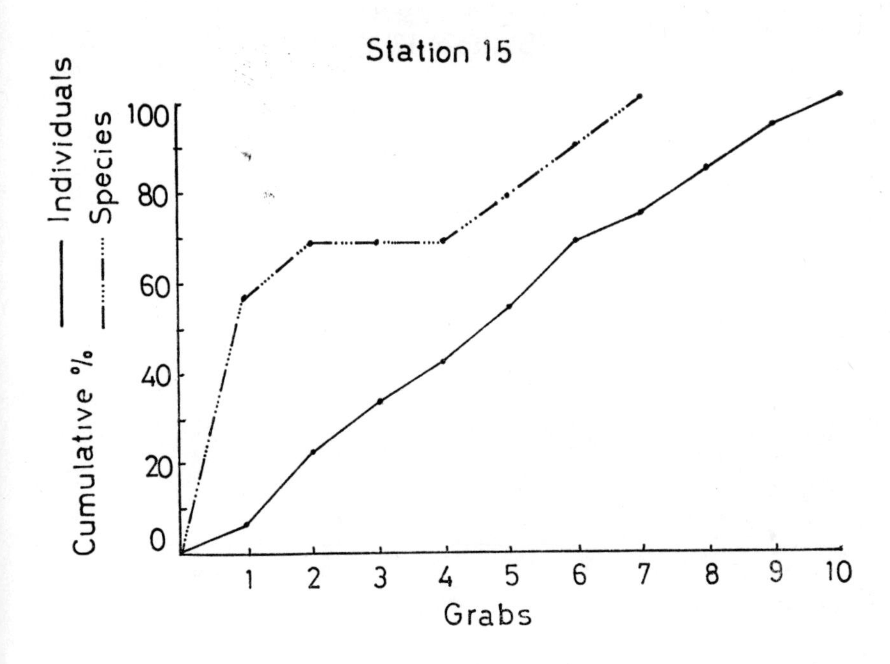

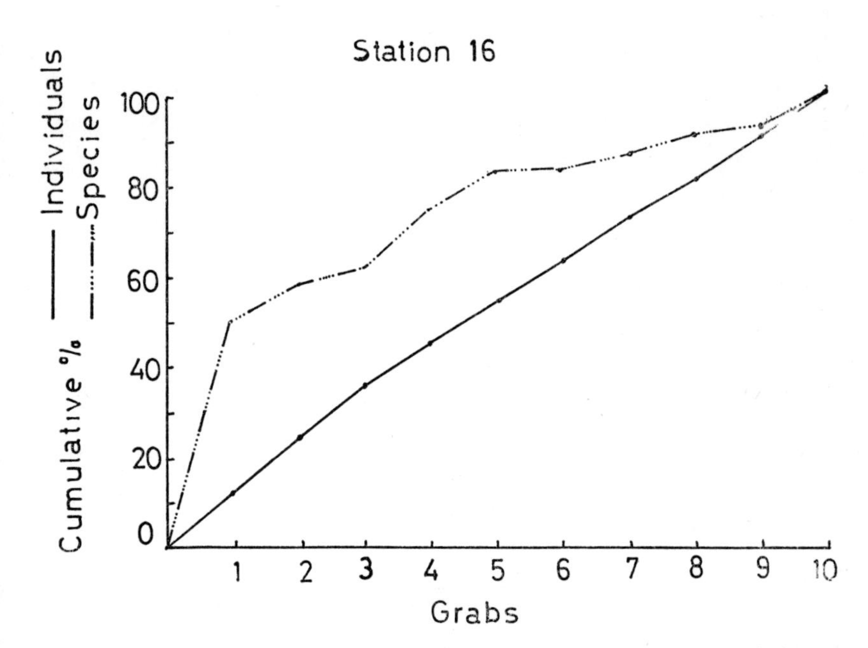

FIG. 32. Graphs of cumulative percentage of new species and cumulative percentage of individuals occurring with ten successive replicate grabs at sampling stations 15 and 16.

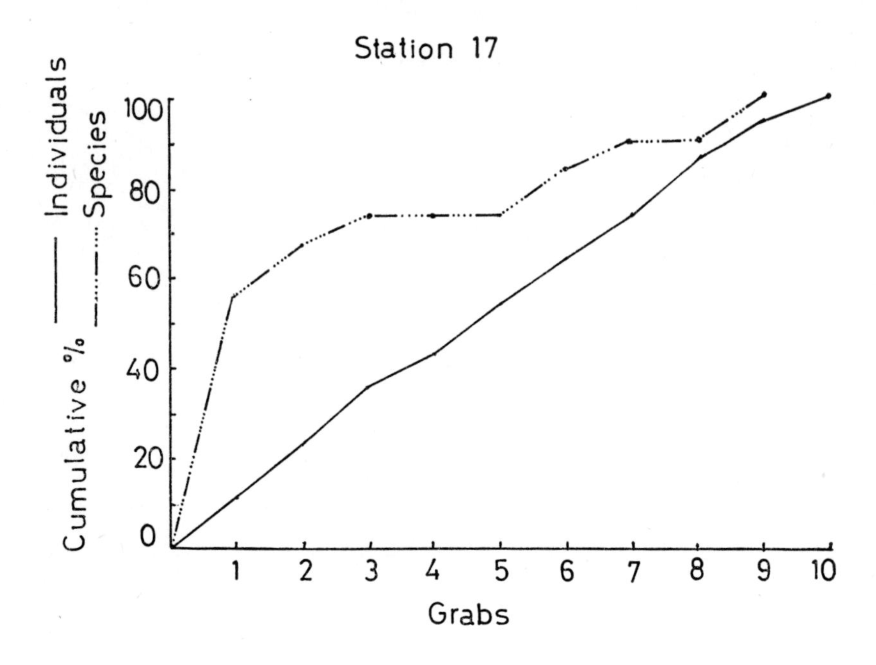

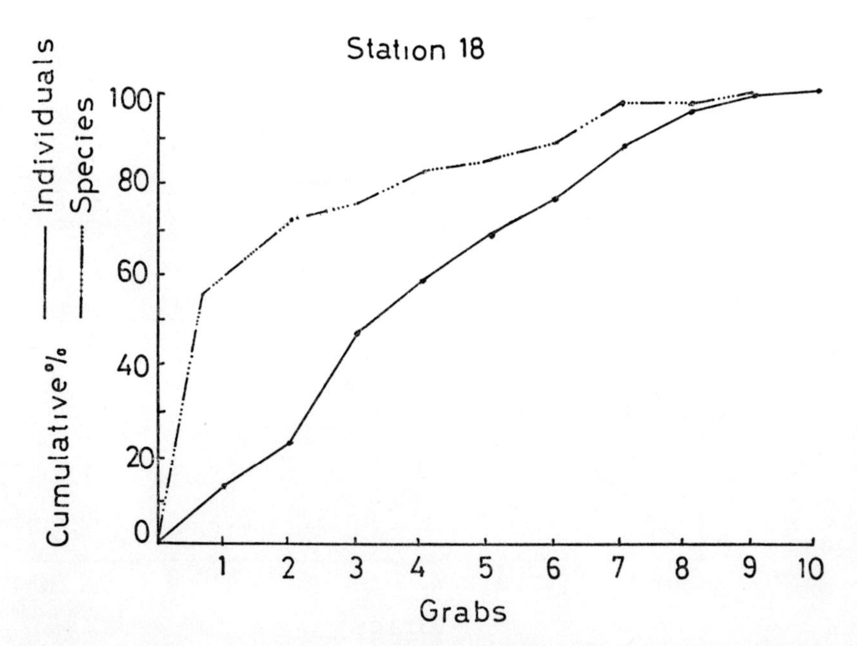

FIG. 33. Graphs of cumulative percentage of new species and cumulative percentage of individuals occurring with ten successive replicate grabs at sampling stations 17 and 18.

It is intended to use at least ten samples at each sampling station in future surveys in order to sample the less dense species more effectively.

PROPOSALS FOR MONITORING

The practical requirements of a monitoring programme are rather different from those of a large-scale biological survey. Having completed the initial survey in 1974, financial considerations and amount of time available restrict the number of samples which can be taken on a routine basis. Having decided on grab-sampling as the most operationally quantitative method available for working from a boat the choice of sampling sites for routine investigation must be influenced by their suitability for grab-sampling. Thus, a number of the stations sampled in this survey would not be revisited on a routine basis due to the inefficiency of grab-sampling on rough ground.

The most uniform area of Milford Haven, both in terms of its sediments, depth and fauna, is Angle Bay (see Figs. 2–26). It is proposed to continue the Angle Bay transect used in this survey up into the littoral regions, and sample the fauna at each station regularly, at least twice a year.

Seasonal variations in species composition and density can be considerable (Parker, 1975), and the assumption that taking samples at the same time of year rules out the effect of temporal variation in population density needs investigation. Angle Bay is an ideal situation in which to study seasonal as well as long-term changes in population density and species composition.

Dale Roads and Dale Shelf provide other fairly uniform areas and a line transect similar to that at Angle Bay could be set up there.

It is generally agreed that change in population density and species make-up due to environmental factors, including abiotic factors such as pollution, will influence the species diversity calculated for the population.

Most of the commonly used diversity indices exhibit a bias, either towards dependence on species number, or on the total number of individuals in the sample.

Scatter diagrams of Gleason's index of diversity (D) plotted against number of species and total number of individuals from which each index was derived show a positive relationship between D and the number of species at a 5% confidence level (Figs. 34–35). The test used was the rapid graphical method outlined in Lewis and Taylor (1967).

The Shannon–Wiener information function ($H(S)$) has been used successfully elsewhere (Lie, 1968; Sanders, 1968; Buchanan *et al.* 1974). Scatter diagrams of $H(S)$ plotted against the species number and total number of individuals from which each index was derived do not show any definite relationship (Figs. 36–37). More than one diversity index should be used when comparing faunas either spatially or temporally removed from each other (Feder *et al.* 1973). Problems of comparison of widely differing faunistic areas in terms of species diversity are not discussed here (see Dicks, 1975). Such problems do not necessarily arise in a monitoring programme where comparisons are temporal rather than spatial (Dicks, 1975).

The values of D, $H(S)$ and ε calculated for the data are given in Table II.

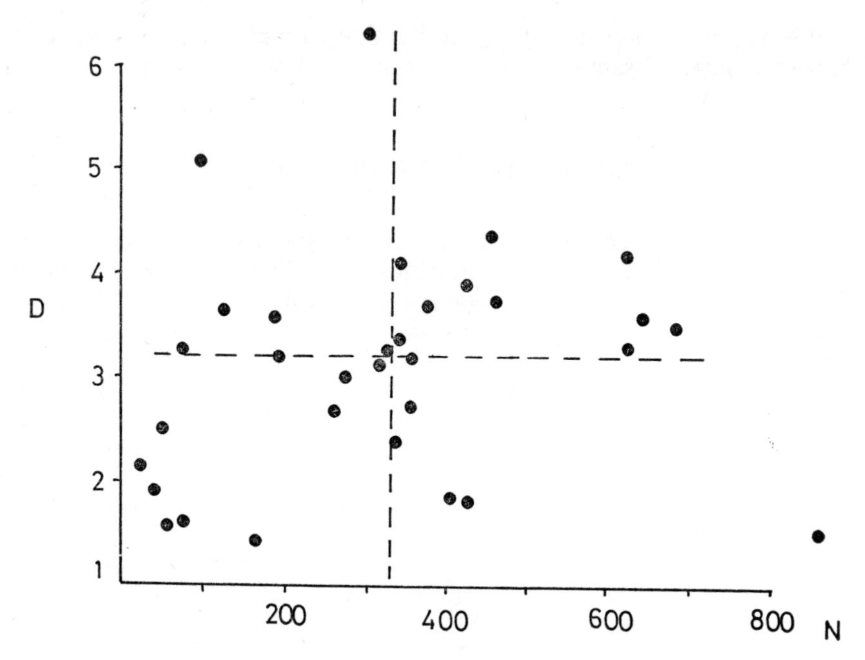

FIG. 34. A scatter diagram of Gleason's index of diversity (D) plotted against the total number of individuals (N) at each sampling station.

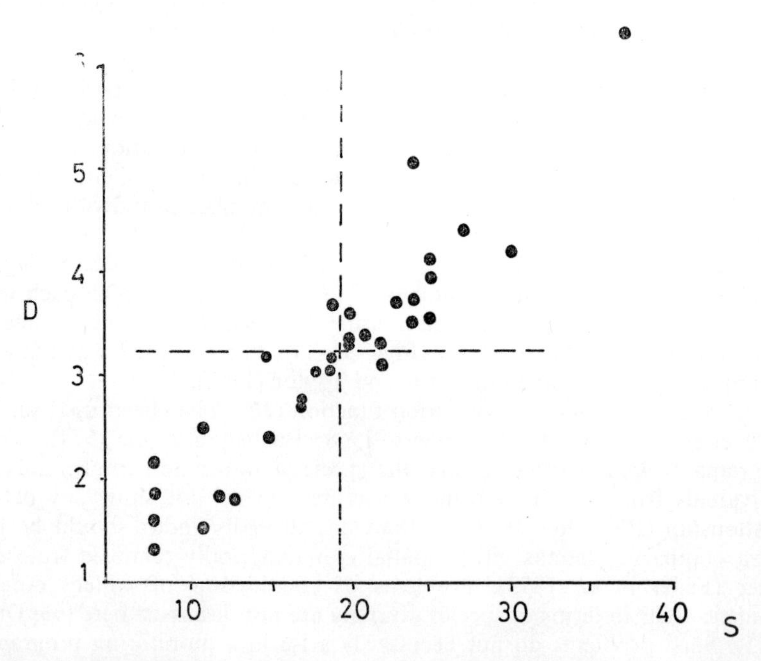

FIG. 35. A scatter diagram of Gleason's index of diversity (D) plotted against the number of species (S) at each sampling station.

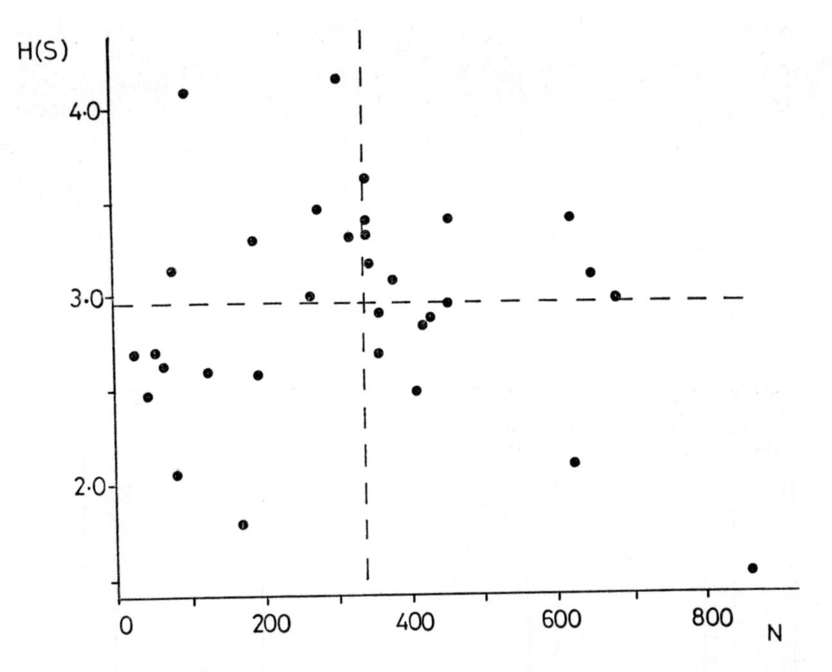

FIG. 36. A scatter diagram of the Shannon–Wiener information function ($H(S)$) plotted against the total number of individuals (N) at each sampling station.

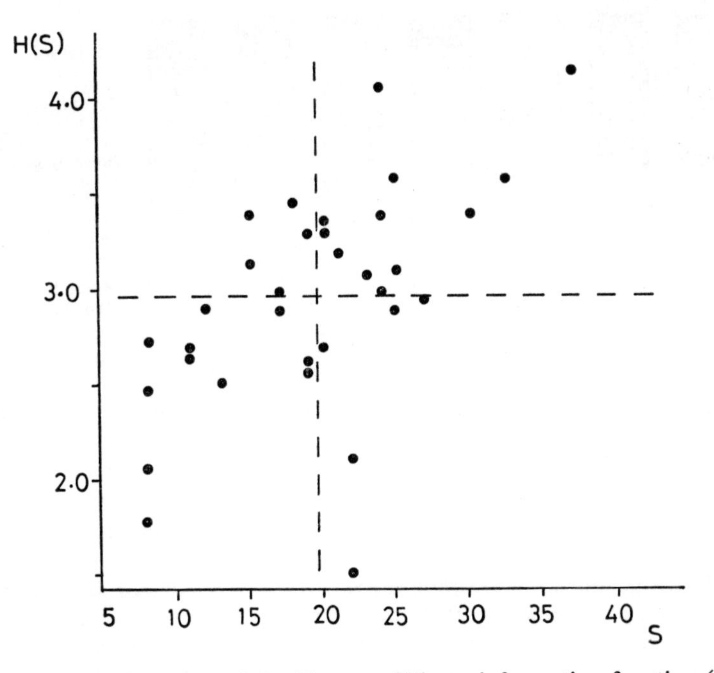

FIG. 37. A scatter diagram of the Shannon–Wiener information function ($H(S)$) plotted against the number of species (S) at each sampling site.

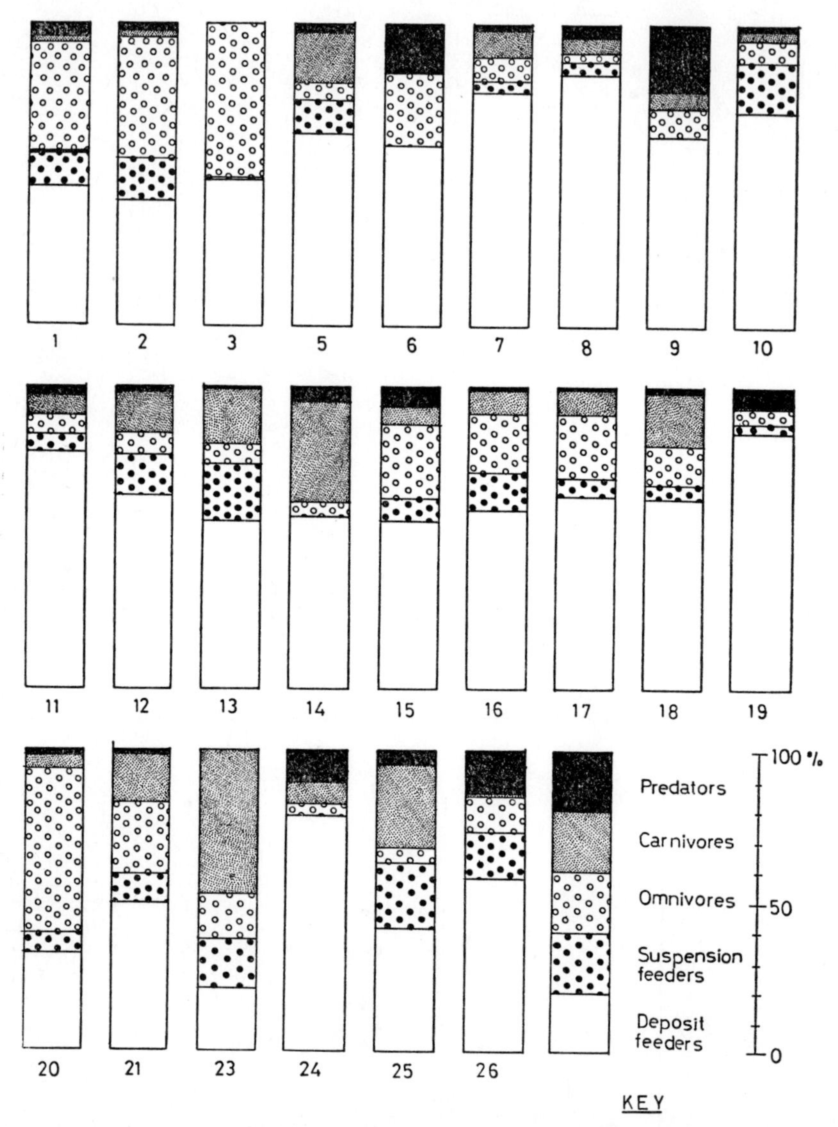

FIG. 38. Histograms showing the species percentage composition, by feeding type at each sampling station.

The highest diversities and information content were found at stations 25 and 26. These are possibly the most 'oceanic' of the sample stations.

In addition to the diversity indices and equitability component described above, it is proposed to examine the relative proportions of different feeding types found at each site within the monitoring programme. Fig. 38 shows the data for this survey. The dominance of deposit feeders in terms of percentage proportion at each station is noticeable. Buchanan *et al.* (1974) have demonstrated the presence of 'opportunistic' species in the macrobenthos,

TABLE II

Number of individuals (N), number of species (S), species diversity (D), Shannon–Wiener function ($H(S)$) and equitability component ε at each station

Sample station	S	N	D	$H(S)$	ε
1.1	19	127	3·70	2·60	0·42
1.2	20	188	3·63	3·32	0·7
2	8	43	1·86	2·48	1·0
3	8	169	1·36	1·77	0·5
5	13	409	1·83	2·52	0·62
6	8	25	2·17	2·73	1·13
7	22	623	3·26	2·13	0·27
8	22	863	3·11	1·53	0·18
9	11	55	2·50	2·70	0·82
10.1	30	623	4·20	3·40	0·50
10.2	24	682	3·52	2·99	0·46
11.1	21	345	3·42	3·18	0·62
11.2	23	378	3·71	3·10	0·52
12.1	25	424	3·97	2·86	0·40
12.2	27	456	4·41	2·96	0·41
13	25	342	4·11	3·62	0·72
14	8	80	1·60	2·07	0·63
15.1	15	340	2·40	3·41	1·00
15.2	18	276	3·02	3·46	0·89
16.1	20	359	3·23	2·67	0·45
16.2	19	318	3·12	3·30	0·74
17.1	12	427	1·82	2·88	0·83
17.2	17	358	2·72	2·95	0·65
18·1	25	645	3·56	3·12	0·52
18.2	24	458	3·75	3·41	0·67
19	17	265	2·69	3·00	0·65
20	19	191	3·24	2·58	0·42
21	20	336	3·27	3·35	0·75
23	11	62	1·56	2·65	0·73
24	15	76	3·23	3·14	0·87
25	37	301	6·31	4·14	0·70
26	24	95	5·07	4·08	1·04

for example the capitellid *Heteromastus*. It is thought that a routine assessment of the proportion of deposit feeders and other categories would be useful as a complement to diversity indices.

The distinction between predation and a carnivorous habit may seem tenuous but a number of polychaetes in particular have been shown to be carnivorous, whilst not necessarily actively hunting their prey. For example, *Lumbrineris* may be scavengers, having lost the prostomial antennae and eyes usually associated with the eunicids (Day, 1967).

The feeding status of many polychaetes is unknown, and can often only be inferred from the anatomy of the animal and observed behaviour of its close relatives. However, this can lead to errors and consequently the broader categories such as 'deposit feeders' and 'suspension feeders' are probably of more value in the context of a monitoring programme than attempts to consider more precisely the habits of individual species.

ACKNOWLEDGEMENTS

I should like to thank the Natural Environment Research Council's Research Vessel Base, who supplied the grab used in this survey.

Sampling was successfully carried out with the assistance of Mrs M. Knight and Miss S. Nicol, who did the initial sieving on the sometimes uncomfortable vessel, and Mr J. Levell, who fixed the sampling positions and assisted with the operation of the grab.

The co-operation of the Milford Haven Conservancy Board during the week of sampling was greatly appreciated.

I am grateful to Dr P. Gibbs for his identification of the Sipunculans and some of the polychaetes. My colleagues in the Oil Pollution Research Unit checked the manuscript and have offered helpful advice.

I also thank Miss A. van der Wel for her assistance throughout this project.

REFERENCES

Allen, J. A. (1967). *Fauna of the Clyde Sea area. Crustacea: Euphausiacea and Decapoda*, Scottish Marine biol. Ass., 116 pp.

Baker, J. M. (1971). The effects of oil pollution on saltmarsh vegetation, Ph.D. thesis, University College of Wales, Swansea.

Barrett, J. and Yonge, C. M. (1958). *Collins Pocket Guide to the Sea Shore*, Collins, 272 pp.

Buchanan, J. B. and Kain, J. M. (1971). In: *Methods for the Study of Marine Benthos*, ed. N. A. Holme and A. D. McIntyre, IBP Handbook No. 16.

Buchanan, J. B. and Warwick, R. M. (1974). An estimate of benthic macrofaunal production in the offshore mud of the Northumberland coast, *J. mar. biol. Ass. U.K.*, **54**, 197–222.

Buchanan, J. B., Kingston, P. F. and Sheader, M. (1974). Long-term population trends of the benthic macrofauna in the offshore mud of the Northumberland coast, *J. mar. biol. Ass. U.K.* **54**, 785–95.

Clark, R. B. (1960). *Fauna of the Clyde Sea area: Polychaeta, with keys to the British genera*, Scottish Marine biol. Ass., 71 pp.

Clark, R. B. and Haderlie, E. C. (1960). The distribution of *Nephtys cirrosa* and *N. hombergi* on the south-western coasts of England and Wales, *J. Anim. Ecol.*, **29**, 117–47.

Clay, E. (1967). Literature Survey of the common fauna of Estuaries. 6. *Nephtys hombergi* Lamark, ICI Brixham Laboratories, 22 pp.

Crapp, G. (1970). The biological effects of marine oil pollution and shore cleaning, Ph.D. thesis, University College of Wales, Swansea.

Crothers, J. H. (ed.) (1966). The Dale Fort Marine Fauna, Supplement to Vol. 2 of *Field Studies*, 169 pp.

Day, J. H. (1967). *Polychaeta of Southern Africa. Part 1, Errantia; Part 2, Sedentaria*, Brit. Mus. (Nat. Hist.), 878 pp.
Dicks, B. (1975). Offshore biological monitoring (this volume).
Fauvel, P. (1923). *Faune de France*, **5**, *Polychètes Errantes*, 488 pp.
Fauvel, P. (1927). *Faune de France*, **16**, *Polychètes Sédentaires*, 494 pp.
Feder, H., Mueller, G. J., Dick, M. H. and Hawkins, D. B. (1973). In *Environmental Studies of Port Valdez*, ed. D. W. Hood, W. E. Shiels and E. J. Kelly. Institute of Marine Science, University of Alaska Occasional Publication No. 3.
Friedrich, H. (1969). *Marine Biology*, Univ. of Washington Press, Seattle, 474 pp.
Gage, J. (1972a). A preliminary survey of the benthic macrofauna and sediments in Lochs Etive and Creran, sea-lochs along the west coast of Scotland, *J. mar. biol. Ass. U.K.* **52**, 237–74.
Gage, J. (1972b), Community structure of the benthos in Scottish sea-lochs I. Introduction and species diversity, *Mar. Biol.*, **14**, 281–97.
Gibbs, P. E. (1973). On the Genus *Golfingia* (Sipuncula) in the Plymouth area, with a description of a new species, *J. mar. biol. Ass. U.K.*, **53**, 73–86.
Graham, A. (1971). *British Prosobranchs*, Synopses of the British Fauna (New Series) No. 2, Linnean Society of London.
Lewis, T. and Taylor, L. R. (1967). *Introduction to Experimental Ecology*, Academic Press, 401 pp.
Lie, U. (1968). A quantitative study of the benthic infauna in Puget Sound, Washington, U.S.A. in 1963–1964, *Fisk. Dir. Skr.* (*Ser Havunders*), **14**, 229–556.
Longhurst, A. R. (1959). The sampling problem in benthic ecology, *Proc. N.Z. Ecological Society*, **6**, 8–12.
McIntyre, A. D. (1971). In: *Methods for the Study of Marine Benthos*, ed. N. A. Holme and A. D. McIntyre, IBP Handbook No. 16.
Morgans, J. F. C. (1956). Notes on the analysis of shallow water soft substrata, *J. Anim. Ecol.*, **25**, 367–87.
Nelson-Smith, A. (1965). Marine biology of Milford Haven: the physical environment, *Field Studies*, **2**, 155–88.
Parker, R. H. (1975). *The Study of Benthic Communities*, Elsevier, 277 pp.
Sanders, H. L. (1968). Marine benthic diversity: a comparative study, *Amer. Naturalist*, **102**, 243–82.
Southwood, T. R. E. (1966). *Ecological Methods*, Methuen, 391 pp.
Stephen, A. C. (1960). *A synopsis of the Echiuroidea, Sipunculoidea and Priapuloidea of British Waters*, Synopsis of the British Fauna No. 12, Linn. Soc. London.
Tebble, N. (1966). *British Bivalve Seashells*, Brit. Mus. (Nat. Hist.), 213 pp.
Thorson, G. (1966). Some factors influencing the recruitment and establishment of marine benthic communities, *Netherlands J. Sea Res.*, **3**, 267–93.

APPENDIX I

Total Numbers of Each Taxa Found at Each Sampling Station. The Feeding Type is also Shown when Information is Available

DF = deposit feeders, SF = suspension feeders, O = omnivores, H = herbivores, C = carnivores, P = predators

Taxa	*Feeding type*	1.1	1.2	2	3	5	6	7	8	9	10.1	10.2	11.1
		Sample station											
Polychaeta:													
Aphrodite aculeata L.	DF												
Harmothoe antilopis McIntosh	P												
Lepidonotus squamatus (L.)	P												
Harmothoe lunulata (Delle Chiaje)	P	1	4				2		1		1		
Harmothoe sp. 1.31	P	1					2	1	3				
Lagisca extenuata (Grube)	P											1	
Sigalion mathildae (Audouin & Milne-Edwards)	P												
Sthenelais boa (Johnston)	P					6		7	9				1
Pholoe minuta (Fabricius)	P										1		
Aphroditid sp. 91											3		
Aphroditid sp. 80		1											
Phyllodoce laminosa Savigny	P?												
Phyllodoce mucosa Oersted	P?							3	2				
Eulalia sanguinea Oersted	P?			1		5			11				
Gyptis capensis Day	P?	1	2									2	
Syllid spp.						7							
Platynereis dumerili (Audouin & Milne-Edwards)	H												
Heteronereis sp.	P?												
Nereid sp. 95	P?											1	
Nereid sp. 46	P?									1			
Nephtys spp.	O?	50	79	17	86	24	6	43	21	5	40	51	5
Glycera sp. 1.24	P												8
Glycera convoluta Keferstein	P									11	2	2	
Goniada maculata (Oersted)	P	3	1								3		
Marphysa sanguinea (Montague)	O/DF												
Marphysa bellii (Audouin & Milne-Edwards)	O/DF												
Nematonereis unicornis (Grube)	O/DF							2	2		2		17
Lumbrineris spp.	C	3		1		67		51	56	1	14	19	19
Arabella iricolor (Montague)	C	1	1								1		
Scoloplos armiger (O. F. Muller)	DF							1					
Orbinid sp. 52	DF?										2		
Aonides oxycephala (Sars)	DF												
Spionid sp.?	DF?												
Magelona spp.	DF	1								22			
Cirriformia tentaculata (Montague)	DF	2	6		18	180	3	389	608	3	1	2	23
Diplocirrus glaucus Haase	DF										1	8	
Scalibregma inflatum Rathke	DF												1
Ophelia limacina	DF											1	
Armandia polyophthalma Kukenthal	DF												
Notomastus latericeus (Sars)	DF	25	21	8	2	66		51	76		6	8	90
Clymene oerstedii (Claparede)	DF	12	17								44	36	17
Maldane sarsi Malmgren	DF												
Owenia fusiformis Delle Chiaje	SF										3	2	2
Sabellaria spinulosa Leuckart	SF								3				
Lygdamis murata Allen	SF												
Pectinaria belgica (Pallas)	DF						1						
Pectinaria koreni Malmgren	DF												
Melinna palmata (Grube)	DF	8	12	6	54		6	1	1	2		15	52
Ampharetid sp. 1.7	DF							1	1				
Lanice conchilega (Pallas)	DF								1				
Pista cristata (O. F. Muller)	DF												
Terebellides stroemi Sars	DF				5		1	2	6				
Terebellid sp. 1.56	DF												
Branchiomma vesiculosum (Montague)	SF												2
Chone infundibuloformis Kroyer	SF										2		
Echone rubrocincta (Sars)	SF												1
Hydroides norvegica Gunnerus	SF												
Pomatoceros triqueter (L.)	SF					1		1	11				
Mollusca:													
Turritella communis Risso	SF												
Calyptraea chinensis (L.)								5					1
Crepidula fornicata (L.)								1					
Philine aperta		3	1										
Chiton asellus (Gmelin)													
Nucula nucleus (L.)	DF					6		31	9	1	14	5	
Nucula turgida Leckenby & Marshall	DF		3	2									
Thyasira flexuosa (Montague)	SF	15	13		1						54	45	16
Mysella bidentata (Montagu)	SF										55	52	
Acanthocardia echinata (L.)	SF												
Acanthocardia tuberculata (L.)	SF												
Parvicardium exiguum (Gmelin)	SF												
Dosinia lupinus (L.)	SF										1		
Dosinia exoleta (L.)	SF												

APPENDIX I—*contd.*

Sample station																			
11.2	**12.1**	**12.2**	**13**	**14**	**15.1**	**15.2**	**16.1**	**16.2**	**17.1**	**17.2**	**18.1**	**18.2**	**19**	**20**	**21**	**23**	**24**	**25**	**26**
															1				
1																			
			1			3		1										1	
														1					
													1						
													4				6		2
3			2	4							3	3							
														1					
																		3	
1	2																		
							2			1	1				2		1		
					21	14		2							1		1	1	
														1					
												2	1						
2							1												
		1																1	
2	6	10	3	4	72	59	82	53	87	80	41	27	13	105	76	9	5	4	11
6																		3	
	3	5									2	3	10						9
1	4		1		1		1		5		2	3						5	2
			1																
							1				1	1							
7	19	21	18								40	30			3			9	
22	63	50	62	26	22	13	23	29	48	11	118	72	2	9	52	26		83	1
	2	7									3	2				1			
					33	38	20	17	26	13	1	1							
																5	2		1
3	3																		
													1						6
	3						1						51	6			20	5	11
13	5	3	45	3	66	47	104	83	168	77	89	65	33	11	29	1	17	1	1
																		1	
2	2		1								5	2						4	
																			5
																			1
96	42	41	32	3	15	10	14	10	6	12	115	75		11	26			52	1
28	29	31	7		23	25	54	42	59	90	73	64		1	28			28	
																		1	
1	3	7	1										7			8	1	7	1
																		2	
																	1		
70	178	217	3		12	10	14	13	18	23	91	52		9	41			4	
										2	8	4							
		1			29	21	1	2	1	4									
		2																	
	2	6									3	2			1				
			1												1				1
2			2																
																		18	
	26	14	48											1				34	
		1																1	
		1	1							1					12			1	
		1																	
		1	6																
	1		66	2									18	1			8		3
18					23	16	36	35	9	26	19	20		7	24			1	
						1	4	5		2	1								
							1	1											
						2													
															3				
																	1		9
																		1	

APPENDIX I—*contd.*

Taxa	*Feeding type*	1.1	1.2	2	3	5	6	7	8	9	10.1	10.2	11.1
							Sample station						
Mollusca—*contd.*													
Venus casina L.	SF		2									2	
Venus ovata Pennant	SF												
Mysia undata (Pennant)											1	1	
Abra alba (W. Wood)	DF	23	12	2	2	1		7	2		142	181	75
Tellina fabula Gmelin	DF									2			
Cultellus pellucidus (Pennant)	SF		1									1	
Thracia convexa (Wood)	SF												
Chaetognatha:													
Spadella cephaloptera													
Ecinodermata:													
Antedon bifida (Pennant)								2					
Ophiura albida Forbes											2	2	
Amphiura filiformis (O. F. Müller)		1									72	98	
Amphipholis squamata (Delle Chiaje)													
Echinocardium cordatum (Pennant)	DF									3			
Holothuroid sp. 1.86	DF										1	1	2
Holothuroid sp. 2.111	DF		7										
Leptosynapta sp.	DF	5			1		4				4		
Crustacea:													
Pagurus prideauxi (Leach)	C/P												
Pagurus bernhardus (L.)	C/P												
Gonoplax rhomboides (L.)	C/P		1										
Macropipus depurator (L.)	C/P								1				
Carcinus maenas (L.)	C/P												
Amphipod spp.						1			8	4	2		1
Anthozoa:													
Cerianthus Lloydi Gosse													
Anemone sp. 1.39	SF	1		6		44		22	31				
Anemone sp. 2.44	SF							1					
Tunicata:													
Tunicate sp. 1.29	SF												1
Tunicate sp. 2.45	SF												
Tunicate sp. 3.77	SF												
Sipuncula:													
Golfingia elongata (Keferstein)	DF												
Golfingia vulgaris (Blainville)	DF												1
Golfingia procera	DF		1								126	146	10
Priapuloidea:													
Priapulus caudatus Lamarck	DF							1					
Nemertean			2										
Sucker fish													

APPENDIX I—*contd.*

Sample station																			
11.2	12.1	12.2	13	14	15.1	15.2	16.1	16.2	17.1	17.2	18.1	18.2	19	20	21	23	24	25	26
		1				4											1	4	2
			1															3	2
96	7	1	12	37	19	9	9	2	6	3	20	19	22	19	18	4	7	6	7
							6	16	8	9	1	5	85				3		9
	1	1					1			3	1	1						1	1
			2																
													1						
1																			
		8		1				1				1	5					2	1
					2	2	4	2											
																			3
1	1																	1	
														1				1	
					1	1				1									
		2														2		1	
											1				1				
								1											
													10	1			2		5
	1													1					
		2	11		1	1		3			5	3			2	2			
															2				
														4	13	2			
1																			
																3			
	1		3															1	
1		3	2									1						2	
	12	15																4	
	2	3															4		
1																			

APPENDIX II

Positions and Approximate Depths of Sampling Stations

Sample station	*Position*		*Approx. depth* (m)
	lat.	*long.*	
1	51°41′·58N	5°05′·82W	4
2	51°42′·00N	5°01′·60W	15
3	51°42′·23N	5°01′·63W	2
5	51°42′·00N	5°00′·76W	16
6	51°41′·06N	4°58′·48W	4
7	51°41′·73N	4°59′·11W	20
8	51°41′·80N	4°59′·72W	13
9	51°41′·08N	5°09′·94W	9
10	51°41′·88N	5°09′·08W	7
11	51°42′·40N	5°08′·97W	5
12	51°42′·47N	5°08′·13W	11
13	51°42′·46N	5°07′·22W	13
14	51°42′·05N	5°06′·77W	15
15	51°41′·03N	5°03′·61W	*0·6**
16	51°41′·20N	5°03′·80W	0·3
17	51°41′·38N	5°03′·94W	1
18	51°41′·60N	5°04′·15W	4
19	51°41′·23N	5°07′·15W	5
20	51°41′·93N	5°02′·88W	20
21	51°42′·20N	5°02′·86W	2
23	51°42′·32N	5°06′·30W	9
24	51°42′·68N	5°06′·50W	5
25	51°41′·90N	5°08′·42W	15
26	51°41′·46N	5°09′·21W	9

* Station 15 is 0·6 m above chart datum.

9

The Effect of Kuwait Crude Oil and the Dispersant BP 1100X on the Lugworm, *Arenicola marina* L.

DAVID LEVELL

(Oil Pollution Research Unit, Orielton Field Centre, Pembroke, Wales)

SUMMARY

This paper describes field experiments on Arenicola marina *L. carried out at Sandyhaven Pill, a small estuary on the northern shore of Milford Haven. The height of the experimental plots above Chart Datum, Milford Haven, varies between 4·5 m and 5 m. The plots are covered at high water for periods of between approximately 3·5 h on neap tides to approximately 5 h on spring tides.*

Kuwait Crude Oil, BP 1100X dispersant, and 1:1 and 5:1 mixtures, by volume, were used as pollutants.

Arenicola *densities were estimated by counting casts. The relationship of cast numbers to animal number was determined as 1:1, although operator efficiency at finding animals in the excavated substrate was found to be a major variant.*

Single spillages were simulated by spraying pollutants over Arenicola *beds at the rate of 0·2 litre/m². A single spillage caused a 25–50% reduction in population density.*

Repeated spillages at two monthly intervals, varying in number from one to six, were carried out. Successive spillages caused progressive reduction of the Arenicola *density. Four successive spillages resulted in eradication of the original population of 20–25/0·25 m².*

Preliminary experiments to determine the influence of surface water and heavy rainfall on the extent of the effect of the pollutant were carried out.

Particle size analysis was carried out for ten stations distributed across the experimental site.

Destructive sampling was carried out at a separate experimental site to determine the effect of pollution by Kuwait Crude Oil on the feeding behaviour, and to determine the fate of polluted animals. It was found that the depression of feeding activity on the day following the oil spillage occurred in up to 75% of the animals. Between 50% and 75% of the original population recovered normal feeding activity. The remainder either died and decomposed in the substrate or quit the substrate. The fate of those leaving the area is not known.

The recolonisation of plots by juvenile Arenicola *was monitored. The recolonisation of recently polluted substrates was inhibited.*

INTRODUCTION

A wide gap in the knowledge concerning the effects of oil pollution and cleaning on littoral communities exists, namely, the effect of pollution on the fauna of soft substrates. This paper is an attempt to begin to fill that gap and is a report, to date, of the work carried out upon the lugworm, *Arenicola marina* L.

Due to the long term nature of monitoring recovery rates, and the inherent difficulties of sampling infaunal species, it was decided that non-destructive sampling methods should be used to restrict the area of experimental site subjected to pollutants, and to reduce the work load involved in sampling. The lugworm was chosen as an experimental animal because its occurrence in high and often uniform densities reduces plot size and sample size necessary for accurate assessment of population density. Its presence can be monitored by non-destructive sampling due to the production of worm casts.

SITE SELECTION

The availability of a suitable experimental site was a major problem. In several areas where sufficient animal densities occurred, other factors which were not compatible with the use of oils and dispersants such as commercial shell fisheries, bait digging, tourist pressure or bird feeding grounds, restricted the choice of experimental site.

A suitable site was found to the north of Sandyhaven Pill, a small estuary on the northern shore of Milford Haven, O.S. grid ref. S.H. 858085. Dense *Arenicola* beds cover the whole length of Sandyhaven Pill. The experimental site was confined to the area of muddy sand between the main drainage creeks at the north of Sandyhaven Pill as it turns to the north east. The changes in site topography throughout the year are minor and occur usually during the winter months when heavy rainfall and flooding during the period of emersion causes erosion and minor changes in the stream bed formation. The height of the experimental plots varies from between 4·5 m and 5 m above Milford Haven Chart Datum, which is 3·17 m O.D. (Newlyn). The area is covered on all tides, the approximate times of emersion to immersion over a tidal cycle are from 7 h : 5 h on extreme spring tides, to $8\frac{1}{2}$ h : $3\frac{1}{2}$ h on extreme neap tides. These times are only approximate and will be dependent upon tidal range and weather conditions.

MATERIALS AND METHODS

Simulated spillages of pollutant were carried out by evenly spraying known volumes of the pollutant over trial areas of *Arenicola* beds. The spraying equipment (Fig. 1) consisted of a hand held lance supplied with pollutant from a 25 litre container by means of a hand pump. A lance fitted with a single plastic flood jet nozzle was used for spraying 1 metre square plots, and three nozzles 12 in apart were used on a boom when 5 metre square plots were sprayed. The hand pump was threaded into the neck of the drum and

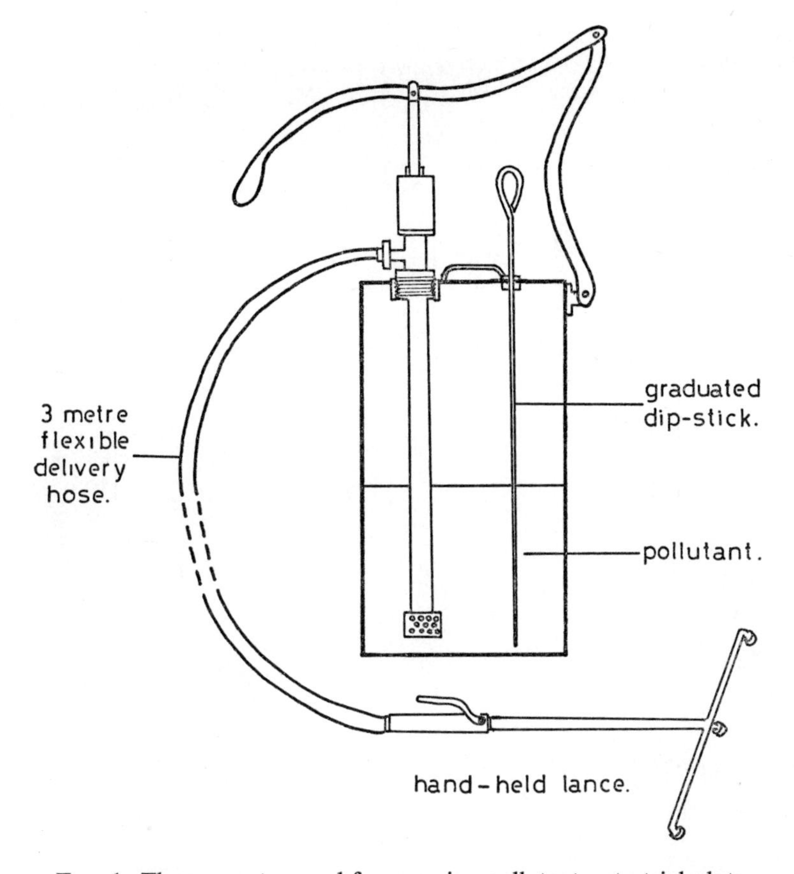

FIG. 1. The apparatus used for spraying pollutant onto trial plots.

could be interchanged between similar sized drums of different pollutant. The volume of pollutant used was determined by a graduated dipstick, inserted through the top of the drum.

Pollutants used were fresh Kuwait Crude Oil, BP1100X dispersant, and mixtures of 1:1 and 5:1 of oil:dispersant. All pollutants were stored in air-tight containers, and fresh mixtures were made up prior to spraying. More detailed specifications of both Kuwait Crude Oil and BP 1100X are given in Table I. BP 1100X is the dispersant used within Milford Haven.

Trial plots were marked out using wooden corner pegs. The arrangement of plots is shown in Fig. 2.

SINGLE SPILLAGES

The first series of plots (site 1) was laid out in April, 1973. This consisted of eight plots 3 m × 5 m. Alternate plots were used as controls. Plot No. 2 was sprayed with 3 litres of fresh Kuwait Crude Oil. Plot No. 4 was sprayed with 3 litres of a 1:1 mixture of oil and dispersant. Plot No. 6 was sprayed with 3

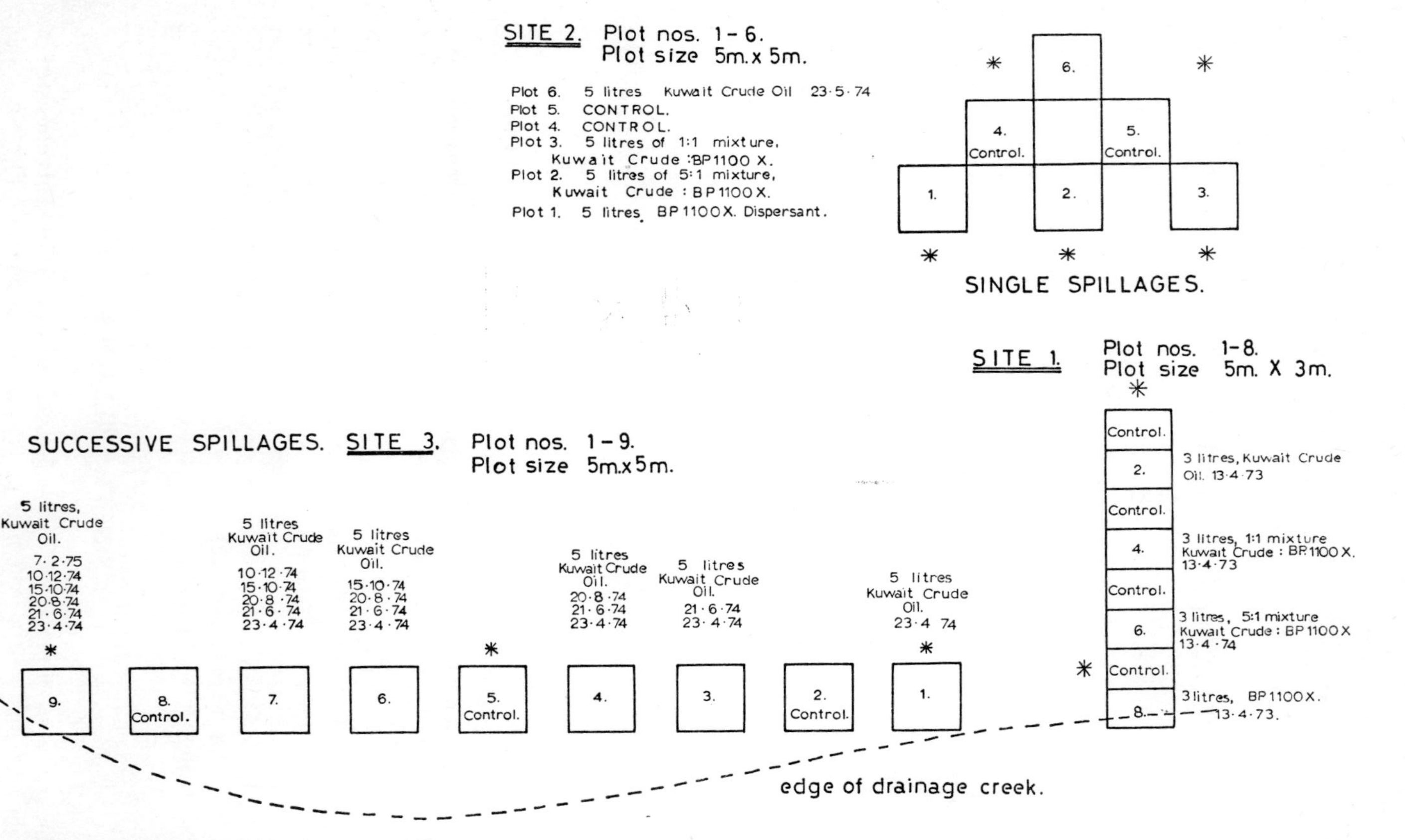

FIG. 2. Arrangement of trial plots at Sandyhaven. * denotes sample sites for particle size, and organic carbon analysis.

TABLE I

Inspection Data and Ash Composition of Kuwait Export Crude Oil (British Petroleum Company, 1966)

Specific gravity 60°F/60°F		0·869
Gravity	°API	31·3
Distillation test:		
Distillate and loss to 150°C	% vol	16
200°C		26
300°C		44
Residues	% vol	56
Sp. gr. 60°F/60°F of residue		0·958
Sulphur content	% wt	2·5
H_2S (dissolved)	% wt	nil
Kinematic viscosity at 50°F	cSt	27·4
70°F		17·0
100°F		9·6
Pour point	°F	−25
Wax content	% wt	5·5
Melting point of wax	°F	120
Carbon residue (Conradson)	% wt	5·2
Asphaltenes	% wt	1·4
Total nitrogen	% wt	0·12
Acidity	mg KOH/g	0·15
Ash content	% wt	0·006
Water	% vol	trace
Water and sediment	% vol	<0·2
Salt content (as NaCl)	lb/1000 bbl	4

These results are typical of the average annual quality of the crude oil. Individual cargoes may show small random and seasonal variations.

Components of BP 1100X Dispersant (British Petroleum Company, 1972)

Fatty acid ester	15% wt
Solvent	84% wt
Additives	1% wt

Solvent has a flash point (closed) above 150°F and contains less than 3% aromatics.

litres of a 5:1 mixture of oil and dispersant. Plot No. 8 was sprayed with 3 litres of BP 1100X dispersant.

A second series of plots (site 2) was established in May, 1973, and population densities within each plot monitored at monthly intervals prior to spraying in May, 1974. Four of the six 5 m × 5 m plots were sprayed with 5 litres of pollutant, two plots acting as controls.

Plot No. 1. 5 litres BP 1100X dispersant.

Plot No. 2. 5 litres of a 5:1 mixture of oil:dispersant.
Plot No. 3. 5 litres of a 1:1 mixture of oil:dispersant.
Plot No. 4. Control—not polluted.
Plot No. 5. Control—not polluted.
Plot No. 6. 5 litres of fresh Kuwait Crude Oil.

REPEATED SPILLAGES

A third series of nine 5 m × 5 m plots (site 3) was established in June, 1973. These consisted of six trial plots and three control plots all separated from each other by a 3 m interval in the arrangement shown in Fig. 2. The population densities of *Arenicola* within these plots was monitored at monthly intervals. Beginning in April, 1974, the six trial plots were sprayed at two-monthly intervals with 5 litres of fresh Kuwait Crude Oil, the number of plots sprayed being reduced by one on each successive occasion. Thus, after ten months (February 1975), the plots had received from one to six spillages.

RECORDING METHODS

The density of *Arenicola* within each trial plot was determined by recording the number of casts within 25 randomly placed $\frac{1}{2}$ m square quadrats. The validity of using cast numbers as an estimate of population density is discussed later. All cast counts were taken at least 1 h after low water, thus having allowed approximately 5 h since the surface of the substrate was exposed by the receding tide. All spraying took place as near as possible to the time of low-water giving approximately 4 h of exposure before the substrate was re-covered by the incoming tide.

In all cases, monitoring was carried out before and after simulated spillages in the following sequence:

1. 24 h before spraying.
2. 24 h after spraying.
3. 3 days after spraying.
4. 7 days after spraying.
5. 14 days after spraying.
6. 28 days after spraying.
7. Thereafter at monthly intervals.

SUBSTRATE ANALYSIS

Particle size analysis was carried out on ten samples of approximately 250 g wet weight of substrate from sample stations shown in Fig. 2. The sample was dried to a constant weight at 100°C, and then shaken through a series of sieves for 30 min, on a mechanical shaker. The weight retained within each sieve, and the bottom pan, was recorded. The results were plotted as cumulative percentage weight against particle size. The median particle diameter, the phi quartile deviation, and the phi quartile skewness were determined:

$$\text{Phi quartile deviation } (QD\phi) = \frac{Q_3\phi - Q_1\phi}{2}$$

$$\text{Phi quartile skewness } (Skq\phi) = \frac{Q_3\phi + Q_1\phi - 2Md\phi}{2}$$

where $Md\phi$ equals the median particle size.

Phi quartile deviation is a measure of the degree of sorting. The phi quartile skewness is a measure of the efficiency of sorting, positive values indicating greater efficiency of sorting of the particles larger than the median particle size, negative values reflecting greater sorting efficiency of the particles smaller than the median particle size.

Hydrocarbon analysis of the polluted substrate. Substrate samples were taken by inserting, horizontally, a 1 cm diameter class specimen tube into the first cm of substrate and another at a depth of 10–11 cm. A 20 g portion of the sample was then added to one litre of distilled water in a stoppered separating funnel and shaken for 5 min. 75 ml of cyclohexane (Spectrosol) was then added, and the funnel again shaken for 5 min. The funnel was then allowed to stand. The cyclohexane and soluble hydrocarbon fraction was separated, and the remainder again shaken well with an additional 75 ml of cyclohexane. After thorough mixing and allowing time for settling out of the two fractions, the cyclohexane and any dissolved hydrocarbons were again separated and added to the first extract. The two cyclohexane extracts were shaken together to form a homogeneous solution, from which a 100 ml aliquot was retained for spectrophotometric analysis at 260 μm wavelength.

Further substrate samples have been taken from plots in site 3 but results are not yet available.

RESULTS

Single spillages. The results of experiments carried out on site 1 are shown in Figs. 3–7. No data is available for December 1973, and January 1974, due to extensive flooding and rainfall making cast density estimates unreliable. The broken portion of the histograms represents the density of small *Arenicola* casts produced by a settlement and colonisation of the plots by juvenile animals in the summer of 1974. Density of casts is expressed in numbers per 0·25 square metre. 95% confidence limits of the mean densities are given. Plot 8 was abandoned in the winter of 1973 because movement of the creek bed had caused extensive flooding and erosion. Figures 8–13 show the results of a replicate set of single spillage experiments (site 2). Again the extremely heavy rainfall and flooding in the winter of 1973–74 made cast counts unreliable for the winter months.

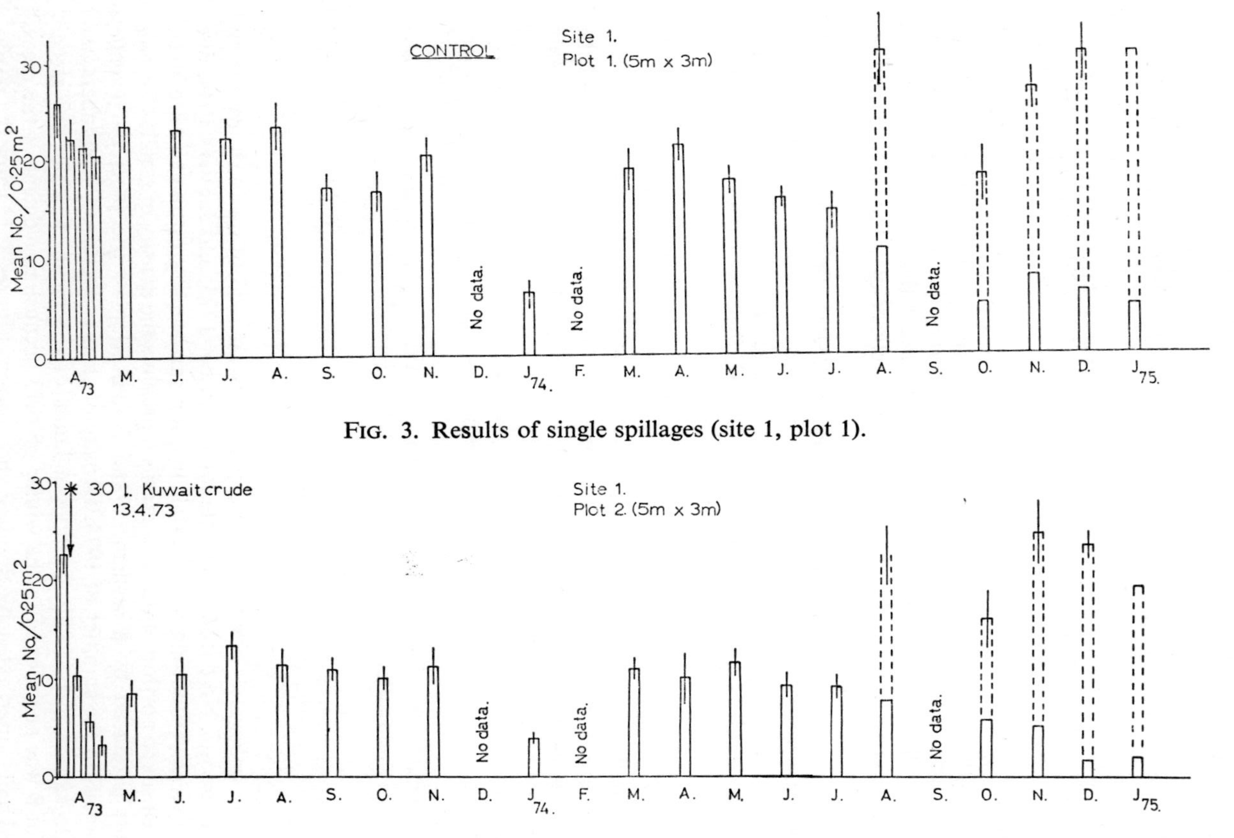

FIG. 3. Results of single spillages (site 1, plot 1).

FIG. 4. Results of single spillages (site 1, plot 2).

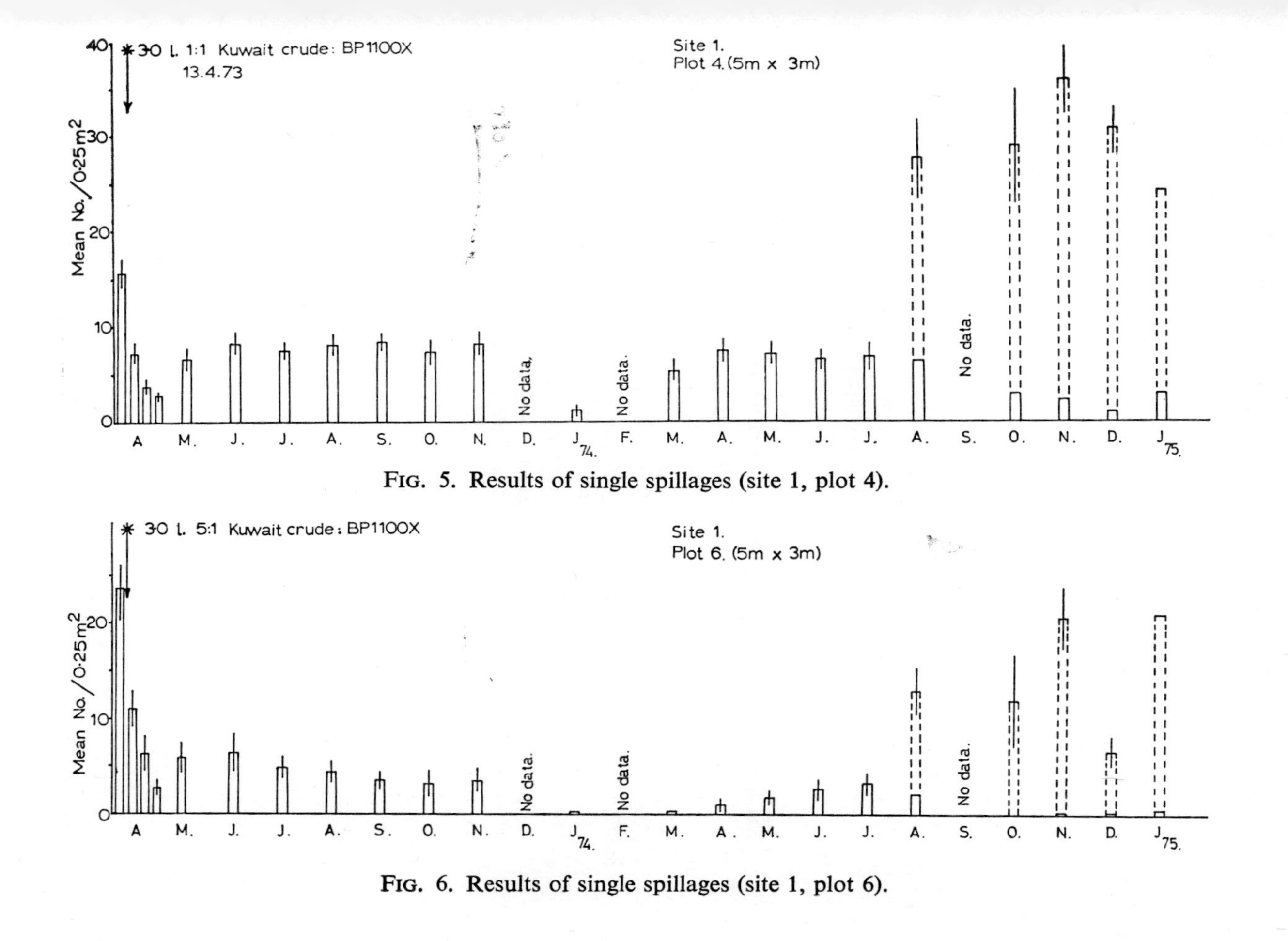

FIG. 5. Results of single spillages (site 1, plot 4).

FIG. 6. Results of single spillages (site 1, plot 6).

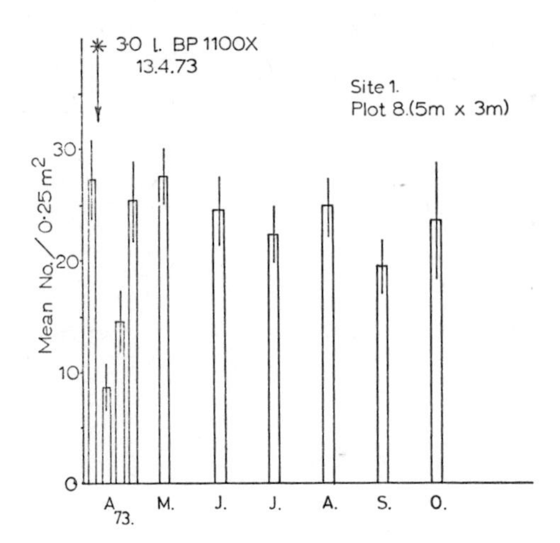

FIG. 7. Results of single spillages (site 1, plot 8).

Successive spillages. Figures 14–22 represent the results obtained from site 3. Again density of casts is expressed in numbers per 0·25 square m. In Figs. 3–5 the symbol * represents the occurrence of a simulated spillage.

Particle size analysis. Figure 23 shows the results of a typical particle size analysis. All ten samples analysed show remarkable uniformity both in median particle size and sorting. The substrate in the upper Sandyhaven area is particularly well sorted with a median particle size ranging from 160 to 175 μm. Data for the analysis is given in the appendix.

Hydrocarbon analysis. Figure 24 shows the results of extraction of cyclohexane soluble hydrocarbons from samples taken from the surface and 10 cm depth of a plot sprayed with 0·2 litres of Kuwait Crude Oil per square metre. Only one sample was taken at each depth and the results are only intended as a guide to the order of scale of removal of the pollutant from the substrate by successive tides. A more comprehensive substrate analysis for hydrocarbons is intended.

DISCUSSION OF METHODS

Acceptability of Spraying Oil as a Method of Replicating a Spill Situation over Sandy Shores or Mud Flats

The oil spill incidents within Milford Haven, to date, have not resulted in oil being deposited as a layer over mud flats. On the occasions when oil slicks have reached sandy/muddy regions before dispersal at sea has been possible, the oil has been deposited as a narrow band trapped either by saltmarsh plants or rocks and boulders which occur at the higher tidal limits of the Milford Haven shores. However, it is conceivable that in a situation where an upper fringe of rocks or saltmarsh plants are not present to act as a trap, then oil could well drain back across a mudflat or sandy shore after being deposited at high water. Similarly a layer of oil could be deposited if accidental release of a large quantity of oil occurred from a storage tank, or pipeline situated above high water mark.

Spraying was used as the only practical means of applying an even layer of pollutant over the substrate surface. The volume of pollutant was determined after trials had been carried out and it was found that 5 litres applied to a 5 m × 5 m plot produced a heavy coating of oil which, after 1–2 h, had penetrated the surface of the substrate, before being covered by the incoming tide. Due to the open nature of the experimental system, the volume of oil present within the substrate is diminished as each successive high tide removes a portion of the oil.

Validity of Using Cast Numbers as an Indication of *Arenicola* Density

Ratio of Cast Number to Animal Number

Linke (1939) and Holme (1948) both considered that a count of the casts gives a reliable indication of the size of the *Arenicola* population, Holme finding that the number of casts in an area corresponded well with the actual number of worms. Newell (1948), however, found this to be an unreliable method of population estimation.

At Sandyhaven a series of preliminary trials were carried out to determine the relationship of cast number to number of animals present. 1 metre squares were marked out and the number of casts present recorded. These metre squares were then fully excavated and the substrate sorted, for animals. At first it was proposed that the substrate be sieved, but as this proved impracticable because of the volume involved and the lack of easily accessible water, the substrate was broken up and sorted by hand. This operation was repeated several times on different occasions by different people. Several factors seem to be important:

(1) Operator efficiency at sorting.
(2) The drainage of the substrate—rapid infilling of holes from water-logged substrates hindered efficient searching.
(3) A systematic removal of substrate to a constant depth was necessary.

A distinct 1:1 ratio was found at Sandyhaven. Variation between number of casts and number of animals of less than 5% was considered acceptable as casts on the edges of the excavated area may have been produced by animals in burrows outside the excavated region and vice versa.

It should be noted that these investigations were carried out on animals larger than 10 cm at least 4 h after the substrate had been uncovered by the receding tide, in natural conditions, i.e. they had not been subjected to pollution.

Time for Production of Casts

It was felt necessary to determine whether *Arenicola* cast production (feeding and defaecation) occurred at a constant rate after the substrate had been

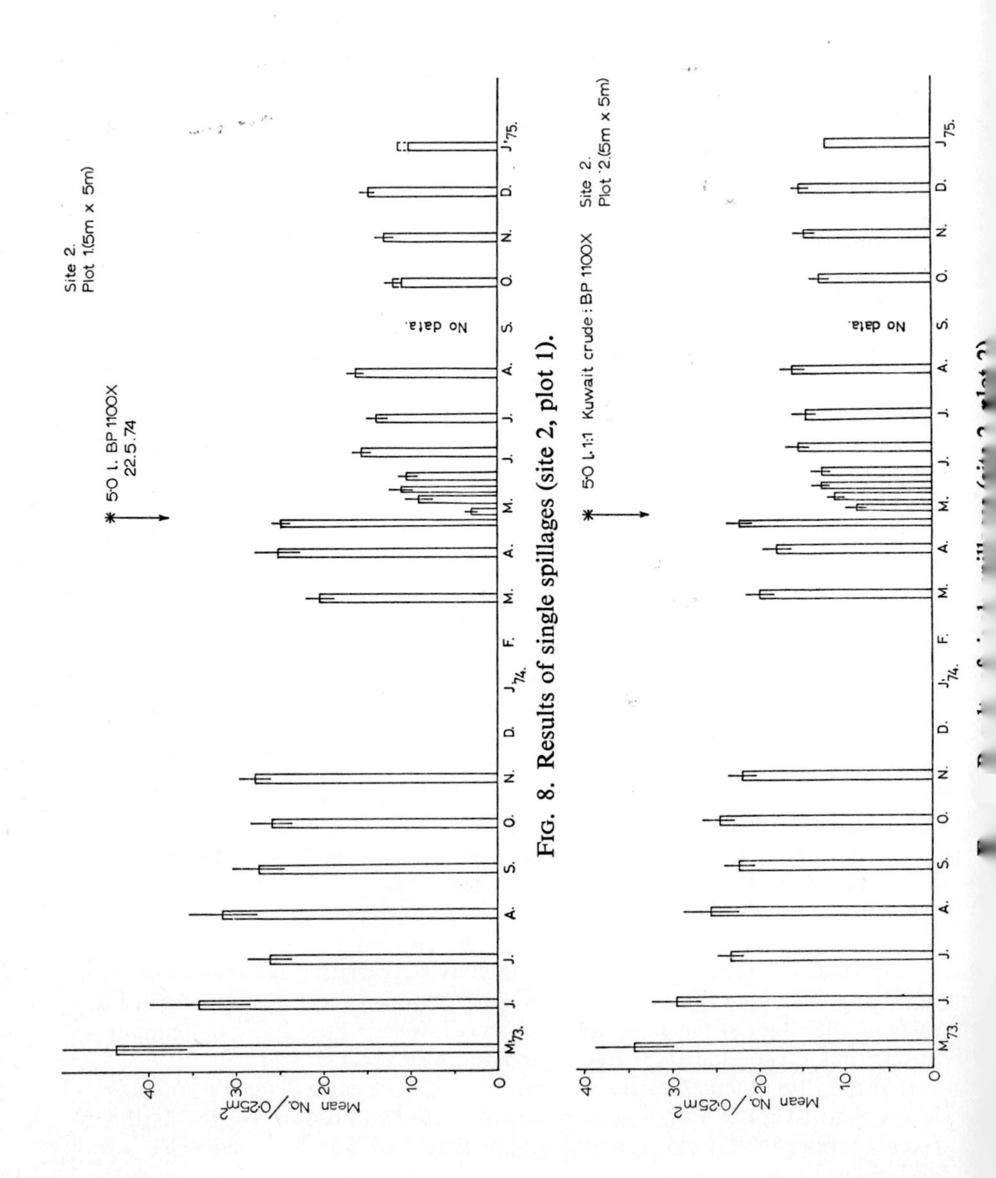

FIG. 8. Results of single spillages (site 2, plot 1).

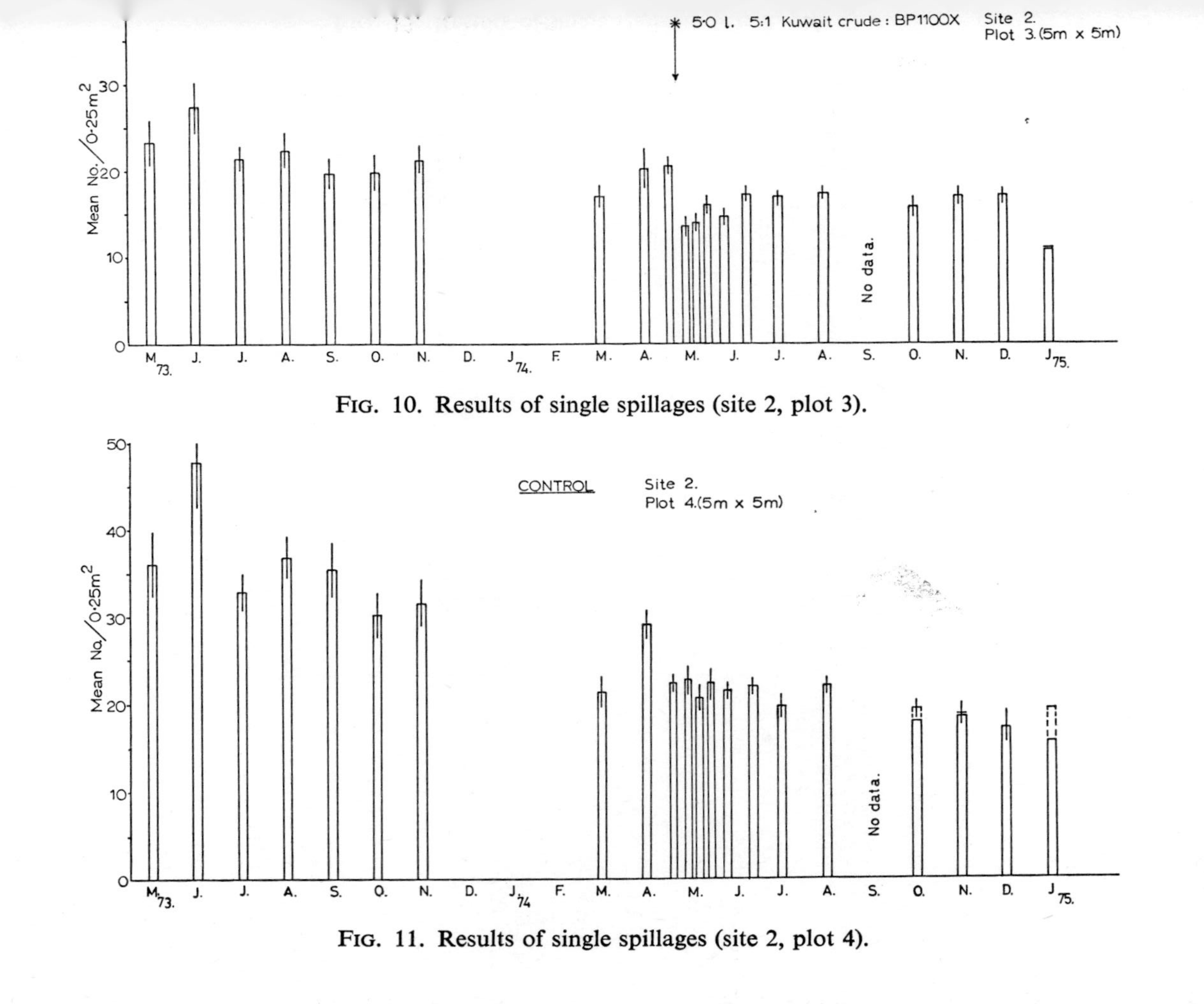

FIG. 10. Results of single spillages (site 2, plot 3).

FIG. 11. Results of single spillages (site 2, plot 4).

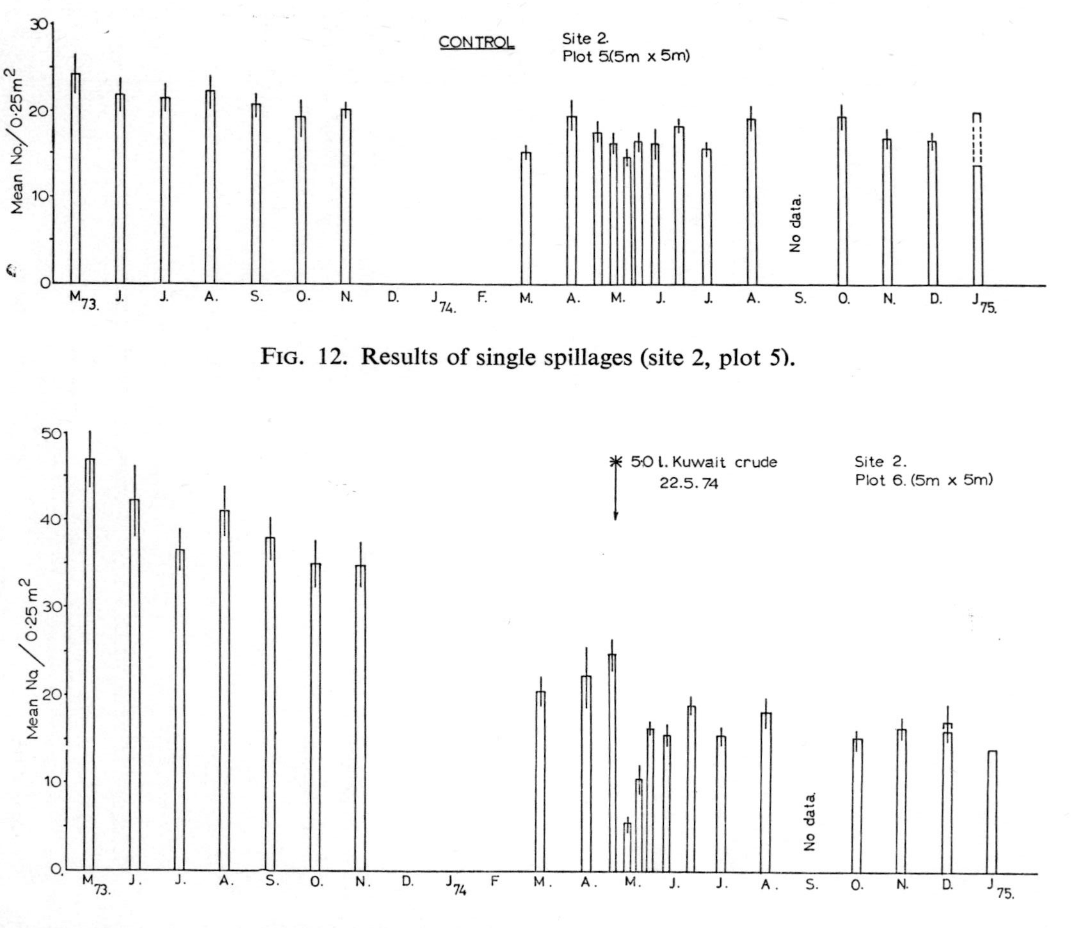

FIG. 12. Results of single spillages (site 2, plot 5).

FIG. 13. Results of single spillages (site 2, plot 6).

exposed by the receding tide. It was found that within 2 h of emersion, a constant number of casts had been produced. The animals continued to eject cast material for several hours, and so increase the size and distinctiveness of the cast. It is thought that once the fluidity of the substrate is lost through drainage, cast production ceases. It was noted that in the gullies and creek beds, cast production continued as long as the substrate remained waterlogged.

The effect of heavy rainfall on the feeding behaviour of *Arenicola* seemed to be to reduce cast production. As rain obscured and eventually washed away casts, particularly the small casts produced by juveniles, cast density counts were only carried out during dry weather.

DISCUSSION OF RESULTS

Single Spillages. Site 1

From Figs. 3–7 it can be seen that the effect which the pollutants have is not significantly different in all the plots sprayed. There is a rapid decline in cast production following the spill. Over the next month, a gradual increase in feeding activity occurs reaching a constant level at about 50–75% of the original *Arenicola* density. The one exception to this pattern was in plot 8 where the dispersant BP 1100X was used. Although there was an initial reduction of cast production, this returned to approximately the original level after a month. Plot 8 was situated on the edge of a shallow creek and it was thought that continual waterlogging and the higher flows of water across the surface of plot 8 may well have aided the rapid recovery of the animals.

The control plots in site 1 were alternate with polluted plots with no separation zone between. It was found that distinct fringe effects were noted on the control plots next to polluted plots. A typical example is plot 1 (Figs. 3–7). The drop in cast production in plot 1 after the adjacent plot was polluted, although not long lasting, was sufficient to raise doubts to its use as an effective control. Therefore, to clarify the situation regarding the siting of controls, and the influence of surface water and waterlogging, a second series of plots (site 2) was established on a more uniform substrate without excessive areas of surface water. The controls at this site were positioned further from the polluted plots (Fig. 2).

In addition, two separate short-term trials were held.

To Determine the Extent to Which Surface Water, or a Poorly Drained Substrate, Influences the Effect of Kuwait Crude Oil on Arenicola

Two sites were selected, one (A) was situated on the shallow sloping edge of a waterlogged creek, over which drainage ran from a higher saltmarsh region. The other site (B) was sprayed after heavy rainfall, the substrate being waterlogged and standing surface water present in the small hollows of the rippled substrate surface. At each site six plots one metre square were each sprayed with 0·4 litres of Kuwait Crude Oil. Table II shows the results. There are no appreciable effects on the production of casts, the pollutant not having penetrated the substrate.

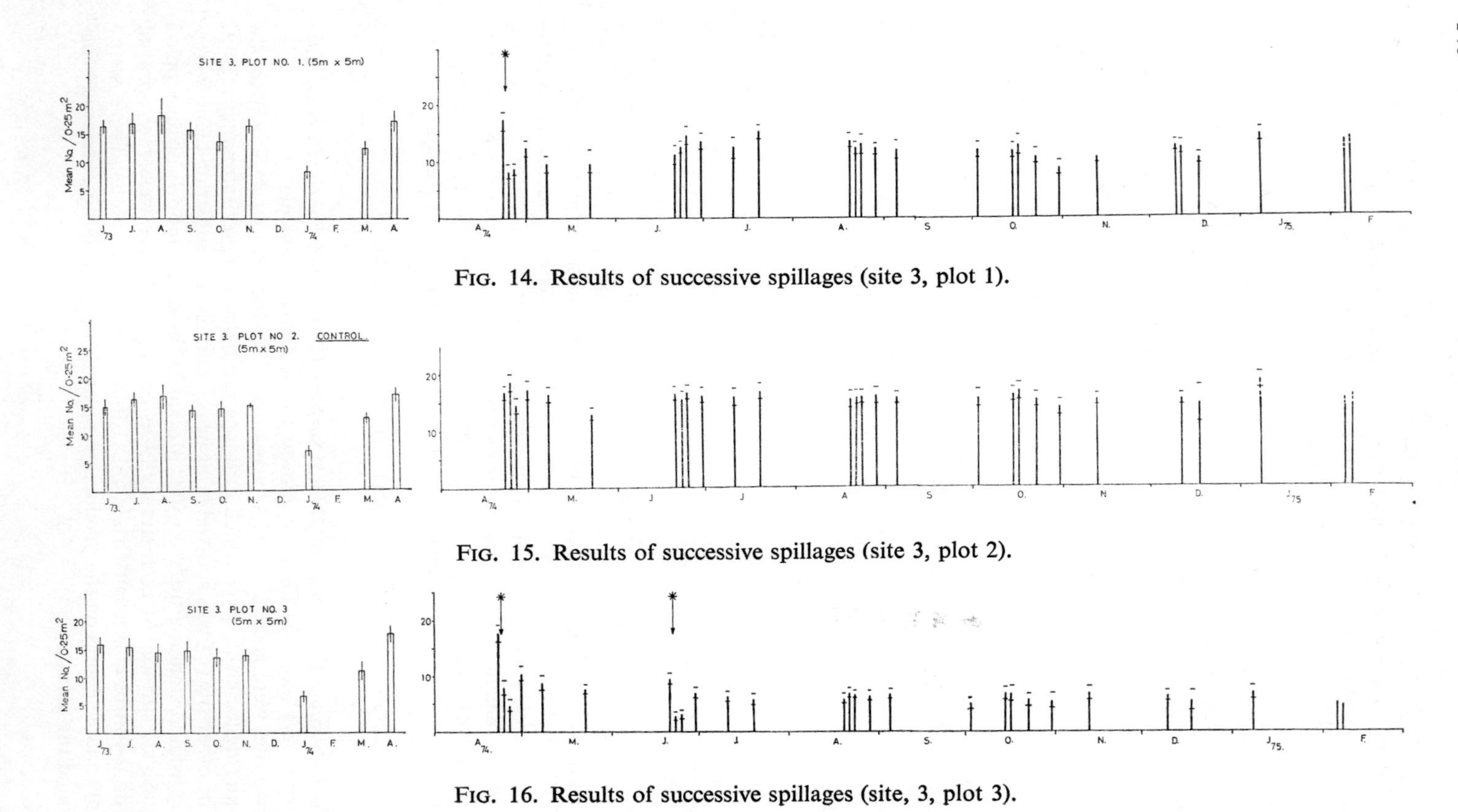

FIG. 14. Results of successive spillages (site 3, plot 1).

FIG. 15. Results of successive spillages (site 3, plot 2).

FIG. 16. Results of successive spillages (site, 3, plot 3).

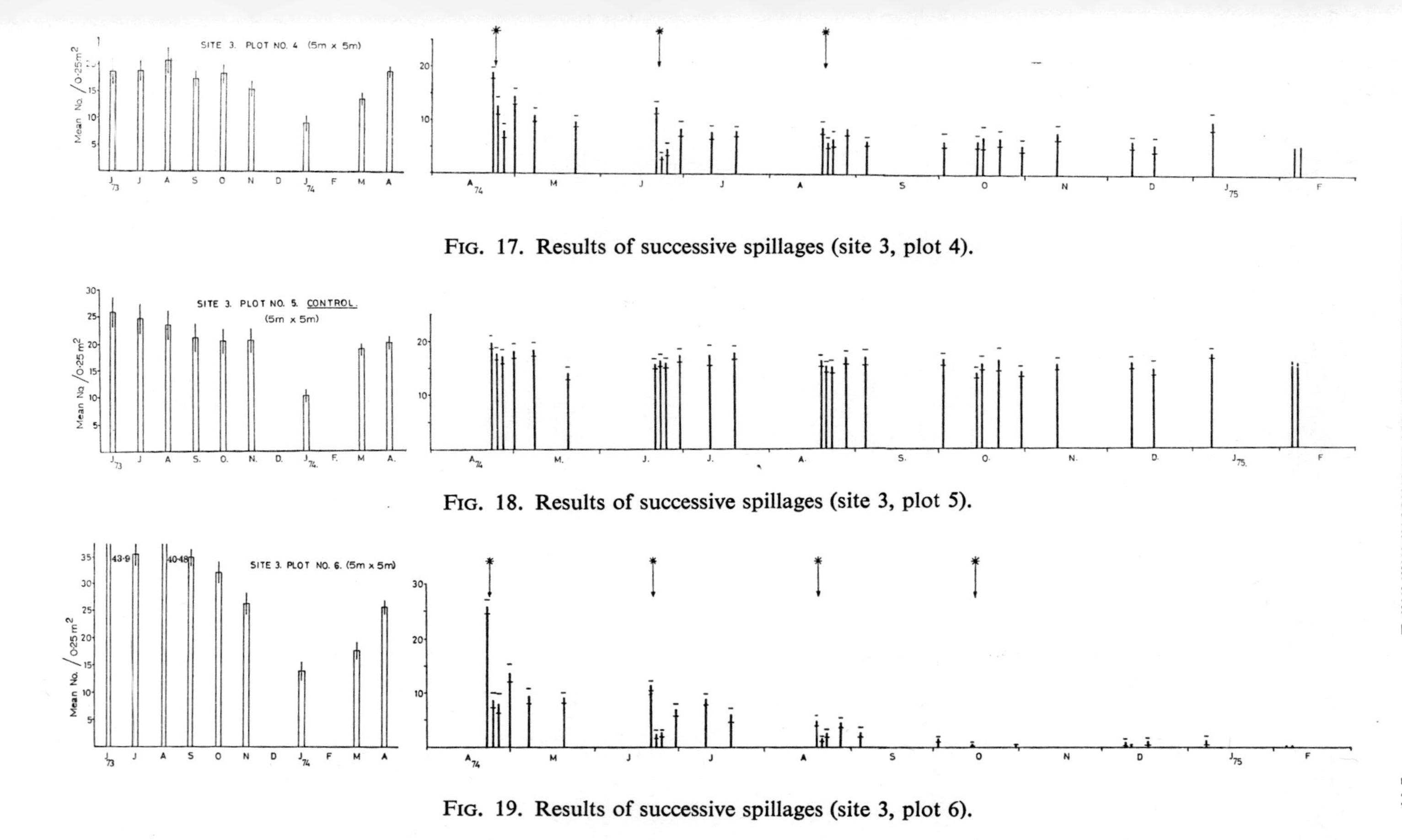

FIG. 17. Results of successive spillages (site 3, plot 4).

FIG. 18. Results of successive spillages (site 3, plot 5).

FIG. 19. Results of successive spillages (site 3, plot 6).

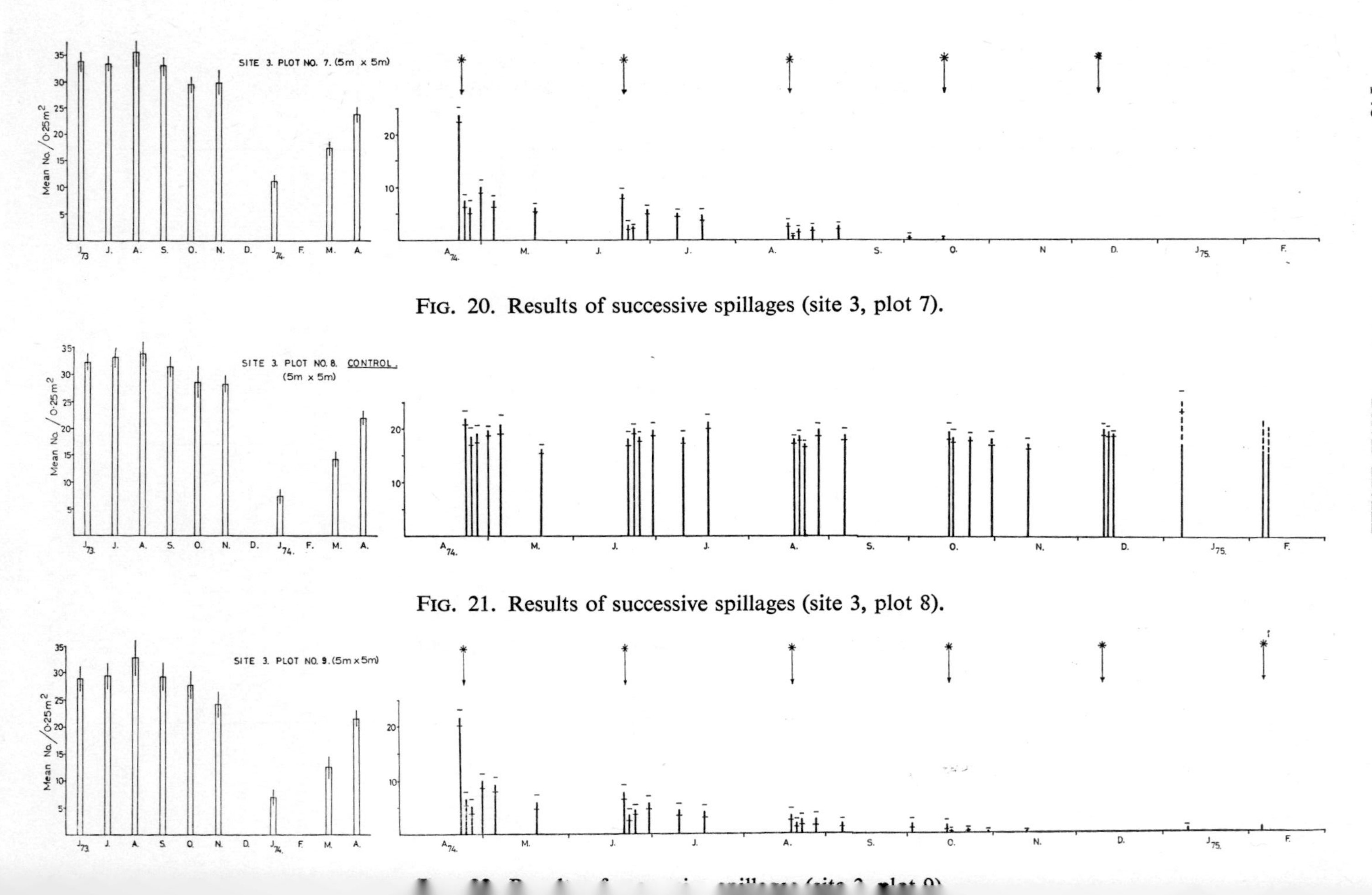

FIG. 20. Results of successive spillages (site 3, plot 7).

FIG. 21. Results of successive spillages (site 3, plot 8).

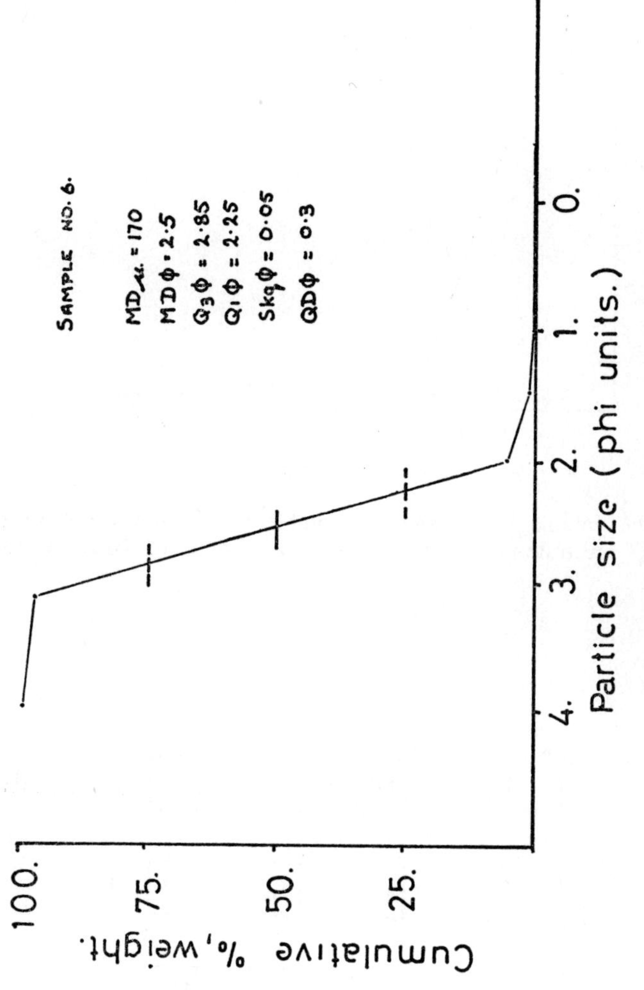

FIG. 23. Results of particle size analysis of sample No. 6.

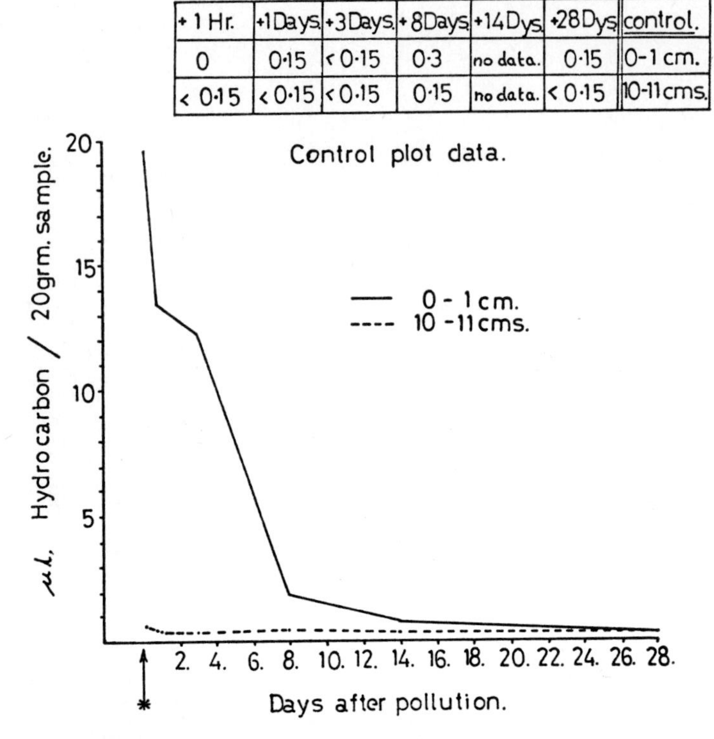

+1 Hr.	+1Days.	+3Days.	+8Days.	+14Dys.	+28Dys.	control.
0	0·15	< 0·15	0·3	no data.	0·15	0-1 cm.
< 0·15	< 0·15	< 0·15	0·15	no data.	< 0·15	10-11cms.

FIG. 24. The volume of cyclohexane soluble hydrocarbons present in 20 g samples of substrate previously oiled with 0·2 litre/m² of Kuwait Crude Oil.

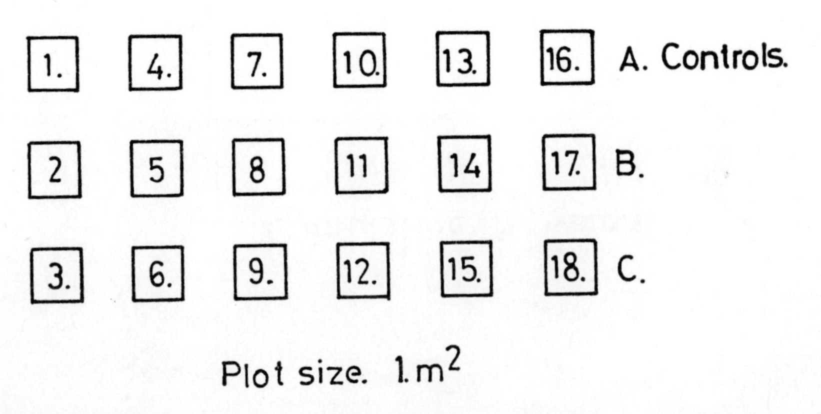

FIG. 25. Arrangement of experimental plots to determine the fate of *Arenicola marina* L. after pollution of substrate by Kuwait Crude Oil.

TABLE II
Results of Polluting Poorly Drained Substrates with Kuwait Crude Oil. The Numbers Are of Casts/m²

	Site A						*Site B*					
Plot No.	*1*	*2*	*3*	*4*	*5*	*6*	*1*	*2*	*3*	*4*	*5*	*6*
Prepollution	16	14	17	13	19	21	24	27	31	25	20	24
+ 1 day	17	15	17	12	15	19	22	24	28	27	20	24
+ 3 days	15	13	16	12	16	19	22	26	29	26	21	23
+ 7 days	16	14	17	14	17	23	23	28	29	20	23	25

To Determine the Fate of Arenicola *Polluted by Kuwait Crude Oil*
The plots were arranged in the manner shown in Fig. 25. The series A were to act as controls and were not polluted. Series B and C were replicates. Each plot in series B and C was sprayed with 0·4 litres of Kuwait Crude Oil. Destructive sampling was used to estimate the number of animals present after the number of casts present had been recorded for each plot.

The sampling procedure was as follows:

(1) One day prior to spraying: record cast numbers in all plots. Record number of animals present within plot nos. 1, 2 and 3.
(2) Spray plots in Series B and C.
(3) One day after spraying record numbers of casts present in all plots. Record number of animals present in plot nos. 4, 5 and 6.
(4) Three days after spraying, record numbers of casts present in all plots. Record numbers of animals present in plots 7, 8 and 9.
(5) Seven days after spraying, record numbers of casts present in all plots. Record numbers of animals present in plots 10, 11 and 12.
(6) Fourteen days after spraying, record numbers of casts present in all plots. Record numbers of animals present in plots 13, 14 and 15.
(7) Twenty-eight days after spraying, record numbers of casts present in all plots. Record numbers of animals present in plots 16, 17 and 18.

The results of this experiment are shown in Table III. The figures included in parentheses are the numbers of animals found after excavation of the plot.

It can be seen from Table III that the cast number follows the same pattern as shown in the results of previous single spillages (Figs. 3–13).

Animals extracted from the substrate 24 h subsequent to pollution all seemed moribund and flaccid although no dead animals were found. During the following week, it appears that a certain percentage of these animals recover and continue to feed (produce casts). The remaining proportion either die and remain in the substrate where they decompose or they quit the substrate. The fate of those which quit the substrate is not known—they may well survive to recolonise other areas or may succumb to the effects of the oil pollution or predators. The only means of testing this would be in a closed system within the laboratory. It was noted that the burrows of all the animals found dead within the substrate were lined with a thick layer of oil, even though the substrate surface showed only traces of an oil film after two

TABLE III
Results of Spraying 0·4 l Kuwait Crude Oil/m² on *Arenicola* Cast Numbers

Plot No.	*Prepollution*	*+1 Day*	*+3 days*	*+7 Days*	*+14 Days*	*+28 Days*
1	31 (30)	A Control				
2	26 (25)	B				
3	28 (29)	C				
4	21	22 (19)	A Control			
5	22	9 (23)	B			
6	22	5 (19)	C			
7	24	22	22 (20)	A Control		
8	22	5	6 (14 + 2 dead)	B		
9	24	4	4(13)	C		
10	25	23	24	24 (25)	A Control	
11	22	4	3	12 (14 + 5 dead)	B	
12	20	3	2	7 (9 + 5 dead)	C	

13	23	20	20	23	20 (20)	A Control	
14	22	5	1	8	9 (10)	B	
15	21	4	1	8	8 (10 + 1 dead)	C	
16	24	19	23	21	20	24 (23)	A Control
17	27	6	5	9	9	11 (13)	B
18	23	3	4	9	10	12 (13)	C

Figures in parentheses refer to the number of animals found within each plot when destructively sampled prior to pollution: at +1 day; +3 days; +7 days; +14 days; and +28 days. Series B and C are replicates of polluted plots.

or three days of tidal submergence. The resumption of feeding by the animals once the polluted substrate surface had been recovered by the tide would tend to draw the oil down into the burrows. It is also possible that the pollutant would naturally trickle down into any head shafts which were open, not having been filled by the disturbance of surface substrate by the receding tide.

Single Spillages. Site 2

The results of single spillages of pollutant on the plots in site 2 (Figs. 8–13) show the same trends as site 1. However, the effect of the dispersant BP 1100X is of the same order as the oil and oil emulsifier mixtures, in its long lasting effects. It is felt that this is due to the more uniform nature in the plots in site 2. There appears to be no significant difference in the extent of effect of a single spillage of emulsifier, oil, or mixtures of both. In all experimental spills, the reduction of animal numbers is in the order of 25–50%.

Repeated Spillages. Site 3

Figures 14–22 are records of the changes in *Arenicola* densities in the plots of site 3 over a period of nearly twenty-two months. Plot nos. 2, 5 and 8 are controls which have received no spillages of Kuwait Crude Oil upon their surface. All three controls show a stable density throughout the whole experiment. The remaining plots have had between one and six spillages of 5 litres Kuwait Crude Oil indicated by the symbol *.

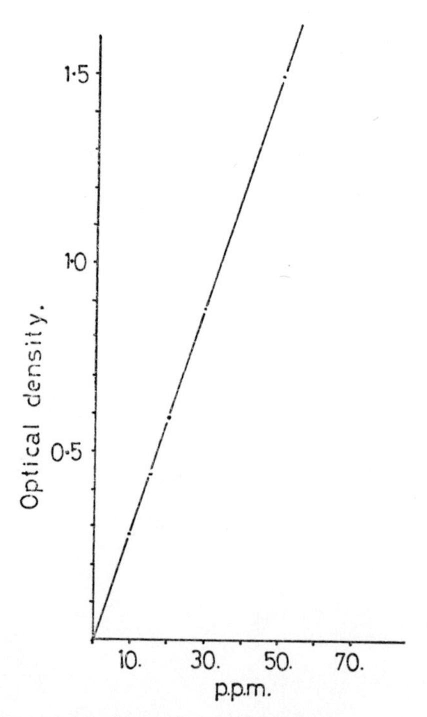

FIG. 26. Spectrophotometric absorption of Kuwait Crude Oil in cyclohexane at 260 μm wavelength.

The monthly pre-pollution densities all show a noticeable reduction for January 1974. It is felt that this was due to the heavy rain and waterlogging of the substrate causing a reduction of cast production rather than a true reduction in *Arenicola* population. Records for December 1973 and January 1974 were unobtainable due to persistent rain and excessive flooding, this making cast estimation unreliable.

On each occasion of a spillage, there appears to be a reduction in cast production followed by a recovery to only a 50–75% of the previous *Arenicola* population as seen in the examples of single spills at sites 1 and 2. The one exception to this appears to be on 20 August 1974, plot no. 4, when the oiling appears to have caused no long-lasting depletion of the *Arenicola* population. It is assumed that this is due to the non-penetration of the substrate by the oil due to excessive water content of the substrate.

The successive oilings cause further reduction of the population until after four spillages the population is almost eradicated. The whole area of site 3 was populated to a density of between 20 and 25/0·25 m^2; a more dense population may well survive more spillages. The wish to restrict pollution to the northern end of Sandyhaven has prevented the testing of this hypothesis. It was noted that in plots 6, 7 and 9 after the depletion of the *Arenicola* population, the additional application of Kuwait Crude Oil had a stabilising effect on the surface of the substrate. Distinctly oily layers were noticeable in the substrate. The reduction of the *Arenicola* population and hence reduction in the turnover of the surface of the substrate, through the action of cast production and substrate ingestion, may well have aided the retention of the oil within the surface layers.

RECOVERY OF POLLUTED PLOTS

The appearance of casts produced by juvenile *Arenicola* is denoted in Figs. 3 to 22 by the broken portions of the histograms. The 95% confidence limits represent the limits of the mean of the sum of Adult and Juvenile densities, although their numbers per 0·25 m^2 were recorded separately. Settlement of juveniles was first recorded in the plots of site 1 in the summer of 1974 during late July and August. The juvenile *Arenicola* are 10–25 mm in length. There was an extensive and heavy settlement or redistribution of post-larval juveniles throughout the Sandyhaven region in summer 1974. The density of settlement in site 1 is heavy, and results in densities equal to, or above, the original prepollution levels of the plots. The size of cast produced by the juveniles is at first very indistinct, only 2–5 mm in height. These are, of course, easily obscured or washed away by surface moisture. Thus the decrease in juvenile cast density in plots 1 and 2 when sampled in October 1974 and in plot 6 in December 1974 may represent a false picture of total numbers produced.

The plots of site 1 had been polluted fifteen months previously, in April 1973, and juveniles occur in all plots. However, in site 2 polluted in April 1974, colonisation is restricted to the non-oiled controls. The presence of second generation casts is noted several months later (October 1974) than in site 1 and is of a much lower density.

Similarly, at site 3, the appearance of juvenile casts occurs in the controls in October (plot 8) and December (plot 2), again in very low densities. By January 1975, the density of juveniles in the control plots seems to be increasing, juveniles having made their appearance in plot 1 which was only oiled on one occasion in April 1974, after a large portion of the substrate had been eroded and new substrate deposited by the movement of the creek sides which borders plots 1 and 9 (see Fig. 2).

It is not known whether settlement by *Arenicola* juveniles is selective or non-selective. There may well occur a complete non-selective 'blanket' settlement of juveniles in the area, survival occurring only where the substrate was unpolluted.

The information available on the redistribution and settlement of larval and post-larval stages is varied and somewhat conflicting.

Newell (1948) suggested that larval stages remain in the top centimetre or two of the sand where the eggs are laid. He considered that dispersal takes place later in the post-larval pelagic stage (Benham Stage).

Laboratory observation of the post-larval stages of *Arenicola* has shown them to be capable of swimming, buoyed up by the mucilaginous sheath in which they are enclosed, but that they did so infrequently and normally restricted their activities to crawling on the bottom of the culture vessel (Newell, 1948; Smidt, 1951).

Field surveys have found the post-larval stages in plankton hauls (Ehlers, 1892; Benham, 1893; Kyle, 1896; Ashworth, 1904; Leschke, 1908; Southern, 1914; Allen, 1915; Blegvad, 1932; M.B.A., 1957).

Thorson (1935) found no larval stages of *Arenicola* in the plankton during his studies near Helsinger and quoted Gustafson as saying that the larval development of *Arenicola* at Fulmars Fjord was similarly non-pelagic.

However, Thamdrup (1935) describes the larvae as pelagic throughout the winter and spring, and considers that it is at this phase that the animals are dispersed.

The occurrence of juvenile worms at a higher tidal level than the adults, has been recorded by Watkin (1942) and Werner (1954, 1956). Werner (1956) believed that migration of the juveniles from the higher tidal levels, occurred in the winter, and was initiated by a drop in temperature.

Spawning of *Arenicola* has not been observed at Sandyhaven, but large (15–20 cm) sexually mature animals of both sexes were abundant in the substrate of the creek bed on 25th September 1973 and were reduced when the same area was investigated on 22nd October, suggesting a large post-spawning mortality or migration (see Newell, 1948). A table of reported spawning times for *Arenicola* has been constructed by Clay (1967). Duncan (1953) reports that the *Arenicola* of the Dale Fort region spawn during the autumn.

CONCLUSIONS

The following general conclusions can be drawn from the results.

(1) Kuwait Crude Oil, BP 1100X and 1:1 and 5:1 mixtures of both all reduce population density of *Arenicola*.

(2) Single spillages of all pollutants result in a 25–50% reduction in the population.
(3) Successive spillages produce a progressive depletion of the population. Four spillages and above result in the virtual eradication of the population.
(4) Waterlogged substrates reduce penetration of the pollutants and thus their effects on *Arenicola* populations.
(5) Pollutants reduced feeding activity of up to 75% of the animals. During the following month recovery of a proportion of these animals occurs. The remainder either die and decompose or quit the substrate. The fate of those which leave the substrate is not known.
(6) Recolonisation of the substrate by juveniles was inhibited in previously polluted experimental areas. Plots polluted in April 1973 showed higher settlement in August 1974 than those polluted in May 1974.

These investigations can only be regarded as preliminary, but they provide a base for future work.

SUGGESTED AREAS FOR FURTHER INVESTIGATION

(1) Continual monthly monitoring of experimental sites to record the recovery of areas subjected to repeated spillages.
(2) A comprehensive investigation into the fate of oil and its retention within polluted substrates.
(3) More extensive field trials to determine the influence of water content of the substrate on penetration of the pollutant.
(4) Laboratory experiments, i.e. a closed system, to determine the fate of total *Arenicola* population.
(5) More extensive field trials involving varying concentration of pollutants.

ACKNOWLEDGEMENTS

I should like to thank the British Petroleum Company for the supply of Kuwait Crude Oil and the dispersant BP 1100X, especially Captain F. H. Johnston at BP Ocean Terminal, Popton Fort. Also the Milford Haven Conservancy Board, particularly Captain G. Dudley for permission to spray oils and dispersant within Milford Haven, and likewise the Pembrokeshire National Park Committee.

REFERENCES

Allen, R. J. (1915). Polychaeta of Plymouth and the south Devon coast, *J. mar. biol. Ass. U.K.*, **10** (4), 592–647.

Ashworth, J. Y. (1904). *Arenicola*, Liverpool Mag. Biol. committee mem, II London.

Benham, W. B. (1893). The post-larval stages of *Arenicola marina, J. mar. biol. Ass. U.K.*, **3**, 48–52.
Blegvad, H. (1932). Investigations of the bottom fauna of drains in the Sound, *Rept. Danish biol. Sta.*, **36**, 1–20.
British Petroleum Company (1966). Inspection data and ash composition of Kuwait Export Crude Oil.
Clay, E. (1967). Literature survey of the common fauna of estuaries, *2. Arenicola marina* Linnaeus, Imperial Chemical Industries Ltd., Brixham Laboratory.
Ehlers, A. (1892). Zur Kenntnis von *Arenicola marina, Nachr. Ges. Wiss. Göttingen.*
Holme, N. A. (1948). The fauna of sand and mud banks near the mouth of the Exe estuary, *J. mar. biol. Ass. U.K.*, **28**, 189.
Kyle, H. M. (1896). On the nephridia, reproductive organs, and post-larval stages of *Arenicola, Ann. Mag. Nat. Hist.*, **18**, 295–300.
Leschke, M. (1908). Beitrage zur Kenntnis der pelagischen Polychaetenlarven der Kieler Fohrde, *Kiel Comm. Wissensch. Meeresuntersuch* (*Kiel*), **I**, 113–16.
Linke, O. (1939). Die Biota des Jadebusenwattes, *Helgol. Wiss. Meeresunters*, **1**, 201–348.
Marine Biological Association (1957). Plymouth marine fauna.
Morgans, J. F. C. (1956). Notes on the analysis of shallow water soft substrata, *J. Anim. Ecol.*, **25**, 367–87.
Newell, G. E. (1948). A contribution to our knowledge of the life history of *Arenicola marina* L, *J. mar. biol. Ass. U.K.*, **17**, 554–80.
Newell, G. E. (1949). The later larval life of *Arenicola marina* L, *J. mar. biol. Ass. U.K.*, **28**, 635–9.
Smidt, E. L. B. (1951). Animal production in the Danish Waddensea, *Medd. Komm. Denmark Fisk.*, **11** (6), 1–151.
Southern, R. (1914). Archiannelida and Polychaeta, Clare Island Survey, *Proc. Roy. Irish Acad.*, **31** (2), 47.
Thorson, G. (1935). Reproduction and larval development of Danish marine bottom invertebrates, *Medd. fra. Komm. Fisk. Haversund, Serie Plankton*, **10** (2), 125.
Thamdrup, H. M. (1935). Beitrage zur Okologie der Waltenfauna auf experimentaler Grundlage, *Med. Komm. Havunders. Kbh. Fisheri*, **10**, 154–87.
Watkin, E. E. (1942). The macrofauna of intertidal sand at Kames Bay, Millport, Buteshire, *Trans. Roy. Soc. Edin.*, **60** (6), 543–61.
Werner, B. (1954). Über der Winterwanderung von *Arenicola marina, Helgol. wiss. Meeresunters*, **5**, 353–78.

APPENDIX I

Single Spillage Data (site 1). Total Cast Number, Mean Density/0·25 m², and 95% Confidence Limits about the Mean are Given. * Denotes Plot has been Subjected to Pollution

Date: 12 April 1973

Site No.	1				
Plot no.	1	2	4	6	8
Total	648	565	393	591	683
Mean ($\bar{x}$)	25·92	22·6	15·72	23·64	27·32
Variance (s^2)	67·66	23·5	13·46	71·57	76·32
Standard error	1·65	0·97	0·73	1·69	1·75
t × S.E.	3·41	2·0	1·51	3·49	3·61
95% Confidence limits	29·33	24·6	17·23	27·13	30·93
	22·51	20·6	14·21	20·15	23·71

Date: 14 April 1973

Site no.	1				
Plot no.	1	2*	4*	6*	8*
Total	556	261	182	274	217
Mean ($\bar{x}$)	22·24	10·44	7·28	10·96	8·68
Variance (s^2)	28·61	14·84	7·54	21·37	27·73
Standard error	1·07	0·77	0·55	0·92	1·05
t × S.E.	2·08	1·59	1·14	1·90	2·17
95% Confidence limits	24·32	12·03	8·42	9·06	10·85
	20·16	8·85	6·14	12·86	6·51

Date: 15 April 1973

Site no.	1				
Plot no.	1	2*	4*	6*	8*
Total	537	142	93	157	365
Mean ($\bar{x}$)	21·48	5·68	3·72	6·28	14·6
Variance (s^2)	37·59	6·23	3·79	17·88	42·25
Standard error	1·23	0·5	0·39	0·85	1·37
t × S.E.	2·24	1·06	0·81	1·75	2·83
95% Confidence limits	23·72	6·74	4·53	8·03	17·43
	19·24	4·62	2·91	4·53	11·77

Date: 18 April 1973

Site no.	1				
Plot no.	1	2*	4*	6*	8*
Total	515	82	72	67	635
Mean ($\bar{x}$)	20·6	3·28	2·88	2·68	25·4
Variance (s^2)	33·83	5·96	1·94	3·56	78·5
Standard error	1·16	0·49	0·28	0·38	1·77
t × S.E.	2·39	1·011	0·58	0·78	3·65
95% Confidence limits	22·99	4·29	3·46	3·46	29·05
	18·21	2·27	2·3	1·9	21·75

Date: 15 May 1973

Site no.	1				
Plot no.	1	2*	4*	6*	8*
Total	589	215	165	146	692
Mean ($\bar{x}$)	23·56	8·6	6·6	5·84	27·68
Variance (s^2)	39·59	11·67	11·08	14·39	37·48
Standard error	1·26	0·68	0·67	0·76	1·22
t × S.E.	2·60	1·40	1·38	1·57	2·52
95% Confidence limits	26·16	10·0	7·98	7·41	30·20
	20·96	7·2	5·22	4·27	25·16

Date: 4 June 1973

Site no.	1				
Plot no.	1	2*	4*	6*	8*
Total	605	264	138	82	422
Mean ($\bar{x}$)	24·2	10·56	5·52	3·28	23·44
Variance (s^2)	17·75	13·92	6·93	4·54	33·32
Standard error	0·84	0·75	0·53	0·43	1·36
t × S.E.	1·73	1·55	1·09	0·89	2·81
95% Confidence limits	25·93	12·11	6·61	4·17	26·25
	22·47	9·01	4·43	2·39	20·63

Date: 11 June 1973

Site no.	1				
Plot no.	1	2*	4*	6*	8*
Total	578	257	207	163	614
Mean ($\bar{x}$)	23·12	10·28	8·28	6·52	24·56
Variance (s^2)	39·19	20·13	7·63	21·32	56·92
Standard error	1·25	0·90	0·55	0·92	1·51
t × S.E.	2·58	1·86	1·14	1·90	3·12
95% Confidence limits	25·70	12·14	9·42	8·42	27·68
	20·54	8·42	7·14	4·52	21·44

Date: 26 July 1973

Site no.	1				
Plot no.	1	2*	4*	6*	8*
Total	555	335	188	122	560
Mean ($\bar{x}$)	22·2	13·4	7·52	4·88	22·4
Variance (s^2)	23·67	11·25	4·09	7·53	37·25
Standard error	0·97	0·67	0·40	0·55	1·22
t × S.E.	2·0	1·38	0·83	1·14	2·52
95% Confidence limits	24·2	14·78	8·35	6·02	24·92
	20·2	12·02	6·69	3·74	19·88

Date: 24 August 1973

Site no.	1				
Plot no.	1	2*	4*	6*	8*
Total	586	286	203	112	621
Mean ($\bar{x}$)	23·44	11·44	8·12	4·48	24·84
Variance (s^2)	32·84	15·59	8·19	7·18	39·97
Standard error	1·15	0·79	0·57	0·54	1·26
$t \times$ S.E.	2·37	1·63	1·18	1·12	2·60
95% Confidence limits	25·81	13·07	9·3	5·60	27·44
	21·07	9·81	6·94	3·36	22·24

Date: 24 September 1973

Site no.	1				
Plot no.	1	2*	4*	6*	8*
Total	429	275	212	93	370
Mean ($\bar{x}$)	17·16	11·0	8·48	3·72	19·47
Variance (s^2)	10·81	8·25	4·93	5·29	28·82
Standard error	0·66	0·57	0·44	0·46	1·20
$t \times$ S.E.	1·36	1·18	0·91	0·95	2·48
95% Confidence limits	18·52	12·18	9·39	4·67	21·95
	15·8	9·82	7·57	2·77	16·99

Date: 22 October 1973

Site no.	1				
Plot No.	1	2*	4*	6*	8*
Total	418	252	183	81	118
Mean ($\bar{x}$)	16·72	10·08	7·32	3·24	23·6
Variance (s^2)	21·29	8·24	11·89	9·36	20·8
Standard error	0·92	0·57	0·69	0·61	2·04
$t \times$ S.E.	1·90	1·18	1·42	1·26	5·25
95% Confidence limits	18·78	11·26	8·74	4·5	28·85
	14·66	8·9	5·90	1·98	18·35

Date: 22 November 1973

Site no.	1			
Plot no.	1	2*	4*	6*
Total	511	285	205	92
Mean ($\bar{x}$)	20·44	11·40	8·20	3·68
Variance (s^2)	17·26	19·67	8·08	9·56
Standard error	0·83	0·89	0·57	0·62
$t \times$ S.E.	1·71	1·84	1·18	1·28
95% Confidence limits	22·15	13·24	9·38	4·96
	18·73	9·56	7·02	2·40

Date: 23 January 1974

Site no.	1			
Plot no.	1	2*	4*	6*
Total	158	103	30	5
Mean ($\bar{x}$)	6·32	4·12	1·2	0·20
Variance (s^2)	10·73	1·94	1·83	0·17
Standard error	0·66	0·28	0·27	0·08
$t \times$ S.E.	1·36	0·58	0·56	0·17
95% Confidence limits	7·68	4·70	1·76	0·37
	4·96	3·54	0·64	0·03

Date: 12 March 1974

Site no.	1			
Plot no.	1	2*	4*	6*
Total	471	279	133	11
Mean ($\bar{x}$)	18·84	11·16	5·32	0·44
Variance (s^2)	26·14	6·72	7·06	0·34
Standard error	1·02	0·52	0·53	0·12
$t \times$ S.E.	2·11	1·07	1·09	0·25
95% Confidence limits	20·95	12·23	6·41	0·69
	16·73	10·09	4·23	0·19

Date: April 22 1974

Site no.	1			
Plot no.	1	2*	4*	6*
Total	536	285	188	28
Mean ($\bar{x}$)	21·44	11·4	7·52	1·12
Variance (s^2)	16·26	11·0	8·51	2·36
Standard error	0·81	0·66	0·58	0·31
$t \times$ S.E.	1·67	1·36	1·20	0·64
95% Confidence limits	23·11	12·76	8·72	1·76
	19·77	10·04	6·32	0·48

Date: 21 May 1974

Site no.	1			
Plot no.	1	2*	4*	6*
Total	445	295	178	47
Mean ($\bar{x}$)	17·8	11·8	7·12	1·88
Variance (s^2)	10·92	11·67	6·03	3·53
Standard error	0·66	0·68	0·49	0·38
$t \times$ S.E.	1·36	1·40	1·01	0·78
95% Confidence limits	19·16	13·2	8·13	2·66
	16·44	10·4	6·11	1·10

Date: 19 June 1974

Site no.	1			
Plot no.	1	2*	4*	6*
Total	399	239	163	68
Mean ($\bar{x}$)	15·96	9·56	6·52	2·72
Variance (s^2)	5·79	8·59	6·59	6·71
Standard error	0·48	0·59	0·51	0·52
t × S.E.	0·99	1·22	1·05	1·07
95% Confidence limits	16·95	10·78	7·57	3·79
	14·97	8·34	5·47	1·65

Date: 22 July 1974

Site no.	1			
Plot no.	1	2*	4*	6*
Total	364	235	170	83
Mean ($\bar{x}$)	14·56	9·4	6·8	3·32
Variance (s^2)	21·76	8·58	12·17	6·64
Standard error	0·93	0·59	0·70	0·52
t × S.E.	1·92	1·22	1·45	1·07
95% Confidence limits	16·48	10·62	8·25	4·39
	12·64	8·18	5·35	2·25

Date: 22 August 1974

Site no.	1			
Plot no.	1	2*	4*	6*
Total	402	294	359	169
Mean ($\bar{x}$)	30·92	22·62	27·62	13·0
Variance (s^2)	32·24	23·76	50·26	14·83
Standard error	1·67	1·35	1·96	1·14
t × S.E.	3·64	2·94	4·27	2·48
95% Confidence limits	34·56	25·56	31·89	15·49
	27·28	19·68	23·35	10·51

Date: 25 October 1974

Site no.	1			
Plot no.	1	2*	4*	6*
Total	237	211	376	157
Mean ($\bar{x}$)	18·23	16·23	28·92	12·08
Variance (s^2)	21·03	21·19	100·41	60·08
Standard error	1·27	1·27	2·78	2·15
t × S.E.	2·78	2·77	6·06	4·69
95% Confidence limits	21·0	19·0	34·98	16·77
	15·46	13·36	22·86	7·39

Date: 19 November 1974

Site no.	1			
Plot no.	1	2*	4*	6*
Total	352	325	468	270
Mean ($\bar{x}$)	27·08	25·0	36·0	20·77
Variance (s^2)	13·74	28·5	34·5	26·69
Standard error	1·03	1·48	1·63	1·43
t × S.E.	2·24	3·23	3·55	3·12
95% Confidence limits	29·32	28·23	39·56	23·89
	24·84	21·77	32·44	17·65

Date: 17 December 1974

Site no.	1			
Plot no.	1	2*	4*	6*
Total	399	310	369	88
Mean ($\bar{x}$)	30·69	23·85	28·38	6·77
Variance (s^2)	21·73	5·64	13·42	7·36
Standard error	1·29	0·66	1·01	0·75
t × S.E.	2·81	1·44	2·20	1·63
95% Confidence limits	33·50	25·29	30·58	8·40
	27·88	22·41	26·18	5·14

Date: 24 January 1975

Site no.	1			
Plot no.	1	2*	4*	6*
Total	399	255	315	274
Mean ($\bar{x}$)	30·69	19·61	24·23	21·07
Variance (s^2)				
Standard error				
t × S.E.				
95% Confidence limits				

APPENDIX II

Single Spillage Data (site 2). Total Cast Number, Mean Density/0·25 m², and 95% Confidence Limits about the Mean are Given. * Denotes Plot has been Subjected to Pollution

Date: 17 May 1973

Site no.	2					
Plot no.	1	2	3	4	5	6
Total	1091	895	580	900	605	1172
Mean ($\bar{x}$)	43·64	34·36	23·2	36	24·2	46·88
Variance (s^2)	375·5	118·49	40·75	77·33	28·75	64·61
Standard error	3·88	2·18	1·28	1·76	1·07	1·61
$t \times$ S.E.	8·01	4·5	2·64	3·63	2·21	3·32
95% Confidence	51·65	38·86	25·84	39·63	26·41	50·2
limits	35·63	29·86	20·56	32·37	21·99	43·56

Date: 6 June 1973

Site no.	2					
Plot no.	1	2	3	4	5	6
Total	878	735	684	1195	546	1054
Mean ($\bar{x}$)	34·12	29·4	27·36	47·8	21·84	42·16
Variance (s^2)	189·55	48·0	49·82	165·08	20·56	93·72
Standard error	2·75	1·39	1·41	2·57	0·91	1·94
$t \times$ S.E.	5·68	2·87	2·91	5·30	1·88	4·00
95% Confidence	39·8	32·27	30·27	53·1	23·72	46·16
limits	28·44	26·53	24·45	42·5	19·96	38·16

Date: 26 July 1973

Site no.	2					
Plot no.	1	2	3	4	5	6
Total	653	582	535	820	356	911
Mean ($\bar{x}$)	26·12	23·28	21·4	32·8	21·44	36·44
Variance (s^2)	36·78	15·79	11·5	27·83	16·01	33·59
Standard error	1·21	0·79	0·68	1·06	0·8	1·16
$t \times$ S.E.	2·5	1·63	1·40	2·19	1·65	2·39
95% Confidence	28·62	24·91	22·8	34·99	23·09	38·83
limits	23·62	21·85	20·0	30·61	19·89	34·05

Date: 24 August 1973

Site no.	2					
Plot no.	1	2	3	4	5	6
Total	787	642	559	922	553	1024
Mean ($\bar{x}$)	31·48	25·68	22·36	36·88	22·13	40·96
Variance (s^2)	85·68	56·98	23·57	31·94	21·61	49·21
Standard error	1·85	1·51	0·97	1·13	0·93	1·40
t × S.E.	3·82	3·12	2·0	2·33	1·92	2·89
95% Confidence	35·3	28·8	24·36	39·21	24·05	43·85
limits	27·66	22·46	20·36	34·55	20·21	38·07

Date: 23 September 1973

Site no.	2					
Plot no.	1	2	3	4	5	6
Total	686	557	491	886	516	948
Mean ($\bar{x}$)	27·44	22·28	19·64	35·44	20·64	37·92
Variance (s^2)	51·84	18·21	16·49	55·67	11·99	33·24
Standard error	1·44	0·85	0·81	1·49	0·69	1·15
t × S.E.	2·97	1·75	1·67	3·08	1·42	2·37
95% Confidence	30·41	24·03	21·31	38·52	22·06	40·29
limits	24·47	20·53	17·97	32·27	19·22	35·55

Date: 22 October 1973

Site no.	2					
Plot no.	1	2	3	4	5	6
Total	649	615	494	754	478	874
Mean ($\bar{x}$)	25·96	24·6	19·76	30·16	19·2	34·96
Variance (s^2)	32·04	18·42	24·27	38·39	26·53	39·96
Standard error	1·13	0·86	0·99	1·24	1·03	1·25
t × S.E.	2·33	1·78	2·04	2·56	2·13	2·58
95% Confidence	28·29	26·38	21·8	32·72	21·25	37·54
limits	23·63	22·82	17·72	27·6	16·99	32·38

Date: 22 November 1973

Site no.	2					
Plot no.	1	2	3	4	5	6
Total	694	550	529	788	502	872
Mean ($\bar{x}$)	27·76	22·0	21·16	31·52	20·08	34·88
Variance (s^2)	21·11	15·42	11·97	42·43	5·66	39·61
Standard error	0·92	0·79	0·69	1·30	0·48	1·26
t × S.E.	1·90	1·63	1·42	2·68	0·99	2·60
95% Confidence	29·66	23·63	22·58	34·2	21·07	37·48
limits	25·86	20·37	19·74	28·84	19·09	32·28

Date: 21 March 1974

Site no.	2					
Plot no.	1	2	3	4	5	6
Total	508	495	423	537	381	512
Mean ($\bar{x}$)	20·32	19·8	16·92	21·48	15·24	20·48
Variance (s^2)	15·81	14·75	9·24	18·26	3·52	16·68
Standard error	0·80	0·77	0·61	0·85	0·38	0·77
$t \times$ S.E.	1·65	1·59	1·26	1·75	0·78	1·59
95% Confidence	21·97	21·39	18·18	23·23	16·02	22·07
limits	18·67	18·21	15·66	19·73	14·46	18·89

Date: 23 April 1974

Site no.	2					
Plot no.	1	2	3	4	5	6
Total	628	446	501	725	488	556
Mean ($\bar{x}$)	25·12	17·84	20·04	29·0	19·52	22·24
Variance (s^2)	36·69	16·89	30·04	16·25	19·51	73·11
Standard error	1·25	0·82	1·10	0·81	0·88	1·71
$t \times$ S.E.	2·58	1·69	2·27	1·67	1·82	3·53
95% Confidence	27·7	19·53	22·31	30·67	21·33	25·77
limits	22·54	16·15	17·77	27·33	17·71	18·71

Date: 21 May 1974

Site no.	2					
Plot no.	1	2	3	4	5	6
Total	622	555	510	558	439	492
Mean ($\bar{x}$)	24·88	22·2	20·4	22·32	17·56	24·6
Variance (s^2)	8·03	11·75	7·33	5·81	10·51	14·57
Standard error	0·57	0·69	0·54	0·48	0·65	0·85
$t \times$ S.E.	1·18	1·42	1·12	0·99	1·34	1·78
95% Confidence	26·06	23·62	21·52	23·31	18·9	26·38
limits	23·7	20·78	19·28	21·33	16·22	22·82

Date: 23 May 1974

Site no.	2					
Plot no.	1*	2*	3*	4	5	6*
Total	79	216	334	557	409	138
Mean ($\bar{x}$)	3·16	8·64	13·36	22·68	16·36	5·52
Variance (s^2)	3·06	7·49	7·82	16·23	8·49	9·93
Standard error	0·35	0·55	0·56	0·81	0·58	0·63
$t \times$ S.E.	0·72	1·14	1·16	1·67	1·20	1·30
95% Confidence	3·88	9·78	14·52	24·35	17·52	6·82
limits	2·44	7·5	12·2	21·01	15·16	4·22

Date: 25 May 1974

Site no.	2					
Plot no.	1*	2*	3*	4	5	6*
Total	224	275	346	518	367	262
Mean ($\bar{x}$)	8·96	11·0	13·84	20·72	14·68	10·48
Variance (s^2)	15·21	3·75	6·39	11·79	6·39	15·26
Standard error	0·78	0·39	0·51	0·69	0·51	0·78
$t \times$ S.E.	1·61	0·81	1·05	1·42	1·05	1·61
95% Confidence	10·57	11·81	14·89	22·14	15·73	12·09
limits	7·35	10·19	12·79	19·3	13·63	8·87

Date: 30 May 1974

Site no.	2					
Plot no.	1*	2*	3*	4	5	6*
Total	277	321	399	560	412	408
Mean ($\bar{x}$)	11·08	12·84	15·96	22·4	16·48	16·32
Variance (s^2)	10·49	5·14	6·87	13·58	7·09	3·81
Standard error	0·65	0·47	0·52	0·74	0·53	0·39
$t \times$ S.E.	1·34	0·97	1·07	1·53	1·09	0·81
95% Confidence	12·42	13·81	17·03	23·93	17·57	17·13
limits	9·74	11·87	14·89	20·47	15·39	15·51

Date: 6 June 1974

Site no.	2					
Plot no.	1*	2*	3*	4	5	6*
Total	262	321	363	539	406	388
Mean ($\bar{x}$)	10·48	12·84	14·52	21·56	16·20	15·52
Variance (s^2)	11·76	5·06	5·18	4·84	18·89	9·26
Standard error	0·69	0·45	0·46	0·44	0·87	0·61
$t \times$ S.E.	1·42	0·93	0·95	0·91	1·80	1·26
95% Confidence	11·40	13·77	15·47	22·47	18·00	16·78
limits	9·06	11·91	13·67	20·65	14·40	14·26

Date: 19 June 1974

Site no.	2					
Plot no.	1*	2*	3*	4	5	6*
Total	391	389	428	550	454	471
Mean ($\bar{x}$)	15·64	15·56	17·12	22·0	18·16	18·84
Variance (s^2)	7·07	10·76	4·94	5·58	4·64	5·81
Standard error	0·53	0·66	0·44	0·47	0·43	0·48
$t \times$ S.E.	1·09	1·36	0·91	0·97	0·89	0·99
95% Confidence	16·73	16·92	18·03	22·97	19·05	19·83
limits	14·55	14·20	16·21	21·03	17·27	17·85

Date: 22 July 1974

Site no.	2					
Plot no.	1*	2*	3*	4	5	6*
Total	345	367	411	493	389	386
Mean ($\bar{x}$)	13·8	14·68	16·44	19·72	15·56	15·44
Variance (s^2)	9·17	10·33	5·09	11·79	4·09	6·76
Standard error	0·61	0·64	0·45	0·69	0·40	0·52
$t \times$ S.E.	1·26	1·32	0·93	1·42	0·83	1·07
95% Confidence	15·06	16·0	17·37	21·14	16·39	16·51
limits	12·54	13·36	15·53	18·3	14·73	14·37

Date: 19 August 1974

Site no.	2					
Plot no.	1*	2*	3*	4	5	6*
Total	410	401	428	551	479	454
Mean ($\bar{x}$)	16·4	16·04	17·12	22·04	19·16	18·16
Variance (s^2)	6·42	11·12	4·11	5·62	12·39	17·06
Standard error	0·51	0·67	0·41	0·47	0·70	0·83
$t \times$ S.E.	1·05	1·38	0·85	0·97	1·45	1·71
95% Confidence	17·45	17·42	17·97	23·01	20·61	19·77
limits	15·35	14·66	16·27	21·07	17·75	16·25

Date: 24 October 1974

Site no.	2					
Plot no.	1*	2*	3*	4	5	6*
Total	297	323	390	423	484	376
Mean ($\bar{x}$)	11·88	12·92	15·6	16·92	19·36	15·04
Variance (s^2)	5·53	7·99	7·83	6·74	11·91	9·29
Standard error	0·47	0·57	0·56	0·52	0·69	0·61
$t \times$ S.E.	0·97	1·18	1·16	1·07	1·42	1·26
95% Confidence	12·85	14·1	16·76	17·99	20·78	16·3
limits	10·91	11·74	14·44	15·85	17·94	13·78

Date: 25 November 1974

Site no.	2					
Plot no.	1*	2*	3*	4	5	6*
Total	325	365	420	473	420	405
Mean ($\bar{x}$)	13·0	14·6	16·8	18·92	16·8	16·2
Variance (s^2)	8·08	9·25	6·75	7·74	7·58	8·0
Standard error	0·57	0·61	0·52	0·56	0·55	0·57
$t \times$ S.E.	1·18	1·26	1·07	1·16	1·14	1·18
95% Confidence	14·18	15·86	17·87	20·08	17·94	17·38
limits	11·82	13·34	15·73	17·76	15·66	15·02

Date: 18 December 1974

Site no.	2					
Plot no.	1*	2*	3*	4	5	6*
Total	372	377	421	447	413	420
Mean ($\bar{x}$)	14·88	15·08	16·84	17·19	16·52	16·8
Variance (s^2)	4·61	5·08	4·56	17·2	5·93	7·0
Standard error	0·43	0·45	0·43	0·83	0·49	0·53
$t \times$ S.E.	0·89	0·93	0·88	1·71	1·01	1·94
95% Confidence	15·76	16·01	17·72	18·9	17·53	18·74
limits	14·0	14·05	15·96	15·48	15·51	14·86

Date: 31 January 1975

Site no.	2					
Plot no.	1*	2*	3*	4	5	6*
Total	285	311	295	252	259	348
Mean ($\bar{x}$)	11·4	12·44	11·8	19·38	19·93	13·92
Variance (s^2)						
Standard error						
$t \times$ S.E.						
95% Confidence						
limits						

APPENDIX III

uccessive Spillage Data. Total Cast Number, Mean Density/0.25m^2, and 25% Confidence Limits bout the Mean are Given. * Denotes the Number of Times Each Plot has been Subjected to Pollution.

Date: 28 June 1973

ite no.	3								
lot no.	1	2	3	4	5	6	7	8	9
otal	407	374	397	465	645	1098	844	807	724
lean ($\bar{x}$)	16·28	14·96	15·88	18·6	25·8	43·92	33·76	32·28	28·96
ariance (s^2)	28·54	10·79	11·44	32·58	45·67	37·41	22·77	12·21	29·62
tandard error	1·07	0·66	0·68	1·14	1·35	1·22	0·95	0·70	1·09
× S.E.	1·07	1·36	1·40	2·35	2·79	2·52	1·96	1·45	2·25
5% Confidence	17·35	16·32	17·28	16·25	28·59	46·44	35·72	33·72	31·21
limits	15·21	13·6	14·48	20·95	23·01	41·40	31·8	30·84	26·71

Date: 26 July 1973

te no.	3								
lot no.	1	2	3	4	5	6	7	8	9
otal	424	408	390	468	613	888	833	828	737
lean ($\bar{x}$)	16·96	16·32	15·6	18·72	24·52	35·52	33·32	33·12	29·48
ariance (s^2)	20·37	10·06	16·08	19·46	43·84	24·76	11·06	19·94	32·68
andard error	0·9	0·63	0·8	0·88	1·32	1·0	0·67	0·89	1·14
× S.E.	1·86	1·3	1·65	1·82	2·72	2·06	1·38	1·84	2·35
% Confidence	18·82	17·6	17·25	20·54	27·24	37·58	34·70	34·95	31·83
limits	15·1	15·02	13·95	16·91	21·8	33·46	31·94	31·29	27·13

Date: 24 August 1973

te no.	3								
ot no.	1	2	3	4	5	6	7	8	9
tal	453	420	361	517	584	1012	885	810	822
ean ($\bar{x}$)	18·28	16·8	14·44	20·68	23·36	40·48	35·4	33·75	32·88
riance (s^2)	61·13	27·33	14·34	33·31	44·66	39·76	35·33	28·20	63·61
andard error	1·56	1·05	0·76	1·15	1·34	1·26	1·19	1·06	1·60
× S.E.	3·22	2·17	1·57	2·37	2·77	2·60	2·46	2·19	3·30
% Confidence	21·5	18·97	16·01	23·05	26·03	43·08	37·86	35·93	36·18
limits	15·06	14·63	12·87	18·31	20·69	37·88	32·94	31·57	29·58

Date: 24 September 1973

e no.	3								
ot no.	1	2	3	4	5	6	7	8	9
tal	394	355	371	416	527	873	819	785	735
ean ($\bar{x}$)	15·76	14·2	14·80	17·33	21·08	34·92	32·76	31·4	29·4
riance (s^2)	14·36	6·0	17·8	10·84	39·41	16·16	16·19	19·42	37·92
ndard error	0·76	0·49	0·84	0·66	1·26	0·80	0·80	0·88	1·23
S.E.	1·57	1·01	1·73	1·36	2·60	1·65	1·65	1·81	2·54
% Confidence	17·33	15·21	16·5	18·69	23·68	36·57	34·41	33·22	31·94
imits	14·19	13·19	13·1	15·97	18·48	33·27	31·11	29·68	26·86

Date: 22 October 1973

Site no.	3								
Plot no.	1	2	3	4	5	6	7	8	9
Total	344	362	340	459	508	804	735	746	697
Mean ($\bar{x}$)	13·76	14·48	13·60	18·36	20·32	32·16	29·40	28·69	27·8
Variance (s^2)	17·26	12·51	15·42	15·66	31·31	23·06	12·33	49·58	38·9
Standard error	0·83	0·71	0·79	0·79	1·12	0·96	0·70	1·41	1·2
$t \times$ S.E.	1·71	1·47	1·63	1·63	2·31	1·98	1·45	2·91	2·5
95% Confidence	15·47	15·95	15·23	19·99	22·63	34·14	30·85	31·6	30·4
limits	12·05	13·01	11·97	16·73	18·01	30·18	27·95	25·78	25·3

Date: 22 November 1973

Site no.	3								
Plot no.	1	2	3	4	5	6	7	8	9
Total	412	377	351	385	513	660	747	709	583
Mean ($\bar{x}$)	16·48	15·08	14·04	15·4	20·52	26·4	29·88	28·36	24·
Variance (s^2)	9·34	11·24	6·46	10·92	30·34	22·67	28·36	14·07	29·
Standard error	0·61	0·13	0·51	0·66	1·10	0·96	1·07	0·75	1·
$t \times$ S.E.	1·26	0·27	1·05	1·36	2·27	1·98	2·21	1·55	2·
95% Confidence	17·74	15·35	15·09	16·96	22·79	28·38	32·09	29·91	26·
limits	15·22	14·81	12·99	14·04	18·25	24·42	27·67	26·81	22·

Date: 23 January 1974

Site no.	3								
Plot no.	1	2	3	4	5	6	7	8	9
Total	212	175	164	231	255	350	279	187	177
Mean ($\bar{x}$)	8·48	7·00	6·56	9·24	10·2	14·0	11·16	7·48	7·
Variance (s^2)	9·26	3·75	5·59	11·11	9·42	18·67	7·31	11·51	12·
Standard error	0·61	0·39	0·47	0·67	0·61	0·86	0·54	0·68	0·
$t \times$ S.E.	1·26	0·81	0·97	1·38	1·26	1·78	1·12	1·40	1·
95% Confidence	9·74	7·81	7·53	10·62	11·46	15·78	12·28	8·88	8·
limits	7·22	6·19	5·57	7·86	8·94	12·24	10·04	6·08	5·

Date: 14 March 1974

Site no.	3								
Plot no.	1	2	3	4	5	6	7	8	9
Total	316	317	278	345	346	445	434	360	316
Mean ($\bar{x}$)	12·61	12·68	11·12	13·8	18·84	17·8	17·36	14·4	12·
Variance (s^2)	9·31	5·48	16·62	7·25	5·81	14·75	8·24	10·67	20·
Standard error	0·61	0·47	0·82	0·54	0·48	0·77	0·57	0·65	0·
$t \times$ S.E.	1·26	0·97	1·69	1·12	0·99	1·59	1·18	1·34	1·
95% Confidence	13·87	13·65	12·81	14·92	19·83	19·39	18·54	15·74	14·
limits	11·35	11·71	9·43	12·68	17·85	16·21	16·18	13·06	10·

Date: 22 April 1974

te no.	3								
ot no.	1	2	3	4	5	6	7	8	9
tal	434	420	444	474	499	645	595	522	542
ean ($\bar{x}$)	17·36	16·8	17·76	18·96	19·96	25·8	23·8	22·08	21·68
riance (s^2)	17·41	9·92	12·86	11·12	8·79	9·67	11·5	9·49	12·56
andard error	0·83	0·63	0·72	0·67	0·59	0·62	0·68	0·62	0·71
× S.E.	1·71	1·3	1·49	1·38	1·22	1·28	1·40	1·28	1·47
% Confidence	19·07	18·1	19·25	20·34	21·18	27·08	25·2	23·36	23·15
limits	15·65	15·5	16·27	17·58	18·74	24·52	22·4	20·8	20·21

Date: 24 April 1974

e no.	3								
ot no.	1*	2	3*	4*	5	6*	7*	8	9*
tal	206	467	200	318	444	219	188	466	165
ean ($\bar{x}$)	8·24	18·28	8·0	12·72	17·76	8·76	7·52	18·64	6·6
riance (s^2)	6·44	13·48	9·42	14·29	7·61	11·94	9·09	15·41	9·92
andard error	0·51	0·73	0·61	0·76	0·55	0·69	0·60	0·79	0·63
S.E.	1·05	1·51	1·26	1·57	1·14	1·42	1·24	1·63	1·3
% Confidence	9·29	20·19	9·26	13·39	18·9	10·18	8·76	20·27	7·9
imits	7·19	17·17	6·74	11·15	16·62	7·34	6·28	17·01	5·3

Date: 26 April 1974

e no.	3								
t no.	1*	2	3*	4*	5	6*	7*	8	9*
tal	220	364	122	201	431	203	156	478	130
an ($\bar{x}$)	8·8	14·56	4·88	8·04	17·24	8·12	6·24	19·12	5·20
riance (s^2)	7·33	10·34	6·19	8·96	10·77	17·11	9·77	15·86	11·00
ndard error	0·54	0·64	0·50	0·60	0·66	0·87	0·63	0·80	0·66
S.E.	1·12	1·32	1·03	1·33	1·36	1·80	1·30	1·65	1·36
% Confidence	9·92	15·88	5·91	9·37	18·6	9·92	7·54	20·77	6·56
imits	7·68	13·24	3·85	6·71	15·88	6·32	4·94	17·47	3·84

Date: 30 April 1974

no.	3								
t no.	1*	2	3*	4*	5	6*	7*	8*	9
al	311	433	265	363	456	343	254	493	249
an ($\bar{x}$)	12·44	17·32	10·6	14·52	18·24	13·72	10·16	19·72	9·96
iance (s^2)	11·42	16·06	10·33	12·18	10·11	15·38	8·72	5·13	10·21
ndard error	0·68	0·80	0·64	0·70	0·64	0·78	0·59	0·45	0·64
S.E.	1·40	1·65	1·32	1·45	1·32	1·61	1·22	0·93	1·32
% Confidence	13·84	18·97	11·92	15·97	19·56	15·33	11·38	20·65	11·28
mits	11·04	15·67	9·28	13·07	16·92	12·11	8·94	18·79	8·64

Date: 7 May 1974

Site no.	3								
Plot no.	1*	2	3*	4*	5	6*	7*	8	9*
Total	230	386	213	265	447	227	179	500	221
Mean ($\bar{x}$)	9·58	16·5	8·88	11·04	18·54	9·46	7·46	20·83	9·
Variance (s^2)	11·64	10·70	8·64	8·39	9·39	10·95	6·87	17·62	10·
Standard error	0·68	0·65	0·59	0·58	0·61	0·66	0·52	0·84	0·
t × S.E.	1·41	1·34	1·22	1·22	1·26	1·36	1·07	1·73	1·
95% Confidence	10·99	17·84	10·1	12·26	19·8	10·82	8·53	22·56	10·
limits	8·17	15·16	7·66	9·82	17·28	8·1	6·39	19·1	7·

Date: 22 May 1974

Site no.	3								
Plot no.	1*	2	3*	4*	5	6*	7*	8	9*
Total	243	325	194	246	352	229	155	408	151
Mean ($\bar{x}$)	9·72	13·0	7·76	9·84	14·08	9·16	6·20	16·32	6·
Variance (s^2)	12·96	7·67	4·94	6·89	8·83	5·39	3·92	4·14	10·
Standard error	0·72	0·55	0·44	0·52	0·59	0·46	0·40	0·41	0·
t × S.E.	1·49	1·14	0·91	1·07	1·22	0·95	0·83	0·85	1·
95% Confidence	12·21	14·14	8·67	10·91	15·3	10·1	7·03	17·17	7·
limits	8·23	11·86	6·85	8·77	12·86	8·25	5·37	15·47	4·

Date: 20 June 1974

Site no.	3								
Plot no.	1*	2	3*	4*	5	6*	7*	8	9
Total	295	415	240	313	396	284	222	456	195
Mean ($\bar{x}$)	11·35	16·6	9·6	12·52	15·84	11·36	8·88	18·24	7
Variance (s^2)	14·8	8·83	5·83	7·01	4·89	4·66	4·86	10·44	10
Standard error	0·77	0·59	0·48	0·53	0·44	0·43	0·44	0·65	0
t × S.E.	1·59	1·22	0·99	1·09	0·91	0·89	0·91	1·34	1
95% Confidence	12·94	17·82	10·59	13·61	16·75	12·25	9·79	19·58	9
limits	9·66	15·38	8·61	11·43	14·93	10·47	7·97	16·9	6

Date: 22 June 1974

Site no.	3								
Plot no.	1*	2	3**	4**	5	6**	7**	8	9
Total	318	392	72	84	411	63	72	504	89
Mean ($\bar{x}$)	12·72	15·68	2·88	3·36	16·44	2·42	2·88	20·16	3
Variance (s^2)	6·38	9·89	2·86	2·57	6·92	3·85	4·44	5·56	6
Standard error	0·51	0·63	0·34	0·32	0·53	0·39	0·42	0·47	0
t × S.E.	1·05	1·3	0·70	0·66	1·09	0·81	0·87	0·97	1
95% Confidence	13·77	16·98	3·58	4·02	17·53	3·22	3·75	21·03	4
limits	11·67	14·38	2·18	2·70	15·35	1·62	2·01	19·19	2

Date: 24 June 1974

te no.	3								
ot no.	1*	2	3**	4**	5	6**	7**	8	9**
otal	368	422	79	116	396	67	66	465	108
ean ($\bar{x}$)	14·72	16·88	3·16	4·64	15·84	2·68	2·64	18·6	4·32
ariance (s^2)	3·96	7·86	3·06	7·91	4·72	2·48	1·07	4·42	7·73
andard error	0·75	0·56	0·35	0·56	0·43	0·31	0·21	0·42	0·56
× S.E.	1·55	1·16	0·72	1·16	0·89	0·64	0·43	0·87	1·16
% Confidence	16·27	18·04	3·89	5·8	16·73	3·24	3·07	17·73	5·48
limits	13·17	15·62	2·43	3·48	14·95	1·96	2·21	19·47	3·16

Date: 28 June 1974

te no.	3								
ot no.	1*	2	3**	4**	5	6**	7**	8	9**
otal	342	405	177	213	433	172	145	499	145
ean ($\bar{x}$)	13·68	16·2	7·08	8·5	17·32	6·88	5·8	19·96	5·8
riance (s^2)	11·06	11·92	5·24	11·59	8·64	7·53	4·25	9·04	8·5
andard error	0·67	0·69	0·46	0·68	0·59	0·55	0·41	0·60	0·58
× S.E.	1·38	1·42	0·95	1·40	1·22	1·14	0·85	1·24	1·20
% Confidence	15·06	17·62	8·03	9·9	18·52	8·02	6·65	21·2	7·0
limits	12·30	14·88	6·13	7·1	16·1	5·74	4·95	18·72	4·6

Date: 5 July 1974

e no.	3								
ot no.	1*	2	3**	4**	5	6**	7**	8	9**
tal	316	404	158	196	433	147	127	463	111
ean ($\bar{x}$)	12·64	16·16	6·32	7·84	17·32	8·88	5·08	18·52	4·44
riance (s^2)	19·16	9·64	4·73	9·47	21·06	6·11	3·24	8·01	7·17
ndard error	0·88	0·62	0·43	0·62	0·92	0·49	0·36	0·57	0·54
× S.E.	1·82	1·28	0·89	1·28	1·90	1·01	0·74	1·18	1·12
% Confidence	14·46	17·44	7·29	9·02	19·22	9·89	5·82	19·7	5·56
limits	10·62	14·88	5·43	6·56	15·42	7·87	4·34	17·34	3·32

Date: 19 July 1974

e no.	3								
ot no.	1*	2	3**	4**	5	6**	7**	8	9**
tal	383	425	145	201	444	149	120	536	102
ean ($\bar{x}$)	15·32	17·0	5·8	8·04	17·75	5·96	4·8	21·44	4·08
riance (s^2)	8·31	11·17	5·42	5·37	9·11	9·54	6·33	10·17	8·24
ndard error	0·58	0·67	0·47	0·46	0·60	0·62	0·50	0·64	0·57
× S.E.	1·20	1·38	0·97	0·95	1·24	1·28	1·03	1·32	1·18
% Confidence	16·52	18·38	6·77	8·99	18·99	7·24	5·83	22·76	5·26
limits	14·12	15·62	4·83	7·09	16·51	4·68	3·77	20·12	2·90

Date: 19 August 1974

Site no.	3								
Plot no.	1*	2	3**	4**	5	6**	7**	8	9**
Total	342	386	148	219	405	121	79	455	88
Mean ($\bar{x}$)	13·68	15·44	5·92	8·76	16·2	4·84	3·16	18·2	3·
Variance (s^2)	8·56	12·59	4·58	8·52	6·17	5·31	2·72	5·08	7·
Standard error	0·59	0·71	0·43	0·58	0·80	0·46	0·33	0·45	0·
$t \times$ S.E.	1·22	1·47	0·89	1·20	1·03	0·95	0·68	0·93	1·
95% Confidence	14·9	16·91	6·81	9·96	17·23	5·79	3·84	19·13	4·
limits	12·46	13·97	5·03	7·56	15·17	3·89	2·48	17·27	2·

Date: 21 August 1974

Site no.	3								
Plot no.	1*	2	3**	4***	5	6***	7***	8	9**
Total	308	394	175	148	377	40	20	472	49
Mean ($\bar{x}$)	12·32	15·76	7·0	5·92	15·08	1·6	0·8	18·88	1·
Variance (s^2)	6·98	8·69	4·42	5·91	6·58	1·42	0·83	5·36	3·
Standard error	0·53	0·59	0·42	0·49	0·51	0·24	0·18	0·46	0·
$t \times$ S.E.	1·09	1·22	0·87	1·01	1·05	0·50	0·37	0·95	0·
95% Confidence	13·41	16·98	7·87	6·93	16·13	2·09	1·17	19·83	2·
limits	11·23	14·54	6·13	4·91	14·03	1·1	0·43	17·93	1·

Date: 23 August 1974

Site no.	3								
Plot no.	1*	2	3**	4***	5	6***	7***	8	9
Total	327	394	168	165	374	64	49	434	64
Mean ($\bar{x}$)	13·08	15·76	6·72	6·6	14·96	2·56	1·96	17·36	2·
Variance (s^2)	12·16	7·69	2·46	12·0	9·71	3·84	2·21	2·24	5·
Standard error	0·70	0·55	0·31	0·69	0·62	0·39	0·30	0·30	0·
$t \times$ S.E.	1·45	1·14	0·64	1·42	1·28	0·81	0·62	0·62	0·
95% Confidence	14·53	16·9	7·36	8·02	16·24	3·37	2·58	17·98	3·
limits	11·53	14·62	6·08	5·18	13·68	1·75	1·34	16·74	1·

Date: 28 August 1974

Site no.	3								
Plot no.	1*	2	3**	4***	5	6***	7***	8	9
Total	302	401	162	212	421	113	57	500	66
Mean ($\bar{x}$)	12·08	16·04	6·48	8·48	16·84	4·52	2·28	20·0	2
Variance (s^2)	5·16	10·12	4·76	7·51	8·81	4·34	3·13	8·25	7·
Standard error	0·45	0·64	0·44	0·55	0·59	0·42	0·35	0·57	0
$t \times$ S.E.	0·93	1·32	0·91	1·14	1·22	0·87	0·72	1·18	1
95% Confidence	13·01	17·36	7·39	9·62	18·06	5·39	3·0	21·18	3
limits	11·15	14·72	5·57	7·34	15·62	3·65	1·56	18·82	1

Date: 4 September 1974

e no.	3								
ot no.	1*	2	3**	4***	5	6***	7***	8	9***
tal	300	393	170	156	422	68	63	478	48
ean ($\bar{x}$)	12·0	15·72	6·8	6·24	16·88	2·72	2·52	19·12	1·92
riance (s^2)	16·58	6·29	4·5	4·02	11·28	4·13	2·01	7·11	3·83
ndard error	0·81	0·50	0·42	0·40	0·67	0·41	0·28	0·53	0·39
< S.E.	1·67	1·03	0·87	0·83	1·38	0·85	0·58	1·09	0·81
% Confidence	13·67	16·75	7·67	7·07	18·26	3·57	3·1	20·21	2·73
imits	10·43	14·69	5·93	5·31	15·5	1·87	1·94	18·03	1·11

Date: 2 October 1974

e no.	3								
t no.	1*	2	3**	4***	5	6***	7***	8	9***
tal	301	393	130	163	413	39	22		50
an ($\bar{x}$)	12·04	15·7	5·2	6·52	16·52	1·56	0·88		2·0
riance (s^2)	10·87	14·13	6·42	9·51	7·26	1·67	1·28		5·33
ndard error	0·66	0·75	0·51	0·62	0·54	0·26	0·23		0·46
S.E.	1·36	1·55	1·05	1·28	1·12	0·54	0·48		0·95
% Confidence	13·4	17·25	6·25	7·8	17·64	2·09	1·36		2·95
imits	10·68	14·15	4·15	5·24	15·40	1·03	0·40		1·05

Date: 14 October 1974

no.	3								
t no.	1*	2	3**	4***	5	6***	7***	8	9***
al	299	407	174	157	351	15	8	495	42
an ($\bar{x}$)	11·96	16·28	6·96	6·28	14·04	0·6	0·31	19·8	1·75
iance (s^2)	8·54	8·96	7·87	8·46	5·87	0·83	0·30	13·75	4·98
ndard error	0·58	0·60	0·56	0·58	0·48	0·18	0·11	0·74	0·45
S.E.	1·20	1·24	1·16	1·20	0·99	0·37	0·23	1·52	0·93
Confidence	13·12	17·52	8·12	7·48	15·04	0·97	0·56	21·42	2·62
mits	10·78	15·04	5·80	5·08	13·04	0·23	0·07	18·28	0·78

Date: 16 October 1974

no.	3								
no.	1*	2	3**	4***	5	6****	7****	8	9****
al	129	169	69	70	158	0	0	474	13
an ($\bar{x}$)	12·9	16·9	6·9	7·0	15·8	0·0	0·0	18·96	0·57
iance (s^2)	5·88	4·77	3·66	8·89	3·29	0·0	0·0	8·21	1·44
ndard error	0·77	0·69	0·61	0·94	0·57	0·0	0·0	0·57	0·24
S.E.	1·74	1·56	1·38	2·13	1·29	0·0	0·0	1·18	0·50
Confidence	14·64	18·46	8·28	9·13	17·03	0·0	0·0	20·14	1·07
mits	11·16	15·34	5·52	4·87	14·57	0·0	0·0	17·78	0·07

Date: 22 October 1974

Site no.	3								
Plot no.	1*	2	3**	4***	5	6****	7****	8	9*
Total	274	384	145	170	214	1	0·0	470	18
Mean ($\bar{x}$)	10·96	15·36	5·8	6·8	16·46	0·04	0·0	18·8	0·
Variance (s^2)	10·79	9·16	7·92	14·42	12·44	0·04	0·0	3·33	1·
Standard error	0·66	0·61	0·56	0·76	0·98	0·04	0·0	0·37	0·
t × S.E.	1·36	1·26	1·16	1·57	2·14	0·08	0·0	0·76	0·
95% Confidence	12·22	16·62	6·97	8·37	18·60	0·12	0·0	19·56	1·
limits	9·6	14·10	4·63	5·43	14·32	0·0	0·0	18·04	0·

Date: 30 October 1974

Site No.	3								
Plot no.	1*	2	3**	4***	5	6****	7****	8	9
Total	220	350	142	137	356	7	0	464	11
Mean ($\bar{x}$)	8·8	14·0	5·68	5·48	14·24	0·28	0·0	18·56	0
Variance (s^2)	9·0	11·17	10·56	8·01	5·86	0·71	0·0	8·92	0
Standard error	0·6	0·67	0·65	0·57	0·48	0·17	0·0	0·60	0
t × S.E.	1·24	1·38	1·34	1·18	0·99	0·35	0·0	1·24	0
95% Confidence	10·04	15·38	7·02	6·66	15·24	0·63	0·0	19·8	0
limits	7·56	12·62	4·34	4·30	13·24	0·0	0·0	17·32	0

Date: 12 November 1974

Site no.	3								
Plot no.	1*	2	3**	4***	5	6****	7****	8	[illegible]
Total	267	387	175	198	391	0	0	438	[illegible]
Mean ($\bar{x}$)	10·68	15·48	7·0	7·92	15·64	0·0	0·0	17·52	[illegible]
Variance (s^2)	6·14	7·68	9·5	11·99	6·49	0·0	0·0	5·76	[illegible]
Standard error	0·50	0·55	0·62	0·69	0·51	0·0	0·0	0·41	[illegible]
t × S.E.	1·03	1·14	1·28	1·42	1·05	0·0	0·0	0·99	[illegible]
95% Confidence	11·71	16·62	8·28	9·34	16·71	0·0	0·0	18·62	[illegible]
limits	9·65	14·33	5·72	6·5	14·59	0·0	0·0	16·52	[illegible]

Date: 9 December 1974

Site no.	3								
Plot no.	1*	2	3**	4***	5	6****	7****	8	[illegible]
Total	314	383	163	154	397	22	0	506	[illegible]
Mean ($\bar{x}$)	12·56	15·32	6·52	6·16	15·88	0·92	0·0	20·24	[illegible]
Variance (s^2)	5·17	6·06	5·26	6·56	6·86	2·78	0·0	8·19	[illegible]
Standard error	0·45	0·49	0·46	0·51	0·52	0·33	0·0	0·57	[illegible]
t × S.E.	0·93	1·01	0·95	1·05	1·07	0·68	0·0	1·18	[illegible]
95% Confidence	13·49	16·33	7·47	7·21	16·95	1·6	0·0	21·42	[illegible]
limits	11·63	14·31	5·57	5·01	14·81	0·24	0·0	19·06	[illegible]

Date: 11 December 1974

e no.	3								
ot no.	1*	2	3**	4***	5	6****	7****	8	9****
tal	306					6	0	495	0
ean ($\bar{x}$)	12·24					0·24	0·0	19·8	0·0
riance (s^2)	9·02					0·46	0·0	5·08	0·0
ndard error	0·6					0·14	0·0	0·45	0·0
< S.E.	1·24					0·29	0·0	0·93	0·0
% Confidence	13·48					0·53	0·0	20·73	0·0
imits	11·0					0·0	0·0	18·87	0·0

Date: 13 December 1974

e no.	3								
t no.	1*	2	3**	4***	5	6****	7*****	8	9*****
tal						0	0	484	0
an ($\bar{x}$)						0·0	0·0	19·36	0·0
riance (s^2)						0·0	0·0	1·99	0·0
ndard error						0·0	0·0	0·28	0·0
S.E.						0·0	0·0	0·58	0·0
% Confidence						0·0	0·0	19·94	0·0
imits						0·0	0·0	18·78	0·0

Date: 17 December 1974

no.	3								
t no.	1*	2	3**	4***	5	6****	7*****	8	9*****
al	94	190	74	74	192	25	0		1
an ($\bar{x}$)	7·23	14·62	5·69	5·69	14·77	1·0	0·0		0·04
iance (s^2)	11·53	30·26	8·90	6·06	4·53	3·16	0·0		0·04
ndard error	0·94	1·52	0·83	0·68	0·59	0·36	0·0		0·04
S.E.	2·05	3·31	1·81	1·48	1·29	0·74	0·0		0·08
% Confidence	9·26	17·93	7·5	7·07	16·05	1·74	0·0		0·12
mits	5·18	11·31	3·88	4·21	13·48	0·26	0·0		0·0

Date: 7 January 1975

no.	3								
no.	1*	2	3**	4***	5	6****	7*****	8	9*****
al	362	466	179	250	436	36	0	634	19
n ($\bar{x}$)	14·48	18·64	7·16	10·0	17·44	1·24	0·0	25·36	0·70
ance (s^2)	9·84	12·82	8·06	15·33	4·84	4·77	0·0	20·16	2·52
dard error	0·63	0·72	0·57	0·78	0·44	0·44	0·0	0·90	0·32
S.E.	1·3	1·49	1·18	1·61	0·91	0·91	0·0	1·86	0·66
Confidence	15·78	20·08	8·34	11·61	18·35	2·15	0·0	27·22	1·36
nits	13·18	17·25	5·98	8·39	16·53	0·33	0·0	23·5	0·04

Date: 5 February 1975

Site no.	3								
Plot no.	1*	2	3**	4***	5	6****	7*****	8	9**
Total	330	388	129	134	395	4	0	545	24
Mean ($\bar{x}$)	13·2	15·52	5·16	5·36	15·8	0·16	0	21·8	0·
Variance (s^2)									
Standard error									
$t \times$ S.E.									
95% Confidence limits									

Date: 7 February 1975

Site no.	3								
Plot no.	1*	2	3**	4***	5	6****	7*****	8	9*
Total	178	212	57	69	203	6	0	267	0
Mean ($\bar{x}$)	13·69	16·34	4·38	5·3	15·61	0·24	0	20·53	0
Variance (s^2)									
Standard error									
$t \times$ S.E.									
95% Confidence limits									

APPENDIX IV

Particle Size Analysis Data

Sample no.	Aperture size (μm)	Weight retained		% Weight	
		Fractional	Cumulative	Fractional	Cumulative
1	>1400	0·0	0·0		
	1400	0·0	0·0		
	710	0·0	0·0		
	500	0·01	0·01	0·005	0·005
	355	0·07	0·08	0·036	0·041
	250	0·27	0·37	0·151	0·192
	125	153·77	154·14	79·890	80·082
	63	37·94	192·08	19·710	99·792
	<63	0·41	192·5	0·210	100·00

Md μm	Mdϕ	$Q_1\phi$	$Q_3\phi$	Skqϕ	QDϕ
160	2·6	2·3	2·9	0·04	0·3

Sample no.	Aperture size (μm)	Weight retained		% Weight	
		Fractional	Cumulative	Fractional	Cumulative
2	>1400	0·10	0·10	0·051	0·051
	1400	0·09	0·19	0·046	0·097
	710	0·34	0·53	0·172	0·224
	500	0·98	1·51	0·496	0·720
	355	2·15	3·66	1·089	1·809
	250	9·20	12·86	4·658	6·467
	125	152·40	165·26	77·160	83·672
	63	31·35	196·61	15·872	99·544
	<63	0·90	197·51	0·455	99·999

Md μm	Mdϕ	$Q_1\phi$	$Q_3\phi$	Skqϕ	QDϕ
168	2·55	2·3	2·9	0·05	0·3

Sample no.	Aperture size (μm)	Weight retained		% Weight	
		Fractional	Cumulative	Fractional	Cumulative
3	>1400	0·01	0·01	0·005	0·005
	1400	0·01	0·02	0·005	0·010
	710	0·02	0·04	0·012	0·022
	500	0·03	0·07	0·015	0·037
	355	0·11	0·18	0·056	0·093
	250	0·69	0·87	0·352	0·445
	125	166·83	167·70	85·122	85·567
	63	26·06	193·76	13·297	98·864
	<63	2·23	195·99	1·138	100·002

Md μm	Mdϕ	$Q_1\phi$	$Q_3\phi$	Skqϕ	QDϕ
165	2·6	2·3	2·85	−0·025	0·275

Sample no.	Aperture size (μm)	Weight retained		% Weight	
		Fractional	*Cumulative*	*Fractional*	*Cumulative*
4	> 1400	0·0			
	1400	0·0			
	710	0·0			
	500	0·04	0·04	0·020	0·020
	355	0·46	0·50	0·235	0·255
	250	1·16	1·66	0·593	0·848
	125	172·91	174·57	88·358	89·206
	63	20·52	195·09	10·485	99·691
	< 63	0·62	195·71	0·317	100·008

Md μm	Mdϕ	$Q_1\phi$	$Q_3\phi$	Skqϕ	QDϕ
168	2·56	2·3	2·8	−0·01	0·25

Sample no.	Aperture size (μm)	Weight retained		% Weight	
		Fractional	*Cumulative*	*Fractional*	*Cumulative*
5	> 1400	0·0			
	1400	0·0			
	710	0·04	0·04	0·020	0·020
	500	0·04	0·08	0·020	0·040
	355	0·24	0·32	0·123	0·163
	250	2·43	2·75	1·244	1·407
	125	180·63	183·38	92·469	93·876
	63	11·25	194·63	5·759	99·635
	< 63	0·71	195·34	0·363	99·998

Md μm	Mdϕ	$Q_1\phi$	$Q_3\phi$	Skqϕ	QDϕ
170	2·55	2·25	2·75	−0·05	0·25

Sample no.	Aperture size (μm)	Weight retained		% Weight	
		Fractional	*Cumulative*	*Fractional*	*Cumulative*
6	> 1400	0·01	0·01	0·005	0·005
	1400	0·08	0·09	0·040	0·045
	710	0·14	0·23	0·071	0·116
	500	0·40	0·63	0·203	0·319
	355	2·17	2·80	1·100	1·419
	250	7·69	10·49	3·899	5·318
	125	181·46	191·95	92·009	97·327
	63	5·04	196·99	2·556	99·883
	< 63	0·23	197·22	0·117	100·00

Md μm	Mdϕ	$Q_1\phi$	$Q_3\phi$	Skqϕ	QDϕ
170	2·55	2·25	2·85	0·0	0·3

Sample no.	Aperture size (μm)	Weight retained		% Weight	
		Fractional	*Cumulative*	*Fractional*	*Cumulative*
7	> 1400	0·0			
	1400	0·06	0·06	0·31	0·031
	710	0·22	0·28	0·113	0·144
	500	0·47	0·75	0·242	0·386
	355	3·20	3·95	1·646	2·032
	250	11·06	15·01	5·688	7·720
	125	171·36	186·31	88·130	95·850
	63	7·55	193·92	3·883	99·733
	< 63	0·52	194·44	0·267	100·00

Md μm	Mdϕ	$Q_1\phi$	$Q_3\phi$	Skqϕ	QDϕ
175	2·5	2·2	2·75	−0·025	0·275

Sample no.	Aperture size (μm)	Weight retained		% Weight	
		Fractional	*Cumulative*	*Fractional*	*Cumulative*
8	> 1400	0·03	0·03	0·015	0·015
	1400	0·06	0·09	0·030	0·045
	710	0·09	0·18	0·045	0·090
	500	0·29	0·47	0·146	0·236
	355	0·96	1·43	0·483	0·719
	250	8·76	10·19	4·410	5·129
	125	184·42	194·61	92·841	97·970
	63	3·77	198·38	1·898	99·868
	< 63	0·26	198·64	0·131	99·999

Md μm	Mdϕ	$Q_1\phi$	$Q_3\phi$	Skqϕ	QDϕ
175	2·5	2·2	2·75	−0·025	0·275

Sample no.	Aperture size (μm)	Weight retained		% Weight	
		Fractional	*Cumulative*	*Fractional*	*Cumulative*
9	> 1400	0·0			
	1400	0·0			
	710	0·01	0·01	0·005	0·005
	500	0·03	0·04	0·015	0·020
	355	0·10	0·14	0·051	0·071
	250	0·36	0·50	0·183	0·254
	125	173·59	174·09	88·422	88·676
	63	21·88	195·97	11·145	99·821
	< 63	0·35	196·32	0·178	99·999

Md μm	Mdϕ	$Q_1\phi$	$Q_3\phi$	Skqϕ	QDϕ
167	2·55	2·25	2·8	−0·025	0·275

Sample no.	Aperture size (μm)	*Weight retained*		*% Weight*	
		Fractional	*Cumulative*	*Fractional*	*Cumulative*
10	> 1400	0·0			
	1400	0·01	0·01	0·005	0·005
	710	0·03	0·04	0·016	0·021
	500	0·09	0·13	0·047	0·068
	355	1·03	1·16	0·533	0·601
	250	5·45	6·61	2·821	3·422
	125	168·09	174·70	86·994	90·416
	63	18·29	192·99	9·466	99·882
	< 63	0·23	193·22	0·119	100·001

Md μm	Mdϕ	$Q_1\phi$	$Q_3\phi$	Skqϕ	QDϕ
170	2·55	2·25	2·8	0·025	0·275

APPENDIX V

Data from Hydrocarbon Analysis of Substrate

Site	Plot no.	Depth (cm)	Date	Time	Transmission	Optical density	ppm in solvent	wt in 20 g sample
			Sprayed with 0·2 litre Kuwait Crude Oil/m²					
2	6	0–1	22.5.74	+1 h	0·2	3·7	130	19·5
		10–11	22.5.74	+1 h	74	0·13	5	0·75
		0–1	23.5.74	+1 day	0·12	2·9	89	13·75
		10–11	23.5.74	+1 day	81	0·08	3	0·45
		0–1	25.5.74	+3 days	0·43	2·4	82	12·30
		10–11	25.5.74	+3 days	88	0·06	2·5	0·30
		0–1	30.5.74	+8 days	46·5	0·33	12	1·8
		10–11	30.5.74	+8 days	85	0·07	3	0·45
		0–1	5.6.74	+14 days	71	0·15	5	0·75
		10–11	5.6.74	+14 days	90	0·045	2	0·3
		0–1	19.6.74	+28 days	90	0·045	2	0·3
		10–11	19.6.74	+28 days	89	0·05	2	0·3
				Control				
2	5	0–1	22.5.74	+1 h	99·5	0·000	0	0
		10–11	22.5.74	+1 h	98	0·01	<1	<0·15
		0–1	23.5.75	+1 day	97·5	0·01	<1	0·15
		10–11	23.5.74	+1 day	96	0·016	<1	<0·15
		0–1	25.5.74	+3 days	98	0·01	<1	<0·15
		10–11	25.5.74	+3 days	98	0·01	<1	<0·15
		0–1	30.5.74	+8 days	91	0·04	2	0·3
		10–11	30.5.74	+8 days	94·5	0·025	1	0·15
		0–1	5.6.74	+14 days		No data		
		10–11	5.6.74	+14 days		No data		
		0–1	19.6.74	+28 days	94	0·025	1	0·15
		10–11	19.6.74	+28 days	99	0·005	<1	<0·15

Discussion

The Sublittoral Macrofauna of Milford Haven

Dr M. W. Holdgate (Institute of Terrestrial Ecology) asked Mr Addy to explain the choice of the Angle Bay transect for a monitoring (or perhaps more strictly, surveillance) project. He could see that repeated sampling of this transect would reveal changes, but clearly research would be needed to elucidate their cause. He did not see how the changes in Angle Bay could readily be related to events in Milford Haven as a whole. **Mr Addy** thought that Angle Bay was suitable for the study of 'natural' population changes because of the uniformity of the sediment throughout most of the bay. There was scarcely anywhere in Milford Haven safe from potential oil spillage and Angle Bay would anyway be the ideal place to assess the effects of a spill on the shallow sublittoral fauna. He agreed that the Angle Bay study should be described as a research rather than a monitoring project.

Mr A. D. McIntyre (Department of Agriculture and Fisheries for Scotland) pointed out that it was important that the methods used should be chosen carefully in relation to the objectives of the work. The 1 mm sieve used in this project to screen off the animals from the sediment would probably retain a large part of the macrobenthic biomass, but it is likely that a substantial proportion of the individual organisms will pass through. Those organisms which are lost, both young individuals and small species, may be the very specimens which would be most and soonest affected by pollution. The fact that the most sensitive individuals have not been included in the study should be clearly recognised in interpreting the results. **Mr Addy** agreed that the use of 1 mm mesh size is likely to miss many individuals both of small species and juvenile forms, and that these may indeed be more sensitive to pollution. However, the aims of the present survey were to provide a background description of the sublittoral macrobenthos of Milford Haven upon which a monitoring scheme might be based, rather than to study the relationships between pollution and community structure. In any subsequent monitoring scheme which may be developed, these relationships must be investigated, probably necessitating the use of smaller mesh sizes. It should be borne in mind that increased sensitivity of sampling can only be achieved at the expense of the area covered by the survey. He felt that the location of the reduced number of sampling stations which could be accommodated in a more sensitive sampling programme can only be decided on the basis of an overall description of the macrobenthic communities in the area of concern.

Mr A. J. O'Sullivan (Atkins Research and Development) asked how important benthic fauna was in the context of oil pollution. Since oil normally floats on the water surface, and reaches the bottom only when dispersed throughout the water column or after sinking, could it not be argued that benthic fauna are only rarely exposed to the direct effects of oil pollution? The major exception to this would be in fairly shallow waters where dispersed oil may cause damage. For example, *Ensis* was affected following the *Torrey Canyon* incident and also in Bantry Bay following the *Universe Leader* spillage of October 1974. **Dr B. Dicks** (Oil Pollution Research Unit) pointed out that spilled oil may have effects on the benthic communities at considerable depths for a number of reasons, two of which sprung to mind immediately:

1. Many benthic organisms have a planktonic larval stage in surface waters which may be affected by oil at the sea surface.
2. A number of spillages have resulted in the natural sinking of considerable quantities of oil which may affect the benthos.

Dr R. G. J. Shelton (Department of Agriculture and Fisheries for Scotland) commented that for highly fecund organisms with pelagic larvae, heavy, density-dependent mortality takes place prior to settlement. For this reason it would require pollution on a massive scale at the early larval stage to affect recruitment to the settled stage in any measurable way. Once the latter has been reached, pollution on a much smaller scale could well have significant effects.

Field Experiments on *Arenicola marina*

Dr T. E. Lester (British Petroleum Co. Ltd) asked if any experiments have been conducted in which the oil and/or dispersant has been applied to test plots as a dispersion in sea water, rather than as a continuous phase. Could Mr Levell also comment on the depth of oil and/or dispersant penetration during the experiments described in his paper? **Mr Levell** said he did not know of any field experiments using oils or dispersants applied as dispersions in sea water. This would perhaps, in some cases, be more representative of the actual spill conditions, but the practicalities of such a procedure precluded its use at Sandyhaven. The penetration of pollutants into the substrate was particularly influenced by substrate water content. In plots which were repeatedly sprayed with oil, the stabilisation of the surface layer by the oil meant a build-up of 2–3 cm of substrate which would normally be fairly mobile sand. Successive spraying added to this build-up of substrate, so after six sprayings a section of the substrate showed six oily layers, the lowest being perhaps 10–15 cm from the surface.

Refinery Effluents and Their Ecological Effects

10
Refinery Effluent Sources, Constituents and Treatments

W. G. ROBERTS
(*British Petroleum Co. Ltd*)

SUMMARY

This paper provides an introduction to the problems of pollution arising at oil refineries, with the main emphasis on effluent water. The major sources of effluent water are described, with an indication of the pollutants which are likely to occur. The treatments available and appropriate for removal of pollutants are briefly described, with comments on their effectiveness and limitations, results which can be achieved with present technology, and an approach to the setting of consent levels.

Air pollution and solid waste disposal are mentioned briefly, insofar as direct or indirect effects on the ecology of water resources may arise.

INTRODUCTION

Since the conference for which this paper is prepared is directed to the effects of oil pollution on marine ecology, atmospheric pollution is considered only in passing and there is also only brief reference to solid waste. The major topic is the avoidance of pollution due to aqueous effluents and the means used for treating waste water within refineries. The major difference between oil refinery waste waters and those from most other manufacturing processes is the continuous presence of oil, and a relatively low biological oxygen demand (BOD). The problems of inland refineries can be substantially different from those on sea-board or estuarine sites, and are mentioned to complete the picture of waste-water treatment as it affects the industry. The statements and comments made are deliberately not confined to UK conditions and practices; the problems are similar world-wide.

AIR POLLUTION

The major atmospheric emissions from refineries come from chimney stacks on process furnaces and large boilers for steam raising. These are similar to those encountered in other industries, and are subject to the same controls on construction, use and quality of fuel (especially in respect of sulphur content). Provided all appropriate regulations and guidance rules are observed to achieve good dispersion of stack gases, there should be no ecological effect of any significance on local waters. The problems of long-range transportation of sulphur and possible acidification of inland waters are still far from being fully understood, let alone resolved. In any event such matters are the concern of all fuel-using industries; any further controls which might be found necessary will apply to oil refineries among many others.

Small quantities of hydrocarbons and odorous material may emanate from refineries. No refiner can afford uncontrolled hydrocarbon loss, to produce a persistent odour nuisance, or to be guilty of unsafe practices. Accordingly hydrocarbon vapour and odour emissions are always under scrutiny and the levels which do occur from refineries are exceedingly unlikely to inflict any damage on the environment, either on land or marine.

EFFLUENT WATER

There are a variety of sources of effluent water in any refinery, though all the sources described below are not necessarily present at any one site. It is convenient to consider these separately, not merely for descriptive convenience, but because their different properties and flow-rates suggest different degrees or kinds of treatment. Wholly or partially segregated systems are commonly used to effect the most economic overall disposal. The aqueous effluent streams which may have to be considered include:

1. Clean surface drainage (storm water).
2. Domestic sewage.
3. Cooling water.
4. Oily water from process and miscellaneous sources.
5. Tanker ballast water and tanker wash water.

Clean Surface Drainage

Most of the rain falling on a refinery will fall on areas which are, or should be, free of oil or other potential pollutants. These areas include the roads, administrative areas, and the greater part of tank farms and pipe tracks. Completely undeveloped areas, free of any risk of contamination, may be drained without treatment to natural watercourses. For the remainder of the clean water, it is usual modern practice to provide a final guard basin with an inverted weir to protect against the possibility of accidental oil spillage or other emergency. The bunded areas in tank farms should always be provided with valved outlets, and before accumulated surface water is drained, the operator must check visually that the water is oil-free. If it is not, then the water must be collected separately or diverted for disposal through the oily water system. Segregation of all clean storm water from water which may regularly contain oil or other contaminants is obviously essential, since flows of storm water can be very large, are essentially intermittent, and any treatment facility must be sized for the maximum short-term flow rate expected.

Domestic Sewage

Domestic sewage from washrooms, canteens, etc., is best disposed of by connection to a public sewer system if such is available, but an isolated site may require its own septic tank or other treatment plant.

Cooling Water

Water cooling for power and process plant can be used only where there is an adequate supply of water, either for once-through use, or for make-up to a recirculating system with cooling towers. Most recent refineries rely mainly on air cooling and have very little water effluent in this category. Where recirculating systems are used, there will be 'blow-down', on an intermittent or continuous basis, of a small part of the total circulating flow to prevent excessive concentration of hardness salts and other dissolved or suspended solids; this blow-down water is normally handled through the refinery oily water system (see below).

Large separate treatment facilities are required to deal with once-through cooling water systems. These are usually found only at coastal refinery sites or on lower reaches of large rivers, particularly at refineries of older construction, and they use salt or brackish water. The flow-rate for a large refinery can be as much as 30 000 m^3/h. The cooling water leaving heat exchangers should be oil-free, but leaks sometimes develop from a variety of causes, and small quantities of oil then appear in the water until such time as it is possible to trace the offending equipment and take it out of service for repair. The cooling water is therefore directed through some form of gravity separator, commonly of the API or similar type (API, 1969). The principles of such separators are discussed below.

Oily Water, Miscellaneous Sources

Oily, or potentially oily, water can accumulate from a variety of sources, including:

(a) Storm water drainage from process units and the immediate surroundings of tanks and pump houses, plus water drained from tanks.
(b) Cooling water from equipment such as pump glands, and from blow-down of recirculating systems.
(c) Process water and chemical drainings.

Items (a) and (b) are unlikely to be significant contributors of contaminants other than oil. It may seem odd to the layman that water has to be drained from oil tanks, but it can appear from several sources in small amounts. Crude oil can contain water all the way from the well, or may pick up small amounts on board tankers. Many products are steam stripped or have aqueous washes during refining and the last trace of water is not separated completely before transfer to tankage. All the more volatile oils are stored in floating roof tanks, and roof drainage can only be handled internally, so that the water may pick up some oil.

It is usual practice to combine all three types of oily water streams mentioned for final treatment, but before describing these processes one must touch on the third item in the list, for this is a major source of such problems as we have with aqueous effluents. The non-oil contaminants which appear, and in particular the total quantities present, depend very much on the types of crude oil processed and the choice of processes for conversion and other post-distillation treatment of products. We are likely to find any or all of the following: sulphides, ammonia, organic acids, phenols, sodium chloride and other inorganic salts from crude oil tankage and desalting, trace sodium hydroxide carried over from washes for removal of sulphide, copper salts from copper chloride sweetening, traces of solvents from lubricating oil treatment—e.g. furfural, ketones and chlorinated hycrocarbons—and occasionally detergents when these have to be used for process unit cleaning. Cooling water and boiler blow-down, and effluent from boiler water treatment may contain, in addition to substances natural to the raw water, added chemicals such as chromates, phosphates and biocides. The most commonly used biocide is chlorine; the dosage used is always the minimum for effective control of unwanted organisms, and the residual quantities in waste water are small enough to be innocuous. Blow-down from boilers should be diverted to the clean water system, since although the total quantities of chemicals present are not large, they can cause flocculation in gravity separators and lead to oil carry-over. These process and chemical contaminants can be dealt with in three phases, the second and third being common to all oily water flows: (a) pretreatment or reuse; (b) primary oily removal; (c) secondary treatments.

'Sour water' containing high concentrations of ammonia and/or hydrogen sulphide is often stripped by steam or by flue- or hydrocarbon-gases to reduce the quantities of these contaminants, and the off-gases incinerated in the process furnaces. Such water appears mainly from distillation sections of the plant, originating from traces of water in feedstock, water added to aid preliminary desalting, and from steam injection to improve fractionation of products. Stripped sour water can be used as make up to desalters or may be passed to the main oily water system where its residual contamination can be dealt with.

Spent caustic soda solutions (usually below 3% NaOH, but containing sulphides and mercaptides and possibly phenolic material, as well as some oil) are reused as far as possible for corrosion control procedures and for desalted pH control on crude oil distillation units. Reuse has obvious economic advantages, and limits the increase in final effluent pH attendant on presence of strong alkali (direct neutralisation with acid has a serious odour problem from release of sulphur-compounds). Means of dealing with the sulphur-bodies from concentrated streams also include direct combustion on a fluidised bed, air oxidation to sulphate and thiosulphate, and flue gas stripping, all of which have operational problems. After any necessary pretreatment, process and chemical wastes rejoin other oily water for the next stage.

The oil removal stage is invariably some form of gravity separator. Gravity separators are designed to allow oil globules of 0·15 mm and greater to be separated from the largest water flow to be treated. The oil content of separated water therefore depends on the number of droplets below this size in the incoming water to the separator.

Pumping tends to disperse oil as very fine droplets; it is therefore important to feed the separator by gravity drainage alone if at all possible. In a modern refinery with gravity drainage, the process effluent after gravity separation usually averages about 25 ppm total oil (including soluble organic materials which show up as oil in the test), but where oily water has been pumped, the figure may be up to twice this. The precise ppm figure is normally less important than whether any remaining oil would cause visible or surface pollution to the receiving water. The basic nature of gravity separation usually affords a good safeguard in this respect.

Suspended solids from such sources as boiler blow-down, and discharges of used alkali which may precipitate such solids, can weigh down oil droplets and carry them through a gravity separator, so such additions to the process water system should be avoided. Bacteria can also flocculate oil, and for this reason domestic sewage should also be excluded from oily waste waters. The volume of oily process water to be treated is typically less than 0·5 ton per ton of crude oil capacity. A once-through cooling water flow may be 20 to 50 times as much.

This difference in volume, plus the difference in likely quality before treatment, usually dictates segregated handling of these two classes of waste water.

We are now at the stage where we have an effluent water which has had the larger part of the oil, and at least part of the other major pollutants, eliminated. This effluent may be of acceptable quality for discharge to the sea, but definitely not good enough for inland waters nor for the more strictly controlled coastal sites. The BOD_5 is typically in the range 100–200 mg/litre for process effluent, quite low by comparison with other industrial effluents. Nearly all the likely contaminants are biodegradable, but some of the remaining oil is likely to be of a type which is relatively slowly oxidised by microorganisms and presence of oil reduces effectiveness of removal of other contaminants. The first part of any secondary treatment should therefore be to reduce oil content still further, preferably to around 10 ppm. Reduction of oil in gravity separator effluent can be achieved by the following processes:

(a) Dissolved air flotation. If used in conjunction with flocculation by polyelectrolyte and/or aluminium sulphate or ferric chloride with pH control, oil contents down to about 15 ppm can be realised.
(b) Flocculation plus sedimentation gives somewhat better results, given adequate settling capacity.

The process selected will depend on the level of oil content which is finally required, and the possible combination with a biological treatment. Large river and upper estuary sites now often require oil contents in the range 5–20 ppm max, the most rigorously controlled inland refineries are required to meet oil levels of less than 5 ppm; both categories will be likely to need biological treatment, especially since BOD levels will also be specified. Refineries in the least sensitive sites may have BOD consent levels which can be met quite consistently without secondary treatment, but the newer developments have commonly to cope with limits of about 20 ppm max, or even substantially lower at some inland sites. Consent levels of BOD are, however, mostly set somewhere in the range 20–50 ppm, according to the type of site involved. Where the quality of effluent from the oily water system after primary treatment demands it, refiners install biological treatment plant, usually one of three types—activated sludge, some form of oxidation lagoon, or percolating filters.

Activated sludge systems have been used in the past, but they have certain drawbacks for refinery duty. The settling tank must be sized for maximum flow conditions, and a refinery flow-rate can vary quite widely. As the BOD of refinery waste water is relatively low, the rate of growth of the sludge is also low, and the sludge will have a relatively high oil content. There is thus not only a difficulty in settling, but also some danger of depletion of the mixed liquor suspended solids if losses in the settling tank exceed the rate of production of excess sludge. To overcome these difficulties it is usual to combine activated sludge with flocculation using iron salts. The ferric hydroxide formed adsorbs oil, forming a convenient structure on which bacteria can grow and also weighting the floc to give good settling. With care, such plants can give a good-quality effluent. The excess sludge, however, containing bacteria, ferric hydroxide and oil, can amount to over 1 ton per hour even after thickening. Because of its oily nature, this sludge must be incinerated. Incinerators are expensive to install, operate and maintain and consume appreciable quantities of auxiliary fuel, besides producing a flue gas which may itself require treatment to meet air pollution standards.

Oxidation lagoons are widely used where an extent of suitable land is available and the system may include mechanical aeration. Sludging by a mixture of humus and settled inorganic matter must occur, and a long-term problem of solids disposal is inevitable. Some refineries use such lagoons only as a tertiary 'polishing' treatment after other biological plant, and the lagoon then also acts as a safety feature in case of a failure elsewhere.

The disadvantages and cost of the conventional combined flocculation/activated sludge treatment are overcome in a new process (BP) based on percolating filtration. Insoluble oil remaining after gravity separation is recovered, and by operating under virtually oil-free conditions, the process makes use of the advantages of modern percolating filters. Because of the low starting BOD of refinery waste waters, a good-quality final effluent can be produced without the accumulation of oily excess sludge, and thus without the need for an incinerator.

An alternative to biological treatment is the use of activated carbon, but it is expensive to install and operate—unacceptably so for a normal refinery effluent readily treatable by other means. It may, however, be useful for petrochemical plant wastes containing materials such as chlorinated hydrocarbons, or as a final 'polishing' treatment to meet especially stringent effluent quality requirements.

Fortunately, biological treatment processes are fairly robust. Many substances which are poisonous in the everyday sense of the word are satisfactorily dealt with at the concentrations likely to occur in a normal refinery effluent. This certainly applies to all hydrocarbons (including the lighter aromatics), phenols, carboxylic acids, cyanides and sulphides. Sudden heavy loadings of oil or chemical wastes such as spent soda can be damaging, at least temporarily; where such problems seem likely to arise they can often be avoided by care in plant housekeeping or by regulated mixing and holding of contaminated water upstream of the biological treatment unit. Special care may be needed where chlorinated hydrocarbon solvents are used. Heavy metals must also be handled warily; copper and zinc, and to a lesser extent chromium, are potentially damaging, both to one's own plant and to organisms in the water ultimately receiving the effluent. Where a significant hazard is known to exist, water authorities may impose consent levels for individual toxic chemicals or metals. It should be emphasised that in oil refinery operations we do not encounter the more toxic and persistent metals such as mercury and cadmium. Zinc and chromium may be present, but are usually precipitated due to the alkalinity of most refinery waste waters.

Tanker Ballast

Tanker ballast water and wash waters have to be cleaned up ashore, after discharge from ships which have made short voyages or which have made their journeys to the refinery entirely in restricted waters where no oily discharges are permitted.

It is usual to provide large reception tanks in which primary gravity separation of oil takes place. The oil is subsequently dewatered and recovered for reuse, and the water containing small amounts of oil passes to the refinery oily water system. Geographical layout sometimes dictates a separate treatment facility in the dock or jetty area, but the processes involved are as described above to reduce oil content to the locally permitted limits for discharge. Normally, nothing other than oil will be present, but if the incoming tanker has washed its tanks, persistent emulsions can be found which are likely to need chemical assistance in order to break. Some refineries have found it necessary to refuse tank washings including detergents, to avoid the risk of emulsion formation in their own effluent. Many detergents, both for ship and shore use, contain aromatic solvents which are highly toxic to marine life, and the use of such material is to be discouraged on these grounds alone.

Recovered Oil

The fate of the recovered oil need not concern us here. Suffice it to say that it is always wet, often partly emulsified, and requires further settling, with the remaining water recycling to the separators. The oil can then be reprocessed or absorbed in fuel oil. One special point should be noted: in areas where leaded motor spirit spills may occur it is usual to try to keep this material separate and pump the gasoline, which separates very readily from water, direct from collecting sumps back to tankage. This minimises the lead hazard and eliminates certain very serious difficulties in reprocessing.

SOLID WASTES

The main problem with solid and semi-solid wastes is to ensure that potentially damaging materials cannot enter watercourses or ground water in significant quantities. To this end most governments exert increasingly close control over all solid disposal, and particularly on its transport and on tipping sites. Oil refiners may use their own selected sites, or, particularly for the more highly toxic and indestructible materials, make use of commercial firms who operate facilities where the risk of water pollution is known to be minimal.

Types of solid waste peculiar to the oil industry include oily sludge from tank cleaning (special rules drawn up by Associated Octel Co. apply to material which may contain traces of lead anti-knock additives) and from the operation of aqueous effluent treatments. Oil contents may be quite low, perhaps only a few per cent, and the proportion of water and inorganic solids possibly high. Incineration is an obvious first step for oily material, but it can be technically difficult, and is costly both in equipment and in auxiliary fuel. Spent catalysts are a problem mainly on account of their heavy metals content, which may include compounds of copper, cobalt and molybdenum. On both economic and environmental grounds there is increasing interest in recovering the metals for reuse (this has always been done with precious-metal catalysts), but processes are not yet universally available, particularly for small intermittent supplies, and some controlled dumping is still unavoidable.

ECOLOGICAL CONSIDERATIONS AND CONTROL LIMITS ON WASTE WATER

The acceptability (or otherwise) of substances emitted to the environment may be judged on the following grounds, applicable to all emissions, though here we are primarily concerned with water as the receiving element and with emissions of a particular industry.

(a) The emission may offer a hazard to human life or health.
(b) There are ecological effects and hazards to other animal and plant life.
(c) It is unpleasant to humans by odour, taste or tainting.
(d) The emission may be aesthetically unpleasing.

The first of these points is obviously critically important, the second and third only slightly less so; the fourth is not likely to give trouble if the others have been properly considered.

All responsible organisations wish to avoid offending against any of these criteria, and to do so they must apply technically sound limitations to the quality of industrial effluents. These limitations should be relevant to the nature and uses of the receiving water and take account of its capacity for dilution. The temptation to apply universal quality control limits to effluents, regardless of their volume relative to the receiving water, because they are thus easier to administer or in the interests of 'harmonisation', should be resisted as inappropriate. It can also be unnecessarily costly in community resources. To be meaningful, such general rules would have to be written to protect the most sensitive locations; in the case of water, to safeguard small rivers used as drinking water sources. Such limits would be neither necessary, nor even meaningful, for a discharge to open sea water. Equally, it is useless to apply stringent controls to one emitter unless others using the same body of water are also controlled and the rules for all take account of natural background in an agreed and concerted scheme to safeguard overall quality. The tests used in applying controls to aqueous effluents from oil refineries may include the following (approximately in order of frequency with which they are stipulated, which is not necessarily the order of importance at any one site):

Oil content
pH
BOD 5-day
Temperature
Suspended solids
Phenols
Sulphides
Ammonia

Other parameters may be included to meet the requirements of a particular site. In all cases, since the tests are apt to be empirical, in the sense that they are chemical and not directly ecological, the test method must be specified along with the permitted limits. All specified methods of test should be in widespread use and be meaningful in respect of both the parameters measured and the limits set.

Direct toxicity tests are applied in some countries. They have serious difficulties of operation and interpretation. They are very imprecise owing to the wide divergencies in toxicity within and between species, and interpretation of limits is bedevilled by the interactions, mostly little understood, between potential poisons, substances naturally occurring in the water, and possible test organisms. Such tests may be better used to try to establish some correlation between overall toxicity and a workable set of 'chemical' tests, bearing in mind the intended use of the receiving water and the species expected to be present.

Numerical limits for tests, individually or in relation to one another, are too wide a subject for proper discussion in a review paper. In many cases they have to be set somewhat arbitrarily, often with a view to best practical means of control at a given time and place. This indeed is probably the correct approach, provided limits have been set with full consideration of all the factors involved. It does, however, lead to obvious difficulties in trying to compare sites unless all the facts are available.

Since it distinguishes refineries from other industries we return to oil content, where most consistency on consent levels has emerged, depending on the type of water to which the discharge is made. The toxicity of oil itself is reasonably low apart from products containing high concentrations of the lighter aromatic compounds. The situation is alleviated by the fact that the lower boiling point fractions are readily lost by evaporation and most susceptible to biological oxidation. Visible oil slicks should not occur at any refinery outfall and the risk of smothering any plant or animal life is accordingly small. Present available treatment methods are adequate to prevent gross damage at any appreciable distance from an outfall. The best ultimate measure of our success or failure is what happens in and beside the water after discharge, and biological monitoring should always be used to back up the day-to-day control measures used by the refinery chemist.

The oil industry accepts the necessity for control over the environmental consequences of its activities. Indeed, much has been done over the years on a voluntary basis, apart from the need to meet specific legislation. Increasingly the controls are mandatory. As citizens, we cannot object, provided we can contribute to a balanced view of the potential dangers, the techniques available to overcome them, the pertinence of particular limitations to our operations and the site under consideration, and the cost to industry and the community.

ACKNOWLEDGEMENTS

Permission for presentation of this paper has been given by the British Petroleum Company Ltd. The author is grateful for the assistance of colleagues and particularly to Dr G. F. Oldham whose unpublished work has been drawn on as well as his 1972 paper.

REFERENCES

API (1969). *Manual on Disposal of Refinery Wastes: Volume on Liquid Wastes*, American Petroleum Institute, New York.

Oldham, G. F. (1972). *Chemical Engineer*, Nov., pp. 418–22.

11
Investigation of Refinery Effluent Effects Through Field Surveys

JENIFER M. BAKER

(Oil Pollution Research Unit, Orielton Field Centre, Pembroke, Wales)

SUMMARY

This paper describes a number of surveys carried out in refinery effluent discharge areas in order to find the species and areas affected. Effluent constituents and volumes vary, and discharge areas range from rocky shores to freshwater swamps. It is possible, however, to classify effluents according to their ecological effects, and the effluent dispersal characteristics form the best basis for such a classification. Well dispersed offshore discharges from modern refineries may not cause any measurable biological changes; in contrast, badly dispersed discharges, for example in mud flat or saltmarsh areas, may greatly reduce the numbers and abundance of species near the effluent. Both salt and freshwater marsh vegetation are capable, however, of surviving at least some chronic oil pollution, and the deliberate management of such communities as a final stage in effluent treatment is an interesting possibility.

INTRODUCTION

Recent estimates (e.g. Jeffery, 1972) indicate that at least as much oil reaches the sea from refinery and petrochemical plant effluents as from tanker and other accidents. Extensive literature is available on the biological effects of oil spillages, but not on the biological effects of oil industry effluents. In the first case, a large amount of oil affects a particular area for a relatively short time. In the second case, the area near an effluent is subjected to continuous but low concentrations of oil and other compounds for a long time. Effluent quality standards are set by regulatory organisations, but are usually arbitrary in the sense that biologically acceptable limits for effluents in different situations have not been worked out. It is important to consider this problem, however, for two reasons. First, new oil industry developments usually involve discharges of ballast water or refinery effluent, often in areas of high biological interest with no previous industrial history. Second, modifications to existing installations are sometimes planned for environmental reasons. In both these cases it is necessary to arrive at some sort of balance between biological and economic considerations. There is now general agreement about the necessity for safeguarding estuarine and coastal flora and fauna; on the other hand, production of very high quality effluents may in some cases involve considerable expenditure with little or no biological benefits.

The effects of effluents depend on many factors other than the obvious ones of effluent constituents and volume. The siting of the outfall, the type of receiving area (rock, mud, sand or saltmarsh) and its associated community of plants and animals, and the movements and quality of the receiving water, all have to be considered.

The primary line of approach used for this work is field surveys of the areas round effluent discharges. These are to find if changes have occurred, and if so, the species and area affected. This paper describes several such surveys and comes to some conclusions about what they show.

LITERATURE REVIEW

The limited published data is summarised in Table I. Data previously published by the Oil Pollution Research Unit are not included in this table, as they are updated and expanded below.

THE MILFORD HAVEN REFINERY EFFLUENTS

There are now four refinery effluents in Milford Haven. Two are normally discharged from jetties, one from a small headland (Wear Point) and one into a small bay (Little Wick Bay). On two occasions, during power crises, one of the jetty discharges has been diverted down a saltmarsh creek (Martinshaven). As the refineries use air cooling, effluent volumes are relatively small (1·5–4·5 million gallons = 6·75–20·25 million litres per day). The water authority quality conditions are given in Table II.

Biological effects have been observed in Little Wick Bay and Martinshaven. Effects have not been observed with the other two effluents.

Little Wick Bay

This area was first described by Crapp (1970, 1971). A photograph of Little Wick taken before industrial development began (reproduced in a Milford Haven Conservancy Board booklet of 1968) shows that the shore was then dominated by limpets and barnacles; however, when Crapp visited it in 1969, the dominant intertidal species was the bladder wrack *Fucus vesiculosus*. A series of six transects (Fig. 1) was established between the older South Hook Point and Gelliswick Bay monitoring sites, in order to study possible effluent effects in more detail. Crapp concluded that:

(1) There is a change in exposure from South Hook Point (grade 3) to Gelliswick (grade 6). These exposure grades are described by Ballantine (1961).

TABLE I

Refinery Effluent Effects: Summary of Data from the Literature

No.	*Effluent characteristics*	*Outfall site and receiving area*	*Biological observations*	*Reference*
1	Effluent from a 0·57 million tonnes per year capacity refinery	Georgia Strait U.S.A., 320 m offshore, 6–9 m below low water	Annual surveys 1954–63 show no effects except absence of winkles (*Littorina planaxis*)	Oglesby and Sylvester (1965)
2	1·25 million litres effluent per day: Min. Mean Max. pH 7·2 8·6 9·5 COD 30 200 475 Oil 2 7 35 Phenols 0 1 2 Amm. N. 5 35 75 (Approx. figures ppm)	Table Bay, S. Africa, 450 m offshore, 7·3 m below low water. Diffuser on pipe end. Current parallel to coast	Barnacle growth on diffuser. No evidence of any effect on bottom fauna at 7 stations	Steere (1970)
3	Oil industry wastes totalling 230 million litres per day	Los Angeles Harbour, U.S.A. Other industrial and domestic wastes	One main bottom species, the polychaete *Capitella capitata*. Removal of oily bottom waste by dredging temporarily increased the number of species	Reish (1965)
4	Min. Max. Temp. 11·8 22·5°C pH 7·2 7·55 COD 98 116 ppm ($KMnO_4$) Oil 0 102·1 ppm (one abnormal figure of 1830 ppm) Phenols 0 1·9 ppm Amm. N. 40·8 74 ppm	North Sea coast, Germany, Creek in salt-marsh and mudflat system	Accumulation of oil in creek mud. Blue green algae and *Nereis* now dominate creeks near discharge	König (1968)

TABLE II

Water Authority Quality Conditions for Refinery Effluents in Milford Haven

Temperature	30°C
pH	5–9
suspended solids	50 ppm
COD ($KMnO_4$)	25 ppm
Oil	25 ppm
Total phenols	3 ppm
Amm. N.	4–6 ppm
Cyanide (as HCN)	0·1 ppm
Sulphide (as H_2S)	1 ppm
Copper	0·3 ppm

(2) Several species of gastropod molluscs were absent or reduced in numbers on the central transects, and this could not be explained in terms of natural environmental factors. *Littorina saxatilis rudis* and *Littorina littoralis* appeared to be unaffected in this way.

(3) Limpets (*Patella vulgata*) and barnacles (*Chthamalus stellatus, Balanus balanoides* and *Elminius modestus* were considerably reduced in numbers on transects 3 and 4.

(4) The seaweeds *Fucus serratus, F. vesiculosus*, and *F. spiralis* were particularly abundant on central transects. *Pelvetia canaliculata* and supralittoral lichens did not appear to be affected.

Crapp's hydrographic studies showed that surface currents away from the outfall are slow, except in the early stages of the ebb and the flood. It is probable that the siting of the outfall leads to some retention of effluent in the bay. However, on a few occasions very thin oil films have been observed moving from the effluent as far as Gelliswick Bay or South Hook Point following the shore. The effluent is thus implicated in the biological gradient of effects observed.

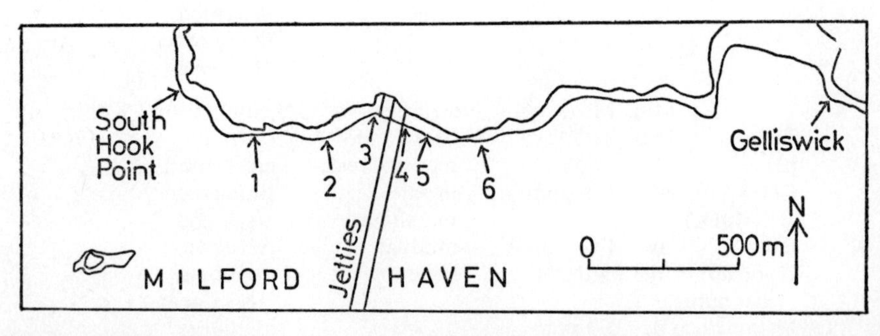

FIG. 1. Positions of South Hook Point, Gelliswick, and Little Wick 1–6 transects, Milford Haven.

TABLE III

Transect Survey Dates

	South Hook Point	*1*	*2*	*3*	*4*	*5*	*6*	*Gellis-wick Bay*
First survey by G. Crapp	4.3.69	21.6.70	6.3.70	28.3.70	9.3.70	8.3.70	20.6.70	30.7.69
Second survey by C. Howe	22.6.74	18.6.74	7.6.74	6.6.74	4.6.74	19.6.74	20.6.74	19.7.74

The transects were re-surveyed by Catharine Howe during 1974, and results for some common species are given in Figs. 2 and 3. Survey dates are given in Table III.

The 1974 results showed that:

(1) The numbers and distribution of some gastropod molluscs correspon ded closely with the earlier data, i.e. *Littorina neritoides*, *L. saxatilis neglecta*, *Monodonta lineata* and *Gibbula umbilicalis* are still absent or rare on central transects. *Littorina saxatilis rudis* appears to have extended its range on some transects, but is otherwise not affected. *Littorina littoralis* numbers are reduced; this correlates well with a reduction of seaweed on some transects.

(2) Results for *Patella vulgata*, *Chthamalus stellatus* and *Elminius modestus* correspond closely with the earlier data, except that the decrease of *Patella* on transect 5 may be significant. There is an increase of *Balanus balanoides* on several transects which could be explained in terms of a good 1974 spat fall; this increase is, however, insignificant on transect 3 nearest the effluent discharge.

(3) There is a decrease of *Fucus vesiculosus* on transects 4 and 5 and a decrease of *Fucus serratus* on transect 5. This does not correlate with increases of limpet or other grazing mollusc abundance, and is difficult to explain. Possibly at the time of the earlier survey *Fucus vesiculosus* was abundant following a localised oil spill, but many young limpets had re-established. Subsequent death of old *Fucus* plants and prevention of re-establishment by the grazing activity of the growing limpets would give the low abundance for *Fucus* recorded on the later survey.

(4) Results for *Pelvetia canaliculata* and supralittoral lichens do not differ greatly from the earlier survey.

The general conclusion, then, is that there have been some changes involving *Littorina saxatilis rudis*, *Littorina littoralis*, *Balanus balanoides*, *Fucus vesiculosus* and *Fucus serratus*. Some of these are as yet unexplained, but there is no obvious increase of the area apparently affected by the effluent, or further decrease in numbers of species within the area already affected.

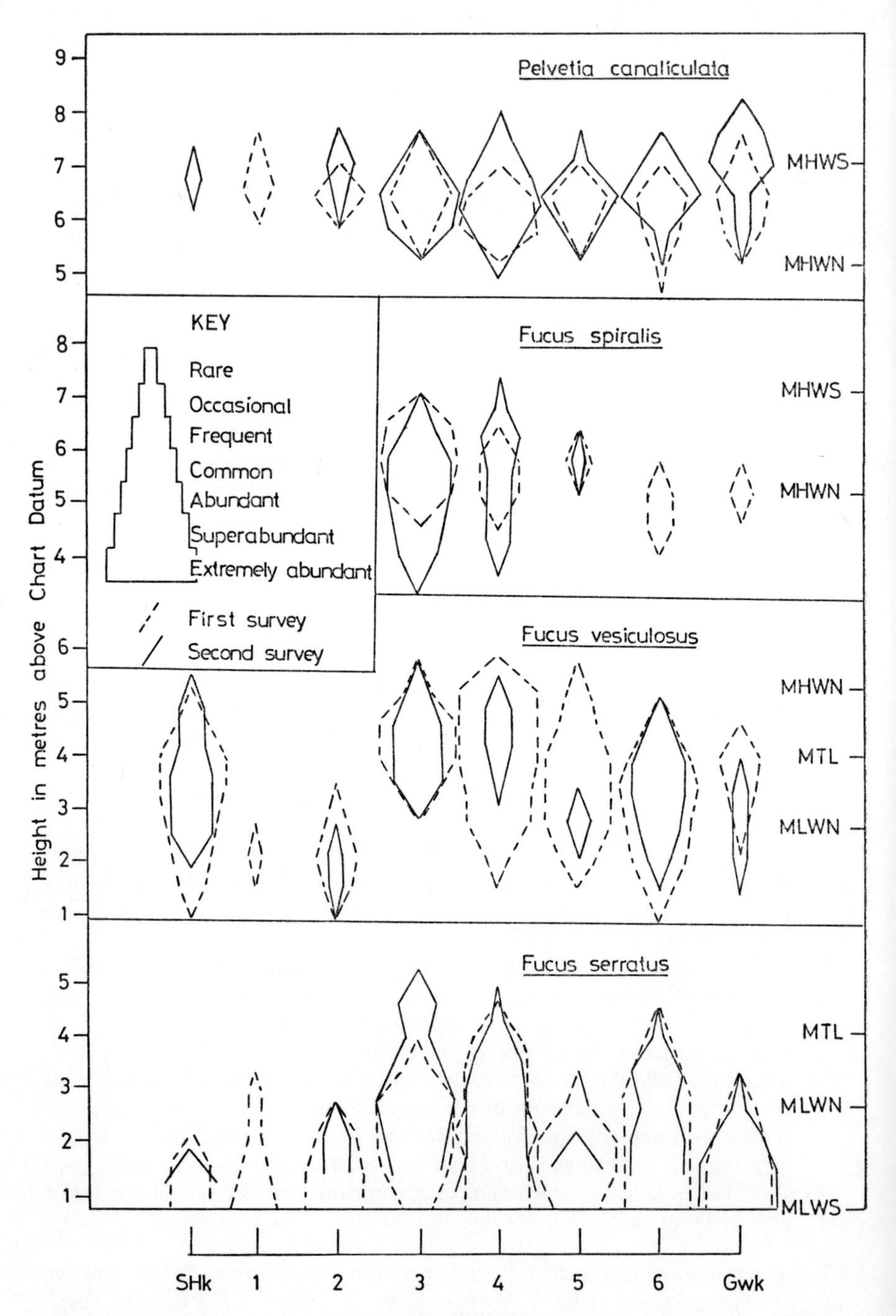

FIG. 2. Transect survey results: some common seaweeds. Abundance scales as in Baker (1975a).

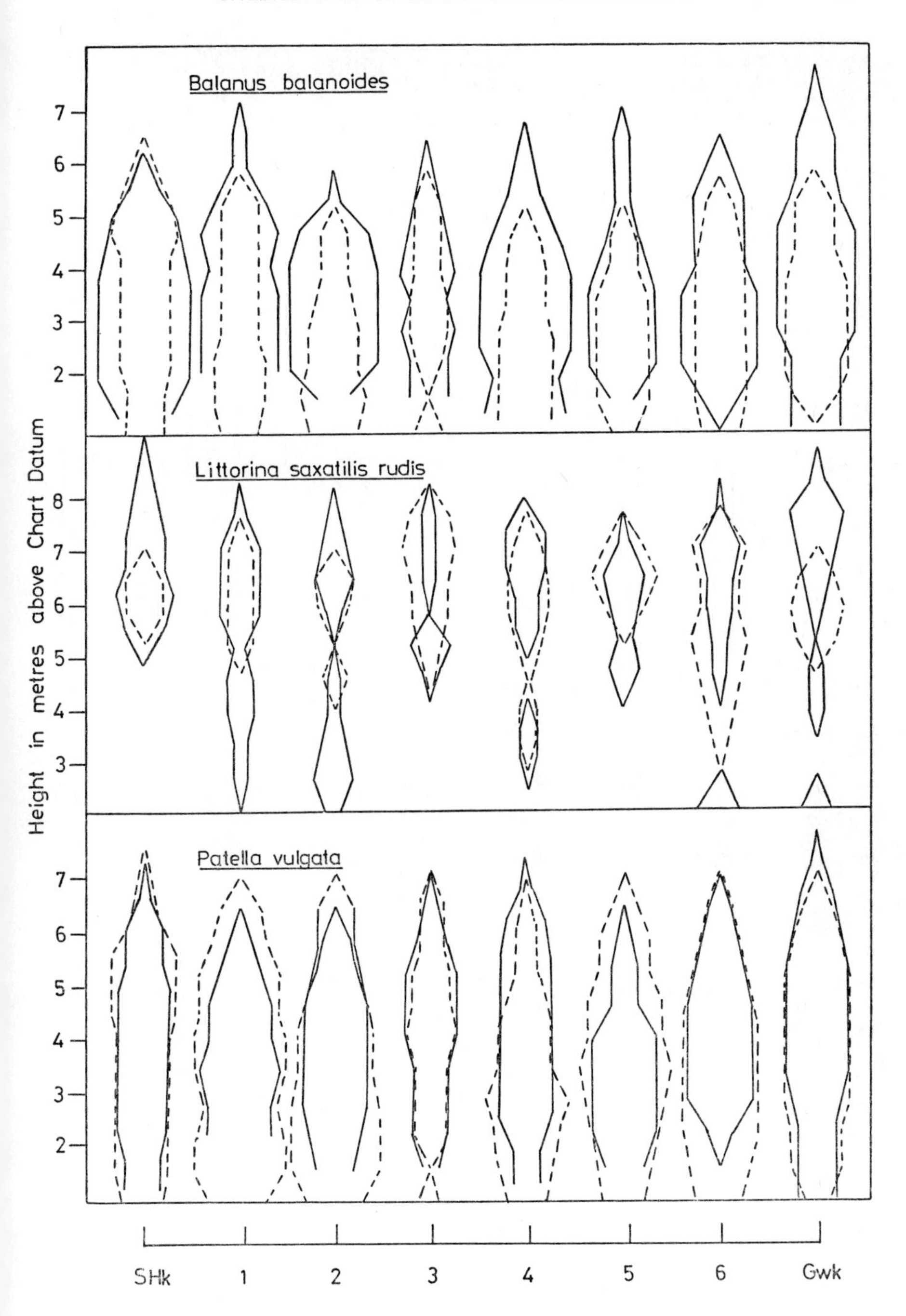

FIG. 3. Transect survey results: some common animals (for key see Fig. 2: abundance scales as in Baker (1957a)).

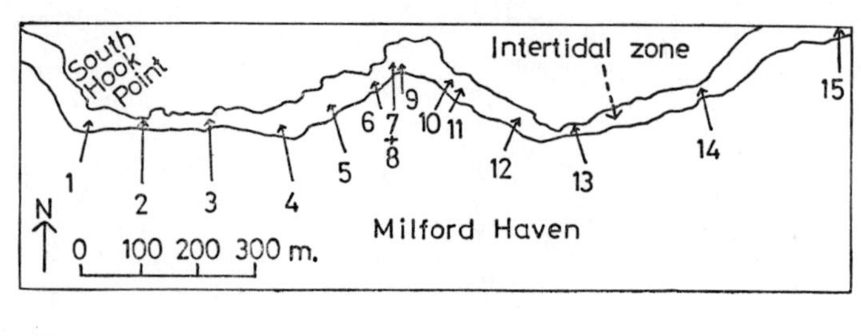

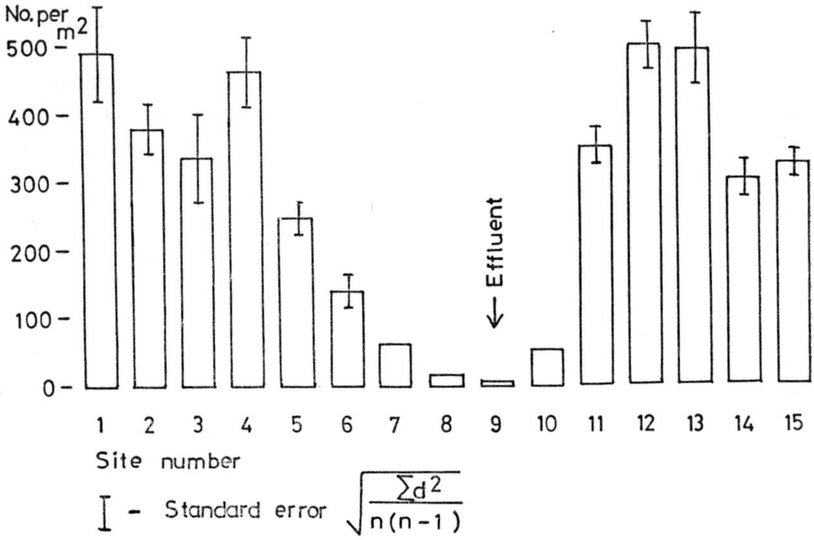

FIG. 4. Limpet sampling sites and densities, May–July 1972.

The limpet (*Patella vulgata*) populations between South Hook Point and Gelliswick have been further studied in some detail. Limpet densities measured at mid-tide level at 15 sites (Fig. 4) show that densities are lowest near the effluent in Little Wick Bay, and measurements of size (Figs. 5–8) show that the largest limpets occur where densities are lowest. The size-frequency histograms of Figs. 5–8 also show that the youngest limpet classes are missing from those areas characterised by low density and large size. Ovary weights measured by Susan Reynard (Fig. 9) indicate, however, that the large limpets near the effluent are healthy.

The large size could indicate:

(1) a fast growth rate. This would follow from the abundant food supply, i.e. the seaweed in Little Wick Bay and the greater shelter afforded by the bay.
(2) greater age. This seems likely because the size-frequency histograms show little or no evidence of recent limpet settlement.

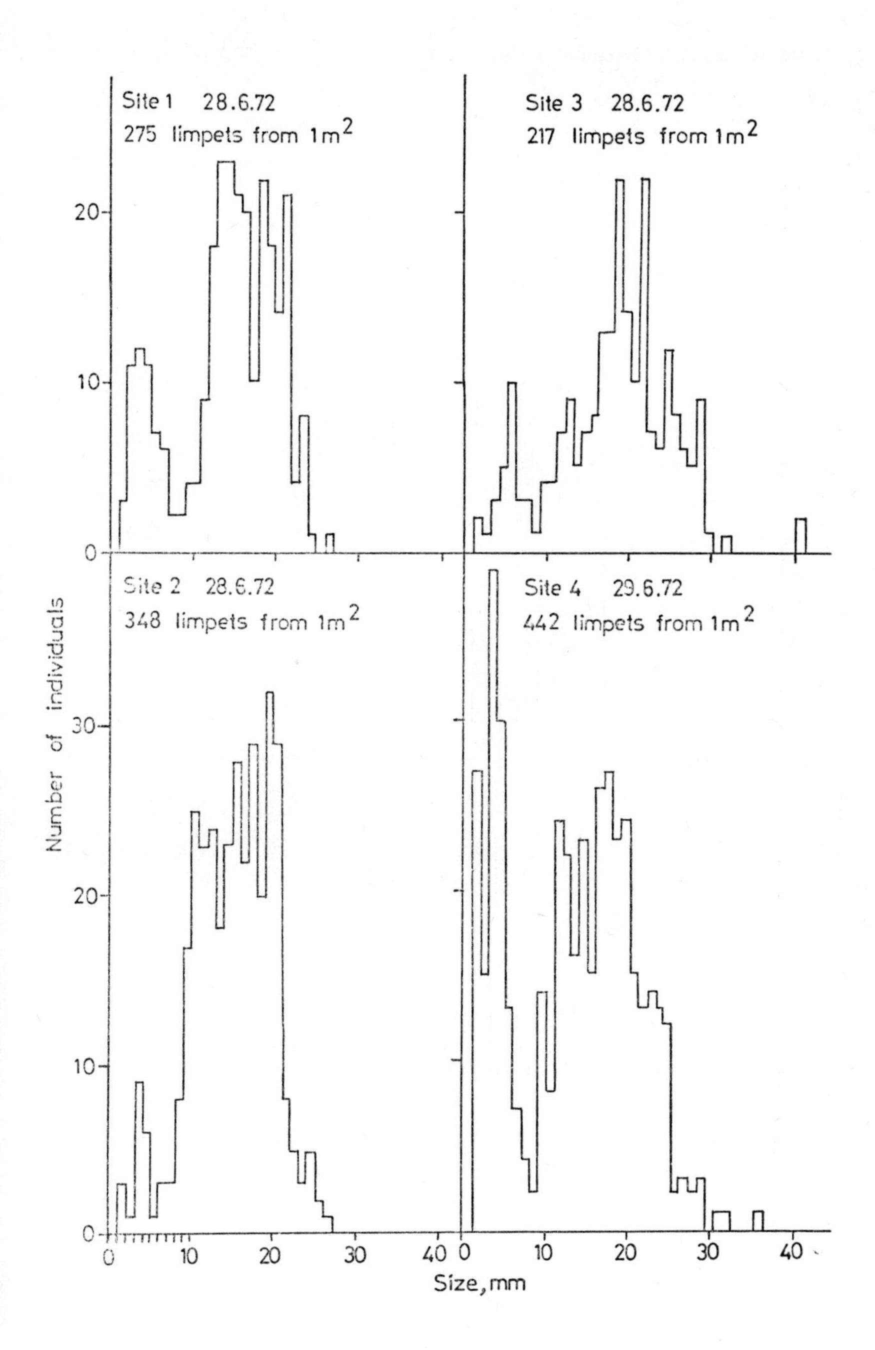

FIG. 5. Limpet size/frequency histograms (sites 1–4). For site locations see Fig. 4.

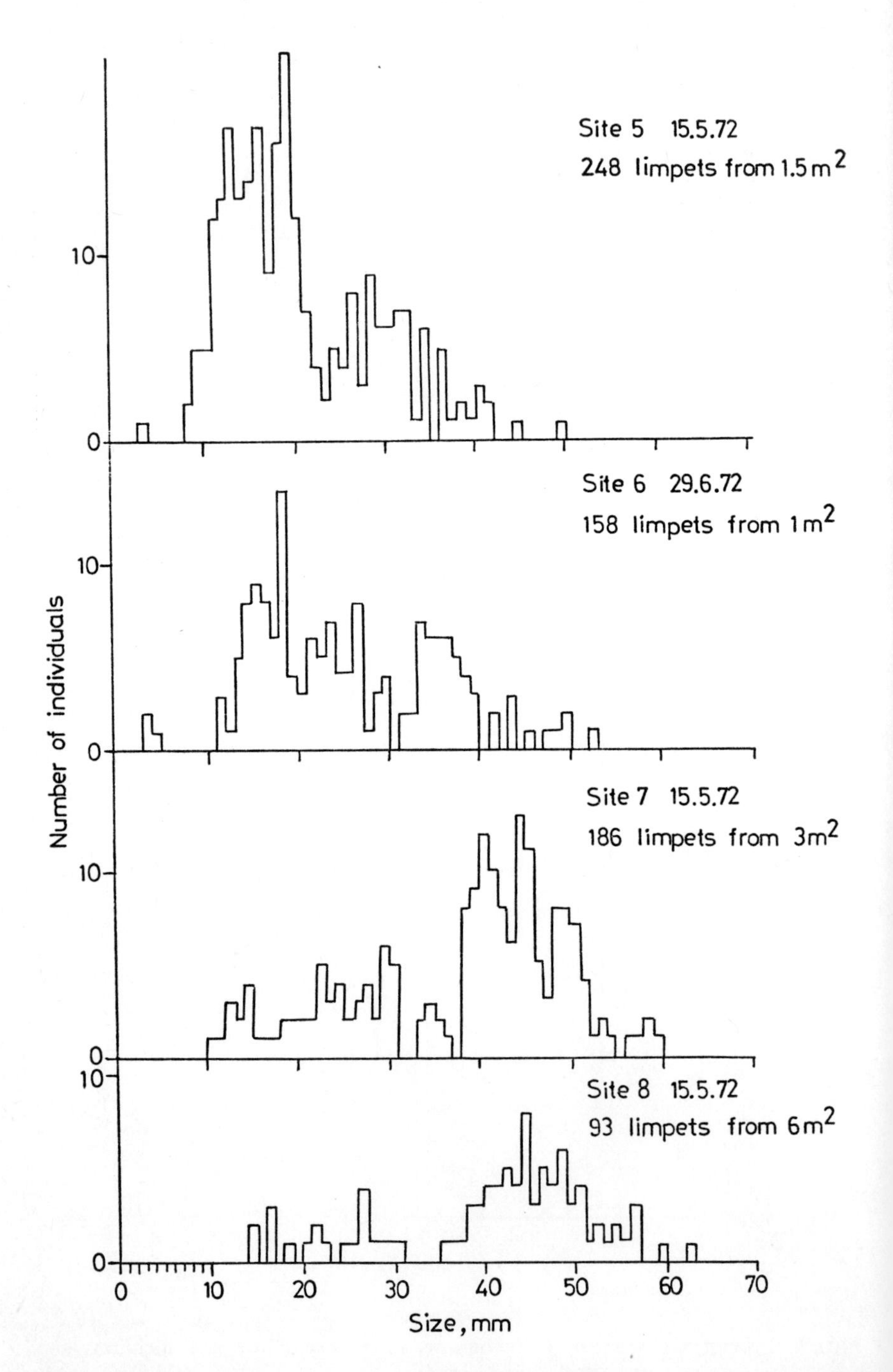

FIG. 6. Limpet size/frequency histograms (sites 5–8). For site locations see Fig. 4.

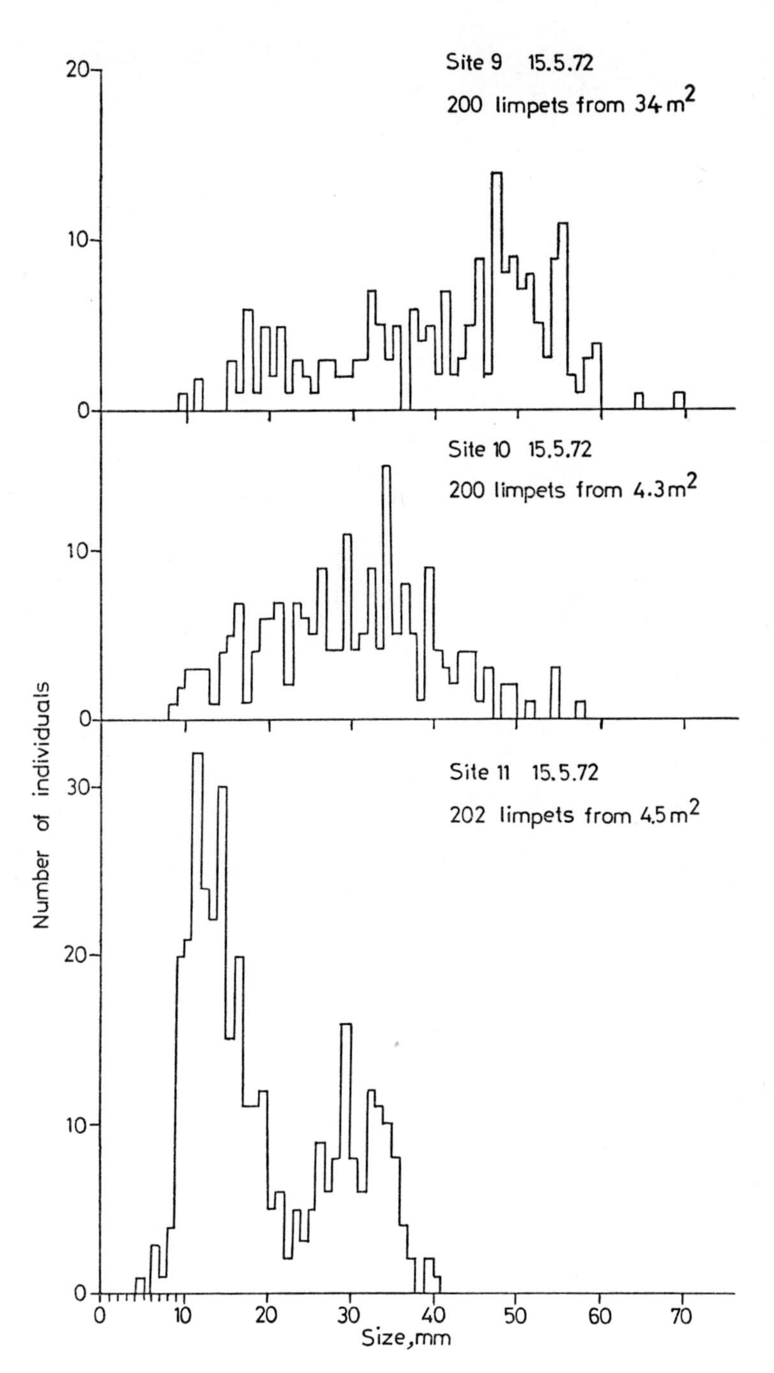

FIG. 7. Limpet size/frequency histograms (sites 9–11). For site locations see Fig. 4.

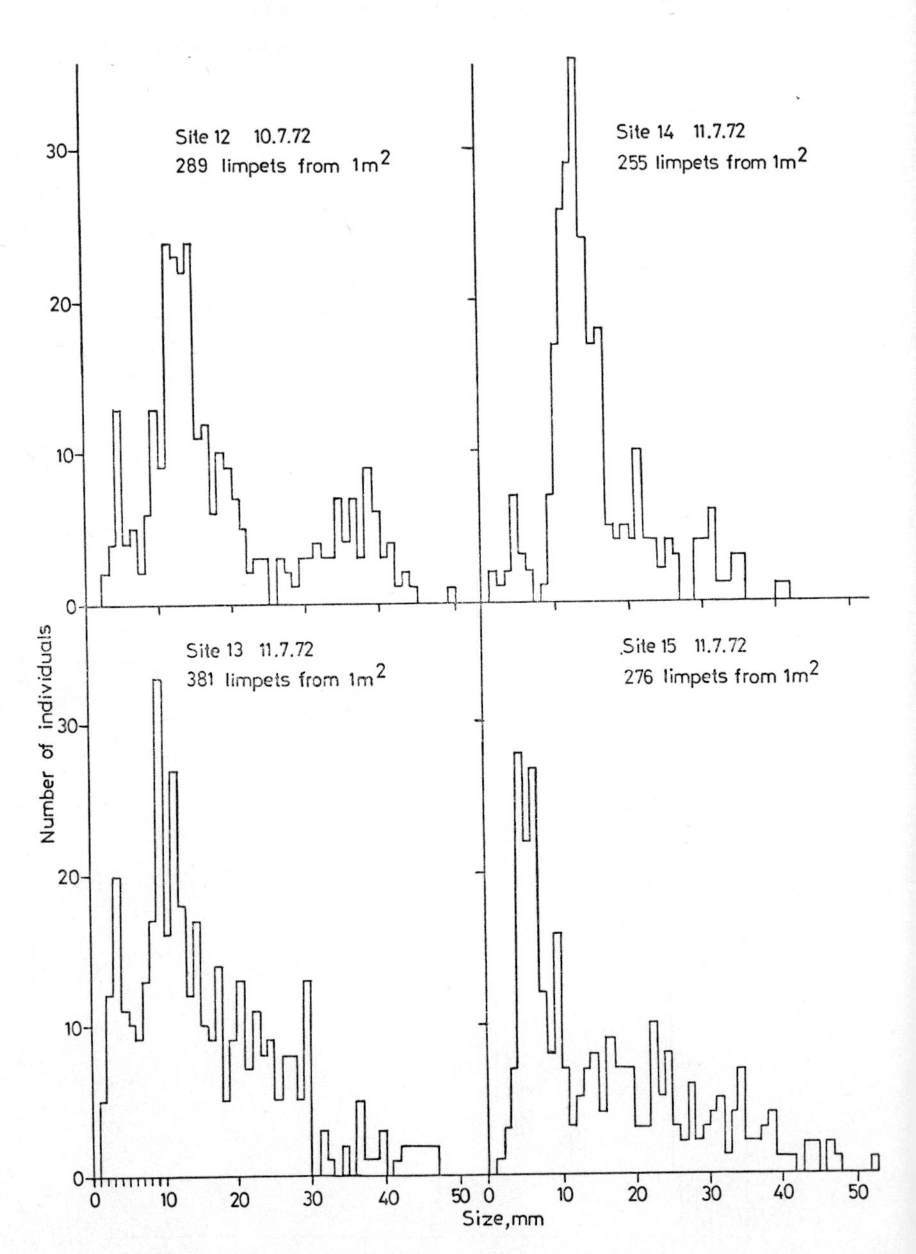

FIG. 8. Limpet size/frequency histograms (sites 12–15). For site locations see Fig. 4.

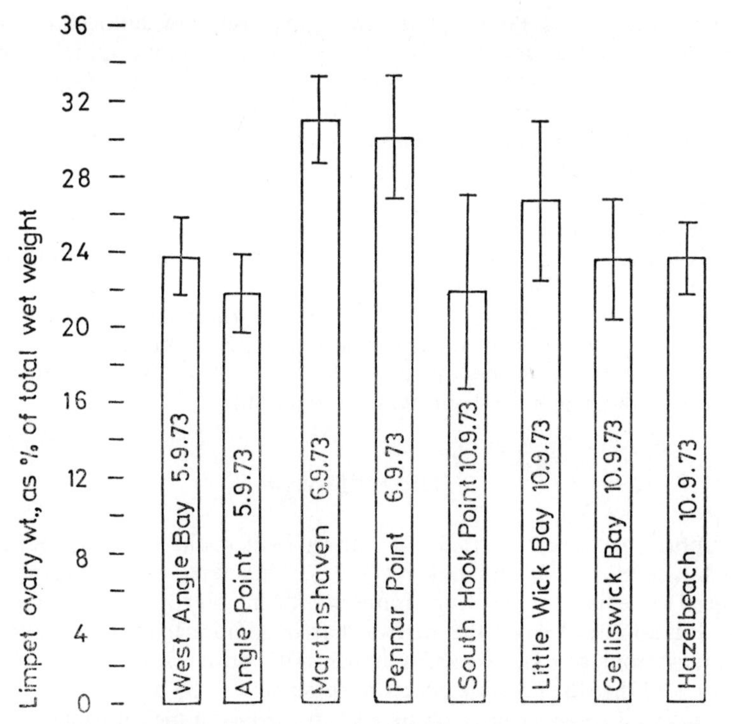

FIG. 9. Weights of limpet ovaries, expressed as a percentage of the wet body weight (from Reynard, 1973). 95% confidence limits are given.

These facts could be explained by two hypotheses:

(1) Limpets on the shore near the effluent are killed by the effluent, the mortality rate being greatest with young limpets. Surviving limpets grow fast because there is abundant space and food.
(2) For several years, settlement of new limpets has been prevented by the effluent. The limpets on this particular shore settled and became established there before 1960. The age of the limpets, perhaps in association with the abundant food, explains their size. It is worth noting here that very little is known about the ages attained by shore organisms, but 16 years is considered feasible for a limpet. It is also worth noting that the settlement of another species, the barnacle *Balanus balanoides* is known to be reduced by this effluent (Crapp, 1971; Dicks, 1975b).

Many questions about these limpets have not yet been answered satisfactorily; the subject is at a stage where field and laboratory experiments are required to provide the answers and these are considered briefly in a subsequent paper in these proceedings (Baker, 1975b).

Martinshaven

There is no real evidence that the normal discharge of the effluent in question has caused any biological changes. Crapp (1970) thought that an area between Bullwell and Popton Point with relatively few limpets and *Fucus vesiculosus* sporelings might have been affected, but following his hydrographic studies of the effluent he concluded that this was not the case.

During the power crises of the 1972 and 1973 winters, the effluent was discharged down the Martinshaven saltmarsh creek instead of being pumped along the jetty. Discharge was during the ebb tide period only as is normally the case.

Effluent in the creek had the following effects.

Large numbers of the ragworm *Nereis diversicolor* left their burrows in the mud, and were swept downstream.

Large numbers of the shore crab *Carcinus maenas* and the goby *Gobius minutus* were likewise swept downstream either sick or dead.

Flocks of gulls fed on the animals near the mouth of the creek, where the water runs slower as it fans out over the mud flats.

THE MEDWAY

Full reports of surveys carried out during 1972 in the Medway, Kent, are given by Levell (1973) and Baker (1973a). They were carried out in order to assess the effects of the Kent refinery effluent on the Medway flora and fauna. It is appreciated that the Medway has, in addition to the refinery effluent, discharges from sewage works, a paper mill, power station and bitumen refinery, and that the possible effects of these have to be borne in mind. Also, many small oil spillages have occurred from jetties in the immediate vicinity of the main refinery discharge, so separation of effluent effects from oil spill effects is not always possible. Effluent characteristics are given in Table IV.

Sample stations (Fig. 10) were selected along both sides of the Medway, and at each site a general qualitative search of the mud was carried out.

TABLE IV

Effluent Water Analyses

Test	*Readings Recorded During Period 1.9.71 to 1.9.72*		
	Average	*Lowest*	*Highest*
Dissolved oxygen (ppm)	6·6 (8·4)	5·6 (6·3)	8·9 (10·3)
Chemical Oxygen Demand (N/80 $KMnO_4$ at 27°C for 4 h) (ppm)	6·7 (2·7)	1·2 (0·8)	12·2 (7·8)
Biochemical Oxygen Demand (5 days at 20°C) (ppm)	35·7 (3·0)	6·1 (0·9)	44·3 (6·9)
pH (°C)	7·6 (7·7)	7·3 (7·6)	7·8 (7·9)
Temperature (°C)	25 (12)	18 (4)	34 (18·5)
Involatile Oil Content (ppm)	5·9	0·8	21
Flow Rate (million gph)	7·1	5·1	7·9

N.B. Figures in parentheses refer to Inlet Water conditions.

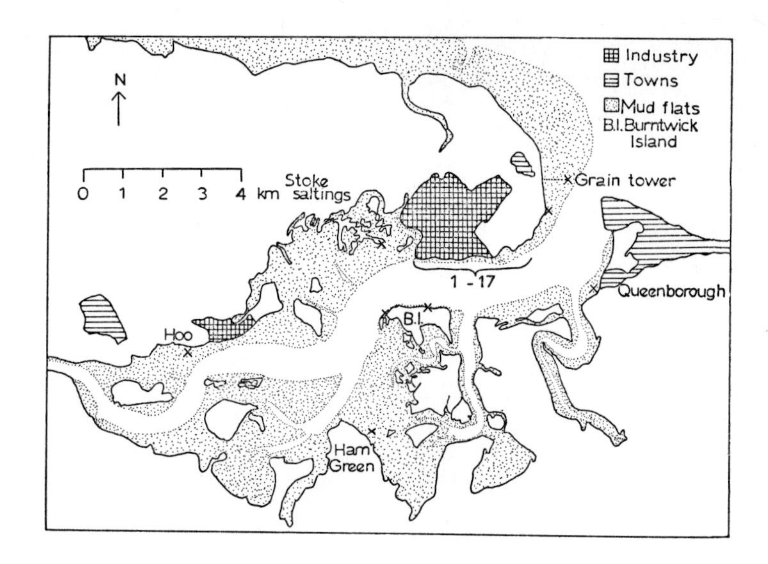

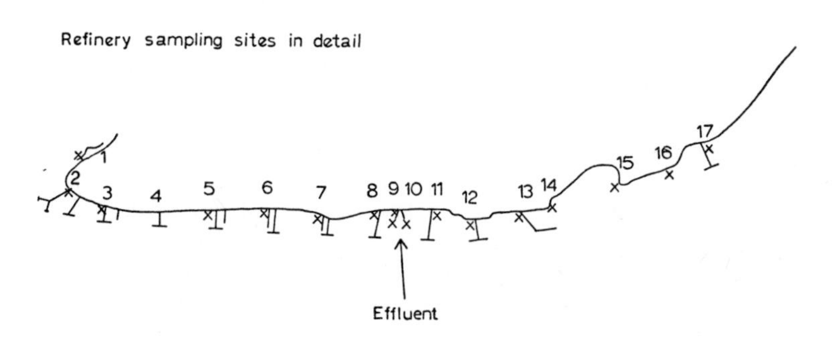

FIG. 10. Positions of Medway sampling sites.

Quantitative sampling was carried out using a $\frac{1}{10}$ m^2 $\times$ 25 cm sampler. This sample had to be sorted by hand as the nature of the mud prevented effective sieving or washing.

Sea-wall transects were also studied: at each site a stone faced sea-wall extended downwards from above extreme high water mark of spring tides, usually to mean tide level or below. All macroscopic species were counted or estimated in contiguous metre squares from approximately mean high water spring tide level to mean tide level.

Water movements were followed from shore or boat by observing the progress of oranges thrown into the main effluent. Water samples were collected for salinity and pH measurements, and the temperature of the water was measured at the sample points. Oil content of mud samples was measured using a relatively crude technique of petroleum spirit extraction followed by evaporation and weighing. All oil contents mentioned subsequently in this paper were measured using this technique. It has to be borne

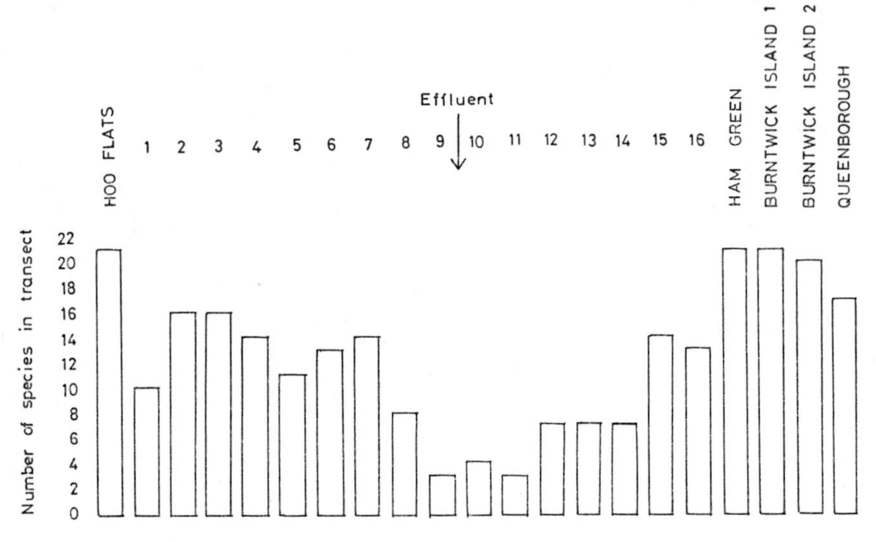

FIG. 11. Numbers of species on Medway sea-wall transects.

in mind that the figures obtained will include all petroleum spirit-extractable materials, whether natural or otherwise.

Numbers of species both of mud fauna and of sea-wall flora and fauna (Fig. 11) were reduced near the effluent. The winkles *Littorina littorea*, *L. saxatilis* and *L. littoralis* were usually rare or absent on the refinery sampling sites, though common elsewhere (Fig. 12). In contrast, fucoid seaweeds were abundant on many of the refinery sampling sites (Fig. 13).

Mud oil contents were generally low, rising to 1·4% near the effluent. The water results indicate a rapid dilution of the effluent to normal estuarine conditions of temperature, salinity and pH, but thin oil films are produced which could become stranded on the shore and drain down the sea-wall. The effluent stream on the ebb follows the shore fairly closely at the eastern end of the refinery area, but on the flood, moves out more quickly into the main channel.

A number of factors may be responsible for the observed distributions of flora and fauna. The fine grained anaerobic silt which is the most common natural substrate of the area tends normally to support few animal species, but it is noticeable that there is a complete absence of any mud dwelling macrofauna in the region of the refinery jetties. At this stage one cannot firmly attribute this to any one cause, although the *Nereis* do seem to decrease in relation to the nearness of the refinery effluent. On the sea-walls, differences in exposure may be partly responsible for the *Ascophyllum* distribution, and *Hydrobia* distribution may be correlated with nearness to salt-marshes where millions of these animals were observed. Some sea-walls were in a better state of repair than others, the refinery sea-wall being in relatively good condition. This means it has less loose stones for animals to live under, and this affects in particular the *Carcinus* distribution. Some transects were necessarily short because the sea-wall gave way to mud above

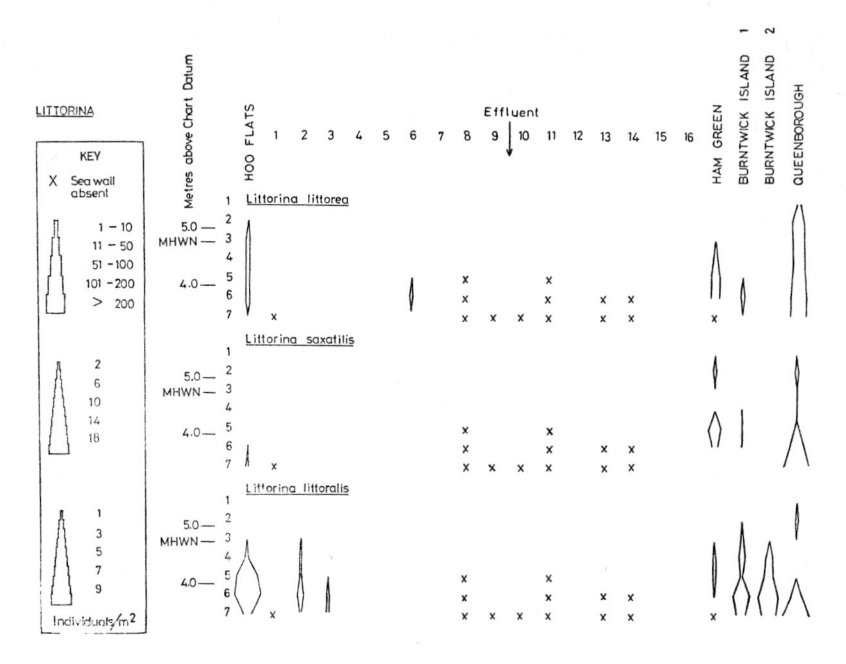

FIG. 12. Distribution of *Littorina* spp. on Medway sea-wall transects.

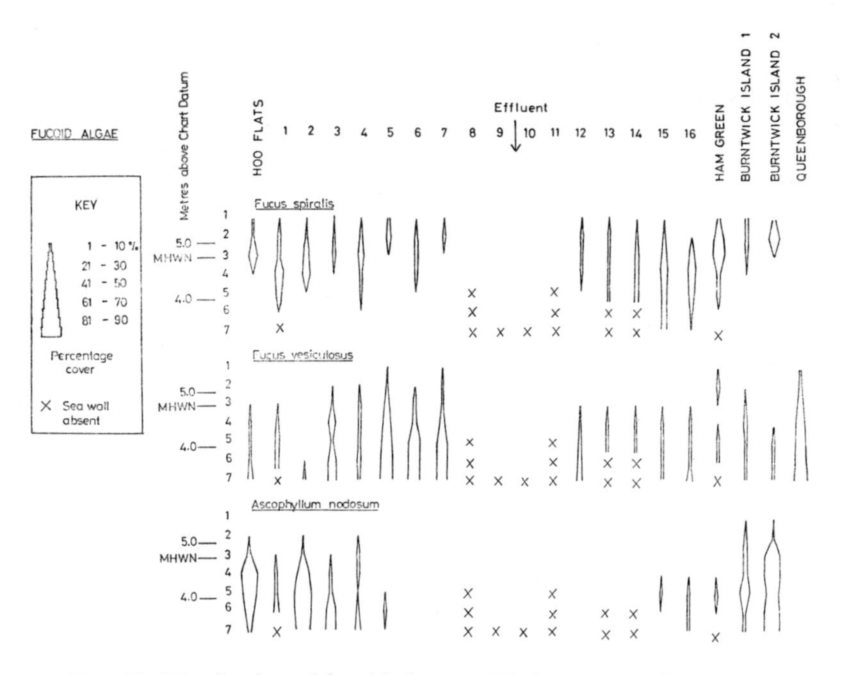

FIG. 13. Distribution of fucoid algae on Medway sea-wall transects.

the mean tide level, and this affects the total numbers of species found per transect.

Bearing all the above factors in mind, there are two main features of distribution which may be explained by refinery operations.

First, a general impoverishment of the area in the immediate vicinity of the main refinery outfall. Here the only organisms present were *Enteromorpha*, diatoms and blue green algae.

Secondly, very low numbers of *Littorina* along the whole of the refinery sea-wall. Emulsifier cleaning and/or effluent, would cause these animals to retract into their shells and they would become dislodged from the sea-wall. Re-establishment probably would not take place in the continued presence of emulsifier or effluent. Relatively strong currents along this part of the Medway shore could be important, either by naturally making it more difficult for winkles to survive, or by intensifying the effects of pollutant-induced retraction. Currents alone do not satisfactorily explain low numbers of *Littorina littoralis* in areas of abundant *Ascophyllum* and *Fucus vesiculosus*.

SOUTHAMPTON WATER

The effects of a large refinery effluent on a saltmarsh in Southampton Water are fully discussed in the paper by Dicks (1975a).

THE CRYMLYN BOG

The refinery at Llandarcy, Glamorgan, Wales, which has been in operation for over 50 years, releases minor effluents into the Crymlyn Bog at three main sites (Fig. 14). Two of these eventually feed the Crymlyn Brook, one a swampy area near Freeman's Bund.

Oil contents are generally below 4 ppm. Oil, however, occasionally enters the bog accidentally, through or over earth banks. The situation is complicated by drainage from old mine workings, which enters the Crymlyn Brook and may have an effect apart from the refinery effluents.

The discharge area is unusual because it is a freshwater swamp—effluents are usually discharged into tidal waters. Surveys were carried out during August 1972 to see how the effluents behaved in this type of habitat and what effect they had on the flora and fauna. Full reports are given by Baker (1973b) and Parsons (1973).

The plant communities show considerable variation along the Crymlyn Brook and its tributaries, and there appear to be two opposing factors determining this. First the brook, in its passage into the bog, becomes broader and slower and is eventually lost in the reed grass *Phragmites*. Under these conditions there is, under normal circumstances, a gradual colonisation by the plants which have been classified here as emergent angiosperms. The second factor, refinery activities and mine drainage, affects the brook after its junction with effluent 1. Effluent is not the only refinery influence, for from this point southwards the brook is also subject to periodic applications of weed killer, and removal of emergent plants to keep the waterway clear.

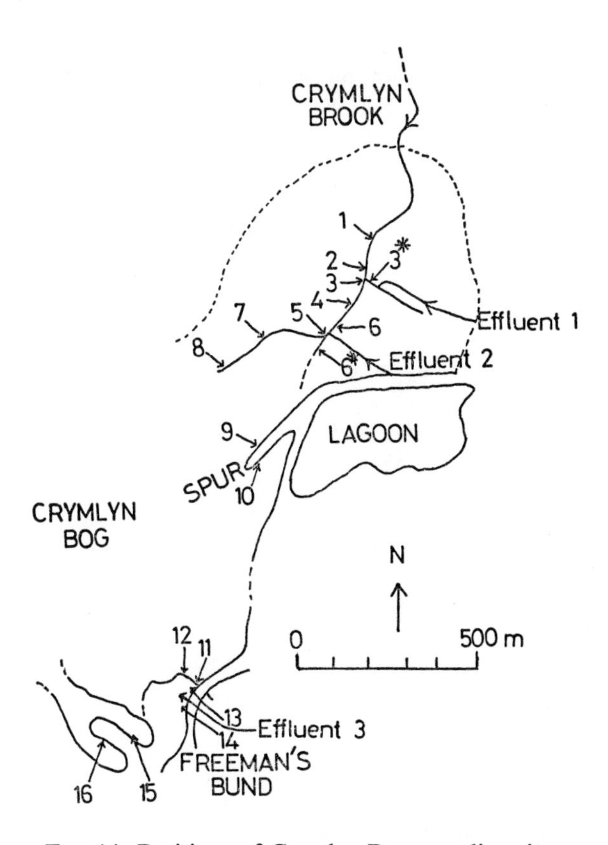

FIG. 14. Positions of Crymlyn Bog sampling sites.

The most obvious effect of the effluent is that the water surface is more or less covered by a thin oily film in the area between discharges 1 and 2. The oil films gradually disappear downstream of discharge 2.

All these influences combine in greatly reducing the emergent angiosperms, and it is of interest to note the corresponding increase of submerged angiosperms (mainly *Callitriche*). These when examined were generally healthy. It was found that under the static oily surface of the stream, cleaner water was flowing at a rate of approximately 6 m/min. The only plant which tolerates oily surface films was found to be the blue green alga *Oscillatoria*, and this was present particularly in effluent streams 1 and 2.

At the southern limit of the stream clearing operations, the Crymlyn Brook is swallowed up in *Phragmites*. It is then difficult to trace either the brook or signs of effluent, as any remaining traces of oil are absorbed among the *Phragmites* plants.

In the area between the spur and the lagoon, *Phragmites* grows in an area of oily mud. The question of how much oil *Phragmites* can tolerate is partly answered by a small transect off the side of the spur. Here the three soil samples taken were from three grades of *Phragmites*—dead, healthy and intermediate. The plants were healthy where the surface soil (0–10 cm) oil content was 32·2%. Owing to the liquid mud nature of the soil, samples could not be obtained from any depth, but it is likely that the highest oil content is near the surface. The underground parts of the *Phragmites* can survive in these conditions owing to oxygen diffusion down the stems.

Soil samples elsewhere, taken mainly from stream sediments, show much lower oil contents (see Table V) than the spur area, though there has been an obvious accumulation of oil in some places. Oil in the soil away from the stream sides could possibly have soaked through from the stream, but is more likely to be a result of flooding in the past. All the low results should be interpreted with caution, because the simple analytical method used does not distinguish between natural and refinery-derived oils. Some of the low results, in other words, may be due entirely or partly to natural oily compounds.

The main results of refinery activities on the plants of the Crymlyn Bog may be summarised as follows:

(1) Death of an area of *Phragmites* between the spur and the lagoon owing to overloading beyond its considerable oil tolerance.
(2) Change of plant type in a part of the Crymlyn Brook, from emergent to submerged angiosperms. This is partly due to weed clearing operations.
(3) Death of all plants other than blue green algae in effluent streams 1 and 2. The Freeman's Bund area does not appear to be affected, though there is some oil in the soil.

There was no animal life at the site between the spur and the lagoon where *Phragmites* was dead and the oil content of the surface mud was about 45%. However, nearby the stands of *Phragmites* and the mats of blue green algae seemed to tolerate, and to some extent contain, the oil which in turn permitted colonisation by some animals, notably Culicid and hoverfly larvae (both air-breathing groups). Differences in the biota were found along the Crymlyn Brook—there was a decrease in the numbers of certain groups near the effluent discharges, notably the Crustacea, Lamellibranchia (the fresh water cockles), water bugs and beetles.

Table V sums up the distribution and abundance of the main animal groups of the Crymlyn Bog as they vary with turbidity and mud oil content.

In the Crymlyn Brook, conditions range from the clear stream that flows through the pasture and water meadow upstream of the bog, to the polluted areas at the mouth of the outfalls. It is difficult to isolate any single factor in the brook that would account for the changes in the structure of the animal communities. Superimposed upon the normal changes that would be found in a stream that flowed into a bog as a result of pH, temperature, oxygen and rate of flow alteration, are other variables such as the outflow of effluent which impose still further variations in these parameters. Added to this the mine-water drainage, the practice of the refinery staff of banking up the sides of the river against flooding, of keeping the water course clear of weed and of using herbicides, a complex of effects is produced which it is not easy to untangle.

From the results in the Crymlyn Brook, it would appear that it is suspended

TABLE V

Summary of Abundance Ratings at Sample Sites

Sample site	1	2	3	3*	4	5	6	6*	7	8	9	10	11	12	13	14	15	16
Isopods	1	5	5	0	2	0	0	0	3	0	0	0	0	0	0	0	5	2
Amphipods	5	4	3	0	0	0	1	0	0	0	0	0	0	0	0	0	0	0
Flatworms	0	0	0	0	0	5	5	0	5	4	0	0	0	0	0	0	0	0
Bivalves	3	4	0	0	0	0	0	3	0	5	0	0	0	0	0	0	0	0
Large gastropods	0	3	3	0	5	4	5	4	5	4	0	0	0	0	0	0	0	0
Small gastropods	1	1	4	0	4	4	5	2	5	2	0	0	0	0	0	0	0	0
Large *Coleoptera*	1	3	3	0	0	0	0	1	0	0	0	0	0	0	0	0	0	5
Small *Coleoptera*	1	5	5	0	3	0	2	0	0	0	0	0	0	0	0	0	0	0
Hemiptera	3	1	0	0	0	0	0	0	0	0	0	0	0	0	0	0	5	5
Diptera (misc.)	1	1	0	0	1	0	1	0	1	1	5	0	4	1	5	1	0	0
Coretura	0	0	0	0	0	0	0	0	0	0	0	0	0	0	0	0	5	3
Rat tailed maggot	0	0	0	0	0	0	0	0	0	0	5	0	0	0	0	0	0	0
Ephemeroptera	0	0	0	0	0	0	0	0	0	0	0	0	0	0	0	0	5	3
Turbidity (silica scale)	0	1	—	75	71	45	40	—	36	—	—	—	0	10	—	—	25	39
% petroleum spirit-extractable material	0·0	0·4	—	2·1/ 2·4	0·6	0·3	—	—	0·7	2·6		32/ 45	2·4	12·6	—	—	7·8	—

Abundance recorded on a five point logarithmic abundance scale where the maximum standing crop (n) scored 5 points, $n/2$ scored 4, $n/4$ scored 3, $n/8$ scored 2, less than $n/8$ scored 1 unless the organism was absent when no points were awarded.

matter that is a critical factor—as indicated by the turbidity. Thus no organisms are found in the first outfall where the mud oil content is 2·0–2·4% and the turbidity is 75, but lower down the stream where the mud oil content is 2·6% and the turbidity is 36 ppm a new community has become established. These areas are comparable in being dredged and banked up to the same extent. Further work is necessary to find the source of the turbidity, i.e. refinery, mine drainage, or other.

The most critical form of pollution would seem therefore to be that where finely divided material remains in suspension for long periods. This would especially affect filter feeders—and the bivalves are limited by this. The fact that no chironomids were found in the first outfall suggests that the bottom mud is also affected in some way.

As mentioned before, colonisation of water between the spur and the lagoon was achieved by certain dipteran larvae where the oil content of the mud was as high as 32–45% but where they were enclosed by the *Phragmites* which acts as a filter and oil absorbent, with the consequent reduction in turbidity.

At Freeman's Bund there was quite a high oil content (2·4–12·6%) but the turbidity was quite low at the outfall; there is a diverse and successful biota in the stretch of water into which the outfall drains.

DISCUSSION

Corresponding to the wide range of effluents described, there is a variety of effects, from no observed damage to death of saltmarsh vegetation over a wide area. These effects, however, fall into relatively few categories which depend mainly on the siting of the outfall.

Category 1. Offshore effluent discharges (see Table I, nos. 1 and 2, and Martinshaven). Though there is evidence to show that the undiluted effluent is toxic, in general the biological effects of this type of discharge seem to be small, presumably owing to rapid dispersion and dilution.

Category 2. Shore discharges with good dispersion (Wear Point, Milford Haven). Discharge from a headland into an area of strong tidal flow has achieved the same good dispersion as category 1.

Category 3. Shore discharges with intermediate dispersion (Little Wick Bay, Milford Haven and the Medway). These are in areas such as small bays where the effluent washes along the shore at some stage of the tide. Reduced numbers of several species may be expected in the immediate vicinity of the discharge, but the effect is localised.

Category 4. Shore discharges with poor dispersion (see Table I, nos. 3 and 4, Southampton Water, and the Crymlyn Bog). Typically, poor dispersion of effluents leads to accumulation of oily sediments and subsequent impoverishment of bottom fauna. Oil from effluents is easily trapped on the marsh vegetation which often grows under the sheltered conditions, and eventually kills it. This applies both to saltmarsh areas and freshwater marsh areas, but in the former the area of damage is likely to be greater owing to spreading of oil films at high tide.

Comments may also be made on the susceptibility of different species. Marsh vegetation is vulnerable because grass species such as *Spartina anglica* and *Phragmites communis* trap oil. In the same way that oil slicks are absorbed by straw, the thin films resulting from effluents are absorbed by these grasses and many such oilings lead to death. The fact that the oil sticks firmly to the plants means that it does not go elsewhere, and as many species can survive successive oilings up to a point there is a possibility here of deliberately managing natural vegetation as an oil trap.

Algae are usually resistant to effluents because they are protected by a mucilage covering. In the case of Little Wick Bay, for example, increase of fucoid algae in the effluent area following decrease of limpets clearly shows that these plants are resistant. In the Medway, fucoid algae were absent from the sea-wall in the immediate vicinity of a large effluent, but the very resistant green algae *Enteromorpha* and blue green algae were present.

Turning to animals, in estuarine mud the animals most resistant to oily sediments appear to be polychaetes such as *Nereis diversicolor*, but even these are absent near some effluents, e.g. in the Medway, and it is known that undiluted effluent can affect *Nereis*. On rocky substrates reduced numbers of limpets (*Patella* spp.) and winkles (*Littorina* spp.) are usual in areas near effluents. Though factors other than effluent may be operating in some cases, as discussed later, it has been shown that limpets moving over oily substrates are likely to drop off owing to a narcotic effect on the foot (Dicks, 1973), and that winkles show reduced activity in undiluted effluent (see Baker, 1975b).

These points may help to explain localised reduced populations, and show that effluent may have an effect not by killing organisms directly, but by causing them to become dislodged and thus liable to predation or movement elsewhere by tidal action.

Interpretation of field survey data is usually a difficult process owing to complicating effects of factors other than effluents. The outfall may be in a relatively sheltered area such as a little bay and this will give rise to a biological gradation relative to the nearby exposed shore, independent of any gradation caused by the effluent. In the case of muddy and sandy shores there may be a gradation of particle size from one end of the sampling area to the other, and this again will produce natural biological gradients. Oil industry operations other than effluent discharge may also account for some biological changes, for example, a shore near an oil terminal may be affected by oil slicks from jetties or tankers, and may be cleaned with dispersants.

The effluent itself can be very variable, even within the limits set by regulatory organisations. Apart from the usual constituents such as oil and traces of phenols, effluents may contain at certain times other substances such as a variety of dispersants used for cleaning within the refinery, dyes used for tracing leaks, and sediment from some cleaning operations. The plants and animals of the area near the discharge will be determined more by maximum levels of constituents than by average effluent composition. If a species cannot tolerate any maximum then it will be eventually eliminated if the maximum occurs at all frequently. Under such conditions, species which can complete their life cycle during the intervals between maxima, for example simple algae, are at an advantage.

Changes in the immediate vicinity of effluent discharges would seem to be unavoidable in sites with the effluent quality standards shown in Tables II and IV, because such effluents, if undiluted or slightly diluted, are toxic in the long term. Does this matter if the area affected is small? It seems more reasonable to trace the wider area where the effluent goes—if damage can be observed here there is cause for concern and a review of the effluent limits. The very localised effects are a different matter—it has to be decided if it is justifiable to impose very strict limits on effluents (with all the expense involved in treatment facilities) to improve conditions on say 200 m of shore. With finite resources there would seem to be greater environmental priorities. It is however important to know whether or not an effluent has a widespread effect and to ascertain if there is any widespread build-up of, for example, oil in organisms or metals in sediments, which could prove deleterious in the long run. It is in these fields that data is needed. It is desirable that refinery effluent areas are regularly surveyed for biological changes, so that long term trends can be determined and the effluent quality, if necessary, changed.

ACKNOWLEDGEMENTS

I would like to thank the Milford Haven Conservancy Board and the many oil company staff who have co-operated in this study. This paper uses results obtained by Catharine Howe, David Levell, Roger Parsons, and Susan Reynard, and I would like to thank them for their help.

REFERENCES

Baker, J. M. (1973a). Biological survey of seawalls in the Medway estuary, O.P.R.U. Annual Rep. for 1973, 17–28.

Baker, J. M. (1973b). Effects of refinery effluents on the plants of the Crymlyn Bog, *ibid.*, 29–35.

Baker, J. M. (1975a). Biological monitoring—principles, methods and difficulties (this volume).

Baker, J. M. (1975b). Experimental investigation of refinery effluents (this volume).

Ballantine, W. J. (1961). A biologically defined exposure scale for the comparative description of rocky shores, *Field Studies*, **1** (3), 1–19.

Crapp, G. B. (1970). The biological effects of marine oil pollution and shore cleansing, Ph.D. thesis, University of Wales.

Crapp, G. B. (1971). Chronic oil pollution, in *The Ecological Effects of Oil Pollution on Littoral Communities* (ed. E. B. Cowell), Inst. Petroleum, London, pp. 187–203.

Dicks, B. (1973). Some effects of Kuwait crude oil on the limpet *Patella vulgata*, *Environ. Pollut.*, **5**, 219–29.

Dicks, B. (1975a). The effects of refinery effluents—the case history of a salt marsh (this volume).

Dicks, B. (1975b). The importance of behavioural patterns in toxicity testing and ecological prediction (this volume).

Jeffery, P. (1972). Oil in the marine environment, Department of Trade and Industry, Warren Spring Laboratory, Stevenage, U.K., 16 pp.
König, D. (1968). Biologische Auswirkungen des Abwassers einer Öl-Raffinerie in einem Vorlandgebiet an der Nordsee, *Helgolander wiss. Meeresunters*, **17**, 321–34.
Levell, D. (1973). Intertidal mud fauna in the Medway estuary, O.P.R.U. Annual Rep. for 1973, 13–16.
Oglesby, R. T. and Sylvester, R. O. (1965). Marine biological monitoring of oil refinery liquid waste emissions, *Engng Bull. Purdue Univ.*, **49**, 167–79.
Parsons, R. (1973). Effects of refinery effluents on the freshwater animals of the Crymlyn Bog, O.P.R.U. Annual Rep. for 1973, 36–43.
Reish, D. J. (1965). The effect of oil refinery wastes on benthic marine animals in Los Angeles Harbour, California, *Sym. Commn. int. Explor. scient. Mer Medit., Monaco*, 355–61.
Reynard, S. (1973). Reproductive potential of limpets near a refinery effluent in Milford Haven, O.P.R.U. Annual Rep. for 1973, 44–6.
Steere, B. F. (1970). Marine disposal of petroleum refinery effluents, paper presented at 1970 Conference of the Institute of Water Pollution Control (Southern African Branch), 16–20 March 1970.

12
The Effects of Refinery Effluents: The Case History of a Saltmarsh

BRIAN DICKS

(*Oil Pollution Research Unit, Orielton Field Centre, Pembroke, Wales*)

SUMMARY

A Spartina *saltmarsh in Southampton Water, which has had a refinery effluent discharged through its creek system since 1951, was surveyed biologically in 1969 and 1970 to assess the extent of ecological damage. The saltmarsh has been re-surveyed twice a year since 1972 to monitor any changes in the distribution of plant species in association with an effluent improvement programme started by the refinery. The results of these surveys are presented and discussed.*

A limited amount of information is available on the condition of the saltmarsh from 1950 up to the start of the recent series of surveys. This information and the monitoring programme results provide a picture of, firstly, severe ecological damage to an area of saltmarsh vegetation resulting in bare mud, followed recently by recolonisation of some parts of this previously denuded area by most of the common saltmarsh species. Though recolonisation appears to be associated with the improvement in effluent quality, the recent mild winters, early springs and late autumns and the virtual absence of acute oiling incidents since 1970 may also have considerably influenced the growth of vegetation. For these reasons, the interim nature of these investigations has been stressed.

If recovery continues, the sequence in which the plants are recolonising the denuded areas could result in a saltmarsh dominated by species other than Spartina anglica.

INTRODUCTION

Extensive saltmarshes dominated by the common cord grass, *Spartina anglica*, occur in Southampton Water, though some areas are now reclaimed for industrial use. This species of grass, with its rhizomatous root system and extensive aerial shoots, stabilises soft marine mud and helps trap further silt to raise the marsh level (Hubbard, 1954; Chapman, 1960, 1964). This process, which allows invasion of the area by other less salt-tolerant plants, thus changes the biology of the area, and is called succession. A typical Southampton Water saltmarsh is therefore composed of, near its landward edge, a mixture of less salt-tolerant plant species or those preferring drier conditions,

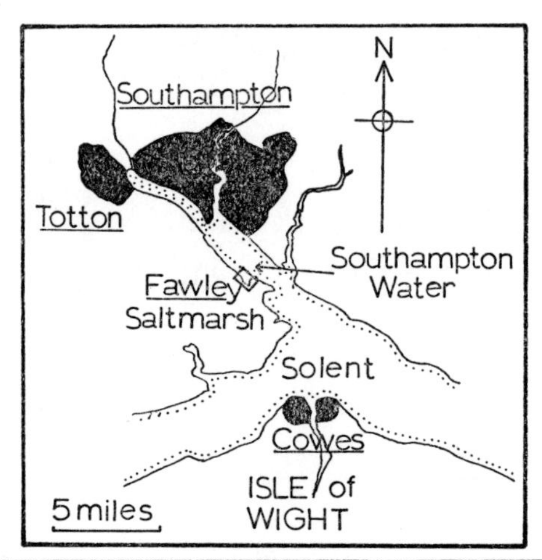

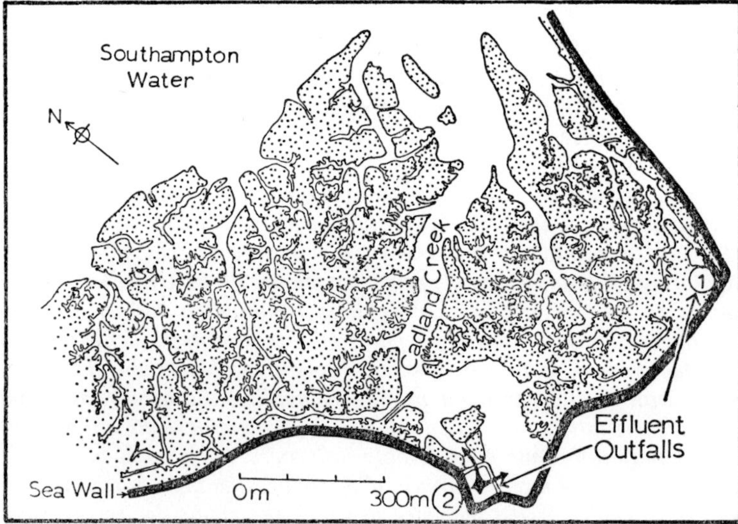

FIG. 1. The saltings and creek system at Fawley, Southampton Water. The location of the saltings in Southampton Water is shown in the upper map, and details of the saltings (stippled) and positions of these and two effluent outfalls in the lower map.

while among the extensive system of creeks and toward the seaward edge more salt-tolerant plants such as *Spartina anglica* and other primary mud colonisers (*Salicornia* spp. and *Suaeda maritima*) occur. Variations in micro-habitat throughout the marsh produced by erosion or changes in drainage patterns provide interruptions to this general successional pattern of change in vegetation from seaward to landward edges of the marsh.

The common marsh plants found at the Fawley monitoring site are: *Salicornia* spp., *Suaeda maritima*, *Spartina anglica*, *Halimione portulacoides*,

Aster tripolium, *Atriplex* sp. and *Juncus maritimus*. A map of the saltings, their creek system and the positions of effluent outfalls is shown in Fig. 1.

The marsh is bounded on its southern edge by reclaimed land, on its landward margin by a sea-wall at the high-tide mark and on its eastward margin by Southampton Water, but extends northwards for a considerable distance out of the area of influence of the refinery effluent along the shore of Southampton Water.

THE EFFECTS OF OIL ON SALTMARSH VEGETATION

All the above species which occur in the Fawley marsh are sensitive to oil pollution, especially to films of oil on the water surface or to oil slicks. *Salicornia* spp. and *Suaeda maritima*, two types of succulent plant, are particularly sensitive, as they have a small underground system from which new growth cannot occur after damage. The reasons for their and other marsh plants' sensitivity are fully discussed by Baker (1970a,b,c,d,e,f) but can be summarised briefly as follows:

(a) There is a high affinity between plant cuticle and oil hydrocarbons. This encourages adhesion of oil to vegetation.
(b) Oil obstructs the leaf pores (stomata) and prevents or restricts gas exchange to root and shoot.
(c) Oil films reduce the light available for photosynthesis.
(d) Some constituents of crude oils are directly toxic to metabolic processes.
(e) Plants with large underground vegetative systems (the perennials) survive oiling better than those without (the annuals), as new growth can occur from these systems after shoots and leaves have been destroyed.

The effects of oil on saltmarshes are also influenced by the number of times incidents occur and, above a certain number, damage can be extensive (Baker, 1970a). Continual (chronic) pollution such as that produced by a refinery effluent probably has its main effect on saltmarsh plants by the almost continual presence of thin oil films on the surface of the water, which, at high tide, reach the plants.

THE FAWLEY EFFLUENT AND ITS DISCHARGE

The main pollutants that occur in refinery effluents are oil, dissolved compounds such as sulphides, phenols and nitrogen compounds, and sludge (Blokker, 1970). Typical refinery waste characteristics of the older type of refinery as produced by the American Petroleum Institute (1963) average around 57 ppm oil. A more recent report (Blokker and Marcinowski, 1970) showed a wide range in effluent quality. The older coastal refineries reported effluent quality to be commonly in the range 11–40 ppm, but approximately 30% reported over 40 ppm. The Fawley effluent, in 1970, averaged 31 ppm oil content with a total discharge volume of 114 000 imp gal/min through outfalls number 1 and 2. Recently these figures have been substantially improved with, in 1972, an average oil content of 25 ppm and 104 000 imp gal/min and, in 1974, an average oil content of 14 ppm and volume discharge of 86 000 imp gal/min.

The effluent is discharged at the high-tide mark to two of the large creeks in the saltings, as illustrated in Fig. 1. Discharge volumes for the two outfalls are roughly equal.

As well as the oil normally dispersed in the effluent, oil from other sources occasionally enters the creek system. These sources include:

(a) accidental spillage within the refinery, which might produce unusually high oil contents in the effluent;
(b) oil spillage from the refinery jetty installations and tankers;
(c) oil spillages from other jetty installations in the vicinity; and
(d) oil spillage from the considerable ship traffic using Southampton Water.

Whilst thin oil films are commonly present in the area, coherent dark slicks appear to be rare. This has been particularly so since 1970, the number of acute incidents both from refinery operations and elsewhere being small.

Whatever the source of the oil, it first affects vegetation near the creek edges, and later, when the initial barrier of the creek edge vegetation has succumbed affects areas further into the marsh.

A HISTORY OF DAMAGE TO THE FAWLEY SALTMARSH 1950–70

Full details of the marsh history in association with refinery developments are given in Baker (1970b) and these can be summarised as follows:

1950 Aerial photograph taken by the RAF (RAF Sortie No. 541/533, print 4005) shows the whole marsh area to be covered by healthy *Spartina anglica* (Fig. 2).

1951 Outfall No. 1 starts operation.

1953 Outfall No. 2 joins No. 1.

1954 Aerial photograph taken by the RAF (RAF Sortie No. 82/895 F21, print 0104) shows what can be interpreted as oily vegetation along the edges of the creeks through which the two outfalls pass (illustrated in Fig. 3).

1962 Photographs of the marsh (taken by Dr D. S. Ranwell, Nature Conservancy Coastal Ecology Research Station) show that large areas of *Spartina anglica* have died and decomposed around the two outfalls.

1966 Ordnance Survey map the area and produce sheets no. SU 4404 and SU 4504 with the boundary of the saltmarsh as shown in Fig. 4.

1969 A series of transects across the marsh show extensive areas of bare mud with the remains of *Spartina anglica* stems and roots (Baker, 1971). At the edge of the denuded area was a belt of oily vegetation behind which the marsh was healthy. The mud level in the denuded areas was 15–25 cm lower than the healthy marsh, presumably due to erosion.

1970 Re-survey of the transects (Baker, 1970b) shows small increases in the amount of oily vegetation at the boundary of the healthy marsh, but otherwise no change from 1969 (Fig. 5). The conclusions reached by Dr Baker after this re-survey were that (a) the damage to the marsh was due to repeated light oilings of the vegetation and shoots from films of oil stemming partly from the effluent and partly from spillages, and (b) recolonisation would be prevented by the continual presence of these films, though the mud itself, whilst containing variable amounts of oil, was not toxic and growth of *Spartina anglica* occurred in samples of the mud in the laboratory.

MONITORING THE FAWLEY SALTMARSH, JULY 1972–DECEMBER 1974

Methods

(i) Vegetation Mapping

Subsequent to the results of the marsh surveys by Baker (1970b), a monitoring programme was started to assess the condition of the marshes twice a year, once in the summer (July), when the plants grow vigorously, and once during the winter (late December or January) when *Spartina anglica* and the annual plants such as *Salicornia* spp. and *Suaeda maritima* die back. The objective of the programme was to locate changes in the extent of marsh vegetation in association with the effluent improvement programme which had been started by the refinery. The series of line transects radiating from the outfall surveyed by Baker (1970b) do not give a complete picture of the plant distribution or effluent effects on the marsh due to unevenness in the effluent distribution caused by the complex creek system and the considerable marsh area involved. This method was replaced by a vegetation mapping technique involving surveying the whole of the marsh area and, using a simple abundance scale based on plant density, roughly quantifying the distribution of all the main marsh plants. Each species was rated as follows:

Abundant (A) The majority of the plants less than 50 cm apart, and often very close to each other.

Common (C) The majority of individual plants between 50 cm and 1 m apart. There may be small clumps of individuals growing closer together within this category, or small patches of less dense vegetation.

Rare (R) Individual plants more than 1 m apart and may be very widely scattered.

Two further categories of vegetation assessment were utilised. These were: bare mud or absent—which was recorded where either no plants or none of that species occurred for 10 m in any direction from the observer—and 'healthy' saltings. The norm for this category was the saltmarsh to the north

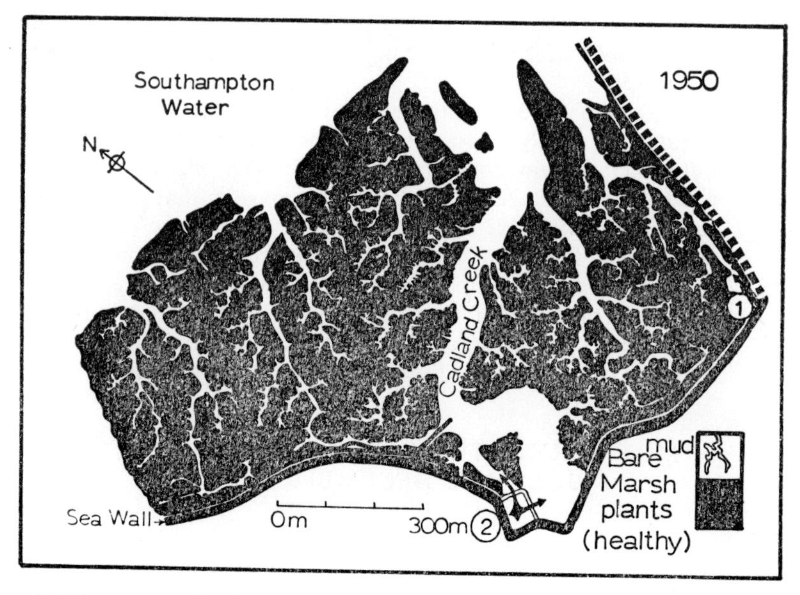

FIG. 2. The extent of saltmarsh vegetation at Fawley in 1950. The information for production of this map was obtained from RAF Photograph Sortie No. 541/533, print 4005.

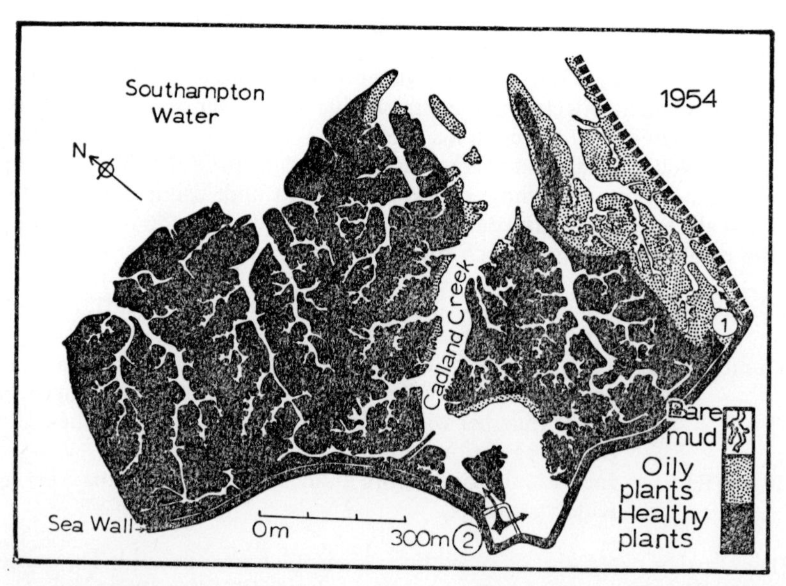

FIG. 3. The extent of saltmarsh vegetation at Fawley in 1954. Two areas of what can be interpreted as oily vegetation are shown around the two effluent-carrying creeks in RAF Photograph Sortie No. 82/895 F21, print 0104.

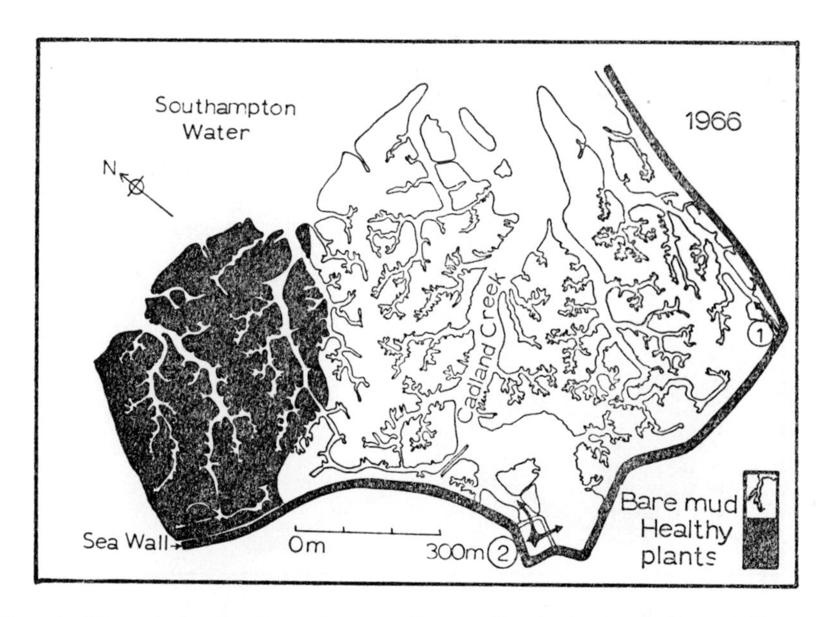

FIG. 4. The extent of saltmarsh vegetation at Fawley in 1966 (from OS charts SU 4404 and SU 4504).

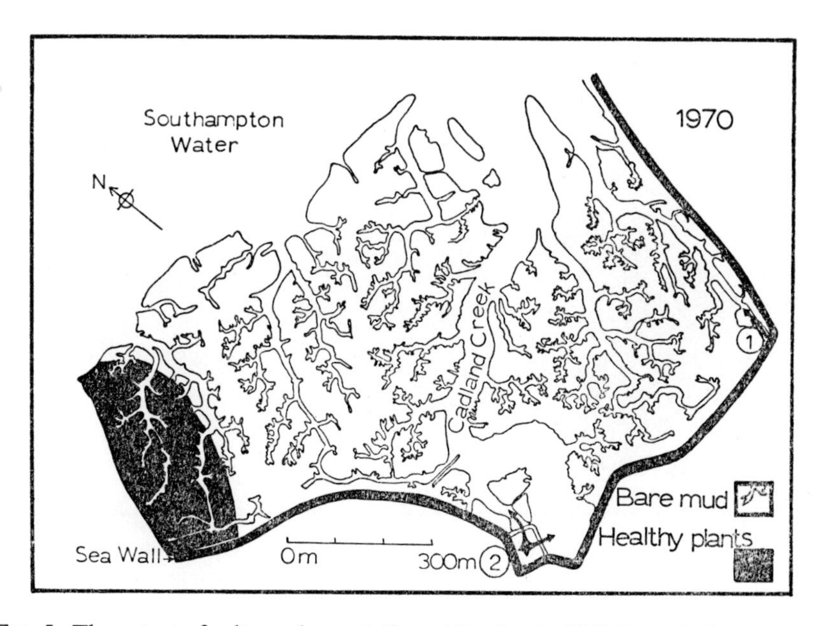

FIG. 5. The extent of saltmarsh vegetation at Fawley in 1970 (from Baker, 1970a).

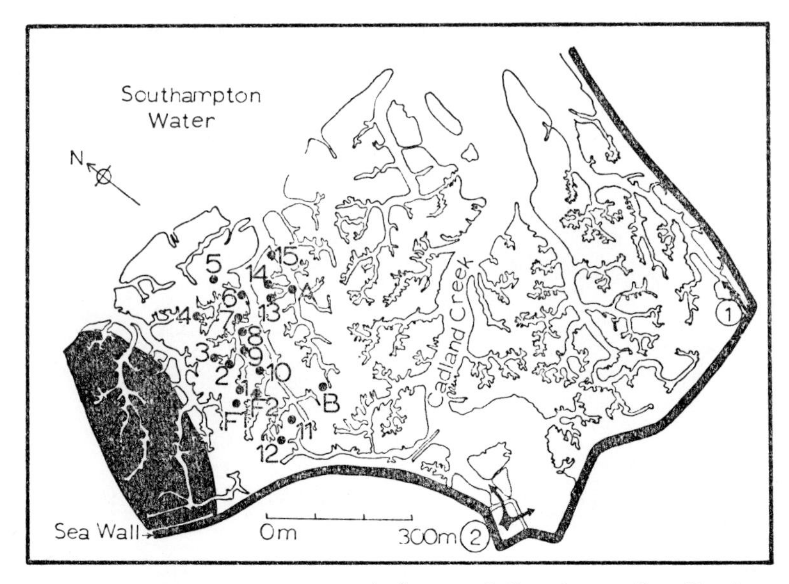

FIG. 6. The location of the measured clumps of *Spartina anglica*, Fawley saltmarsh. Measurement started in December 1972 and has continued to the present time. The extent of saltmarsh vegetation in 1970 is shown (shaded).

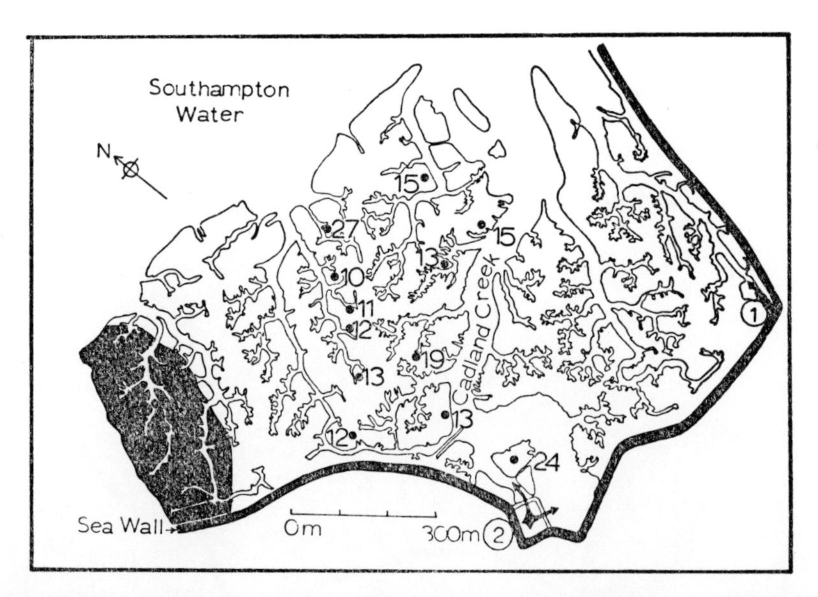

FIG. 7. The location of groups of *Spartina anglica* plants transplanted from the healthy marsh to denuded areas in July 1973 at Fawley. The numbers of shoots in each transplant is shown. The extent of saltmarsh vegetation in 1970 is also shown (shaded).

of the outfall area where there were no manifest effects of pollution damage. This category was extended to cover areas previously affected by oil but now containing all the main marsh plants growing vigorously, either commonly or abundantly, but at lower densities or in slightly different proportions to the unaffected marsh. To fit into this category, vegetational cover of the mud should be complete with no sign of oil or pollution damage to the existing population of plants or shoots.

The categories for vegetation classification are arbitrary, and were chosen for speed of use and to fit with the observed plant distributions. They conveniently describe the broad differences in observed density, and provide a clear picture of the present plant community. Quantification of vegetational measurement was not possible because of the large areas of marsh involved.

In addition to the assessment of distribution of the main marsh plants, the extent of other mud colonisers, e.g. blue green algae, diatomaceous films, and filamentous green algae, were noted, particularly in the winter months when these organisms grow well due to the wetter conditions.

Two further experimental schemes were initiated to monitor vegetational changes in association with mapping.

(ii) Spartina *Clump Measurement*

Detailed measurement was made of the shape and size of individual clumps of *Spartina* growing in the areas which had been denuded by pollution. It was expected that spread or decline of these clumps would act as a pollution indicator. The location of these clumps is shown in Fig. 6. All occur in the area where no living plants were recorded in 1970 (Baker, 1970b), and have therefore grown since 1970. Some (13, 14, 15, A and B) appeared as clumps of shoots (presumably from viable root stock) between the surveys in July and December 1972.

(iii) Transplants

As a follow-up to the *Spartina* clump measurements, transplants of groups of shoots and root stock of *Spartina* were made in July 1973 from the healthy marsh to areas near to and further from Cadland Creek. The progress of these transplants was recorded by the counting of healthy shoots during subsequent surveys. The location of the transplants and the number of healthy shoots at that time is shown in Fig. 7.

Results

(i) Change in Marsh Plant Distribution, July 1972–December 1974

The vegetational mapping shows change in the distribution of all the dominant marsh species between July 1972 and December 1974. For comparative purposes, all maps show the boundary of saltmarsh vegetation found by Baker in 1970. This boundary, with the exception of its strand line end, has proved to be the limit of pollution effects on plants to date. The distribution of *Salicornia* is illustrated in July 1972, July 1973 and July 1974 (Fig. 8); of *Spartina* in July 1972, July 1973, July 1974 and December 1974 (Fig. 9); and of *Halimione* in July 1972 and December 1974 (Fig. 10), as examples of change in the extent of saltmarsh vegetation over the monitoring period.

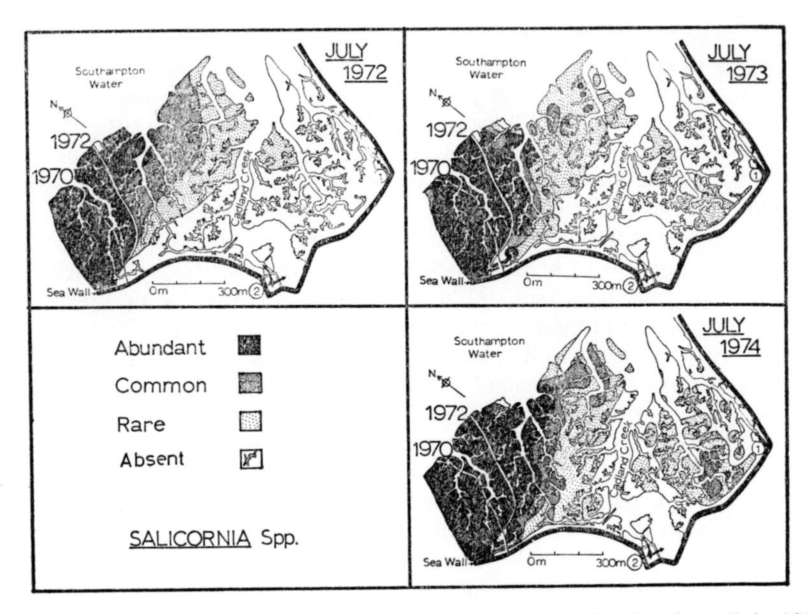

FIG. 8. The changes in distribution of *Salicornia* spp. at Fawley from July 1972 to July 1974. The boundaries of 'healthy' saltings in 1970 and 1972 are marked.

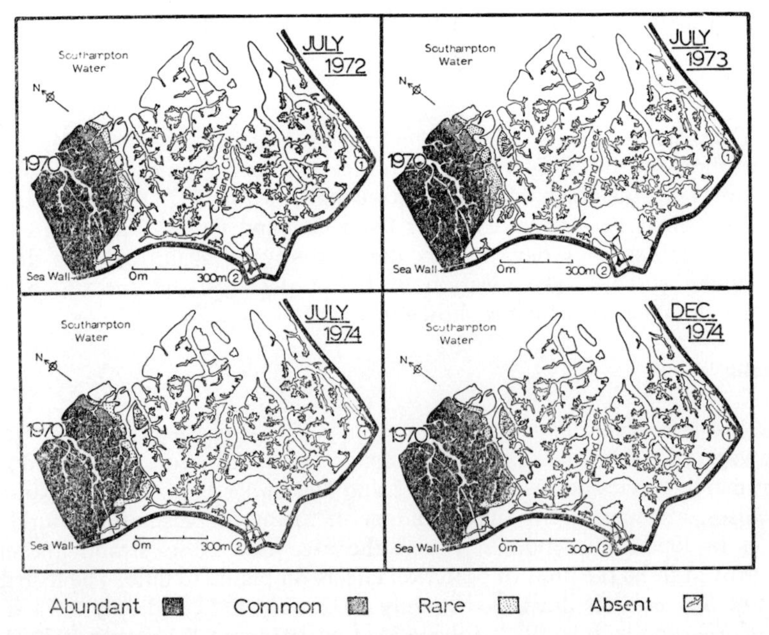

FIG. 9. Changes in the distribution of *Spartina anglica* at Fawley between July 1972 and December 1974. The boundary of the vegetation in 1970 is marked on each map.

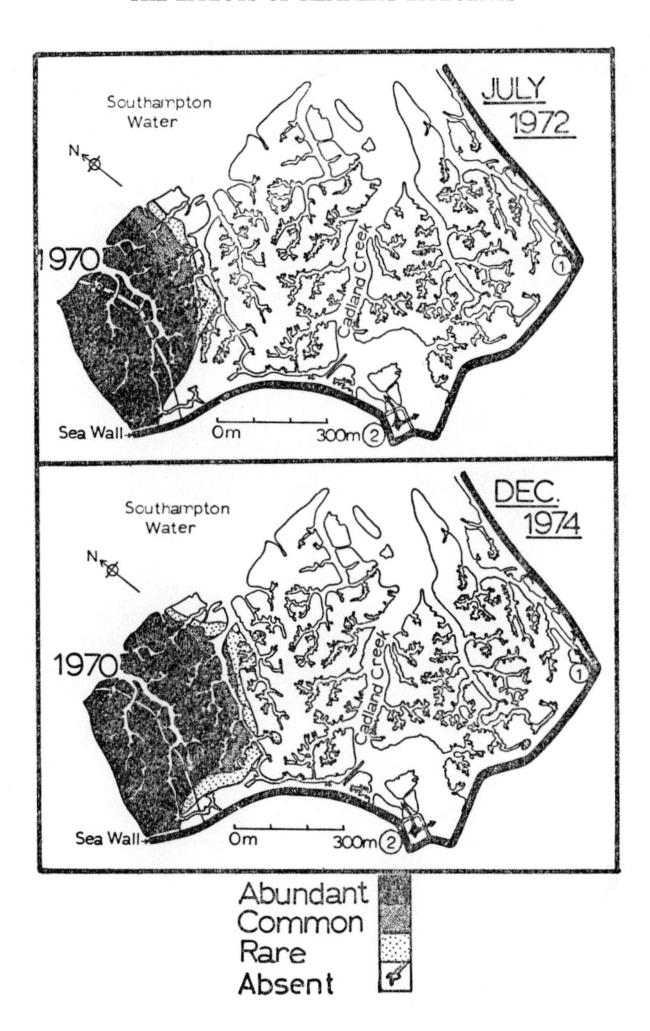

FIG. 10. The distribution of *Halimione portulacoides* at Fawley in July 1972 and December 1974. The boundary of the vegetation in 1970 is also marked.

Suaeda has shown very similar change in distribution to *Salicornia* and *Aster* to *Halimione*. All species have clearly begun recolonisation of those areas previously denuded by pollution, again with the exception of the area of the strand line. Oil was found only occasionally in small areas of the marsh during this period. In the strand line area (high-tide mark), however, further regression of all species had occurred between 1970 and 1972 (Figs. 8–10), although between 1972 and 1974 no further damage was noted. Seedlings of *Salicornia* and *Suaeda* established at the edge of this area between 1973 and 1974.

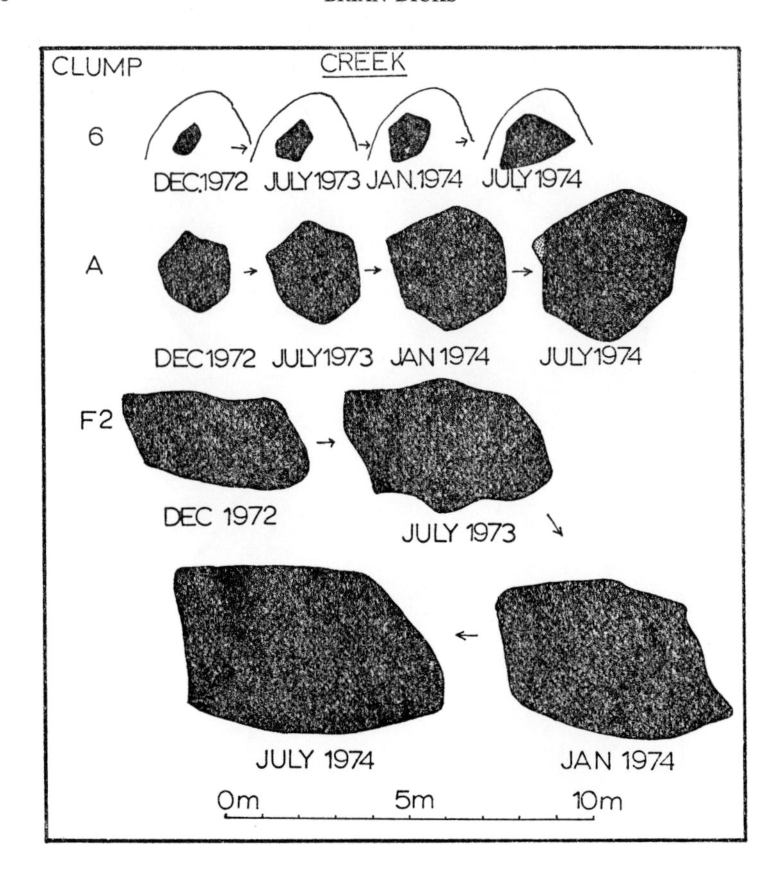

FIG. 11. Change in size of three of the 19 measured clumps of *Spartina anglica* between December 1972 and July 1974. Stippled areas have died back after growth.

(*ii*) *Measured Clumps of* Spartina anglica

The progress of the measured *Spartina* clumps between December 1972 and July 1974 is illustrated with selected examples of the 19 monitored clumps in Fig. 11. Every clump has grown in size over the period monitored, some considerably, although some small parts of some clumps died back apparently not related to oiling. Some of the clumps are no longer mapped because they have been enveloped by surrounding *Spartina* growth and are no longer distinguishable.

(*iii*) *Transplanted* Spartina anglica

The numbers of healthy shoots in the transplants were counted at transplantation in 1973 and during the three surveys of January, July and December 1974. The results are shown in Fig. 12. Whilst those transplants made to previously denuded areas but well away from the creeks carrying effluent have all flourished, those transplanted to near these creeks have all died.

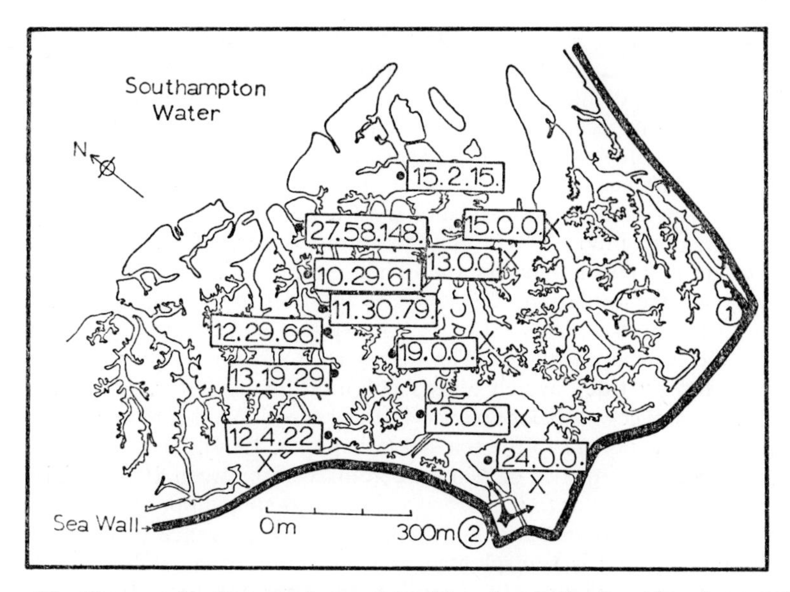

FIG. 12. Changes in the numbers of healthy shoots in the *Spartina anglica* transplants between July 1973 and December 1974 at Fawley. The three figures at each transplant site show numbers of shoots at transplanting in July 1973 (left-hand figure), in July 1974 (middle figure) and in December 1974 (right-hand figure). Transplants marked × were oily on the shoots in January 1973, and the majority of these subsequently died.

(*iv*) *Diatomaceous Films, Blue-green Algae and Filamentous Green Algae*
All these types of plant have flourished, particularly during the winter months of the survey. In December 1972 these organisms occurred over most of the bare mud area with the exception of approximately a 50 m bare area around each outfall and along the edges of the creeks taking the main effluent streams. Their distribution has not changed since, but the wet mild winters have produced very dense growths in most areas.

(*v*) *A Disease Affecting* Spartina anglica
During the survey of December 1974, *Spartina* throughout the experimental area was shown to be suffering from extensive seed infection by an ergot fungus *Claviceps purpurea.* This fungus seriously reduces the seeding ability of *Spartina*, the infection attacking the *Spartina* seeds and using the seed food store to produce its spores. Though the infection is unlikely to kill plants, it can cause decrease in vigorous growth.

DISCUSSION

The series of marsh vegetation distribution maps (Figs. 2–5 and 8–10) clearly show two phases in the progress of the Fawley marsh. In 1950 the area was presumably a flourishing *Spartinetum* similar to the unaffected areas of marsh to the north of the refinery. This was followed by a phase of extensive ecological damage between 1951 and 1970 when large areas of vegetation were killed and decomposed to leave bare mud (Figs. 2 and 3). Subsequent to this date the denuded area has entered a recolonisation phase when all the main marsh species have re-established in various areas of the bare mud. The initial colonisers have been *Salicornia* spp. and *Suaeda*, followed by *Aster* and *Halimione*, and, with the least spread into the denuded areas, by *Spartina*. This process appears similar to a normal successional process of colonisation of mud flats in Southampton Water, where *Salicornia* and *Suaeda* are primary colonisers and, when established, start raising the mud level by silt accumulation. This stage is followed by a *Spartinetum*, which stabilises the mud with its extensive root system and accumulates more silt, raising the level even higher. The marsh then becomes better drained and less affected by tides, which allows colonisation by *Aster*, *Halimione* and *Juncus*. This successional sequence is determined by the rising mud level, drainage and exposure to the sea salt (Chapman, 1964).

The observed sequence of plant recolonisation between 1970 and 1974 at Fawley, though similar to normal succession, has some important differences. For example, *Halimione* and *Aster* appear out of sequence before *Spartina*, and when *Spartina* is transplanted to some areas where it has not spread naturally it grows vigorously. Therefore the appearance of *Aster* and *Halimione* before *Spartina* in the recolonisation sequence is not determined by a successional factor like attainment of a suitable mud level, but presumably by factors which influence the spreading of the plants. It is likely that mud level is still suitable for growth of most of the marsh species over a large area of the marsh, for the whole area was previously covered by healthy marsh, the mud is still bound and stabilised to some extent by the old *Spartina* root systems and has eroded only a small amount (15–25 cm, Baker, 1970b). In some localised areas, erosion may be greater or less than average. Those with greater erosion may only be suitable for recolonisation by *Salicornia*. The main difference between those plants which spread to bare areas first and those which come later seems to be the success with which the plants spread by seed. *Salicornia*, *Suaeda* and *Aster* are annual plants reproducing primarily by seed. The abundance of *Salicornia* and *Suaeda* in the healthy marsh means that large numbers of seeds will be available to establish in any new areas where the seeds can germinate and grow. Hence they will be among the first to enter any suitable areas. *Halimione* spreads both by seed and vegetative means. *Spartina*, however, spreads mainly by vegetative methods and seeds very poorly (Dalby, 1970). This seeding may further be reduced by the heavy ergot infection noted during the latest survey.

It is feasible, therefore, that if recovery continues, large areas of the denuded mud will end up dominated by *Aster*, *Halimione*, *Salicornia* and *Suaeda* rather than *Spartina*, and the original *Spartinetum* of the marsh may not return. Hence the recolonisation pattern and its species composition for the present will probably be determined by seed distribution and seedling success. Observation of the marsh plants during surveys supports this. *Salicornia*, during its spread from July 1972 to July 1974, first established in the area between Nos. 1 and 2 outfalls all around the heads of the small creeks off the main creeks, This fits well with distribution of the seeds by the rising and falling tides. Around these few established individuals, the following year's

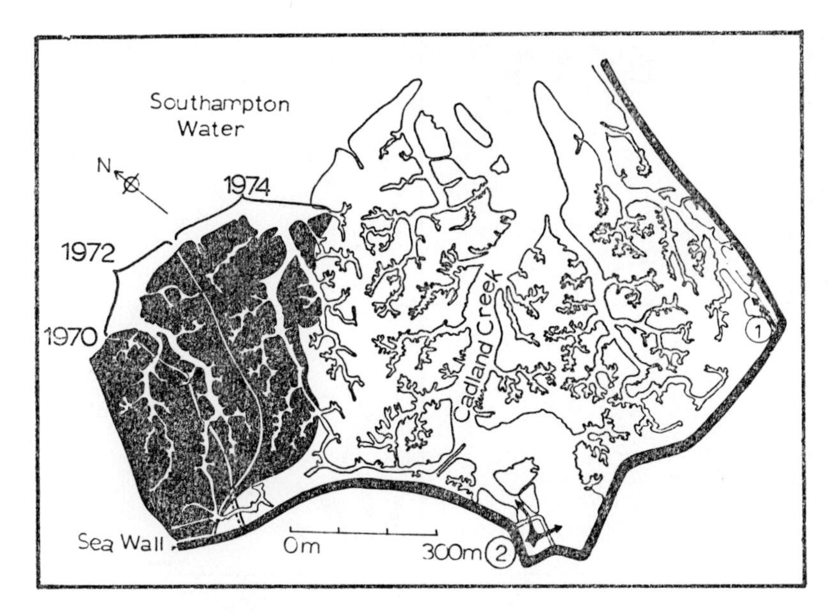

FIG. 13. Extent of 'healthy' saltings at Fawley in 1970, 1972 and 1974. The areas marked 1972 and 1974 were determined as 'healthy' saltings after the surveys of July 1972 and July 1974.

seeds from these few established plants greatly increased the density and spread of this species in that area. Similar seedlings spread have been observed in *Suaeda* and *Halimione*. The slow spread of *Spartina*, therefore, seems due to its predominantly vegetative reproduction. Most interestingly, for the first observed time since 1972, *Spartina* has produced seedlings which have established in new areas between July and December 1974, again primarily in small creeks.

As the *Spartinetum* stage may not be the end-result of recolonisation over the whole marsh (though it is in some of the areas near the 1970 boundary which have already recovered almost completely—see below), the definition given for 'healthy' saltings may not be applicable. A revision of the definition to include the return of the marsh to 100% vegetational cover, though not necessarily dominated by *Spartina*, should be made.

As the limiting factors for the spread of *Spartina* seem to be its poor seeding and slow vegetative reproduction method, and as transplants have established in some areas, an extensive transplantation scheme may well speed up recolonisation of denuded areas by this species, and affect a return to a *Spartinetum*. More important, such a transplanting operation will show in those areas where the effluent is still exerting an effect, or areas where marsh level has eroded sufficiently to prevent recolonisation. An extensive transplant programme has been started for these reasons.

In the two areas marked in Fig. 13, vigorous growth of all the main marsh species has occurred, and there is no evidence of fresh oil damage. The area nearest the 1970 boundary was in this 'healthy' condition in 1972, whilst

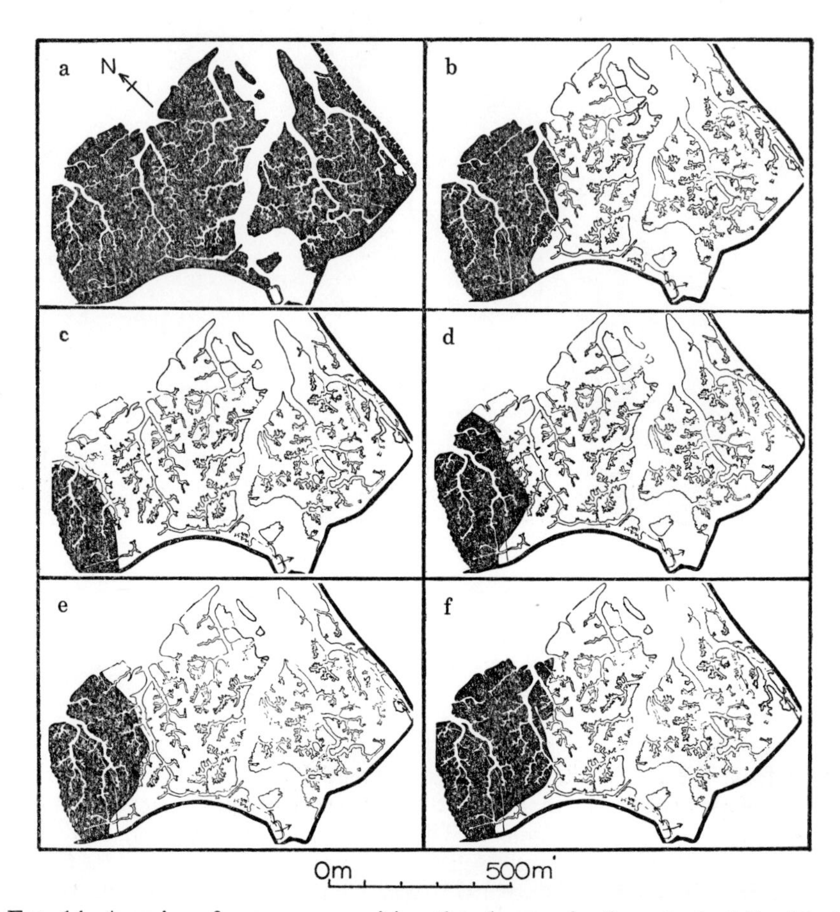

FIG. 14. A series of maps summarising the changes in the extent of 'healthy' saltings (shaded areas) at Fawley from 1950 to July 1974. (a) 1950; (b) 1966; (c) 1970; (d) 1972; (e) 1973; (f) July 1974.

the second area reached this state in 1974. The monitored *Spartina* clumps in this area have grown vigorously, and many are no longer measurable as they have been incorporated into extensive spreading *Spartina*. Whilst there are occasional patches of dead oiled vegetation (pre-1970) in these areas, the overall condition of the salting vegetation is healthy and recovered to a *Spartinetum*. Though growth in these areas is not quite as dense as that in the nearby unaffected marsh, they can now be considered as 'healthy' saltings. Changes in the extent of the 'healthy' saltings between 1950 and December 1974 are summarised in Fig. 14, and in the extent of saltmarsh plants in Fig. 15.

Whilst recolonisation of large areas of denuded mud has occurred, there is an interruption to this pattern in the strand line area (Figs. 8–10, 14 and 15). Any new oil usually ends up in this area after deposition by high tides. This has probably been the case since damage to the marsh started, and has resulted in very oily mud in this region (Baker, 1971). The occasional accu-

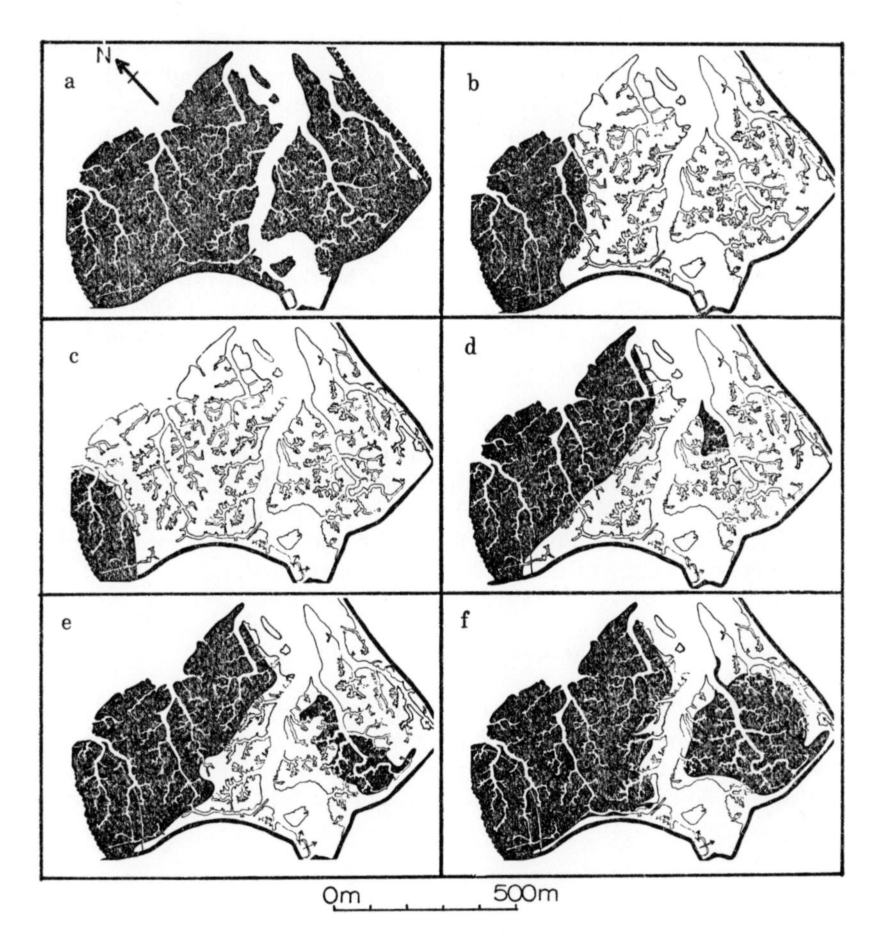

FIG. 15. A series of maps summarising the changes in areas containing living marsh plants (shaded areas) at Fawley from 1950 to July 1974. (a) 1950; (b) 1966; (c) 1970; (d) 1972; (e) 1973; (f) 1974.

mulation of new oil in this area, and the refloating of old oil at high tides, produced further regression of the marsh plants between 1970 and 1972, although since then no further damage has been observed. Some seedlings of *Salicornia* and *Suaeda* have established nearby among dead oily vegetation, but none have established in an area between 10 and 20 m wide immediately below the strand area. This area may well not recover for a considerable period of time, either until all oiling stops or until the oil there is stabilised in the sediment and can no longer float into and on to the vegetation.

An interesting point on recolonisation is that neither mud containing old oil (Baker, 1970a) nor dead oily *Spartina* seems to prevent growth of seedlings, particularly of *Salicornia* and *Suaeda*. There are areas of dead oily *Spartina* vegetation (oiled previous to 1970) in the boundary region of the healthy marsh, 1970, which are colonised now by *Salicornia*, *Suaeda*, *Halimione*, *Aster* and even by vegetative shoots of *Spartina*.

The reasons for the observed recolonisation patterns are not certain at present. It is likely that improvement of effluent quality and reduction in the effluent volume have played a part. Little fresh oil has been observed on saltmarsh plants since monitoring started in 1972. However, the mildness of winters over the last three years accompanied by early springs and late autumns may well have influenced marsh plant growth and reproduction, causing early germination and a very long growing season. A severe winter or cold spell after early germination may have considerable deleterious effects. A further point of importance may be the relatively few oil slicks which have affected the marsh since 1970. The sensitivity of *Salicornia* spp. to oil on the water surface and its role as a primary recoloniser may mean that one serious spillage could considerably alter the present recolonisation pattern.

For these reasons, and to watch the situation in the strand line area where oil which occasionally occurs on the sea surface usually ends up, continued monitoring is necessary. To determine accurately the relative importance of effluent improvement, climatic change or oil spillages to the recolonisation of denuded areas, study of the marsh for some time to come is required, and the interim nature of this report should be stressed.

ACKNOWLEDGEMENTS

I am grateful to Esso Petroleum Company for sponsoring this work, and to Peter Sutton for his assistance during survey work and in checking the manuscript. I am also grateful to members of the Oil Pollution Research Unit and to Andrew Arnold for assistance in the field during these surveys.

REFERENCES

American Petroleum Institute (1963). *Manual on Disposal of Refinery Wastes Vol. 1: Waste Water Containing Oil*, 7th ed., 104 pp.

Baker, J. M. (1970a). Studies on saltmarsh communities—Successive spillages, in: *The Ecological Effects of Oil Pollution on Littoral Communities*, ed. E. B. Cowell, Elsevier, 1971, pp. 21–32.

Baker, J. M. (1970b). Studies on saltmarsh communities—Refinery effluent, *ibid.*, pp. 33–43.

Baker, J. M. (1970c). Studies on saltmarsh communities—Seasonal effects, *ibid.*, pp. 44–51.

Baker, J. M. (1970d). Studies on saltmarsh communities—Oil and saltmarsh soil, *ibid.*, pp. 62–71.

Baker, J. M. (1970e). Studies on saltmarsh communities—Comparative toxicities of oils, oil fractions and emulsifiers, *ibid.*, pp. 78–87.

Baker, J. M. (1970f). Studies on saltmarsh communities—The effects of oils on plant physiology, *ibid.*, pp. 88–98.

Baker, J. M. (1971). The effects of oil pollution and cleaning on the ecology of saltmarshes. Ph.D. thesis, Univ. Coll. S. Wales, Swansea.

Blokker, P. C. (1970). Prevention of water pollution from refineries, in: *Water Pollution by Oil*, ed. P. Hepple, Applied Science Publishers for Institute of Petroleum, 1971, pp. 21–36.

Blokker, P. C. and Marcinowski, H. J. F. (1970). *Survey on Quality of Refinery Effluents in Western Europe*, Stichting CONCAWE, Report 17/70.
Chapman, V. J. (1960). *Saltmarshes and Salt Deserts of the World*, Leonard Hill, London, 392 pp.
Chapman, V. J. (1964). *Coastal Vegetation*, Pergamon Press, Oxford, 245 pp.
Dalby, D. H. (1970). The saltmarshes of Milford Haven, Pembrokeshire, *Field Studies*, **3**(2), 297–330.
Hubbard, C. E. (1954). *Grasses*, Penguin Books, 463 pp.

13
Experimental Investigation of Refinery Effluents

JENIFER M. BAKER
(Oil Pollution Research Unit, Orielton Field Centre, Pembroke, Wales)

SUMMARY

A number of different types of experiment used for the investigation of refinery effluents are described with examples. On the one hand are settlement plate and transplant experiments useful for monitoring and the further study of effects observed during field surveys; on the other hand are laboratory tests for the toxicity ranking of effluents or effluent constituents. The latter type are useful for control purposes and for finding why effluents are toxic. It is difficult to arrange realistic field experiments useful for making predictions about the ecological effects of new effluents in new communities, so monitoring is particularly important in such cases.

INTRODUCTION

There are many different types of experiment useful for the investigation of refinery effluents, none of these, however, can replace field surveys and monitoring though some are suitable for inclusion in monitoring programmes. Experiments range from transplant and settlement plate techniques used in the field with whole effluents, to laboratory toxicity tests on effluent constituents.

The different types of experimental approach are outlined below, with examples from our own work and from the literature given when possible.

SETTLEMENT PLATES

The settlement plate technique involves the fixing of uniform substrates, for example squares of asbestos or glass microscope slides, on jetty piles, buoys, or other suitable fixtures over a wide area. Growth of organisms is recorded over a period of time, and this should show up effluent gradient effects uncomplicated by local variations in type of substrate. This method is likely to be useful in some monitoring schemes for delimiting areas of effect and finding changes in such areas from year to year.

TRANSPLANTS

Transplants, or 'detector species' can be used in the same sort of way as settlement plates for measuring the area of effect of an effluent. Suitable species show responses such as change in behaviour, change in growth or reproduction, or death, when influenced by the effluent.

An experiment involving the transplanting of the saltmarsh grass *Spartina anglica* is described in the paper by Dicks (1975a). In an investigation of the Kent refinery effluent, groups of winkles (*Littorina littorea*) were transplanted along the foreshore (for map see Fig. 10 in Baker, 1975).

In the first experiment, groups of 100 winkles marked with spots of cellulose paint were released on 20.9.73 on the refinery sea wall near the effluent and at the western and eastern extremes of the refinery area. Controls were established near the mouth of the Medway. On 31.10.73, the following numbers were recaptured: 0 near the effluent, 25 and 22 at the extremes of the refinery area, 58 and 45 near the mouth of the Medway. Winkles near the effluent were observed to be in a sick, flaccid condition very soon after transplanting, and were presumably then washed away. There is an indication of more widespread effect, though this cannot of course be proved without replication of the experiment.

To overcome the problem of wandering winkles, a further experiment was conducted using plastic mesh cages tied down to the foreshore. These each contained 20 winkles plus a seaweed food supply, and were set out on 8.7.74, at the sites previously mentioned plus some intermediate sites in the refinery area. On 11.7.74 they were collected—all winkles were healthy including those from near the effluent. The difference between the two results may reflect overall effluent improvement or variation in toxicity of the effluent over periods of a few days. In this context, transplants near the effluent serve as a field bioassay.

Transplants were used for another purpose in the case of the Little Wick effluent, Milford Haven. The limpet (*Patella vulgata*) populations of this area (described in Baker, 1975) are unusual because the overall density is low but individual limpets are large. This could happen if adult limpets were affected by the effluent, leaving only a few particularly robust individuals with an abundant food supply. Limpets (in the size range 20–40 mm) were therefore transplanted onto the end of the effluent pipe, and to control areas, to find if there was an effluent effect on adult animals. In all cases there were initial losses which can be ascribed to the trauma of moving; otherwise, however, limpets on the effluent pipe survived as well as the controls (Fig. 1). This particular experiment therefore provides evidence to suggest that it is limpet settlement or the young stages which are affected.

EXPERIMENTAL DISCHARGES IN THE FIELD

Experimental discharges of effluent on otherwise unpolluted shores are limited by the problem of transporting effluent, and of course by the undesirability of causing damage. A few short term (half hour or less) discharges have been carried out in isolated areas with Milford Haven effluents—on

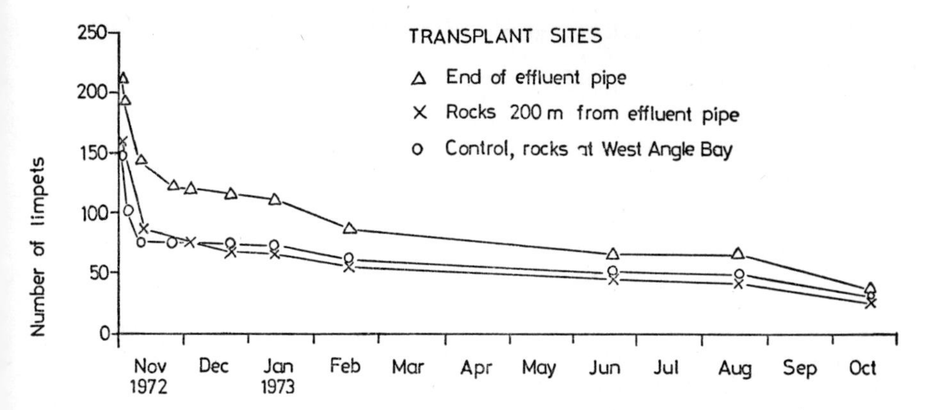

FIG. 1. Survival of transplanted limpets at different sites.

Patella vulgata on a rocky shore, *Littorina saxatilis* in a rock pool, and filamentous red algae (*Griffithsia*, *Polysiphonia* and *Ceramium*) on a mud flat. These short discharges of undiluted effluent did not cause any mortality but a behavioural response was noticed in the case of *Littorina saxatilis*—50% crawled up the sides of the rock pool out of the effluent.

The temporary discharge of a refinery effluent down a saltmarsh creek (see 'Martinshaven' in Baker, 1975) could be regarded as an incidental extension of this type of experiment. In this case, a few days discharge was sufficient to cause behavioural changes, sickness and mortality in ragworms, crabs and gobies.

The advantage of field experiments is that it is possible to know the state of a community before it is treated, to regulate the treatment, and to follow interspecific reactions after treatment. Field experiments involving oils and dispersants have been invaluable, however those involving effluents, and therefore necessitating continuous flow arrangements, are much more difficult to set up.

LABORATORY INVESTIGATION OF SHORT TERM BEHAVIOURAL RESPONSES

Common types of laboratory toxicity test use death of the test animal as the criterion for toxicity. This is not usually a convenient criterion for tests involving effluents because effluents are not usually acutely toxic—death of test organisms may not occur for days or weeks if it occurs at all. As Cowell (1974) has pointed out, the long term continuous flow experiments necessary for this type of test suffer from almost insuperable problems of differential absorption of hydrocarbons onto tubing and tank sides and stripping of volatile compounds through aeration. Easily observed short term behavioural responses are more promising as a basis for bioassay or toxicity ranking tests.

Responses which have been used so far are:

Drop-off reaction of limpets.
Non-activity of barnacle nauplii.
Retraction of winkles into their shells (Parsons, 1972a).
Escape reaction of ragworms from their burrows.

Such responses are described more fully in the paper by Dicks (1975b).

Measurement of the respiration rates of 'convenience organisms' such as yeast is another possibility for effluent bioassays—it has been used for dispersant ranking by Nelson-Smith (1969) and Reynard (1973). There is, however, scope for using a much wider range of responses than is indicated here, for within the limited scope of toxicity ranking (e.g. comparing one effluent constituent with another or the whole effluent from day to day) it probably does not matter which species are used (see Cowell, 1974). Standardisation has bureaucratic rather than biological advantages.

The other use of laboratory investigation of behavioural response is in the interpretation of field survey data. For example, barnacle settlement patterns near the Little Wick effluent may be explained in terms of behavioural responses of nauplii to effluents (see Dicks, 1975b). We would very much like to investigate the effects of the Little Wick effluent on limpet larvae, to see if the peculiarities of the population there can be explained in terms of prevention of settlement caused by an effluent induced response. It has so far, however, proved impossible either to catch sufficient limpet larvae for an experiment or to induce adult limpets to spawn in the laboratory.

TOLERANCE RANKING

Tolerance ranking of species from rocky shores and saltmarshes has been worked out for oils and dispersants (Baker, 1971; Crapp, 1971a) but not for refinery effluents, though similar orders of tolerance might be expected. Less is known about other communities, though freshwater fish have been ranked in a search for suitable bioassay test animals for refinery effluents (Douglas, 1963; Gould, 1962; Ward, 1962). Clearly, neither extremely tolerant nor extremely sensitive species are suitable. Tests on the tolerance of a range of species are necessary before sensible decisions can be made concerning bioassay species, and this process is inevitably linked with observation of behavioural responses as previously described.

BIOASSAYS ON WHOLE EFFLUENTS

The purpose of routine bioassays on effluents must be to determine whether they are more or less toxic than (a) a standard or (b) than at some time in the past. Bioassay results can be extremely useful in showing trends, sudden increases of toxicity caused by malfunction, or, in the case of tests on different effluent streams, for the location of main sources of toxicity within a refinery. However, it must be stressed that ecological predictions cannot be made entirely from bioassays, and that meeting a standard, as defined by a relatively short term test, does not guarantee freedom from long term ecological effects.

Most countries do not require routine bioassays on refinery effluent, however it is worth recording that Canadian regulations as from November 1973 require regular tests using rainbow trout as the standard organism (Environment Canada, 1974).

Bioassays of the Kent refinery effluent, using winkle activity as an indicator, were carried out during August 1974 in a preliminary attempt to locate sources of toxicity. Test winkles (*Littorina littorea*) were collected from Grain tower, near the mouth of the Medway, and used to assay effluent samples collected over 24-hour periods from different effluent streams. Effluent samples were all kept in a refrigerator until needed, then raised to test room temperature before being tested simultaneously on tanks of winkles. (It is not valid to test different sets of samples at different times of day, because winkles have a diurnal rhythm of activity.)

In the case of cooling water from the main plant and the lubricating oil plant, winkle activity was not depressed significantly below that of the controls; however, there was a depression in the case of the process water. Further process water tests, using samples taken from the weirs of two separators, confirmed this result and also indicated a considerable difference between the two separators. One had a very variable pH and salinity, and had a general depressing effect on winkle activity. The other had less variable pH and salinity, and a less depressing effect on activity; however, there appeared to be an inverse relationship between pH and activity. Since the lowest activity was at the two values closest to the pH of sea water, it is unlikely that pH alone is the causal factor, but it may reflect other effluent characteristics that were not analysed.

Though these results do not go very far in elucidating the problem of effluent toxicity, they do illustrate the possible use of bioassays for this purpose.

TOXICITY RANKING OF EFFLUENT CONSTITUENTS

The problem of effluent toxicity can also be approached from another direction, the study of individual known constituents such as sulphides or other factors such as temperature, and their interactions. Here again, the approach is one of toxicity ranking and a number of test organisms are suitable. Jenkins (1964) using fish, studied the toxicity of effluent components (excluding oil) and their interactions. He concluded that ammonia, phenols and sulphides are important toxicants, but could find no consistent relationship between toxicity and chemical oxygen demand or alkalinity. Interactions occurred between pH and ammonia, pH and sulphide, and ammonia and phenol. Interaction between ammonia and pH (increase of pH resulting in reduced survival in ammonia) appeared to be the most important.

Oil, apart from being a very variable constituent, is difficult to study for reasons already given. However, the following work is relevant to the effluent problem. The effects of prolonged exposure to low concentrations of Kuwait crude oil, uncomplicated by other toxicants, was investigated by Crapp (1971b) using starfishes (*Asterias rubens*). Starfishes kept and fed under these conditions were eventually killed, although several weeks elapsed before signs of damage were noted. The work of Mackin and Hopkins (1962) using bleedwater (brine effluent from salt dome oilfields, in this case containing about 30 ppm oil) is also of interest. Laboratory experiments showed that bleedwater, crude oil, water extracts or emulsions of crude oil had no effect on survival of oysters over periods of several months. In physiological experiments, bleedwater and water extracts of crude oil slowed water pumping and filtering rates of oysters, but these effects were temporary and reversible. A discharge of 6600 barrels of bleedwater per day at Lake Barre increased mortality of oysters as far as 50–75 ft from the point of discharge and apparently caused decrease of shell growth and glycogen storage in oysters as far as 150 ft away, but had no detectable effect at greater distances.

The salinity of refinery effluents is of importance. Any freshwater stream discharging into estuarine or coastal waters produces salinity gradients which are reflected in the biota, and the same will apply to a low salinity effluent. Low salinity is implicated, for example, in the non-activity response of barnacle nauplii to an effluent in Milford Haven (Dicks, 1975b). There may, however, be interactions between salinity and other factors.

Interaction between oil and salinity and between total effluent and salinity has been investigated by Ottway (1972) and Parsons (1972b) respectively. Crude oils and effluents are more toxic to animals at the extreme limits of their salinity tolerance range. However, *Littorina saxatilis* is capable of recovering from short term exposure to effluent in a range of salinities 20–55% (Parsons, 1972b). Effluent toxicity followed by this range of osmotic stress was not sufficient to stop the animals crawling clear. A size related response has been noticed in the case of the pinfish *Lagodon rhomboides* (Kloth and Wohlschlag, 1972). In this case sub-lethal effluent pollution depressed fish metabolism, and this was affected by salinity in the case of the larger fish.

An experiment using the ragworm escape reaction (following the Martinshaven incident described by Baker, 1975) demonstrates that low salinity alone is not enough to explain some effects. The effluent in question had at the time of the experiment the low salinity of $3{\cdot}25\,^0/_{00}$. Tanks of ragworms (*Nereis diversicolor*) in mud were subjected to effluent, to clean sea water of salinity $28\,^0/_{00}$, and to clean diluted sea water of salinity $3{\cdot}25\,^0/_{00}$. Only the effluent produced an escape reaction; the worms otherwise remained in their burrows. Extraction of worms from the mud at the end of the experiment showed that they remained healthy when subjected to the low salinity sea water. Worms which escaped from their burrow under the influence of effluent recovered when put into clean sea water, but under natural conditions they would of course be quickly eaten by predators such as gulls and would not have a chance to recover.

CONCLUSIONS

In conclusion, experimental work on refinery effluents falls into two main categories. On the one hand are experiments used for monitoring or explaining ecological effects observed in the field; on the other are experiments used for ranking toxicity of whole effluents or effluent constituents. Field experiments to study the effects of effluents on hitherto unpolluted communities are practically impossible to organise in a realistic fashion, and it is difficult to make ecological predictions without such experiments. The monitoring of new effluents is therefore particularly important because it replaces the field experiment stage of the investigatory and prediction-making process.

ACKNOWLEDGEMENTS

This work would not be possible without the fullest co-operation from many oil companies for which we are most grateful. I would like to thank John Addy, Catharine Howe and Roger Parsons for assistance with experiments.

REFERENCES

Baker, J. M. (1971). Successive spillages, in *The Ecological Effects of Oil Pollution on Littoral Communities* (ed. E. B. Cowell), I.P., London, pp. 21–32.

Baker, J. M. (1975). Investigation of refinery effluent effects through field surveys (this volume).

Cowell, E. B. (1974). A critical examination of present practice, in Beynon, L. R. and Cowell, E. B. (eds.) *Ecological Aspects of Toxicity Testing of Oils and Dispersants*, I.P., London, pp. 97–104.

Crapp. G. B. (1971a). Laboratory experiments with emulsifiers, in *The Ecological Effects of Oil Pollution on Littoral Communities* (ed. E. B. Cowell), I.P., London, pp. 129–49.

Crapp, G. B. (1971b). Chronic oil pollution, *ibid.*, pp. 187–203.

Dicks, B. (1975a). The effects of refinery effluents—the case history of a saltmarsh (this volume).

Dicks, B. (1975b). The importance of behavioural patterns in toxicity testing and ecological prediction (this volume).

Douglas, N. H. (1963). Evaluation and relative resistance of sixteen species of fish as test animals in toxicity bioassays of petroleum refinery effluents, Ph.D. thesis, Oklahoma State University.

Environment Canada (1974). Petroleum Refinery Effluent Regulations and Guidelines, Environmental Protection Service Report, EPS 1-WP-74 1, Water Pollution Control Directorate.

Gould, W. R. (1962). The suitabilities and relative resistances of twelve species of fish as bioassay animals for oil refinery effluents, Ph.D. thesis, Oklahoma State University.

Jenkins, C. R. (1964). A study of some toxic components in oil refinery effluents, Ph.D. thesis, Oklahoma State University.

Kloth, T. C. and Wohlschlag, D. E. (1972). Size-related metabolic responses of the pinfish *Lagodon rhomboides* to salinity variations and sub-lethal petrochemical pollution, *Contributions in Marine Science*, **16**, 125–37.

Mackin, J. G. and Hopkins, S. H. (1962). Studies on oyster mortality in relation to natural environments and to oil fields in Louisiana, *Publ. Inst. mar. sci. Univ. Texas*, **7**, 1–131.

Nelson-Smith, A. (1969) Micro-respirometry and emulsifier toxicity, O.P.R.U. Annual Report 1969, pp. 43–7.

Ottway, S. (1972). Some effects of oil pollution on the life of rocky shores, M.Sc. thesis, University of Wales.

Parsons, R. (1972a). Some sub-lethal effects of refinery effluent upon the winkle *Littorina saxatilis*, O.P.R.U. Annual Report 1972, pp. 21–3.

Parsons, R. (1972b). Note on the recovery of *Littorina saxatilis* from pollution with refinery effluent when under osmotic stress, *ibid.*, p. 31.

Reynard, S. (1973). An evaluation of respirometry for toxicity testing, O.P.R.U. Annual Report 1973, pp. 74–7.

Ward, C. M. (1962). The relative resistance of 15 species of fishes to petroleum refinery effluent and the suitability of the species as test animals. Ph.D. thesis, Oklahoma State University.

14
Some Physical and Biological Effects of Oil Films Floating on Water

SHEILA VAN GELDER-OTTWAY
(formerly with Oil Pollution Research Unit)

SUMMARY

This paper describes investigations made on some physical and biological effects of oil films floating on water, with special reference to the rock-pool environment. Experiments with simulated rock pools demonstrate that photosynthesis and animal activity may be reduced, oxygen levels decreased and temperature increased if oil films are present. Experimental plants and animals generally survived during the period of the experiments and later recovered in clean sea water.

From a discussion of the literature, it appears that oil films on water do allow some degree of gas exchange, although the rate of exchange may be complicated by oxygen uptake in the oil in the process of abiotic or microbial oxidation. Effects on rate of exchange are minimal or non-measurable on the open sea and maximal in still water such as rock pools or under various experimental conditions. Temperature and light-reduction effects follow a similar pattern.

INTRODUCTION

Oil spilt at sea rapidly spreads as a thin film over the water surface. In the open sea, wind and wave action break up oil films and increase the rate of physical emulsification of oil into droplets. Oil washed ashore into rocky pools or spilt in sheltered inland or coastal waters, however, may remain as a continuous film of oil on the water surface for some time, thus forming a physical barrier between the air and water. Such a physical barrier will affect the rates of gas exchange and heat exchange between the air and water, and also the transmission of light. Each one of these physical factors is of importance to aquatic life, which may therefore be adversely affected by an oil spill whether or not the oil itself is toxic.

This paper describes investigations made on some physical and biological effects of oil films floating on water, with special reference to the rock-pool environment. An investigation on the light-transmitting properties of different types of crude oil films was reported in a previous paper (Ottway, 1971); this will be further discussed below.

OIL FILMS AND GAS EXCHANGE

Effects of Oil Films over Water on Photosynthesis

Two artificial rock pools, (a) and (b), were set up by putting 13·6 litres of sea water and 280 g (wet weight) of freshly collected *Enteromorpha* into each of two fibreglass tanks (capacity 45 litres); the depth of water in each tank was then 11 cm. Each tank was placed on a laboratory bench away from direct sunlight and beneath a 100 watt tungsten lamp as light source 40 cm above the surface of the water. The light was switched on from 0900 h until 1900 h daily. Over the following 3 days measurements were made, at 4 h intervals, of the oxygen diffusion rate (ODR) and the pH level in the water at a depth of 1–2 cm. The ODR was measured using a platinum cathode (surface area of platinum = 5·5 mm^2) with a calomel reference electrode, while the pH values were measured using chemical indicators in a BDH 'capillator' outfit. On the fourth day a film of fresh Kuwait crude oil 0·75 cm thick was carefully introduced on to the water surface of tank (a) in such a way that it had no or very little contact with the alga. Subsequently ODR readings were made as before, carefully plunging the ODR measuring electrode quickly through the oil film so as not to disturb it. The electrode was cleaned carefully each time before use. The pH readings were taken as before except that a small sample of water from beneath the oil was removed first with a hypodermic syringe. Three days after the oiling, the crude oil on the water surface of tank (a) was carefully removed by absorption on to paper tissues; the alga underneath was washed free of oil and put into clean sea water. Over the following 4 days the ODR was again measured at 4 h intervals, using the alga from tank (b) placed in clean sea water as a control.

Figures 1 and 2 show the diurnal fluctuation of ODR and pH levels in tanks (a) and (b) before and after oiling. Fig. 3 shows the recovery of the alga from tank (a) after 3 days under an oil film. (ODR values are proportional to oxygen concentration: thus the diurnal ODR fluctuations observed

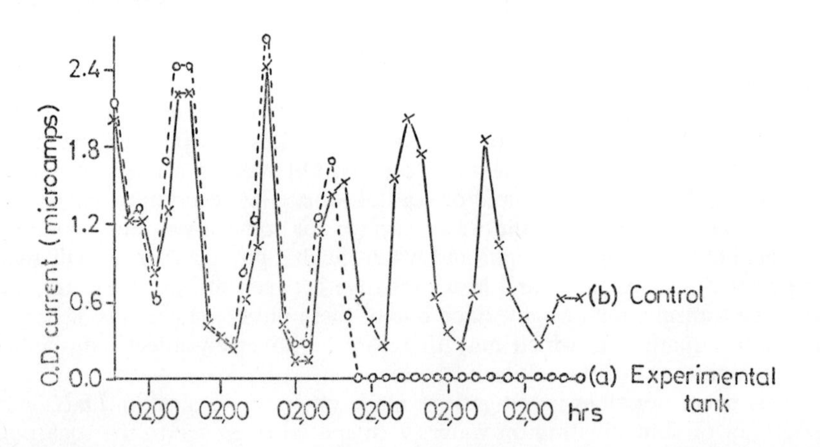

FIG. 1. Diurnal fluctuation in ODR (oxygen diffusion rate) in artificial rock pools containing *Enteromorpha* in sea water: (a) in experimental tank before and after addition of a crude oil film 0·75 cm thick; (b) in control tank with no oil.

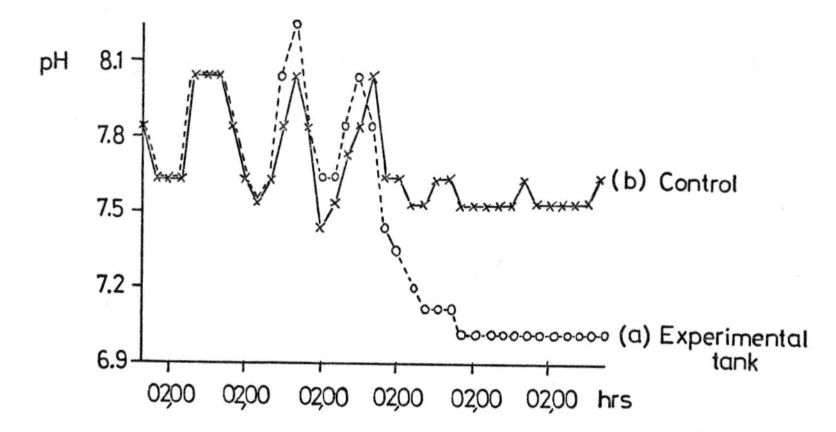

FIG. 2. Diurnal fluctuation in pH in artificial rock pools containing *Enteromorpha* in sea water: (a) in experimental tank before and after addition of a crude oil film 0·75 cm thick; (b) in control tank with no oil.

in tanks (a) and (b) are a measure of the fluctuation of oxygen concentration in the water.) The appearance of the *Enteromorpha* retrieved from under the oil film was dull in colour and somewhat shrivelled and flaccid compared with the control; a few small dead crustaceans were also found in it. After the 4 day recovery period, however, the alga had become a brighter green and more closely resembled the control in appearance.

The diurnal pattern of fluctuations in oxygen concentration in the experimental tanks (a) and (b) indicates the photosynthetic activity of *Enteromorpha.* It is clear from Fig. 1 that the introduction of a film of crude oil 0·75 cm thick on the water surface results in an immediate cessation of this diurnal fluctuation, with a very rapid fall in the oxygen concentration of the water to zero. It appears that the presence of the black oil film on the water surface firstly inhibits photosynthesis so that oxygen production ceases, and secondly acts as a physical barrier between the air and water so that dissolved oxygen available in the water is rapidly used up by respiration of the alga (and any other small organisms present with it). It is possible that oxygen from the air may pass through this thickness (0·75 cm) of crude oil, but not in sufficient quantity or at a fast enough rate to meet the normal respiratory demands of the plant. The rapid fall in pH of water under the oil film shown in Fig. 2 is similarly indicative of a corresponding increase in carbon dioxide concentration as a result of the oil acting as a barrier between the water and air. The recovery of the *Enteromorpha* which had been under the oil film, illustrated in Fig. 3, shows that the alga was capable of regaining its normal photosynthetic activity after several days.

Effects of Oil Films over Water on Animal Behaviour

The animals chosen for this study were three species of intertidal gastropod common on rocky shores in Britain: the dog whelk *Nucella lapillus*, the edible periwinkle *Littorina littorea* and the rough periwinkle *Littorina saxatilis*. In the case of each species, 125 individuals were collected and divided

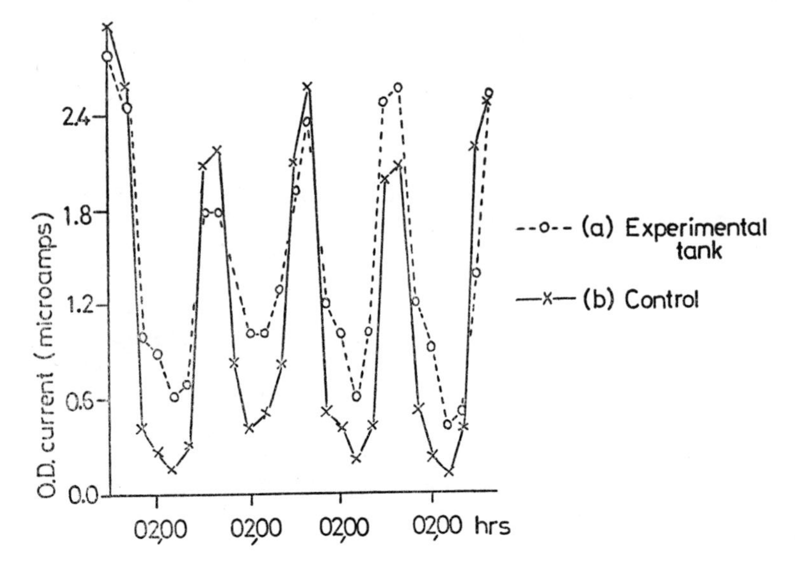

FIG. 3. Diurnal fluctuations in ODR (oxygen diffusion rate) in artificial rock pools containing *Enteromorpha* in sea water: (a) with *Enteromorpha* recovering from 4 days under a crude oil film; (b) with normal *Enteromorpha* in control tank.

into 5 groups of 25 animals. Each group of 25 animals was placed in a glass jar with vertical sides, and sea water was then poured into the jars in suitable volumes to fill the jars approximately two-thirds full: 1000 ml in the cases of *N. lapillus* and *L. littorea* and 250 ml in the case of the smaller species, *L. saxatilis*. Kuwait crude oil was then carefully added to four out of five jars for each species to produce surface films of the following thicknesses:

1. 0·0 cm (control).
2. 0·1 cm.
3. 0·2 cm.
4. 0·4 cm.
5. 0·6 cm.

In the case of jar no. 2, the oil film was discontinuous over the water surface. No aeration or food was supplied, and the jars were left for 24 h in a constant-temperature room at 12°C. After this period, all 25 animals had emerged from the water in the non-oiled jar no. 1 in the case of each species, but none at all had emerged from the water in the other jars. This behaviour of animals of each species in the control jar is typical of intertidal gastropods which normally climb out of water up the sides of tanks under laboratory conditions. Animals in jars 2–5 for each species had been observed attempting to crawl through the oil film up the glass side, but with no success. A long rectangular piece of slate was therefore introduced into each jar to provide a rough surface at a less steep angle. Although some of the gastropods climbed

TABLE I

Effects of Oil Films over Water on Three Species of Intertidal Gastropod

Species	*Jar*	*Volume of sea water in each jar* (ml)	*Thickness of oil film on water* (cm)	*ODR value of water*	*pH value of water*	*Percentage final mortality*
Nucella lapillus	1	1000	0·0	2·6	7·4	0
	2	1000	0·1	0·9	7·2	40
	3	1000	0·2	0·9	7·2	36
	4	1000	0·4	0·8	7·1	20
	5	1000	0·6	0·6	7·1	8
Littorina littorea	1	1000	0·0	2·9	7·6	0
	2	1000	0·1	0·9	7·4	4
	3	1000	0·2	0·6	7·4	0
	4	1000	0·4	0·6	7·3	4
	5	1000	0·6	0·5	7·1	0
Littorina saxatilis	1	250	0·0	2·5	8·8	0
	2	250	0·1	2·0	8·4	12
	3	250	0·2	1·4	8·0	44
	4	250	0·4	1·1	7·4	28
	5	250	0·6	1·0	7·4	52

ODR (oxygen diffusion rate) and pH values are given for sea water containing 25 individuals of each species of mollusc after 5 days of being covered by crude oil films of various thicknesses. The final column shows percentage mortality figures for each species after a 5 day recovery period following the 5 day experimental period in water under oil films.

on to the slate provided, they were still unable to crawl through the oil film on the water surface.

Oxygen diffusion rate (ODR, which is proportional to oxygen concentration) and pH readings of the water under the oil film in each jar were taken 5 days after the addition of the oil films. ODR values were measured using a platinum cathode and calomel reference electrode, while pH values were measured using indicators, as described in the previous section. After these readings had been made, the animals in each jar were retrieved, washed free of oil and put back into the same water as that in which they had been under the oil films, but now filtered clean of surface oil and vigorously aerated. The animals were then observed over the following 5 day period to see if they recovered.

Table I shows the results of the ODR and pH readings for the water in each jar for each species after 5 days under oil films. After this 5 day period under oil films, the animals of each species appeared to be very lethargic, and in many of them the soft parts of the body were extruded to a great extent. After the animals had been washed and put back into aerated, filtered sea water, they regained normal appearance, proceeded to move around actively and began to emerge out of the water up the sides of the tanks in which they had been placed. Mortalities for each species at the end of the 5 day recovery period are also shown in Table I. Figures 4–6 show the

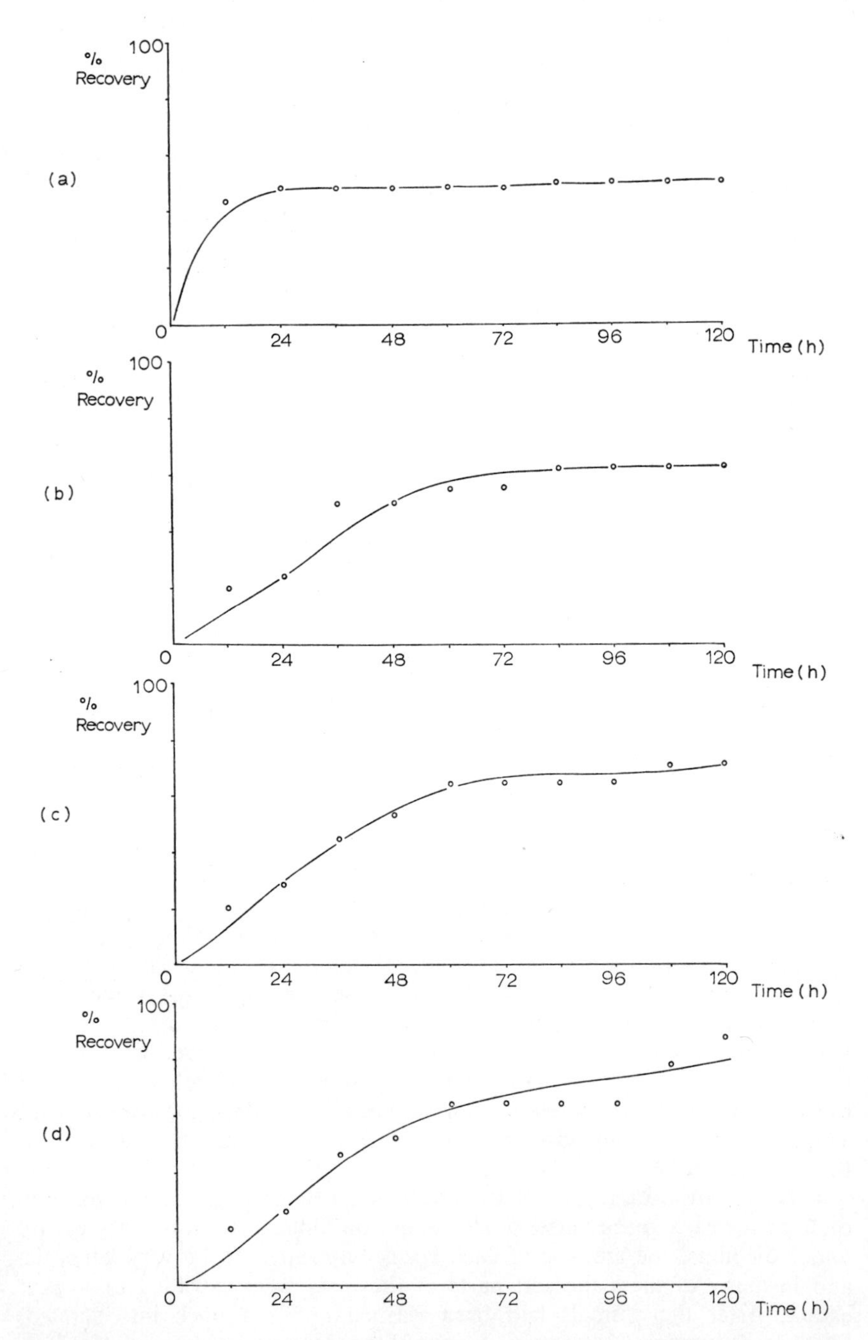

FIG. 4. Recovery of *Nucella lapillus* in aerated sea water, after being in the same water for 5 days under crude oil films of thickness: (a) 0·1 cm; (b) 0·2 cm; (c) 0·4 cm; (d) 0·6 cm.

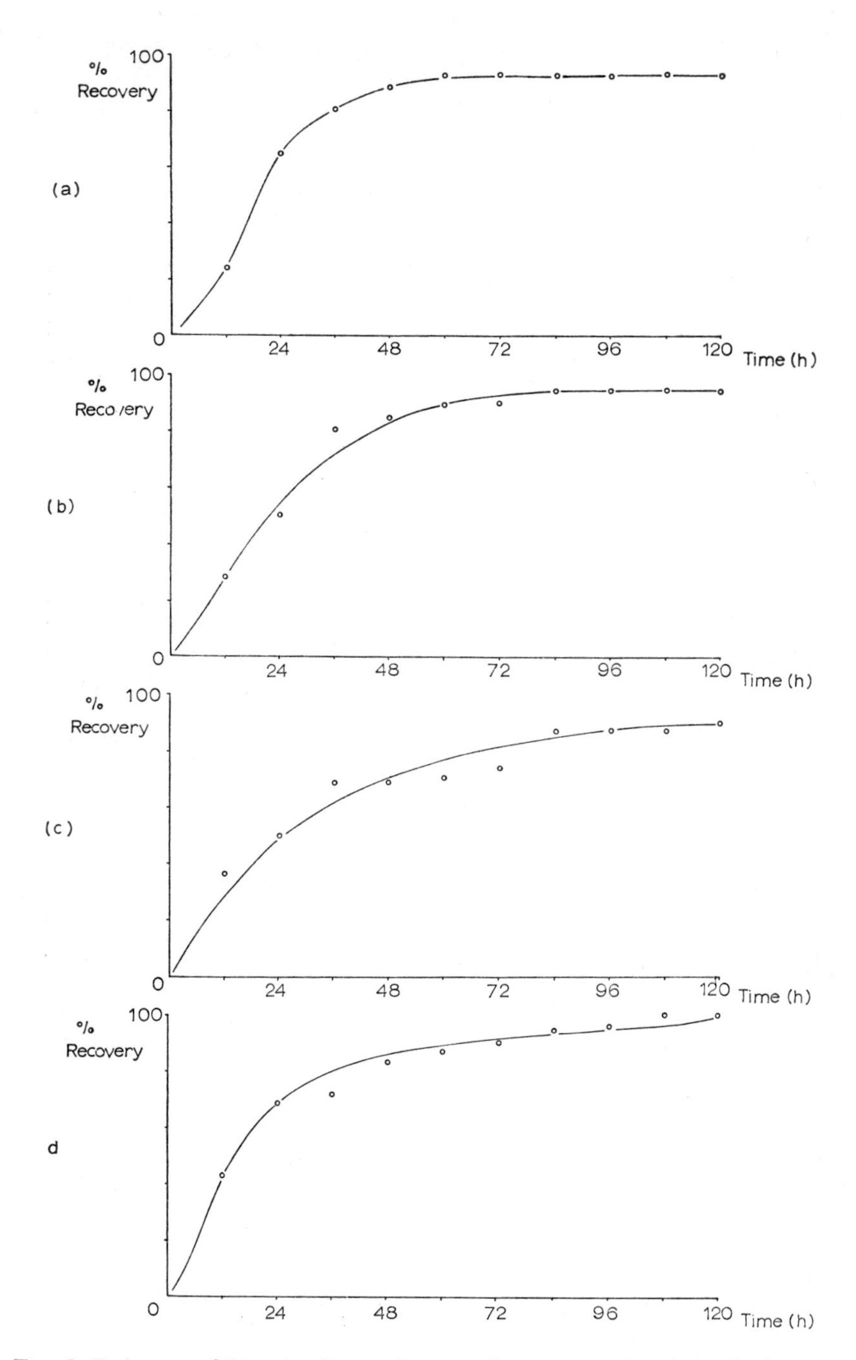

FIG. 5. Recovery of *Littorina littorea* in aerated sea water, after being in the same water for 5 days under crude oil films of thickness: (a) 0·1 cm; (b) 0·2 cm; (c) 0·4 cm; (d) 0·6 cm.

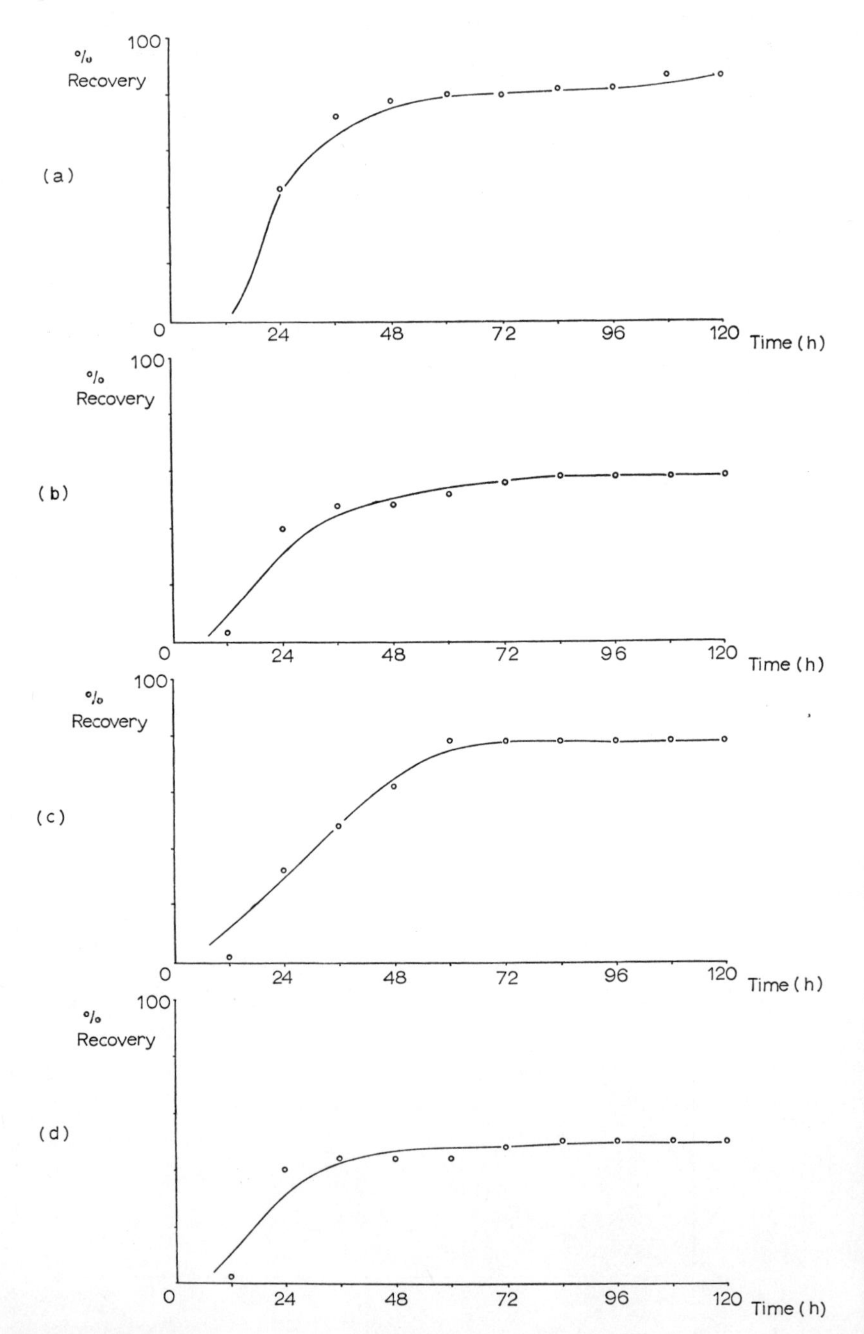

FIG. 6. Recovery of *Littorina saxatilis* in aerated sea water, after being in the same water for 5 days under crude oil films of thickness: (a) 0·1 cm; (b) 0·2 cm; (c) 0·4 cm; (d) 0·6 cm.

recovery rate for each species having been in water under crude oil films of various thicknesses.

From these experiments, it appears that the gastropod species investigated are unable to crawl out of water through oil films 0·1–0·6 cm thick on the water surface. Kinsey (1973), however, during the course of field experiments on an Australian coral reef, noticed that molluscs of the species *Haliotis* crawled to and fro through a crude oil film 0·1–0·7 mm thick introduced on to the surface of an enclosed body of water.

The results of the experiments described above indicate that the animals which had been in water under oil films for several days were in a state of respiratory stress. As shown in Table I, in the case of each species of animal, after 5 days under an oil film the ODR and pH values of the water decrease with increasing thickness of oil film. Since the ODR values are directly proportional to oxygen concentration, and pH values inversely proportional to carbon dioxide concentration of the water, it therefore appears that Kuwait crude oil films of up to at least 0·6 cm in thickness do allow some exchange of gases between air and water. Oil films of all thicknesses used in these experiments nevertheless appear to induce respiratory stress in these animals, judging from their observed extreme lethargy and the extrusion of large areas of their bodies, presumably in order to expose as large an area of the body surface as possible to facilitate gas exchange between the water and body tissues: this would be a most dangerous state of exposure to predators in the wild. Percentage mortality figures for each species shown in Table I indicate very good recovery for *L. littorea.* Mortality figures for *N. lapillus* and *L. saxatilis*, however, are more variable, with overall average mortalities of 26% and 34% respectively for the animals kept in water under oil films. For these two latter-mentioned species some contamination of the water from the decomposition of animals which had died early on in the experiment was suspected: it is very difficult indeed to tell exactly when a gastropod mollusc is dead, especially if it has been very lethargic beforehand. Figures 4–6 show that for all species recovery rates did not differ greatly for groups of animals having been in water under oil films of increasing thickness and thus with decreasing amounts of oxygen available for respiration. Considering that none of the animals were fed at any time during the experiments, and that they were kept in the same small volumes of sea water for 10 days, the percentage mortality figures are not excessively high. It therefore appears that damage suffered by the animals on account of the oil was of a physical rather than a toxic nature.

OIL FILMS AND HEAT TRANSFER

Experiments with Artificial Rock Pools

Two artificial rock pools were set up using two fibreglass tanks (capacity 45 litres) each containing 13·6 litres of sea water to a depth of 11 cm. These two tanks, I and II, were set up in the open air with tank I containing sea water only while tank II contained in addition a dense cover of the green alga *Enteromorpha.* The tanks were left to settle for 18 h, and thereafter the water temperature at the top and bottom of each tank was measured at 2 h intervals from early morning until late evening, using a mercury-in-glass thermometer, for an initial 2 day period. On the third day, 250 ml of Kuwait crude oil were added to each tank, producing a total cover of oil 0·22 cm thick. Water temperatures immediately below the oil layer and at the bottom of the tank were subsequently measured as before over the following 3 day period. Weather conditions over the whole 5 day period (1–5 June 1971) were consistently fine, with clear skies, hot sun, very little wind and no rain. Air temperatures in the immediate vicinity of the tanks were recorded at the same time as the sea-water temperatures.

Figures 7 and 8 show the diurnal temperature fluctuations of sea water in tanks I and II both before and after oiling, together with the diurnal fluctuation of air temperature in sunshine. From these figures it can be seen how the surface sea-water temperatures are higher after a layer of crude oil 0·22 cm thick has been added, attaining a maximum temperature of 34°C; in the absence of the oil layer the surface sea water heats up more slowly than the air and attains a lower maximum temperature under similar weather conditions. The presence of an oil film, at least in conditions of warm sunshine, appears to increase the temperature difference between the surface water at a depth of 1 cm and the bottom water at a depth of 10 cm. The covering of oil on the water, however, does not appear to reduce the rate of heat loss at night, presumably as a result of re-radiation from the surface of the black oil. The presence of the alga in tank II does not seem to have any significant effect on the temperature of the surface or bottom water either before or after oiling.

Heat Absorption Properties of Oil

Experimental determinations of the rate of heat absorption by oil in warm sunshine were also made on a small scale. Equal volumes (170 ml) of Kuwait crude oil (black in colour), gasoline (colourless) and tap water were put into identical glass vessels and exposed all together in sunlight for 2 days (4–5 June 1971), when weather conditions were consistently warm, dry and sunny with very little wind. Over this 2 day period, at 2 h intervals during the daytime the liquid in each vessel was stirred and its temperature measured with a mercury-in-glass thermometer. After 2 days the final volumes of the liquids remaining were measured.

Figure 9 shows the different rates of heat absorption and loss by the small volumes of Kuwait crude oil and gasoline, with tap water for comparison, when these samples were left out of doors in sunshine for 2 days. On the second day there was more cloud cover, accounting for the lower temperatures attained. Table II shows the reduction in volume in each sample after 2 days owing to evaporation: this gives some idea of the relative amounts of heat absorbed by each sample which was used up as latent heat of evaporation rather than increasing the temperature of the sample.

From Fig. 9 it can be seen that the temperature of a small volume of crude oil can rise to 45°C in a few hours in summer sunshine, while equal volumes of water and gasoline attain a temperature of 34°C under the same conditions.

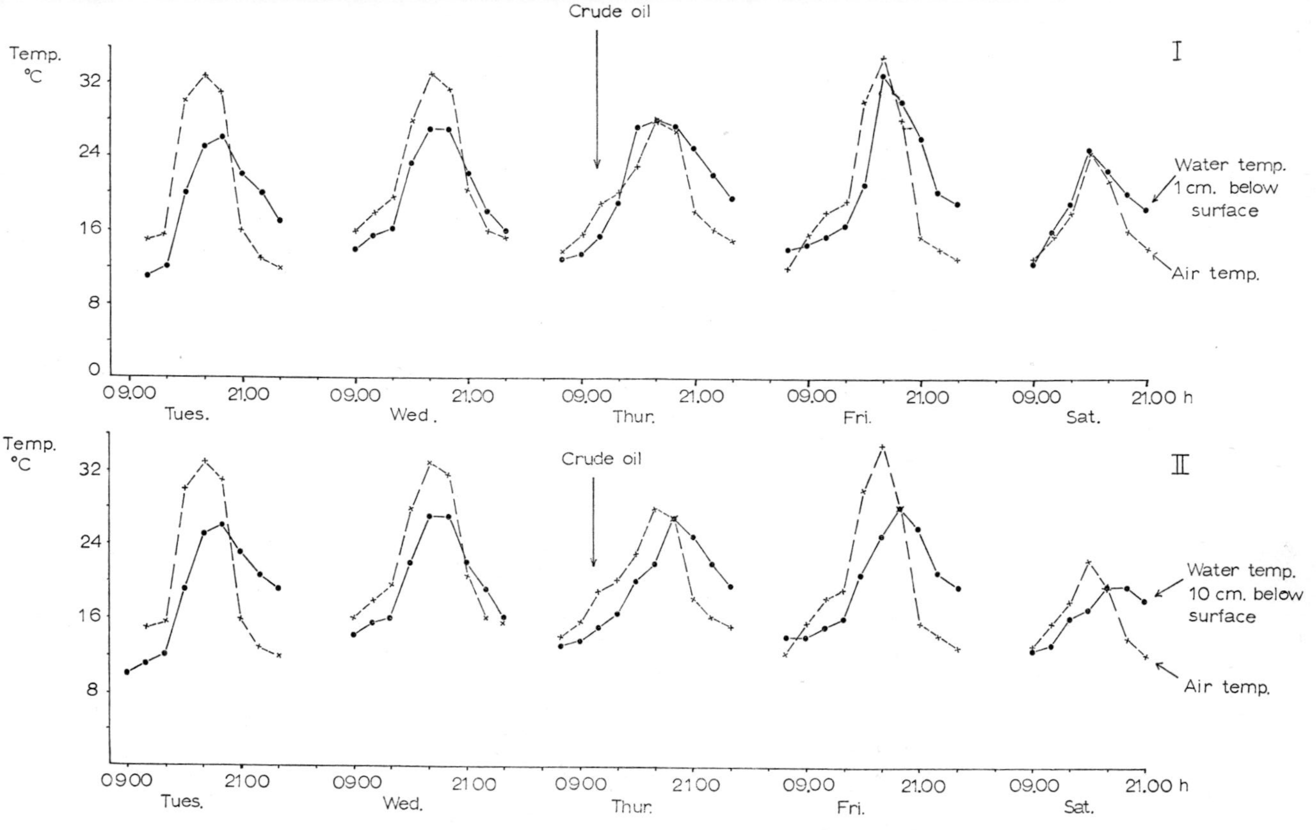

FIG. 7. Diurnal temperature fluctuations in artificial rock pools before and after addition of crude oil films 0·22 cm thick. Tank I contains sea water only; tank II contains sea water and *Enteromorpha*.

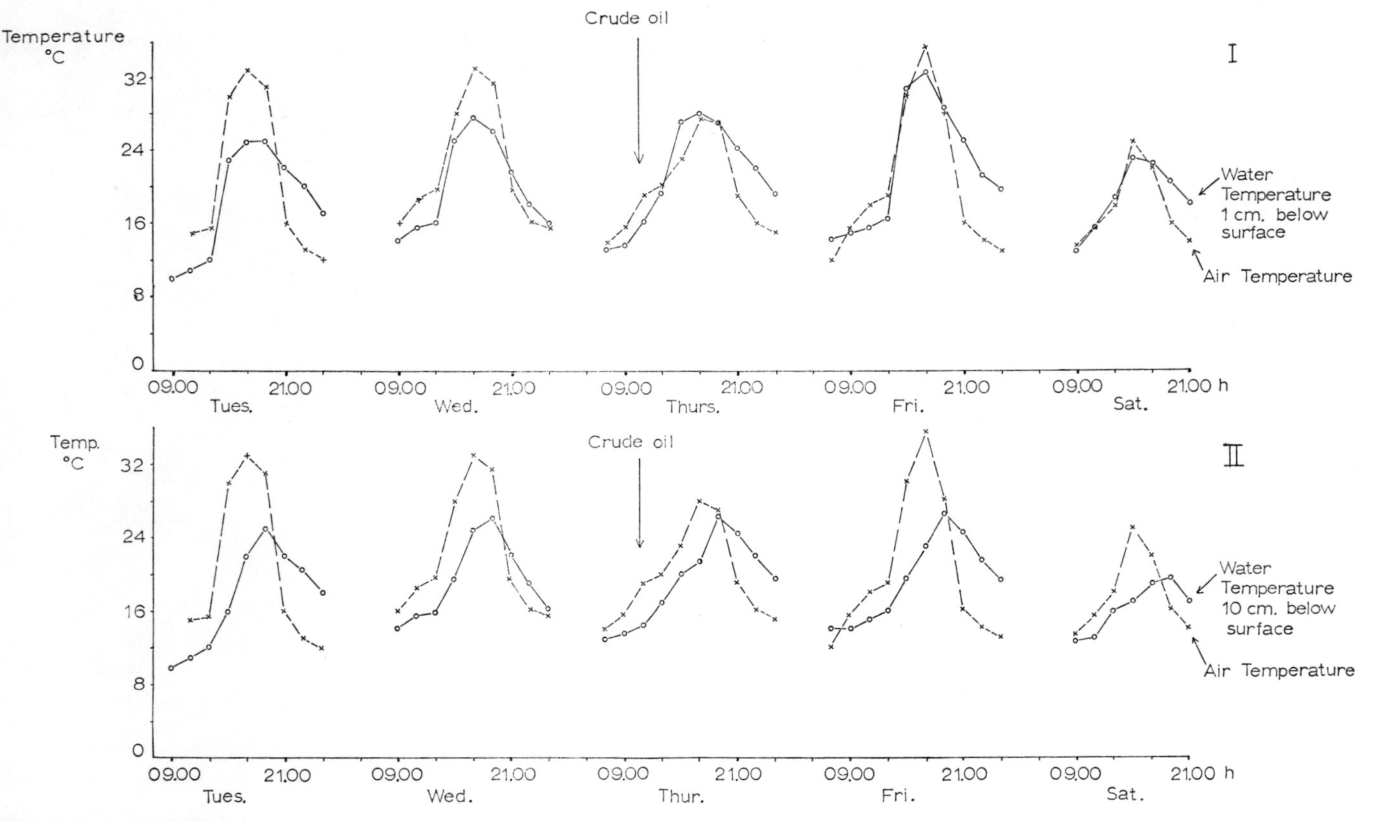

FIG. 8. Diurnal temperature fluctuations in artificial rock pools before and after addition of crude oil films 0·22 cm thick. Tank I contains sea water only; tank II contains sea water and *Enteromorpha.*

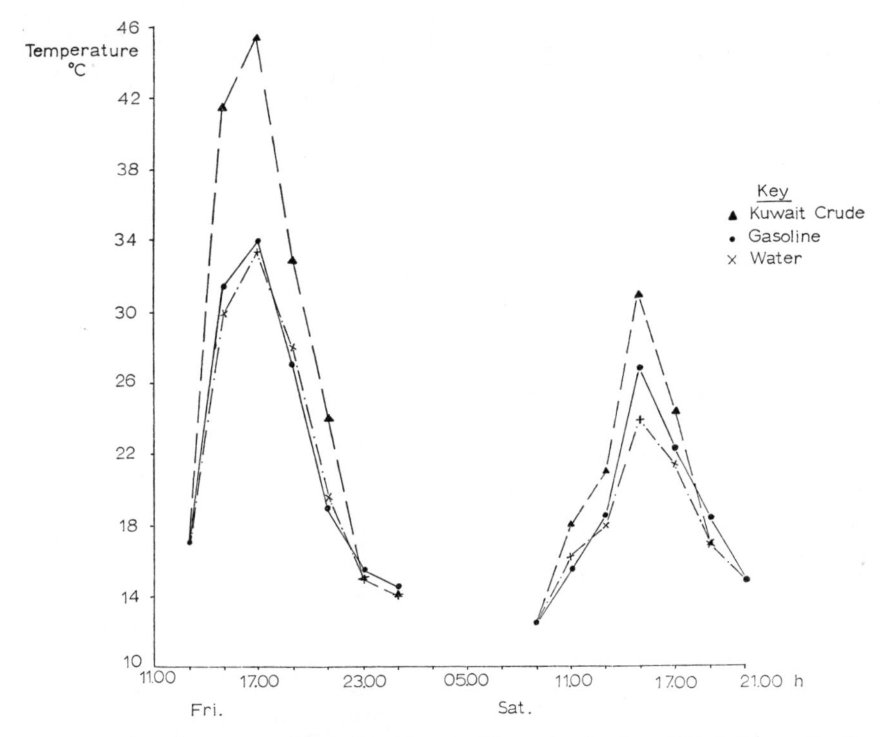

FIG. 9. Rate of heat uptake and loss by small equal volumes of Kuwait crude oil, gasoline and water. Each liquid sample (170 cm^3) was left outside in sunlight for 2 days (4–5 June 1971).

Evaporation Rates of Different Types of Oil

The relative rates of evaporation of different types of oil were measured away from direct sunlight in a laboratory at a mean temperature of 25°C over a 3 day period. The oils used were four different crude oils and five different petroleum fractions, with distilled water for comparison. The five petroleum fractions used were all derived from the same crude oil mixture containing

TABLE II

Reduction in Volume by Evaporation of Oil Samples left Outside in Sunshine for Two Consecutive Days (4–5 June 1971)

Liquid	*Original volume* (ml)	*Final volume* (ml)	*Percentage loss in volume*
Kuwait crude	170	120	29
Gasoline	170	32	81
Distilled water	170	143	16

TABLE III

Evaporation Rates of Different Types of Oil

Oil code no.	*Type of oil*	*Percentage volume of oil evaporated in 72 h from 50 ml sample at 25°C*
A	Fraction: Dieselect	0·0
B	Fraction: No. 2 Fuel	0·0
C	Fraction: Gasoline	45·0
D	Fraction: 3500 Fuel	0·0
E	Fraction: Kerosene	7·0
CT6	Crude (Kuwait)	10·0
CT3	Crude	8·5
CT7	Crude	11·0
CT10	Crude	0·0
Control	Distilled water	4·5

75% Kuwait crude and 25% Nigerian crude: these fractions were dieselect (a diesel fuel), No. 2 fuel oil, gasoline, 3500 fuel oil (a heavy fuel) and kerosene. A volume of 50 ml of each oil was poured into a glass boiling tube of cross-sectional area 3·85 cm^2. The percentage volume of oil evaporated in each case was calculated after 72 h.

Table III shows the results of these calculations. It should be noted that whereas oils A, B, C, D and E are each distinct oil fractions with homogeneous properties, oils CT3, CT7, CT10 and CT6 (Kuwait crude) are all crude oils of varying origins, and are thus mixtures of a great number of hydrocarbons with very different properties. The evaporation rates of these crude oils therefore cannot be constant. From Table III it can be seen that there is a wide range in evaporation rates for the different types of crude oil and refined products investigated. The evaporation rates of these oils in sunshine would probably differ further still as a result of greater heat absorption by the blacker oils.

DISCUSSION

From the experiments on the survival of plants and animals in water under oil films, it appears that living organisms of the species investigated are able to survive in water under oil films up to 0·6–0·75 cm thick for a period of 3–5 days. The rate of gas exchange between air and water, however, does seem to be greatly decreased due to the presence of the oil film since the animals tested appeared to be in a severe state of respiratory stress while in water under oil films. The extent to which gas exchange is reduced by the presence of the oil film must depend upon a number of factors, including the thickness of the oil film, the fraction of the water surface covered by oil, and the volume and surface area of the body of water concerned.

Rock pools are the most likely situation in the marine or maritime environment where in the event of an oil spill a relatively small body of water may be covered completely by a layer of oil for a period of several days. Under such conditions, the oxygen and carbon dioxide concentrations of water in rock pools would certainly be affected by the presence of an oil film on the surface. In rock pools the oxygen and carbon dioxide concentrations of the water fluctuate much more than in the open sea. Pyefinch (1943), whilst working on Bardsey Island, found a regular diurnal variation in oxygen concentration and pH in rock pools while uncovered by the sea; the highest values of both occurred in pools with the densest growth of algae. From the moment the tide ebbs from the pool, the oxygen concentration and pH rise to levels determined by the amount of sunlight available for photosynthesis and the temperature of the pool; a sudden drop in oxygen concentration and pH occurs when the rising tide floods the pools and mixing of the waters takes place. Clearly the greatest diurnal fluctuation in oxygen concentration and pH will occur in pools higher on the shore, since these will be uncovered by the tide for longer periods, and especially in coastal areas (such as south-west Britain) where the time of low water at a spring tide is during the middle of the day. Ganning (1967), investigating a rock pool just above sea level in the Baltic (where tidal action is minimal), found a similar diurnal fluctuation in oxygen concentration and pH, with the greatest fluctuation at a depth of 1 cm in floating algae, and the smallest fluctuation at the maximum depth measured (40 cm). According to Lewis (1964) a low oxygen content in bottom waters of pools may be caused by decomposition of accumulations of algal debris, common in high-level pools, possibly in conjunction with a thermal or salinity discontinuity layer.

In the open sea, surface waters are normally saturated with oxygen (from 8 cm^3/litre in high latitudes to 4·5 cm^3/litre in the tropics) and are often slightly supersaturated due to photosynthesis (Harvey, 1963). Detectable diurnal variation in oxygen content of surface waters has been reported. Immediately beneath the surface waters is an oxygen-poor layer, attributed to the respiration of animals and bacteria. The rate of oxygen absorption into water from the air is much greater when the surface is agitated, for example by wind and wave action, than under still conditions. The concentration of total carbon dioxide in solution in oceanic waters bears roughly an inverse relation to their pH. Near-surface waters have a low carbon dioxide content, while in the lower oxygen-poor layer there is some 5–6 cm^3/litre more carbon dioxide in the water than at the surface. A seasonal change in the total carbon dioxide in the water occurs in continental shelf waters, with a reduction in carbon dioxide concentration as a result of the activity of photosynthesising phytoplankton. On the open sea an oil slick is unlikely to cause any great depletion in oxygen in the underlying water, especially in turbulent conditions. During the Santa Barbara channel spill, the United States Bureau of Commercial Fisheries (quoted by Straughan, 1971a) were unable to detect any oxygen depletion under the oil slick, while another source (Smithsonian Institution, 1969) reported the oxygen saturation beneath a 'heavy oil slick' of unspecified thickness to be 98·5% of that in clear water nearby. Kinsey (1973), during the course of field experiments on a fenced-off part of an Australian coral reef, found that under conditions of light wind, oil films of 0·1 and 0·7 mm thickness caused no significant interference with oxygen and carbon dioxide transfer through the water surface other than that associated directly with the calming effect of the oil films.

Slight oxygen depletion in water under oil films, although by itself not a serious threat to living organisms, may well increase the toxicity of the oil still further to certain species. Tagatz (1961), investigating the toxicity of petroleum products to juvenile American shad, found that the lethality of gasoline, fuel oil and bunker oil was increased when accompanied by low dissolved oxygen. Cairns and Scheier (1957) found that periodic low dissolved oxygen decreased the tolerance of blue-gill sunfish (*Lepomis macrochirus*) and the pond snail (*Physa heterostropha*) to potassium cyanide and naphthenic acids. Wiebe (1935) found that bream, bass and catfish died within 4 h when kept in water under a 0·2 cm thick layer of crude oil; during this time the dissolved oxygen content fell from an initial 1·4 ppm to 0·35 ppm, suggesting that the mortalities were at least partly attributable to oxygen depletion.

A number of workers have investigated the rate of absorption of oxygen by water through oil films with somewhat differing results. Roberts (1926) found that oxygen-depleted water under thin films of diesel, gas and fuel oils became oxygenated to varying degrees depending on the type and thickness of oil film: an oil film thickness of 0·0002 cm resulted in 24 h in subsequent percentage oxygen saturation in the water of 99% (diesel oil), 91% (gas oil), 98% (600 sec fuel oil) and 94% (1500 sec fuel oil), while a 0·003 cm thick film of diesel oil resulted in 60% saturation. Agitation was found to increase the rate of oxygen absorption. Boswell (1950) found that a layer of crude oil 0·05 cm thick reduced the rate of oxygen absorption by boiled sea water to 85% of the control in 6 days. Brown and Reid (1950), however, had more variable results with crude oil: they found that in some cases deoxygenated water covered by a 1·4 mm thick layer of crude oil absorbed as much oxygen as the water without oil, while in other cases the same thickness of oil absorbed up to 25% less oxygen than the water without oil over a 2 day period. Layers of oil 1·7 and 2·00 mm thick also permitted oxygen absorption by underlying water from the air to varying degrees. Downing and Truesdale (1955) investigated the effects of the addition of oil films to slowly stirred distilled water on the rate of oxygen solution. They found that the addition of films of light spindle oil caused a reduction in the rate of oxygen solution from 2·65 cm/h to 2·2 cm/h as the thickness of the oil film was increased from 5×10^{-6} to 5×10^{-3} cm; further addition of oil resulted in a more rapid change in the rate of oxygen solution which was reduced to 0·8 cm/h at an oil film thickness of 1 mm. The addition of films of heavy marine fuel oil resulted in little change in the rate of oxygen solution with oil film thicknesses less than 10^{-4} cm; increasing the thickness from 10^{-4} to 5×10^{-2} cm, however, reduced the rate of oxygen from 4·15 to 0·7 cm/h, i.e. a decrease of 83%. Mazmanidi and Kovaleva (1972), in experiments using Anastasiyevkaya crude oil and oil extracts, observed that the addition of oil to sea water (oil film thickness unspecified) has a specific effect on the oxygen regime. They found that in the absence of aeration the dissolved oxygen content decreases as the oil concentration increases; decreases in dissolved oxygen content of 2·5–3·0 mg/litre in 6 h and up to 1–2 mg/litre per day were observed which resulted in the death of fish from asphyxia.

Under conditions of aeration the oxygen content was found to remain approximately constant with a slight tendency to decrease.

Oil films of certain thicknesses therefore do seem to permit some degree of gas exchange; however, once oil has penetrated animal or plant tissues its properties of facilitating gas exchange may be altered. McMillan and Riedhart (1964) found that the duration of photosynthesis inhibition in oiled citrus leaves correlated with the dissipation time of the oil, indicating that the oil does not allow or inhibits gas exchange. Schramm (1972) found that the presence of crude oil films 0·01 mm thick on the surface of the alga *Porphyra umbilicalis* caused lowering of diffusion rates of photosynthetic gases. Schramm tested three different crude oils with varying results, and concluded that the composition of the oil was an important factor.

The 'solubility' of oxygen in oil is considered by Schoch (quoted by Brown and Reid, 1950) to be greater than that in water, although no type of oil is specified and no figures are given. From a number of sources it appears that oil has the capacity to take up and contain oxygen, although the oxygenation of oil appears to result inevitably in immediate changes in chemical composition of the oil in such a way that it seems misleading to speak of the 'solubility' of oxygen in oil in absolute terms. Berridge *et al.* (1968) state that oxygen attacks hydrocarbons in the liquid phase, autoxidation being a free radical chain process; the products of these oxidation reactions include acids, carbonyl compounds, alcohols, peroxides and sulphoxides. A higher rate of oxidation occurs in the presence of catalysts in the oil (e.g. vanadium), or in sunlight inducing photo-oxidation. ZoBell (1962) states that free or dissolved oxygen can oxidise oil, although substances such as nitrate can act as oxygen donors; figures are supplied for the BOD (biochemical oxygen demand) values of a number of different types of hydrocarbon. McKee (1956) discussed the BOD of hydrocarbons and their derivatives, quoting a number of sources giving varying values of BOD for certain hydrocarbons; he concluded that hydrocarbons are not readily attacked by micro-organisms, but once the hydrocarbon molecule is invaded by oxygen, nitrogen or other atoms it becomes much more susceptible to biological degradation. Moreover, biochemical decomposition of oil appears to occur primarily at the air/water interface, as in the case of an oil slick at sea. It appears that more than 100 species of micro-organisms are capable of oxidising oil (ZoBell, quoted by McKee, 1956). Young (1934) found that aerobic fungi and bacteria could grow under oil, indicating that oils must contain sufficient oxygen for growth. It was found that oxygen was available to different extents within the oils tested (varying from 0–44% sulphonatable). Young also quoted Washburn (1928), who gave the following figures for the concentration of oxygen in oils: paraffin oil 0·0185% (wt), transformer oil 0·025% and Russian petroleum, 0·03%.

It therefore appears that oil films on water do allow some degree of gas exchange between air and water, although the rate of exchange may be complicated by oxygen uptake in the oil in the process of biotic or microbial oxidation.

The investigations carried out on the thermal effects of oil show that a black crude oil film 0·22 cm thick, in conditions of warm sunshine, permits the temperature of the immediately underlying layer of water to rise to a higher degree than it would do in the absence of the oil film. This heating effect of an oil film on the surface of a body of water would probably be appreciable only in very still conditions, e.g. as in a rock pool, and in the event of the oil being black in colour, thus absorbing more heat. Oil stranded on an exposed rock surface must have the same heating effect but probably results in a higher maximum temperature being attained owing to the low thermal conductivity and high specific heat of solid rock compared with water. Straughan (1971a) reported on preliminary research indicating that oil raises temperatures in the intertidal area at low tide. Nicholson and Climberg (1971) observed that oil adheres more readily to dry and warmed areas of the upper intertidal zone than to cooler, wet sites in zones lower on the shore. Straughan (in discussion, 1971b) further reported on preliminary work which showed that the surface temperature of a rock covered with neat oil rose faster and reached a higher level than that of a clean rock; some of the temperature differences involved were up to 5°C above normal temperature fluctuations. At the same time, the surface temperatures of wet and dry asphalt were found to rise at different rates but to the same level as that of a clean rock. In the same discussion, White (1971) referred to the established practice of using a covering bitumen emulsion to stimulate the germination and growth of certain crops. The main factor here appears to be the heat-absorbing properties of this treatment of the soil surface. Soil temperatures may fluctuate widely during the day, however, under normal conditions. Spencer (1970) reported on daily ranges of up to 10°C recorded in littoral soil on mud flats at a depth of 1 cm; a thin covering of black oil could possibly increase this range still further on account of its heat-absorbing properties.

From the experiments carried out on the heat-absorbing properties of oil, it can be seen that a black oil can absorb a great amount of heat within several hours in warm sunshine, and can attain a temperature at least 11°C above that of an equal volume of water. A heavily oiled rock or rock pool surface might therefore reach a temperature of 40°C on a warm sunny day in a British climate, judging from the temperature levels attained in the experiments. This temperature of 40°C is in many cases in excess of the lethal temperatures and points of 'heat coma' for a number of common intertidal gastropods, as shown in Table IV (from Lewis, 1964). Therefore the heating effects of oil could have substantial effects on rocky shore life even at sublethal temperatures by rendering certain species lethargic in a state of 'heat coma', and thus more susceptible to predation or other adverse physiological influences such as desiccation or osmotic stress. Some plants are also susceptible to small temperature rises of a similar order. Biebl (quoted by Lewis, 1964) found that for certain Rhodophyceae a change of only 2°C proves critical. Some algae occupying the upper and middle shore are also very sensitive to high water temperatures because they have delicate lipid membranes easily distorted by heat (Boney and Corner, 1959); a film of black oil over algae in water in sunshine could therefore appreciably raise the water temperature to a point at which the lipid membrane structure is endangered.

The experiments on evaporation rates of different types of oil not in sunshine show a wide range of evaporation rate, as illustrated in Table III. Crude oils and refined oil products spilt at sea are therefore likely to persist

TABLE IV

Lethal Temperatures and Points of Heat Coma for *Nucella lapillus* and *Littorina* species

Species	*Air*		*Water*	
	Heat coma	*Lethal temp.*	*Heat coma*	*Lethal temp.*
Nucella lapillus	28	36	27–28	32–33
L. littorea	32	42–43	31	39
L. littoralis	30–32	38–39	30–31	37
L. saxatilis	36–38	40	30–31	39–41
L. neritoides	ca. 40	42–43	35	42

Values are expressed in °C, and are taken from Lewis (1964).

for different lengths of time depending on weather conditions and the evaporation rate of the oil. Berridge *et al.* (1968) state that the rate of evaporation of an oil at atmospheric pressure depends primarily on the vapour pressure of the oil at the ambient temperature, but rapid evaporation of, say, an oil slick at sea will be favoured by high wind speeds. The spreading of an oil slick on water accelerates evaporation as a result of increase in surface area, at the same time leading to increased viscosity and pour point of the oil. Brunnock *et al.* (1968) studied the effects of weathering on the physical and chemical properties of oil by analysing samples of Kuwait crude oil collected after the *Torrey Canyon* spill; the first sample, collected after less than one day's exposure, had lost approximately a quarter of the original crude, probably mostly by evaporation, but also by a variety of other processes including solution, dispersion, sinking on particles, etc. This figure of *ca.* 25% loss in one day compared with the value of 10% loss for Kuwait crude in three days under the experimental conditions illustrates to some extent the increase in evaporation rate induced by exposure to sunshine, air and water turbulence, and by virtue of the spreading properties of the oil at sea.

The toxicity of oils varies at different temperatures, as described in a previous paper (Ottway, 1971). This may partly account for the seasonal variation in susceptibility of certain organisms to pollution by oil, as shown by Baker (1971a) in the case of saltmarsh plants. Crapp (1971) found a similar seasonal variation in the susceptibility of several species of intertidal gastropods to dispersants. The physical properties of oil, such as evaporation, viscosity and dissolution of water-soluble fractions, vary with temperature, and this probably accounts for the variation in toxicity at different temperatures. Baker (1971a) observed that Kuwait crude oil was more toxic to saltmarsh plants in hot sunny weather, which may be due either to a decrease in viscosity of the oil or to an increase in the formation of toxic acids or peroxides as photo-oxidation products: either of these possibilities would result in greater penetration of the oil components into the plant tissues. Mazmanidi and Kovaleva (1972), in their experiments using Anastasiyevkaya crude oil and oil extracts, observed that when oil is added to sea water, increase in temperature results in increases of the amount of oil products going into solution, the oxidisability of the oil and the carbon dioxide content of the water; increase in temperature could therefore intensify the damaging effect of spilt oil on aquatic organisms.

Investigations on the light-transmitting properties of different types of crude oil have been reported in a previous paper (Ottway, 1971). It was shown that films of different types of crude oil on water absorb light to a widely varying degree; the absorption spectra of all the oils examined showed maximum absorption in the region of 0·325–0·35 μm. This indicates that the same group of compounds in all oils is responsible for light-absorption properties. Harva and Somersalo (quoted by Nelson-Smith, 1970) found that ultraviolet light at 0·27–0·4 μm is strongly absorbed by petroleum oils, due to the presence of aromatic rings. The United States Bureau of Commercial Fisheries (quoted by Straughan, 1971c) found a reduction in penetration by ultraviolet light under oil slicks following the Santa Barbara Channel spill. An oil film on water thus absorbs more light towards the blue end of the spectrum. Certain photosynthetic pigments of marine algae have absorption spectra showing maximum absorption towards the blue end of the spectrum (e.g. chlorophyll b and β-carotene). The photosynthetic activity of these pigments would therefore be impeded by the presence of an oil film over the surface of a stretch of water or a rock pool in which the algae live. Although photosynthesis in these algae might be severely inhibited or cease altogether, the plants would not necessarily die. Burrows (in discussion, 1971) reported that her experimental work with *Fucus* and *Laminaria* species showed that these can continue to exist in a non-growing condition for long periods varying up to 9 months in extremely low light intensities; growth was found to continue when light intensity was increased. Burrows concluded that lack of light alone would be unlikely to have a detrimental effect on the algae in a rock pool. Clendenning (1964, quoted by Nicholson and Climberg, 1971) found that particulate matter in suspension alters the light-transmitting properties of water and significantly reduces photosynthetic rates in *Macrocystis*, the giant bladder kelp. Fine emulsions of water in oil could possibly also have the same effect. North *et al.* (1964), during laboratory toxicity studies following the *Tampico Maru* spill on the west coast of Mexico, found that young kelp blades of *Macrocystis*, collected from the surface canopy and kept under surface films of diesel oil of initial film thickness of about 0·02 mm, showed 67% reduction in photosynthetic capacity; young kelp blades from fronds collected at a depth of 60 ft showed total lack of photosynthesis in 3 days. Both groups of kelp blades, however, were brought into contact with the surface film several times a day, which probably resulted in penetration of the plant tissues by oil; such penetration of oil, rather than the reduction of light intensity, may have caused the photosynthesis inhibition through cell injury, as discussed by Baker (1971b), or by a lowering of the diffusion rate of carbon dioxide, as discussed by Schramm (1972).

Roberts (1926) found that the fresh-water plants *Groenlandia densa* (called *Potamogeton densus*), *Ranunculus aquatilis* and *Callitriche platycarpa* (called *C. verna*) grow luxuriantly when kept in running water under thin films (0·004 cm) of gas oil, diesel oil and oils of 600 and 1500 sec viscosity.

Under the same conditions with stagnant water, however, although all the plants remained alive there was very little growth. Such stagnant conditions would prevail in an oiled rock pool, within which growth inhibition might therefore occur in the case of algae.

In the open sea, reductions in light intensity under oil slicks may inhibit the photosynthesis of phytoplankton, but such an effect is likely to be very short-lived on account of the spreading tendency of oil into very thin films and the relative movement of oil slicks with respect to underlying masses of water. Some light intensities were measured under a heavy slick following the Santa Barbara Channel pollution by crude oil from a well blow-out; light intensities beneath the oil were found to be generally 1% of the surface intensity and, at best 5–10% (Smithsonian Institution, 1969).

Crude oils are mixtures of a large number of different hydrocarbons with different chemical and physical properties; the relative proportions of these compounds change substantially during the weathering of crude oil owing to differential evaporation and degradation rates. The light-transmitting properties of oil are therefore subject to change as the oil is weathered, although the blackest fractions of crude oil which transmit least light are also the most persistent. In contrast to crude oils, petroleum fractions are relatively homogeneous in their chemical composition; thus they have constant evaporation rates and weather in a more consistent fashion. Oil fractions vary widely in their light-transmitting properties, from the transparent, more volatile fractions such as gasoline and kerosene, to the black persistent fractions, such as heavy fuel oil.

A significant effect of sunlight falling on oil is the photo-oxidation of oil, resulting in the oxidation of hydrocarbons to give such products as acids, carbonyl compounds, alcohols, peroxides and sulphoxides, many of which are soluble in water (Berridge *et al.* 1968); sunlight may therefore have the effect of increasing the toxicity of oil to marine organisms in the immediate vicinity of a spill. Different oils exhibit a wide range of oxidation rates on the sea; for example in bright sunlight a high-paraffin, low-sulphur crude oil would oxidise far more rapidly than a less paraffinic, highly sulphurous crude. Pilpel (1968) stated that these photo-oxidation products of oil may break down even further into carbon dioxide and water, or polymerise into dense lumps. Freegarde and Hatchard (1970) report that fluorescent aromatic compounds in crude oil decompose more rapidly as a result of photo-oxidation than non-volatile material as a whole; thus photo-oxidation of aromatic components of crude oil proceeds more rapidly than the overall rate of conversion to volatile or water-soluble products.

Oil slicks at sea have distinctive optical properties, better revealed by infra-red photography than to the human eye (Worsley, 1969). Such optical effects could deceive birds on account of the resemblance of oil slicks to fish shoals or floating debris; this could explain the observation by Lemmetyinen (quoted by Nelson-Smith, 1970) and others (ICBP, 1960) that the long-tailed duck *Clangula hyemalis* L. settles preferentially in oil patches. In the same way, shoals of pelagic fishes observed to avoid areas of water under oil slicks (North *et al.* 1964; Smithsonian Institution, 1969) may be reacting simply to the shadowing effects of the slicks above them, possibly resembling the shadow of a large predator.

REFERENCES

Baker, J. M. (1971a). Seasonal effects, in: *The Ecological Effects of Oil Pollution on Littoral Communities*, ed. E. B. Cowell, Institute of Petroleum, London, pp. 44–51.

Baker, J. M. (1971b). The effects of oil on plant physiology, in: *The Ecological Effects of Oil Pollution on Littoral Communities*, ed. E. B. Cowell, Institute of Petroleum, London, pp. 88–98.

Berridge, S. A., Dean, R. A., Fallows, R. G. and Fish, A. (1968). *Scientific Aspects of Pollution of the Sea by Oil*, Institute of Petroleum, London, pp. 2–9.

Boney, A. D. and Corner, E. D. S. (1959). Application of toxic agents in the study of the ecological resistance of intertidal red algae, *J. mar. biol. ass. U.K.*, **38**, 267–77.

Boswell, J. L. (1950). Experiments to determine the effect of a surface film of crude oil on the absorption of atmospheric oxygen and water, Texas A & M Research Foundation Project 9, 6 pp.

Brown, S. O. and Reid, B. L. (1950). Experiments to test the diffusion of oxygen through a surface layer of oil, Texas A & M Research Foundation Project 9, 5 pp.

Brunnock, J. V., Duckworth, D. F. and Stephens, G. G. (1968). *Scientific Aspects of Pollution of the Sea by Oil*, Institute of Petroleum, London, pp. 12–27.

Burrows, E. M. (1971). In: Discussion, in *The Ecological Effects of Oil Pollution of Littoral Communities*, ed. E. B. Cowell, Institute of Petroleum, London, p. 205.

Cairns, J. and Scheier, A. (1957). The effects of periodic low oxygen upon the toxicity of various chemicals to aquatic organisms, *Proc. 12th Indust. Waste Conf., 1957, Purdue Univ.* (*Eng. Bull.*, **94**, 165–76).

Crapp, G. B. (1971). Laboratory experiments with emulsifiers, in: *The Ecological Effects of Oil Pollution on Littoral Communities*, ed. E. B. Cowell, Institute of Petroleum, London, pp. 129–49.

Downing, A. L. and Truesdale, G. A. (1955). Some factors affecting the rate of solution of oxygen in water, *J. appl. Chem.* (*Lond.*), **5**, 570–81.

Freegarde, M. and Hatchard, C. G. (1970). The ultimate fate of oil at sea. Interim Report No. 7: An investigation of the photodecomposition of crude oil, Admiralty Materials Laboratory, Report No. AML/5/70.

Harvey, H. W. (1963). *The Chemistry and Fertility of Sea Waters*, Camb. Univ. Press, 240 pp.

ICBP (International Committee for Bird Preservation) (1960). *Annual Report of British Section*, British Museum (Nat. Hist.), London, pp. 16–21.

Kinsey, D. W. (1973). Small-scale experiments to determine the effects of crude oil films on gas exchange over the coral back-reef at Heron Island, *Environmental Pollution*, **4**, 167–82.

Lewis, J. R. (1964). *The Ecology of Rocky Shores*, English Univ. Press, 323 pp.

McKee, J. E. (1956). Oily substances and their effects on the beneficial uses of water, Calif. State Water Pollution Control Board, Sacramento.

McMillan, R. T. and Riedhart, J. M. (1964). The influence of hydrocarbons on the photosynthesis of citrus leaves. *Proc. Florida State Hort. Soc.*, **77**, 15–21.

Mazmanidi, N. D. and Kovaleva, G. I. (1972). Experimental data on the effect of oil on some chemical properties of seawater, Ext: *Oceanology, Moscow*, **12**(5), 684–9 (English translation).

Nelson-Smith, A. (1970). The problem of oil pollution of the sea, *Adv. mar. Biol.*, **8**, 215–306.

Nicholson, N. L. and Climberg, R. L. (1971). The Santa Barbara oil spills of 1969: a post-spill survey of the rocky intertidal, in: *Biological and Oceanographical Survey of the Santa Barbara Channel Oil Spill*, ed. D. Straughan, Allan Hancock Foundation, Univ. of S. Calif., Vol. 1, pp. 325–400.

North, W. J., Neushell, M. and Clendenning, K. A. (1964). Successive biological changes observed in a marine cove exposed to a large spillage of mineral oil, *Symp. Pollut. mar. Micro-org. Prod. petrol., Monaco*, pp. 335–54.

O'Sullivan, A. J. (1971). In: Discussion, in *The Ecological Effects of Oil Pollution on Littoral Communities*, ed. E. B. Cowell, Institute of Petroleum, London, p. 204.

Ottway, S. (1971). The comparative toxicities of crude oils, in: *The Ecological Effects of Oil Pollution on Littoral Communities*, ed. E. B. Cowell, Institute of Petroleum, London, pp. 172–80.

Pilpel, N. (1968). The natural fate of oil on the sea, *Endeavour*, **27**, 11–13.

Pyefinch, K. A. (1943). The intertidal ecology of Bardsey Island, N. Wales, with special reference to the re-colonisation of rock surfaces and the rock pool environment, *J. Anim. Ecol.*, **12**, 82–108.

Roberts, C. H. (1926). The effect of oil pollution upon certain forms of aquatic life and experiments upon the rate of absorption through films of various fuel oils of atmospheric oxygen by sea water, *J. Cons. perm. int. Explor. Mer.*, **1**, 245–75.

Schramm, W. (1972). The effects of oil pollution on gas exchange in *Porphyra umbilicalis* when exposed to air, *Proc. 7th Int. Seaweed Symp., 1972*, pp. 309–15.

Smithsonian Institution (1968). Center for short-lived phenomena: information reports, Event No. 55–68.

Spencer, J. F. (1970). Diurnal and seasonal temperature changes in the littoral soil on Pwllchrochan Flats, Milford Haven, in relation to Pembroke Power Station, in: *The Effects of Industry on the Environment* (*Proc. Symp.*), ed. E. B. Cowell, Field Studies Council, pp. 11–25.

Straughan, D. (1971a). What has been the effect of the spill on the ecology in the Santa Barbara Channel? in: *Biological and Oceanographical Survey of the Santa Barbara Channel Oil Spill*, ed. D. Straughan, Allan Hancock Foundation, Univ. of S. Calif., pp. 401–19.

Straughan, D. (1971b). in: Discussion, in *The Ecological Effects of Oil Pollution on Littoral Communities*, ed. E. B. Cowell, Institute of Petroleum, London, p. 100.

Straughan, D. (1971c). In: *Biological and Oceanographical Survey of the Santa Barbara Channel Oil Spill*, ed. D. Straughan, Allan Hancock Foundation, Univ. of S. Calif.

Tagatz, M. E. (1961). Reduced oxygen tolerance and toxicity of petroleum products to juvenile American shad, *Chesapeake Sci.*, **2**, 65–71.

White, D. W. (1971). In: Discussion, in *The Ecological Effects of Oil Pollution on Littoral Communities*, ed. E. B. Cowell, Institute of Petroleum, London, p. 100.

Wiebe, A. H. (1935). The effect of crude oil on freshwater fish, *Trans. Amer. Fish. Soc.*, **65**, 324–31.

Worsley, R. (1969). Aerial photography of oil slicks, *Field Studies Council Oil Pollution Research Unit Annual Report, 1969*, pp. 48–50.

Young, P. A. (1934). Fungi and bacteria as indicators of the effects of petroleum oils on apple leaves, *Phytopath.*, **24**(3), 266–75.

ZoBell, C. E. (1962). The occurrence, effects and fate of oil polluting the sea, *Int. Conf. on Water Pollut. Res., London, Sept. 1962*, Section 3, paper no. 48, 27 pp.

Discussion

Effluent Treatment and Standards

Miss S. Hainsworth (Oil Pollution Research Unit) commented that API separators and skimmers seem to work efficiently if properly operated, but some of the ones she had seen have to be manually adjusted so that their level is correct to skim off the top 1 in or so containing most of the oil. Could these not be modified fairly simply to work automatically, thus keeping pace with changing levels due to differences in flow through the system? **Mr Roberts** said that the standard skimmer mechanism is too coarse to follow small changes in level with the sensitivity required and he doubted if it could be successfully automated. Some attempts were being made to produce a skimmer which can follow level changes accurately or which will operate efficiently over a fairly wide range without adjustment. It would have to be of low cost and high reliability.

Mr Roberts agreed with **Mr A. J. O'Sullivan** (Atkins Research and Development) that site selection should take into account the capacity of receiving waters and the proximity of ecologically sensitive areas. Mr O'Sullivan wondered if levels of 1 ppm of oil in the final effluent could be achieved economically at the present time and whether the recirculation of water within the refinery process (e.g. spent caustic and sour water) has much potential for reducing the volume of waste water requiring treatment and discharge. Reducing the volume of the discharge and reducing the levels of biologically active substances present can both be equally effective in preventing damage. **Mr Roberts** said that levels of 1 ppm were technically feasible but demanded tertiary treatment such as osmosis, carbon adsorption or a large aeration lagoon. Economic viability could only be assessed in relation to a specific site. Specifications as low as 1 ppm might be accepted in a completely new development, but modification of existing plant could be prohibitively expensive. Recirculation and re-use could certainly be increased; it was again a matter of economics at a particular site. If smaller volumes of much more concentrated waste were produced, it might be necessary to consider completely different clean-up processes for what is left.

Mr P. J. Osbaldeston (North West Water Authority) agreed with the concept of standards based on an agreed method of analysis, particularly as regards oil. He was surprised that many statutory bodies and oil companies are still content to use the phraseology 'no visible oil'. In the case of a dispute, did one consult a panel of magistrates or a panel of opticians? On the subject of quantification he asked:

1. How does one analyse for 'oil'?
2. If one uses an extraction method, is one analysing for 'oil' or for a range of soluble organics?
3. Are oil refineries exempt from the implications of the Oil in Navigable Waters Act?

Mr Roberts replied that the commonly used analysis methods rely on a first extraction step. This would show up non-oil organics as 'oil' but he thought the quantities were likely to be very small in most situations. The second step was either (a) evaporation and weighing, which does not measure volatiles but is simpler to carry out, or (b) an IR or similar technique aimed at measuring total organics. The Oil in Navigable Waters Act was succeeded by the Prevention of Oil Pollution Act, 1971. It appears that a refinery effluent is exempt from the provisions of the Act, although an oil spill originating from a refinery is not. However, oil in effluent does not go uncontrolled; it is normal for consent levels to be imposed by the appropriate authorities under other Acts and regulations.

Mr N. C. Morgan (Nature Conservancy Council) asked Mr Roberts whether, when he referred to the levels of oil in water in ppm achieved by different stages of separation for refinery effluent water, these were mean levels or maximum levels above which the oil level would never go? Secondly, did not the latest Warren Spring gravity separator reduce the oil content down to a maximum of 10 ppm? **Mr Roberts** said that the oil content levels he used were guidelines only, typical of what should be achieved during normal operation. Absolute guarantees for any process were virtually impossible. 10 ppm is about the level of oil which is soluble in water, so he did not see how any gravity separator could consistently achieve such a figure. A well-operated separator of good design should achieve 10 ppm suspended oil (i.e. about 20 ppm total) for much of the time, but flow rates, amount of oil and other materials present will all influence performance.

On the question of costs, **Mr H. Jagger** (Esso Petroleum Co. Ltd) said he thought Mr Roberts would agree that his data referred to a fairly small (say 100 000 b/d) hydroskimming refinery. For a larger and more complex refinery these costs would be considerably higher. On oil contents of effluents it was worth noting that with four refineries, a major crude oil terminal and a large oil-fired power station all located in Milford Haven and with consents in each case permitting up to 25 ppm oil in effluent, there is no overall ecological damage in the Haven. **Professor A. R. Halliwell** (Heriot-Watt University) developed this point made by Mr Jagger and referred back to earlier papers. He said that the requirements for any effluent must be related to the area into which it is discharging, in particular to the quantity of water which is available for further dilution and diffusion and to the sensitivity of the environment. To quote a figure of 25 ppm without reference to the total quantities discharged and the water available for further dilution was both misleading and dangerous. The only satisfactory method was to quote the volume of water or areas affected by different concentrations. For example, 1 ppm may appear to be a high standard of effluent, but if it is for a considerable volume of effluent discharging into a comparatively small river it can result in many miles of river having concentrations of the order of say 0·01–0·1 ppm. On the other hand, 50 ppm discharged from a similar-size effluent into a fast-moving part of an estuary may result in concentrations of much less than 0·01 ppm in all areas except a few hundred feet around the effluent discharge points. Only by quoting the volumes of water affected at various concentrations can different areas and schemes be compared and extremes be avoided. **Dr M. W. Holdgate** (Institute of Terrestrial Ecology) said that recently a number of Western European countries signed the Paris Convention (which the UK has not yet ratified). This states that Contracting Parties will use 'best practical means' (a technical term implying a blend of what is technically feasible and economically acceptable) to prevent the pollution of the sea by a number of substances including oil. Pollution is defined in the Convention as the release by man into the marine environment of substances or energy liable to cause hazard to human health, damage to living resources or ecosystems, loss of amenity or interference with legitimate uses of the sea. It is important to define pollution in terms of damage and for us to recall that our controls are designed to prevent unacceptable effects. Generally effects are related to the concentrations of a toxic substance at the target. Ideally our standards should be set in terms of the concentration we will tolerate at a particular target, and if we are to relate this to the standards in ppm to apply to an effluent, we obviously must take account of the pathway from source to target.

Natural Vegetation as a Tertiary Treatment

Mr O'Sullivan said he liked the suggestion of deliberately managing natural vegetation as an oil trap. It is in line with a similar suggestion made nearly ten years ago by Dr Eugene Odum that, in order to metabolise most efficiently the waste products of society and industry, ponds or lagoons should be constructed into which a variety of organisms would be introduced. Natural selection would then operate to result in a community of organisms hopefully capable of breaking down the waste, and this process could be aided by some degree of genetic manipulation. Mr O'Sullivan could see the possibility of refinery effluents receiving a final treatment by slow passage through specially established beds of *Phragmites communis*, *Spartina anglica* or similar vegetation. **Mr E. B. Cowell** (British Petroleum Co. Ltd) added that at the BP Refinery, Gothenburg, *Phragmites communis* grows in a tertiary holding lagoon. On the two occasions when oil has reached the lagoon system it has been effectively trapped by the *Phragmites*. There is now a policy of encouraging the growth of the *Phragmites* to increase the oil-trapping potential of the lagoon system. **Mr C. A. Sinker** (Field Studies Council) pointed out that *Phragmites* is also a potential cash crop of some value, being much in demand for thatching in some parts of England. Natural reed-beds, an important wildlife habitat, are a scarce and decreasing resource in this country. The industry could thus perform a double service to environmental conservation by managing lagoons of this kind, and make a modest annual profit out of the operation. Cut reed currently sells at up to £1 per standard bundle, and a well-grown crop may yield 400 bundles to the acre (approximately 1000 bundles per hectare).

Effects on Marine Organisms

Dr R. H. Cook (Environmental Protection Service, Canada) referred to the large limpets in the vicinity of the Little Wick refinery outfall described by Dr Baker and asked if the thermal component of the effluent contributed to the increased size, and if there were any other environmental effects observed that could be attributable to the thermal nature of refinery effluents. **Dr Baker** replied that in this particular case she thought that low salinity was more important as the heating in the area under consideration was in the order of 1°C. **Dr I. C. White** (Ministry of Agriculture, Fisheries and Food) wondered if the large limpets at Little Wick were an example of giganticism caused by stress, as has been described for winkles (*Littorina littorea*) subjected to parasites. He believed such animals no longer reproduce but show increased growth. **Dr R. Mitchell** (Nature Conservancy Council) had further observations on the subject of 'monster limpets'. One of Dr Baker's hypotheses was that these animals occurring at refinery effluent outfalls represent relict animals surviving from the population that existed before the release of effluent. She postulated that these animals were large because they were old (about 15 years) and had abundant food. In 1968 he and Mrs Mitchell had observed 'monster limpets' on the concrete shores of the cooling-water outfall of Marchwood Power Station in Southampton Water. The density of these limpets was low (less than $1/m^2$) and there were no small limpets in this population; each animal had its own distinct grazing area surrounded by ungrazed algae. He suggested that these limpets possibly represented a remnant of the population that became established on the outfall structure before the effluent was first discharged. It would seem unlikely that any limpet veligers could settle in the outfall since effluent discharge began owing to elevated temperatures, high velocities and chlorine residues. As Marchwood Power Station first came into operation in the mid-1950s, the limpets would have been about 10–15 years old. This observation could therefore reinforce Dr Baker's hypothesis that these animals are large because they are old and food is abundant. The elevated outfall temperatures would no doubt also cause an increased growth rate contributing to a large size if food was not limiting. **Miss Hainsworth** said that she had started to measure the limpets near the effluent outfall at Little Wick, and at various distances from the outfall, to try to establish whether or not growth rates were responsible for the large size of the limpets near the outfall. A definite clear ring on the shell of about 1 mm which was obviously new growth had been observed during April.

Mr O'Sullivan referred to Dr Baker's observation that in Little Wick Bay the youngest age classes are missing, and to Dr Crapp's work on limpet populations affected by oil pollution which indicated that younger limpets were more sensitive. Did she feel that both sets of results point to the use of limpet population structure as an indication of the extent of pollution by oil? **Dr Baker** said she thought the use of limpet population structure was very promising, but populations were of course also greatly affected by natural factors such as exposure and more information was needed on this. There was some evidence that younger limpets were more sensitive to oil pollution, but not all experiments had given this result.

Mr K. Hiscock (Coastal Surveillance Unit, Menai Bridge), referring to the effects of refinery effluents on limpets, said he had found limpets to be affected in a similar manner in relation to a halogenated acidified effluent. In addition, many other common, widely distributed littoral species were excluded from the area of the effluent which he studied. Which littoral species other than *Patella* were excluded from the area of refinery effluents? **Dr Baker** said that near the Little Wick refinery effluent, numbers of barnacles and gastropods were also reduced and there appeared to be some similarity between the effects of the two effluents.

Effluent Toxicity

Dr G. Robinson (Gulf Oil) said that to the refiners the assessment of individual components of industrial effluent for toxic effects was important. It would enable realistic consent limits to be set on refinery effluents and this information would help both the industry and the water authorities. He thought that the use of detergents in the refineries around Milford Haven would have little effect on the toxic properties of effluent. Their use is kept to a minimum because of trouble that they might cause in separators. He added that experiments using refinery effluents must be carefully designed because the effluent is not stable and cannot be stored for any length of time. Sulphide, for example, can be oxidised rapidly, possibly to sulphate or thiosulphate, in the presence of sunlight (UV light), and unless the effluent is stabilised bacterial growth can take place rapidly. **Mr P. D. Holmes** (British Petroleum Co. Ltd) and **Dr White** emphasised the difference between dispersants and detergents, and **Mr Holmes** and **Mr Roberts** added that Dr Baker's observations concerning the difference between night and day effects of a refinery effluent were probably due to the use of detergents during the daytime. About one year ago the Kent refinery changed its detergent to a 'quick-break' type which reduced oil levels in the effluent, and already a significant effect has been observed on the ecology of the area adjacent to the effluent discharge. Detergents in refineries are increasingly rigorously controlled, and a change to 'quick-break' types can improve the effluent oil content by 50%.

Dr M. Spooner (Marine Biological Association of the UK) thought that much of the more volatile and therefore the major part of the immediately toxic fractions would be lost from the oil in the API separators and in later treatment. If so, the oil droplets which emerge would be more like a weathered oil which field experience suggests is fairly inert. This makes it seem probable that the noticeable toxicity of a refinery effluent may perhaps chiefly reside in the other unknown ingredients, although some toxic water-soluble oil fractions could still be present. **Dr T. E. Lester** (British Petroleum Co. Ltd) thought that in view of the difficulties in unravelling the toxic effects of effluent components acting either alone or in combination, a simple comparison of refinery effluent and treated ballast water might be worth consideration. Ballast water alone does not contain certain of the constituents present in refinery process water, but its properties are of considerable interest in the context of North Sea oil production developments.

Herbicides and halogenated hydrocarbons were briefly discussed. **Dr Baker** had commented that within refineries, vegetation around oil tanks may be treated with herbicides and excess would probably be washed into effluent streams by heavy rainfall. **Mr Roberts** thought this should be kept in perspective compared with agricultural use. **Mr T. Bokn** (Norwegian Institute for Water Research) pointed out that halogenated hydrocarbons are quite oil-soluble and thus an oil slick may have much higher concentrations of these compounds than the surrounding water. **Dr White** added that there was some evidence that some organochlorine compounds appear to be concentrated in oil films on the sea surface, but he did not know of any evidence that demonstrated that this facilitates their entry into marine organisms.

Toxicity Testing of Oils, Effluents and Dispersants

15

The Comparative Toxicities of Crude Oils, Refined Oil Products and Oil Emulsions

SHEILA VAN GELDER-OTTWAY

(formerly with Oil Pollution Research Unit)

SUMMARY

This paper describes toxicity texts carried out in the laboratory to determine the comparative toxicities of crude oils, refined oil products, and oil emulsions to several species of intertidal molluscs common on British rocky shores. Generally oil products with low boiling points appear to be more toxic than the heavier fuel oils while crude oils are intermediate with respect to toxicity. Heavier oils are, however, more of a physical hazard.

There appears to be no simple relationship between, on the one hand the toxicity of mixtures of oil and dispersants, and on the other the toxicity of oil or dispersants alone. In the case of one widely used low toxicity dispersant, the oil and dispersant mixture was more toxic than either the crude oil or the dispersant.

INTRODUCTION

This paper describes toxicity tests carried out in the laboratory to determine the comparative toxicities of crude oils, refined oil products and oil emulsions to several species of intertidal mollusc common on British rocky shores. The oil emulsions investigated include both oil-in-water emulsions and oil emulsified by chemical emulsifying agents developed for the purpose of cleaning up oil spilt at sea or on the shore.

CRUDE OILS

Crude oils vary widely in chemical composition and physical properties. The toxicity of different types of crude oil has been found to vary in experiments carried out on several species of gastropod mollusc, as reported in a previous paper (Ottway, 1971). Similar variations in toxicity among different types of crude oil have also been found by Baker (1971a), in experiments on turves of salt marsh grasses *Puccinellia maritima* and *Festuca rubra*, and by Kühnhold (1969), in experiments on fertilised eggs and yolk sac larvae of herring. From these and other experiments it is evident that the most volatile fractions of crude oil are the most toxic, and that fresh crude oil is more toxic than residue.

REFINED OIL PRODUCTS

Experiments were carried out to determine the toxicity of several different types of refined oil products compared to that of crude oil. The refined oil products investigated were as follows:

A: Dieselect (a diesel fuel).
B: No. 2 Fuel.
C: Gasoline (leaded).
D: 3500 Fuel (a heavy fuel oil).
E: Kerosene.

All of these fractions (A–E) were derived from a crude oil mixture consisting of 75% Kuwait crude and 25% Nigerian crude. Three types of household oil products were also investigated: 20/50 motor oil (BP Super Viscostatic), premium graded paraffin heater fuel and medicinal paraffin. The toxicity of all these fractions was tested simultaneously with that of Kuwait crude oil for comparative purposes.

Seven species of littoral molluscs were used in the toxicity tests. These were the gastropods *Littorina littorea*, *Littorina saxatilis*, *Littorina neritoides*, *Nucella lapillus*, *Gibbula umbilicalis* and the bivalve *Mytilus edulis*. The experimental procedure adopted for toxicity determination was based on the method established by Perkins (1968) and subsequently modified by Crapp (1969 and 1971a). Batches of 50 animals of each species were exposed to each type of oil for a period of 1 h or 6 h, followed by a 5-day recovery period in clean aerated sea water. The recovery rate of each species of gastropod was recorded by noting the number of individual animals which crawled out of the water and up the sides of the tank. Mortalities in the gastropod species were taken as the number of animals still immobile and remaining in the water after 5 days; in the case of *Mytilus edulis* mortalities were taken as the number of individuals showing no signs of life and with permanently gaping shells. These experiments were all carried out during June–August 1971.

Mortalities occurring as a result of the toxicity tests are shown as percentages in Table I, while the recovery rates of *Littorina littorea*, *L. neritoides* and *N. lapillus* are illustrated in Figs. 1–3. It can be seen from Table I that the different types of oil vary widely in their toxicity to the species tested. Figures 1–3 show the variation in the pattern of recovery rates of each species from exposure to the oils. The order of toxicity of the oils tested is not exactly the same for all species, but some trends are clearly evident. Considering Kuwait crude oil and the refined products (A–E), gasoline is clearly overall the most toxic fraction. This particular gasoline was leaded, however, which may have contributed to its high toxicity. Kerosene is the next most toxic fraction, and is slightly less toxic than Kuwait crude oil. The remaining fractions dieselect and No. 2 fuel oil are relatively less toxic to the species tested, while the heavy black 3500 fuel oil is almost completely non-toxic.

TABLE I

Percentage mortality figures for several species of intertidal gastropod after exposure to different types of oil, followed by a 5-day recovery period in sea water. Laboratory temperature: 17–20°C.

			Type of oil								
Species	*n*	*Exposure period (h)*	*Kuwait crude*	*Dieselect (A)*	*No. 2 fuel (B)*	*Gasoline (C)*	*3500 fuel (D)*	*Kerosene (E)*	*Household paraffin*	*Medicinal paraffin*	*Motor oil*
Littorina littorea	50	6	58	2	6	28	2	66	82	0	10
Littorina littoralis	50	1	16	0	18	72	4	20	18	0	0
Littorina saxatilis	50	1	6	0	6	2	0	4	76	0	0
Littorina neritoides	50	1	30	14	4	64	4	24	10	2	2
Gibbula umbilicalis	50	1	4	6	4	24	0	2	26	2	32
Nucella lapillus	50	6	18	2	4	54	4	10	46	0	—
Mytilus edulis	50	1	14	6	2	10	0	0	2	0	10

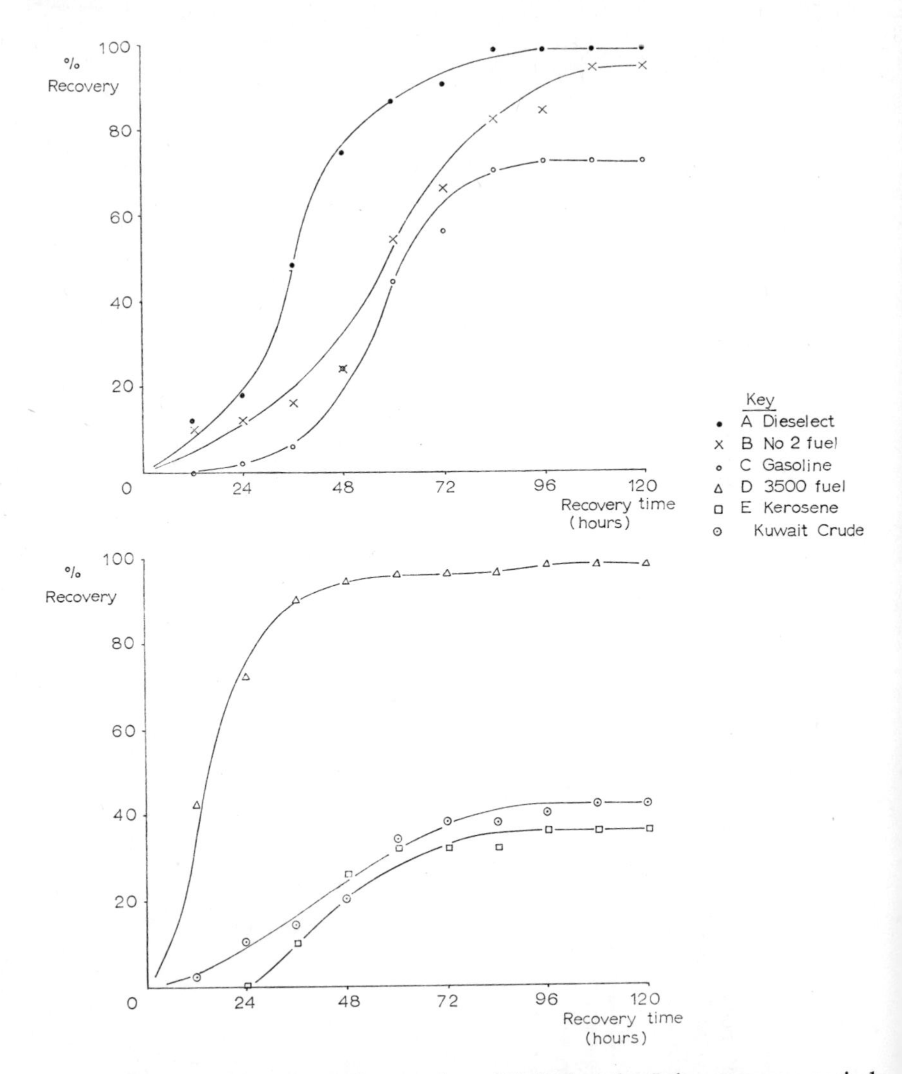

FIG. 1. Recovery of *Littorina littorea* (n = 50) during the 5-day recovery period following exposure for 6 h to crude oil and various oil fractions.

This order of toxicity of oil fractions is similar to that found by Howe and Ottway (1971) when investigating the toxicity of these same fractions to the prawn *Leander squilla*, and parallels that found by Crapp (1971b) who tested the toxicity of these same fractions on the gastropods *Littorina littorea*, *Littorina littoralis* and *Patella vulgata*.

The three household oil fractions investigated similarly show a variation in toxicity with the different species. Medicinal paraffin, as one might expect, is non-toxic to all species, while the 20/50 motor oil and household fuel paraffin are toxic to the different species to varying degrees. Both the 20/50 motor oil and the domestic paraffin undoubtedly contain various additives,

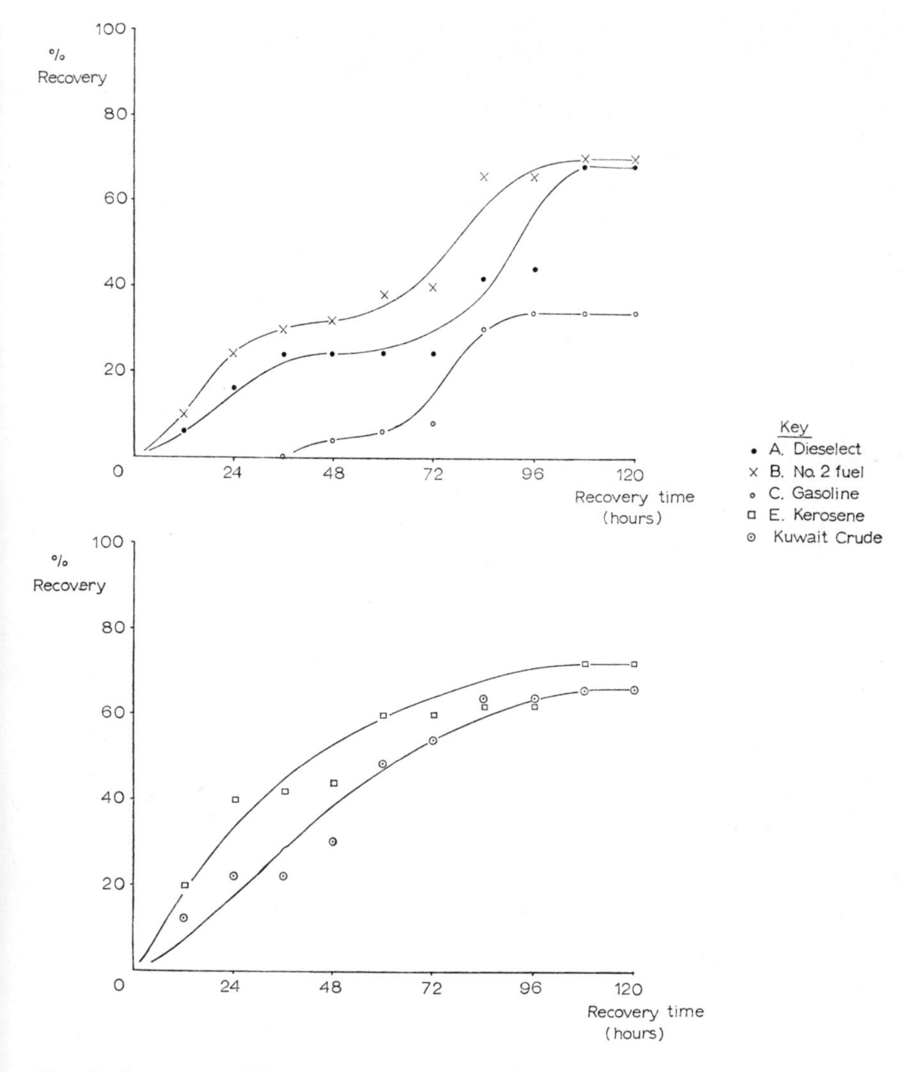

FIG. 2. Recovery of *Littorina neritoides* ($n = 50$) during the 5-day recovery period following exposure for 1 h to crude oil and various oil fractions.

while the latter is distilled to remove the majority of the more toxic aromatic components which occur in commercial kerosene.

The toxicity of refined oil products has been previously investigated in a variety of aquatic fauna and flora. Tagatz (1961), in his experiments on the toxicity of petroleum products to juvenile American shad (*Alosa sapidissima*) found that of the fractions tested gasoline was the most toxic, fuel oil somewhat less toxic and bunker oil least toxic. Chipman and Galtsoff (1949) investigated the effects of oil mixed with sand on aquatic animals, and found in experiments on the hydrozoan *Tubularia crocea* and the toadfish *Opsanus*

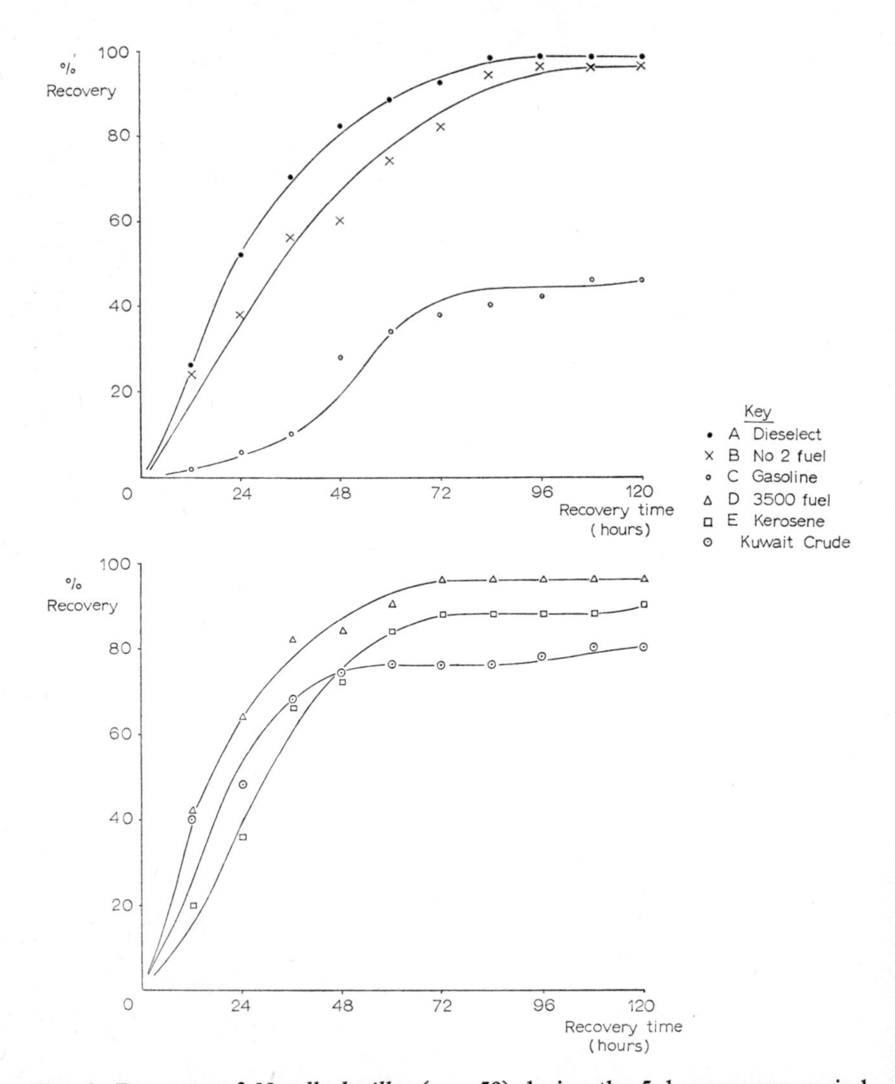

FIG. 3. Recovery of *Nucella lapillus* ($n = 50$) during the 5-day recovery period following exposure for 6 h to crude oil and various oil fractions.

tau (L.) that of the oils tested crude oil was the most toxic and lubricating oil the least toxic, while fuel oil and diesel oil were of intermediate toxicity; their experiments on the oyster *Crassostrea virginica* showed diesel oil to be slightly less toxic than crude oil. Mironov (1967) found kerosene to be less toxic than Russian crude oil, while heavy fuel oil was non-toxic to the intertidal gastropods of the Black Sea *Bittium reticulatum* (da Costa), *Rissoa euxinia* (Mil.) and *Gibbula divaricata* (L.). Mironov and Lanskaja (1967) also found several species of diatoms to be sensitive to both kerosene and fuel oil at concentrations from 100 ppm to 1%, with some suppression of growth induced by sub-lethal concentrations. Crapp (1971b) found Kuwait residue to be less toxic than fresh Kuwait crude oil to a number of intertidal species of mollusc. Baker (1971a) similarly found fresh Kuwait crude oil to be more toxic than Kuwait residue to saltmarsh turves (*Puccinellia/Festuca*), while low-boiling distillate fractions of Kuwait crude oil (10, 20 and 30% by volume) were found to be more toxic still.

The oil products, with low boiling points, thus appear to be considerably more toxic than the heavier fuel oils and bunker oils, while crude oils are intermediate with respect to toxicity. Allen (1971), however, while investigating the effects of water-soluble components of 16 different petroleum fractions on the early development of the sea urchin *Strongylocentrotus purpuratus*, found that cleavage was very sensitive to these fractions to varying degrees, but that the lighter products were less toxic than heavier crude and bunker oils. This may be partly explained by the observation of Kühnhold (1969) that the damaging influence of petroleum fractions on the development of herring eggs and larvae decreased with increasing boiling points of these fractions, while the phenomenon was partially inverted when the fraction had been emulsified. The low-boiling hydrocarbons present in oil include aromatics such as benzene, toluene, xylene, naphthalene and phenanthrene, all of which are water-soluble to some extent, and which are known to be toxic to a wide variety of marine organisms (see Nelson-Smith, 1970, for a review). Among the higher-boiling crude oil fractions, Blumer (1969) considered the polycyclic aromatic hydrocarbons to be seriously damaging long-term poisons, inducing carcinogenic activity, although the significance of this in the marine environment is still disputed. It is evident, however, that some compound(s) present in oil may interfere with cell division, and thus cause the abnormalities in cleavage of sea urchin eggs as described by Allen (1971) and perhaps also the distorted larvae hatching from eggs of the fish *Rhombus maeoticus* in sea water containing oil and oil products, observed by Mironov (1970). These polynuclear aromatic hydrocarbons, however, are present at least in minute concentrations in all marine waters because they are formed by aqueous flora (Suess, 1970), and it is unlikely that they would be present in sufficient concentration in marine waters to cause very much biological damage, except perhaps in the event of an oil spill in a very confined body of water.

In addition to toxicity, the physical properties of refined oil products may cause biological damage. The degree of persistence of an oil is dependent on the physical nature of the oil. 'Persistent' oils were defined by the Ministry of Transport (1953) as including crude, residual fuel and lubricating oils, also tar oils and creosote etc., while non-persistent oils include all other distilled hydrocarbon oils, such as motor spirit, kerosene and gas oil, and also animal and vegetable oils. The persistent types of oil are more likely to cause biological damage by virtue of their physical properties, possibly for a considerable length of time, while the non-persistent types of oil are more likely to cause biological damage by virtue of their toxic nature for a relatively short period of time. Exposure of oil by spillage at sea results in the fairly rapid disappearance of non-persistent oils by evaporation, while persistent oils become weathered in characteristic ways. The physical and chemical changes undergone by crude oil when exposed include an increase in asphaltene content, a slight increase in wax melting point, a decrease in sulphur content and increase in both viscosity and specific gravity (Brunnock *et al.*, 1968). The changes occurring in an exposed fuel oil are similar to those in exposed crude oil. As persistent crude oils and fuel oils are weathered, they become more of a physical hazard to marine organisms, for example by becoming attached in lumps to algae and animals, thereby overweighting and dislodging them. Birds are also much more susceptible to persistent oils, since at most stages of weathering these oils adhere readily to plumage and thus destroy its insulating properties. Crapp (1971b) and Baker (1971a) both investigated the biological effects of Kuwait atmospheric residue and certain residue blends, similar in physical properties to weathered crude oils. These residues were found to be effectively non-toxic, but their relatively high viscosity renders them physically dangerous to marine life. Lubricating oils, like fuel and crude oils, are also of a persistent nature, but differ in being often very thin, unlike the weathered crude and fuel oils which are thick and heavy at sea temperatures (Nelson-Smith, 1970).

The more persistent types of oil may also be toxic to marine animals, especially when in the form of an emulsion. Zitko (1971) found that large quantities of Bunker 'C' oil are taken up by a variety of invertebrates and distributed through their tissues: spectrofluorometric methods revealed the initial accumulation of this oil in the guts of flounders and lobsters. Zitko does not discuss the effects of this uptake on the health of the animals concerned.

Refined oil products thus vary widely in toxicity. In general, the low-boiling fractions are more toxic than crude oil, which is in turn more toxic than the heavier, high-boiling residual fractions; some degree of reversal of this order of toxicity is apparent, however, when the oil is in the form of an emulsion. From the review of some of the major spillages in recent years it appears that the biological damage resulting from spillages of refined oil products is determined more by the degree of persistence of the oil concerned than the absolute toxicity of the oil when fresh.

OIL EMULSIONS

Oil-in-water Emulsions

The toxic effects of oil-in-water emulsions were investigated by making emulsions of Kuwait crude oil in sea water using a hand-operated emulsifying apparatus (supplied by Ormerod Engineers Ltd.).

Emulsions are made with this type of instrument by first pouring the liquid constituents into an aluminium cup (capacity 0·56 litre) from which liquid is drawn into a cylinder beneath through a ball-valve housed within a piston. Power supplied by a hand-operated lever effects the movement of the piston with the cylinder. Pressures up to $13{\cdot}8 \times 10^7$ dyn/cm^2 can be achieved with the type of model used. A 2% (20 ppt) emulsion was initially made by passing a mixture of 10 ml Kuwait crude oil and 490 ml sea water through the emulsifier, the mixture in the cup meanwhile being stirred vigorously and continually by hand. The resulting emulsion was cloudy brown in colour and stable for at least 12 h. This emulsion was in practice somewhat less concentrated than 2%, since a film of oil was always unavoidably left on the sides of the emulsifier cup. In order to compensate for this, less concentrated emulsions were made by immediate repeated dilution of the theoretically 2% emulsion. Five dilutions in all were made in this way, so that the final six emulsions tested for toxicity were of the following concentrations:

(1) 20 ppt emulsion of crude oil in sea water
(2) 10 ppt emulsion of crude oil in sea water
(3) 5 ppt emulsion of crude oil in sea water
(4) 2·5 ppt emulsion of crude oil in sea water
(5) 1·25 ppt emulsion of crude oil in sea water
(6) 0·625 ppt emulsion of crude oil in sea water

These emulsions varied in colour from cloudy pale brown in the first to almost clear and colourless in the sixth, and were tested for toxicity immediately after being made. The experimental procedure was the same as that in the previous section ('Refined oil products').

The species used in these experiments were the intertidal gastropods *Littorina littorea*, *Littorina littoralis* and *Patella vulgata*. The toxicity of Kuwait crude oil (neat) was tested at the same time as that of its emulsions; a control tank with clean sea water was also set up. The activity of the animals

TABLE II

Percentage mortality figures for *Littorina littoralis, Littorina littorea* and *Patella vulgata* at the end of a 5-day recovery period following exposure to Kuwait crude oil and oil-in-water emulsions of different concentrations. Laboratory temperature: 12–14°C. Number of animals in each experimental group given in parentheses.

Exposure period (h)	*Concentration of oil-in-water emulsion* (ppt)	*Percentage mortality*		
		L. littoralis ($n = 25$)	*L. littorea* ($n = 25$)	*P. vulgata* ($n = 50$)
6	20	8	0	58
6	10	0	0	34
6	5	4	0	36
6	2·5	4	2	4
6	1·25	0	0	8
6	0·625	0	0	10
6	Crude oil (neat)	64	28	100
6	Sea water (control)	4	0	6
24	20	8	16	—
24	10	6	6	—
24	5	0	12	—
24	2·5	2	0	—
24	1·25	2	0	—
24	0·625	0	6	—
24	Crude oil (neat)	92	42	—
24	Sea water (control)	2	0	—

was observed while they were immersed in the emulsions. These experiments were carried out during January and February, 1971.

Table II shows the percentage mortality figures for the three species of gastropod at the end of the 5-day recovery period following exposure for 6 h and 24 h to crude oil and oil-in-water emulsions of different concentrations. Figure 4 illustrates the activity of *L. littoralis* in emulsions of different concentrations during the first 6 h of exposure of 50 individual animals; an animal was considered to be active if it emerged from its shell and moved around. From Table II it can be seen that mortalities of *L. littorea* and *L. littoralis* are negligible following exposure for 6 h to emulsions at all concentrations used, and only very low following exposure for 24 h at the higher concentrations (5–20 ppt). *Patella vulgata*, however, found by Crapp (1971b) to be one of the most sensitive rocky shore species to oil pollution, suffered substantial mortalities from a 6-h exposure to the oil-in-water emulsions,

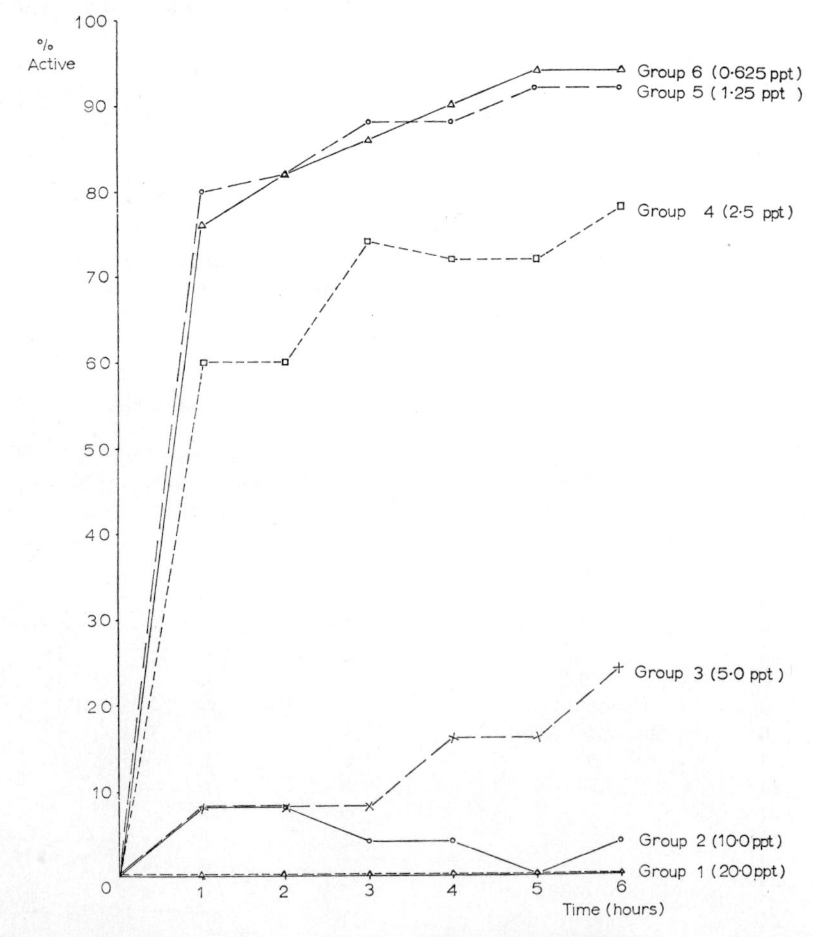

FIG. 4. Percentage activity of *Littorina littoralis* ($n = 50$) during a 6-h period of immersion in emulsions of crude oil in sea water, at various concentrations.

with mortality increasing with concentration of the emulsions. From Fig. 4 it can be seen that *L. littoralis* remains inactive while immersed in emulsions more concentrated than 2·5 ppt, while concentrations of 1·25 ppt and less do not inhibit the animal from emerging from its shell and moving around.

The physical and chemical characteristics of oil-in-water emulsions were described by Pilpel (1968) as water, in the continuous phase, containing droplets of oil up to a few millimetres in diameter, stabilised by the presence on their surface of hydrophilic groups such as $—COO^-$ (acid), —OH (alcohol), —CHO (aldehyde), $—OSO_3^-$ (sulphate) and $—SO_3H^-$ (sulphonate). Such oil-in-water emulsions are readily miscible in sea water, leading to rapid dispersion at sea. Parker *et al.* (1971) investigated the stability of fine dispersions of Kuwait crude oil in water; they found that residual fine dispersions containing about 1·0 ppm of oil can persist for several weeks. It therefore seems that in turbulent seas significant amounts of oil could remain dispersed long enough to be carried quite deep or far away from the original slick. Oil dispersed in water in the form of droplets with diameters of 1 μm is invisible; such a degree of dispersion results in a million-fold increase in surface area (Berridge *et al.*, 1968a).

Investigations on the biological effects of oil in the form of oil-in-water emulsions have shown varying degrees of toxicity and physical damage to be caused by such emulsions. Spooner (1968a) found that physically emulsified Kuwait crude oil droplets were able to pass through the gut of mussels (*Mytilus edulis*) undigested. Furthermore, considerable numbers of ciliates were observed ingesting minute oil globules (2–3 μm in diameter) after these had passed through the gut of mussels (Spooner, 1968b, in discussion following a paper by Wardley Smith). Oil droplets in emulsion in sea water were found by Parket *et al.* (1971) to pass unchanged through the gut and into the faecal pellets of copepods and barnacle larvae (the nauplius stage of *Balanus balanoides*), indicating an unselective process of food filtration in these animals. Simpson (1968) reports that oil emulsified by wave action is ingested in the form of droplets by filter feeders such as mussels, cockles and oysters; although oil in this form appears to have no serious effect on these animals, it taints their flesh and thus renders them commercially worthless. North *et al.* (1964) observed that after the *Tampico Maru* spill extreme surf and wave action resulted in the formation of diesel fuel emulsions which eliminated sea stars of the genus *Pisaster*, and caused high mortalities in sea urchins (*Strongylocentrotus* spp.). Subsequent laboratory toxicity experiments showed certain damaging biological effects of diesel oil emulsions, including reduction of photosynthetic capacity by the kelp *Macrocystis*, and inactivation of the tube feet of sea urchins (*Strongylocentrotus purpuratus*) resulting in the loss of grip by these animals on the substrate. Cole (1941) reported the damage of emulsified fuel oil to fish gills, the adherence of these droplets leading to death by asphyxiation. Similar findings were made by Thomas (quoted by Gutsell, 1921) with emulsions of petroleum residue and light fuel oil, and by Mironov (1970) who found emulsified oil to be much more damaging to fish than an oil film on the water surface. Mironov also found that emulsions of diesel oil in water caused the death of zooplankton.

Oil in the form of an emulsion of dispersed droplets in water may contribute to the effects of chronic pollution caused, for example, by oil refinery effluent water. Baker (1971b) found that excessive damage was suffered by colonies of *Spartina anglica* on salt marshes bordering Southampton Water, which she attributed to the effects of effluent water of a nearby oil refinery. Crapp (1971c) similarly found local abnormalities in the shore ecosystem in the immediate neighbourhood of the outfall site of the effluent water of an oil refinery in Milford Haven.

Under natural conditions at sea, the formation of water-in-oil emulsions, or 'chocolate mousse', may occur to a greater extent than the formation of oil-in-water emulsions. These water-in-oil emulsions may contain 30–80% water, and are very stable, additional stability possibly being conferred by external agents such as organic debris (Berridge *et al.* 1968a). Mousses vary in stability according to the chemical composition of the oil; their microbial degradation takes place at a very slow rate, if at all (Berridge *et al.* 1968b). Mousses are thus likely to be far more persistent than oil-in-water emulsions. The biological effects of mousse were examined by Baker (1971a), who found that mousse containing Kuwait crude oil and 88% sea water had only a blanketing effect on salt marsh grasses, the shoots of which eventually penetrated the mousse. Nelson-Smith (1970) stated that mousses cling more readily to algae, especially when dried out at the top of the shore, and thus cause the overweighted plants to be torn off by waves. It therefore appears that water-in-oil emulsions are likely to cause biological damage by physical effects, while oil-in-water emulsions probably cause more biological damage by toxic effects.

Oil and Dispersant Mixtures

This set of experiments was designed to investigate the toxicity of mixtures of crude oil and dispersants compared to that of (1) the neat crude oil and (2) the neat dispersants alone. Mixtures (500 ml) of Kuwait crude oil and three different dispersants were made by shaking together manually for 5 s different proportions of oil, dispersant and sea water; the relative proportions of each liquid in each mixture are shown in Table III. The dispersants used were BP 1002, Test Blend X and BP 1100X. The neat crude oil alone and neat dispersants alone were also tested for toxicity simultaneously.

The experimental procedure was the same as that in the previous section. The species *Littorina littorea* and *Littorina littoralis* were used. Batches of 50 animals of each species were exposed to each liquid being tested for toxicity for a period of 1 h (*L. littorea*) or 30 min (*L. littoralis*), after which the animals were put into clean aerated sea water and allowed to recover over a 5-day period. Mortalities were taken as the number of animals in each group still immobile and remaining in the water at the end of this 5-day period. These experiments were performed during August, 1971.

Table III shows the percentage mortality figures for *L. littorea* and *L. littoralis* at the end of these experiments. From this table it can be seen that BP 1100X, the more recently developed 'low-toxicity' dispersant, differs in its toxic effects from the other two dispersants investigated. In the case of BP 1100X the oil and dispersant mixtures are more toxic to both *Littorina* species than either neat crude oil or dispersant, neat or diluted. In the case of BP 1002 and Test Blend X, however, the oil and dispersant mixtures are more toxic to *L. littorea* than the oil alone and just as toxic as the dispersant

TABLE III

Percentage mortality figures for *Littorina littorea* and *Littorina littoralis* after periods of 1 h and 30 min respectively of exposure to crude oil, emulsifiers and mixtures of these in various concentrations and dilutions, followed by a 5-day recovery period. Laboratory temperature: 19–21°C. 50 animals were in each experimental group for each species.

	Constituents of toxic liquid			*Percentage mortality*	
Emulsifier	*Kuwait crude oil* (ml)	*Emulsifier* (ml)	*Sea water* (ml)	*L. littorea*	*L. littoralis*
BP 1002	—	500 (neat)	—	26	100
	100	400	—	24	100
	100	200	200	0	96
	100	100	300	4	92
	—	250	250	8	100
	—	125	375	2	82
Test blend X	—	500 (neat)	—	100	100
	100	400	—	96	100
	100	200	200	4	96
	100	100	300	6	92
	—	250	250	6	100
	—	125	375	2	82
BP 1100X	—	500 (neat)	—	0	54
	100	400	—	38	80
	100	200	200	26	94
	100	100	300	42	76
	—	250	250	10	76
	—	125	375	4	54
	500 (neat)	—	—	12	64
	—	—	500 (control)	0	2

alone when undiluted; when the dispersant in the mixtures is diluted, however (i.e. with sea water), then the mixtures are relatively very low in toxicity. For *L. littoralis* with BP 1002 and Test Blend X, the oil and dispersant mixtures are more toxic than oil alone and just as toxic as the neat or diluted dispersant. For both species, the lower the concentration of dispersant in sea water (without any oil), the lower the mortalities: an exception to this, however, is BP 1100X which resulted in a lower mortality when neat than when diluted for both species of gastropod. This may be explained, however, by complete retraction of the animals into their shells during exposure to neat dispersant.

From the results of these experiments it can be seen that mixtures of crude oil and dispersants are often more toxic than the oil by itself, and that, in the presence of oil, dilution of these dispersants is vital for a reduction in toxicity. Investigations made by other workers on the toxicity of oil and dispersant mixtures have given somewhat conflicting results. Perkins (1968) found that the median tolerance level of the shore crab (*Carcinus maenas*) to mixtures of BP 1002 and crude oil at a 96 h LC_{50} was 15·0, compared to 29·0 for BP 1002 alone, indicating that the addition of oil increases the toxicity of the dispersant. Spooner (1968a) found that mixtures of Kuwait crude oil and the dispersant Corexit were more toxic to barnacle larvae (*Elminius modestus*) and young adult mussels (*Mytilus edulis*) than Kuwait crude oil alone; some of the toxicity of these mixtures is still attributable to the oil, however, since weathered Kuwait crude + Corexit was less toxic than fresh Kuwait crude + Corexit to the barnacle larvae. With a larger size range of mussels (3·0–3·5 cm in length), however, Spooner found that both bunker C and Kuwait crude oil (at 1000 ppm), and Corexit (at 100 ppm) caused only very low mortalities (3–7%) after a 20 h exposure, while mixtures of these oils (individually) with Corexit caused no increase in toxicity over that of each oil alone. Portmann and Connor (1968) found that the addition of twice its volume of oil to Polyclens increases substantially its toxicity to the cockle (*Cardium edule*) and renders it slightly more toxic to brown shrimps (*Crangon crangon*), although BP 1002 and Gamlen become less toxic when treated in this way. Nelson-Smith (1970), in preliminary experiments using yeast in a micro-respirometer, found that the addition of equal amounts of fresh crude oil to BP 1002 approximately halves its toxicity; he suggested that oil may bind some of the toxic elements of the dispersant and thus effectively remove them from culture solution. Van de Wiele (1968) found that emulsions of Finasol are just as toxic to the brine shrimp (*Artemia salina* L.) and the guppy (*Lebistes reticulatus* Peters) whether or not crude oil is added; two types of Finasol and four crude oils were investigated, of which the dispersants alone were far more toxic than any of the crude oils alone. Field experiments by Crapp (1971d) with Kuwait crude oil and BP 1002 showed that for a number of rocky shore species the oil and dispersant mixtures (or 'runoff' from areas oiled and cleaned on the shore) were more toxic than oil alone or oil cleaned by dispersant 30 min after its application.

Mixtures of oil and dispersants on the open sea are likely to cause relatively little biological damage since such emulsions appear to disperse and become degraded fairly rapidly. In the event of spilt oil coming ashore and then being treated with dispersants, however, severe biological effects may occur. In such a case, oil and dispersant mixtures would become dispersed in rock pools which would otherwise have only a surface covering of oil until the next tide washed this away. The toxicity of such mixtures accounts for the number of barren pools reported on Cornish shores following their treatment with dispersant after the *Torrey Canyon* spill (Smith 1968); empty limpet seats, or 'scars', were especially conspicuous in pools, and extensive mortalities were suffered by several species of rock pool algae, including *Corallina officinalis* and *Lithothamnion*.

There thus appears to be no simple relationship between on the one hand the toxicity of mixtures of oil and dispersants and on the other the toxicity of oil or dispersant alone. The conflicting results described above indicate that a considerable number of factors are of importance in determining the overall toxicity of such mixtures; such factors include the type of dispersant and oil, the relative proportions and dilutions of these and the degree of emulsification.

REFERENCES

Allen, H. (1971). Effects of petroleum fractions on the early development of a sea urchin, *Mar. Poll. Bull., Newcastle*, **2**, No. 9.

Baker, J. M. (1971a). Comparative toxicities of oils, oil fractions and emulsifiers, in (ed. E. B. Cowell), *The Ecological Effects of Oil Pollution on Littoral Communities*, Proc. Symp., Institute of Petroleum, London, pp. 78–87.

Baker, J. M. (1971b). Refinery effluent, in (ed. E. B. Cowell), *The Ecological Effects of Oil Pollution on Littoral Communities*, Proc. Symp., Institute of Petroleum, London, pp. 33–43.

Berridge, S. A., Dean, R. A., Fallows, R. G. and Fish, A. (1968a). In *Scientific Aspects of Pollution of the Sea by Oil*, Institute of Petroleum, London, pp. 2–9.

Berridge, S. A., Thew, M. T. and Loriston-Clarke, A. G. (1968b). In *Scientific Aspects of Pollution of the Sea by Oil*, Institute of Petroleum, London, pp. 35–57.

Blumer, M. (1969). Oil pollution of the ocean, in (ed. D. P. Hoult), *Oil on the Sea*, Proc. Symp., Plenum Press, pp. 5–13.

Brunnock, J. V., Duckworth, D. F. and Stephens, G. G. (1968). In *Scientific Aspects of Pollution of the Sea by Oil*, Institute of Petroleum, London, pp. 12–27.

Chipman, W. A. and Galtsoff, P. S. (1949). Effects of oil mixed with carbonized sand on aquatic animals, Spec. scient. Rep. U.S. Fish. Wildl. Serv., 1:1–53.

Cole, A. E. (1941). Effects of pollutional waste on fish life, *Symp. Hydrobiol.*, Wisconsin, pp. 241–259.

Crapp, G. B. (1969). The biological effects of marine oil pollution and shore cleansing, O.P.R.U. Annual Report, 1969, pp. 27–42.

Crapp, G. B. (1971a). Laboratory experiments with emulsifiers, in (ed. E. B. Cowell), *The Ecological Effects of Oil Pollution on Littoral Communities*, Proc. Symp., Institute of Petroleum, London, pp. 129–49.

Crapp, G. B. (1971b). The ecological effects of stranded oil, *ibid.*, pp. 181–6.

Crapp, G. B. (1971c). Chronic oil pollution, *ibid.*, pp. 187–203.

Crapp, G. B. (1971d). Field experiments with oil and emulsifiers, *ibid.*, pp. 114–28.

Gutsell, J. S. (1921). Danger to fisheries from oil and tar pollution of water, Rep. U.S. Commnr. Fish. (Append. 7).

Howe, C. and Ottway, S. (1971). Some effects of crude oil, oil fractions and products on the prawn *Leander squilla*, O.P.R.U. Annual Report, 1971, pp. 22–8.

Kühnhold, W. H. (1969). Der Einfluss wasserlöslicher Bestandteile von Rohölen und Rohölfraktionen auf die Entwicklung von Heringsbrut, *Ber. Dt. Wiss. Komm. Meeresforsch.*, **20**, H.2, 165–71.

Ministry of Transport (1953). Report of the Committee on the prevention of pollution of the sea by oil, H.M.S.O., London.

Mironov, O. G. (1967). The effect of oil and oil products upon some molluscs in the littoral zone of the Black Sea, *Zool. Zh.*, **46**, 134–6 (in Russian).

Mironov, O. G. (1970). The effect of oil pollution on flora and fauna of the Black Sea, F.A.O. Tech. Conf. on Mar. Poll. and its effects on living resources and fishing, Rome, Dec., 1970.

Mironov, O. G. and Lanskaja, L. A. (1967). Biology and distribution of plankton of the southern seas, Oceanographical Commission, Moscow, pp. 31–4 (in Russian).

Nelson-Smith, A. (1970). The problem of oil pollution of the sea, *Adv. mar. Biol.*, **8**, 215–306.

North, W. J., Neushell, M. and Clendenning, K. A. (1964). Successive biological changes observed in a marine cove exposed to a large spillage of mineral oil, *Symp. Pollut. mar. Micro-org. Prod. petrol., Monaco*, pp. 335–54.

Ottway, S. (1971). The comparative toxicity of crude oils, in (ed. E. B. Cowell), *The Ecological Effects of Oil Pollution on Littoral Communities*, Proc. Symp., Institute of Petroleum, London, pp. 172–80.

Parker, C. A., Freegarde, M. and Hatchard, C. G. (1971). The effect of some chemical and biological factors on the degradation of crude oil at sea, in (ed. P. Hepple), *Water Pollution by Oil*, Proc. Symp., Institute of Petroleum, London.

Perkins, E. J. (1968). The toxicity of oil emulsifiers to some inshore fauna, *Field Studies*, **2** (suppl.), 81–90.

Pilpel, N. (1968). The natural fate of oil on the sea, *Endeavour*, **27**, 11–13.

Portmann, J. E. and Connor, P. M. (1968). The toxicity of several oil spill removers to some species of fish and shellfish, *Mar. Biol.*, **1**, 322–9.

Simpson, A. C. (1968). Oil, emulsifiers and commercial shellfish, *Field Studies*, **2** (suppl.), 91–8.

Smith, J. E. (Ed.) (1968). '*Torrey Canyon*', *Pollution and Marine Life*, Cambridge University Press, 196 pp.

Spooner, M. F. (1968a). Preliminary work on comparative toxicities of some oil spill dispersants and a few tests with oil and Corexit, Marine Biology Association, Plymouth.

Spooner, M. F. (1968b). In discussion, in *Scientific Aspects of Pollution of the Sea by Oil*, Institute of Petroleum, London, p. 67.

Suess, M. J. (1970). Polynuclear aromatic hydrocarbon pollution of the marine environment, F.A.O. Tech. Conf. on Mar. Pollut. and its effects on living resources and fishing, Rome, Dec., 1970.

Tagatz, M. E. (1970). Reduced oxygen tolerance and toxicity of petroleum products to juvenile American shad, *Chesapeake Sci.*, **2**, 65–71.

Van de Wiele, C. (1968). Toxicité des détergents et des pétroles. 4. Toxicité de l'émulsion, Institut Royal des Sciences Naturelle de Belgique.

Zitko, V. (1971). Determination of residual fuel oil contamination of aquatic animals, *Bull. Environ. Contam. Toxicol.*, **5**(6), 559–64.

16

The Importance of Behavioural Patterns in Toxicity Testing and Ecological Prediction

BRIAN DICKS

(*Oil Pollution Research Unit, Orielton Field Centre, Pembroke, Wales*)

SUMMARY

Many toxicity tests for oil pollutants rely on mortality of a test organism, both for ranking compounds in order of toxicity and for making ecological predictions. These tests often do not take into account two important aspects of animal behaviour which can substantially influence the effect of a pollutant on an organism. These are (a) that changes in behaviour in response to sub-lethal doses of pollutant may have far-reaching ecological effects, and (b) that natural rhythm in an animal's activity may substantially influence its susceptibility to pollutants.

Examples are described which illustrate these points. The first involves a study of settlement of an acorn barnacle, Balanus balanoides, *around a refinery effluent outfall, and a study of some effluent effects on a larval stage of this species. The second conerns the responses of the limpet,* Patella vulgata, *to crude oil. Also included are notes on escape reactions of* Nereis diversicolor *produced by sub-lethal doses of refinery effluent, and the reaction of* Littorina saxatilis *to a sub-lethal dose of refinery effluent.*

It is concluded from the results of these studies:

(*a*) *that rhythm in an animal's activity can substantially influence its susceptibility to pollutants, and that this should be taken into account during toxicity testing for comparative purposes;*

(*b*) *that mortality tests may produce misleading results for making ecological predictions, as sub-lethal doses can result in mortality on the shore by inducing an unfavourable behavioural response;*

(*c*) *that tests using such behavioural responses are more suitable than mortality tests for making ecological predictions because they take into account criteria of importance to organisms in their environment; and*

(*d*) *that the low salinity of a refinery effluent was at least as important as other effluent characteristics in producing a response in barnacle nauplii, and this response has resulted in a reduced population density of adult barnacles in the immediate area of the effluent outfall.*

INTRODUCTION

Many toxicity tests for oil pollutants rely on the criterion of mortality of a sensitive test organism, with the aim of assessing relative toxicities of different compounds and to make ecological predictions of short- or long-term pollutant effects. A useful discussion of the aims and problems of toxicity testing can be found in Beynon and Cowell (1974). However, such tests often do not take into account two important aspects of animal behaviour which can substantially affect test results and the validity of ecological prediction from them. The first of these is the observation that pollutants in sub-lethal doses may elicit behavioural responses in some organisms which can have far-reaching ecological implications, either for that organism or for others as well by upsetting carefully balanced communities (Dicks, 1973; Baker and Crapp, 1974; Swedmark, 1974). If such responses occur, then mortality tests with these organisms may not be suitable for ecological prediction. The second aspect of behaviour concerns the fact that many marine organisms show a circadian rhythm in their activity (Rao, 1954; Brown *et al.* 1956; Salanki, 1966; Dicks, 1972), and this may considerably affect test results depending on the state of activity of the organism at the start of a test or during the test (Dicks, 1973). A third point worth noting in the context of making ecological predictions of pollutant effects on shore life is that it is common to use a 48 or 96 h toxicity test (Portman and Wilson, 1971; Swedmark, 1974; Wilson, 1974), whereas the tidal cycle ensures considerable variation in pollutant concentration over a much shorter time period, and may mean the removal of a pollutant after only a few hours. Short-duration tests of pollutants may, therefore, be more useful for ecological prediction than longer ones.

Examples are described here which illustrate these points. The first involves a study of settlement of acorn barnacles (*Balanus balanoides*) around a refinery effluent outfall, and a study of effluent effects on the larval forms of this species. The second, which concerns the responses of the limpet, *Patella vulgata*, to the effects of crude oil, is described fully elsewhere (Dicks, 1973), but summarised here. Notes on escape reactions of *Nereis diversicolor* from a refinery effluent and the reaction of *Littorina saxatilis* to effluent (Parsons, 1972) are also included.

SETTLEMENT OF *BALANUS BALANOIDES* AROUND A REFINERY EFFLUENT OUTFALL

The Effects of Refinery Effluent on the Settlement of *Balanus balanoides* on a Rocky Shore

The common acorn barnacle, *B. balanoides,* which is abundant in Milford Haven, occurs over a large part of the rocky shore from near the low tide mark to near high tide mark, but most commonly in the middle of the shore (mean tide level). It starts life with a larval stage which spends two or three weeks in the plankton in early spring feeding on the abundant phytoplankton growth. The larval stage is spent in two forms, an early swimming stage (nauplius) which is attracted towards light (and therefore the sea surface), which then metamorphoses to a second cyprid stage which is the settling stage, and searches the shore for a suitable place to settle and change to the adult form. During regular shore surveys around the Little Wick refinery outfall in Milford Haven (Baker, 1973), it was noticed that settlement of the young barnacles was less dense on rocks near to the outfall point than on those farther away. This situation was investigated by detailed measurement of barnacle density at sites varying in distance from the outfall point.

(i) Methods

The survey was carried out in June 1974. At this time of year the newly settled barnacle spat are easily distinguished from older barnacles by their small size and clean white colour.

To overcome problems of comparison of results from site to site, all measurements of barnacle density were made at the same height on the shore (mid tide level, where *B. balanoides* is abundant) and on areas of rock with the same aspect and similar slope. As the effluent discharge pipe runs through a small sandy bay, there were no suitable rocks close to the outfall point for experimental work. However, as the pipe is concrete-coated and this has been settled by barnacles, measurements were made at three sites on the pipe itself—at the outfall point and at 18 and 30 m up the pipe from this point. The position of sample sites is illustrated in Fig. 1.

Density was measured by counting the numbers of *B. balanoides* (both newly settled spat and older barnacles) in ten randomly placed 5 cm × 5 cm quadrats at each site. 0·05 probability limits of the mean density per quadrat were calculated for each site by finding standard error (σ_n) from

$$\sigma_n = \left(\frac{\sum (x)^2 - (\sum x)^2/n}{n(n-1)}\right)^{1/2}$$

where x is each observed quadrat density and n is the total number of observations (10 in this case).

Values for t were found from the appropriate statistical table for $p = 0{\cdot}05$ and the appropriate number of degrees of freedom $(n - 1)$. 0·05 limits for each site are then $\bar{x} \pm n^t$, where $\bar{x}$ = mean density per quadrat at that site.

(ii) Shore Survey Results

These are summarised graphically in Fig. 2 as mean densities of barnacle spat per quadrat and mean density of *B. balanoides* per quadrat plotted against the distance from the outfall point. It is evident that fewer barnacles have settled near to the outfall than further away, with no barnacles at all settling on the end of the pipe, though they settle readily further up the pipe. At site 1 on the western side of the pipe where settlement was very low, those which have settled show no obvious mortality since settlement. It is logical to assume that as settlement has been influenced, the effect of the effluent is on the larval stages of the organism and not on the settled stage. This has also been indicated by the experiments of Baker (1973, 1975) on the limpet distribution around this outfall. For this reason, experiments were conducted in the laboratory on the larvae of *B. balanoides*.

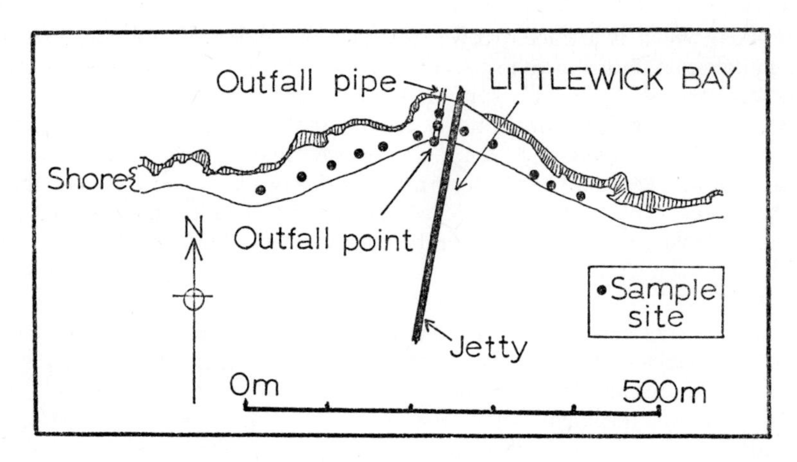

FIG. 1. The location of sample sites around the Little Wick Bay refinery effluent outfall.

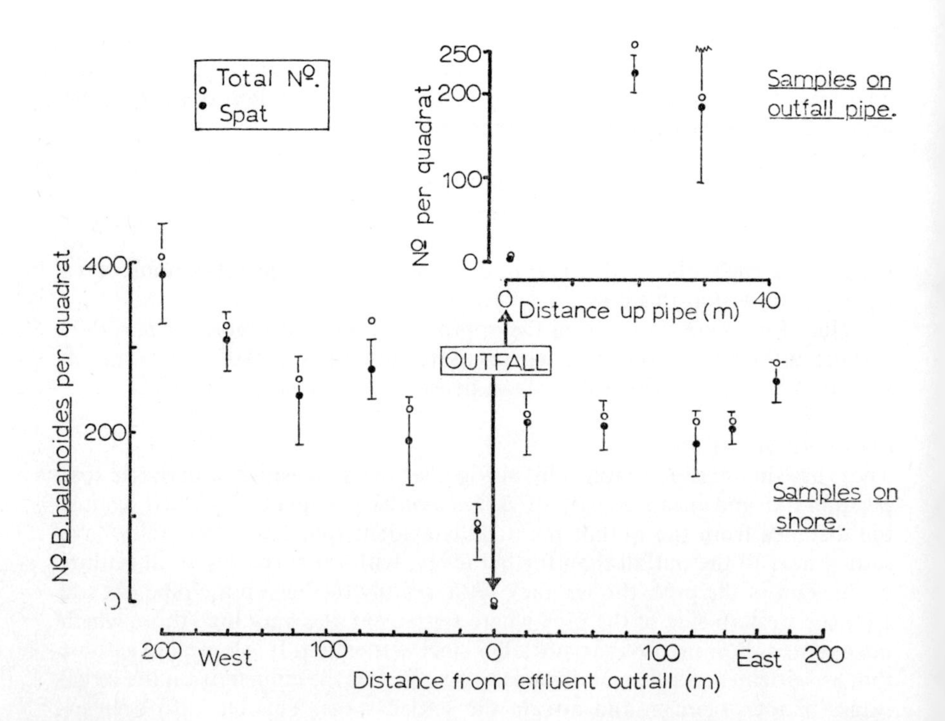

FIG. 2. The density of *Balanus balanoides* at 14 sampling sites of different distance from Little Wick Bay refinery effluent outfall. 10 quadrats of 5 × 5 cm were counted at each site. 95% confidence limits for the means of spat density are shown.

Laboratory Tests with Larvae of *B. balanoides*

(*i*) *Methods*

Due to difficulties in obtaining sufficient numbers of the settling larval stages (cyprids), experiments were made on the earlier larval stages (nauplii). These were obtained with a standard Plymouth MBA zooplankton net (mesh aperture 174 μm). Identification of the nauplii was made under a low-power binocular microscope using Pyefinch (1948) and Newell and Newell (1963). By sampling during the reproductive period of *B. balanoides* (early spring), the greatest number of nauplii in the samples were of this species. All nauplii were acclimatised to laboratory conditions for 12–18 h before being used experimentally.

Two types of test were undertaken, all for very short time periods (up to 2 h exposure): static tests, where groups of 10 larvae were placed in experimental solutions in petri dishes; and layer tests, where experimental solutions (which were of lower density than sea water) were carefully floated on normal sea water in test tubes using a fine pipette. Batches of larvae were introduced into either layer by pipette, and their reactions noted. The tubes were illuminated from above, thus causing the nauplii to swim towards the test solution. Their reactions on encountering the test solutions were noted.

After each experiment the larvae were returned to normal sea water and their recovery (or otherwise) watched. No attempt was made to assess any longer-term effects on those which recovered.

All experiments were carried out at the sea temperature from which the nauplii were collected (12°C $\mp$ 1°C). Effluent for use in the laboratory was obtained from the Texaco Refinery in Milford Haven.

(*ii*) *Results of Static Tests*

Initial experiments showed that placing nauplii in dishes of effluent resulted in immediate cessation of activity and sinking to the dish floor. Groups of nauplii were left in the effluent for periods of up to 2 h, and then returned to dishes of fresh sea water. All recovered within 15 min and were actively swimming in the dish. Measurement of the effluent salinity revealed it to be, at 9‰, considerably less than the normal Milford Haven sea salinity of 34‰. To distinguish whether this or other of the effluent characteristics was producing the cessation in swimming activity, sea salt, obtained by evaporating sea water to dryness, was added to the effluent to make it up to 34‰ and the test repeated. Under these conditions, the nauplii did not cease swimming even after 2 h. To test the effects of both effluent and salinity more thoroughly, series of tests and controls were made. A range of solutions from salinity 9–34‰ were made up and the percentages of nauplii swimming after time intervals from 5 to 60 min were noted. The test was repeated five times with 10 nauplii per dish. After the experiments, the nauplii were returned to fresh sea water and their recovery (if the solution had prevented them from swimming) was noted. The results are illustrated graphically in Fig. 3.

Salinities below 18‰ produced immediate cessation of swimming activity, which was not recovered during the experimental period. At 18‰ and 26‰ salinity, though swimming was initially inhibited in most nauplii, over a period of time an increasing number became active, presumably due to some

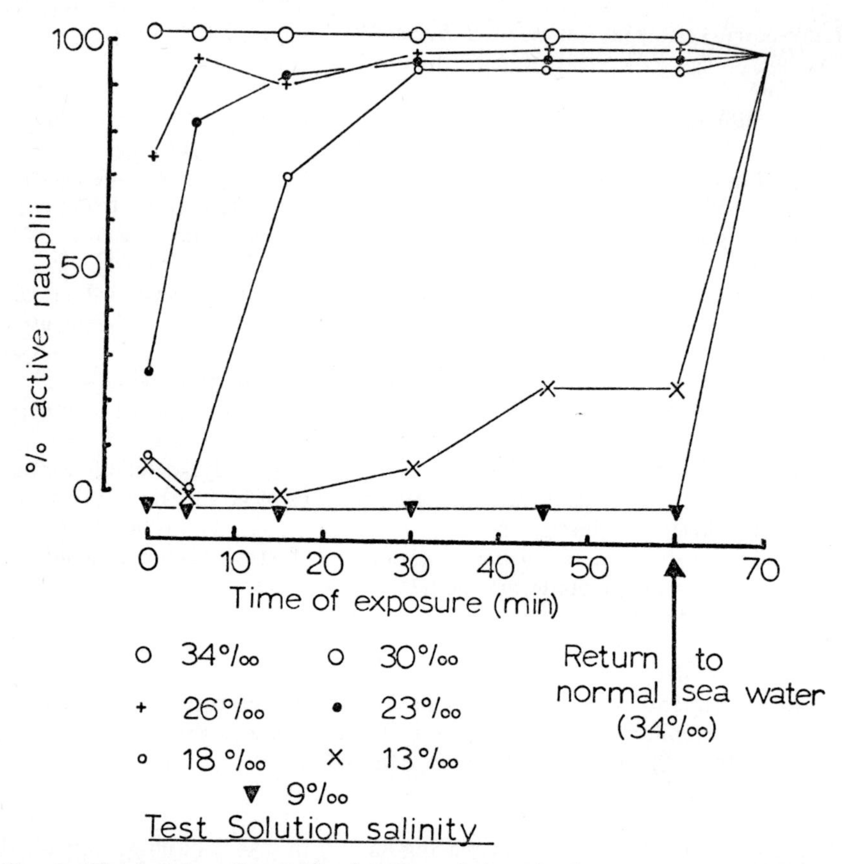

FIG. 3. The activity of nauplii of *Balanus balanoides* in sea waters of different salinity. Each curve represents experiments with 50 animals.

sort of equilibration process. Above 26‰ there was little effect on naupliar activity. All nauplii recovered quickly to normal activity when returned to full-strength sea water.

To compare the effects of effluent at various salinities with sea water at various salinities, two series of test solutions were used. These were: sea water at 34‰, 23‰ and 9‰ salinity made by diluting sea water with distilled water; effluent at 34‰, 23‰ and 9‰ salinity made by adding sea salt to effluent at 9‰ salinity. To act as a control of the sea salt addition, sea water was first diluted to 9‰ with distilled water and then made up again to 34‰ with sea salt to make a seventh experimental solution. Numbers of nauplii actively swimming in the test solution at various times from 5 min to 1 h were noted, and then returned to full-strength sea water to watch recovery. The results are illustrated graphically in Fig. 4. Both effluent and sea water at 9‰ salinity produced total cessation of swimming, but whilst sea water at 23‰ salinity initially stopped 25% of the nauplii swimming, effluent at the same strength stopped 75%. After 15 min almost all the nauplii which stopped swimming in 23‰ salinity had recovered swimming ability, but those in the

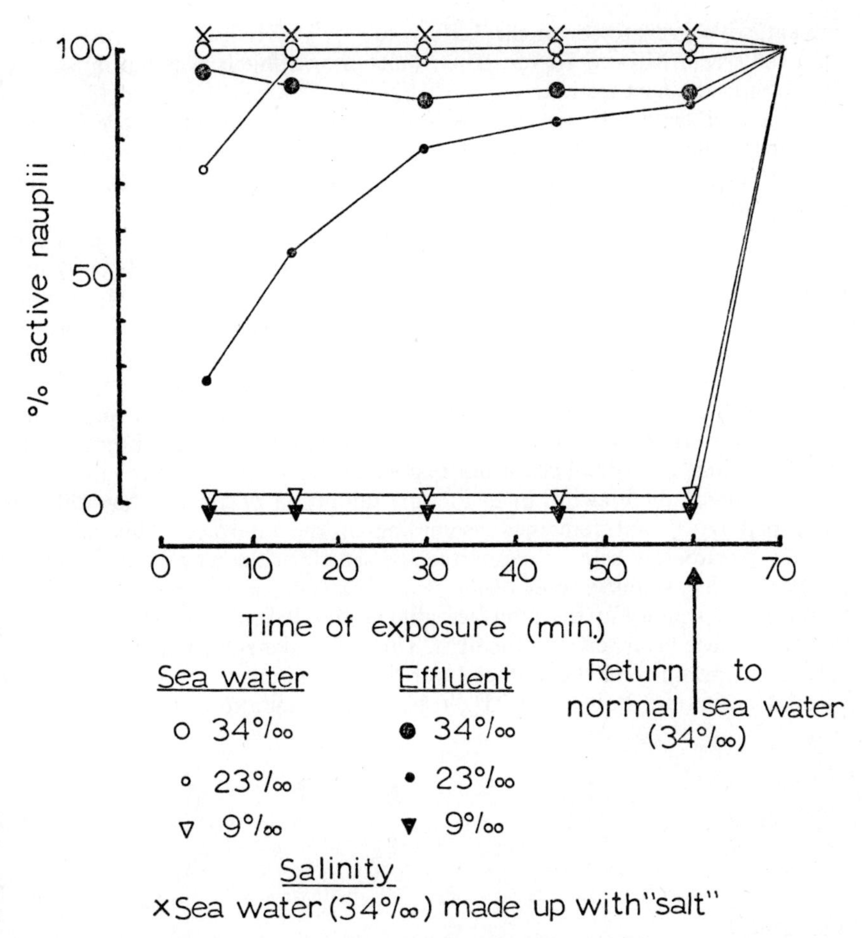

FIG. 4. A comparison of the percentage of nauplii of *Balanus balanoides* actively swimming in sea water and refinery effluent of three different salinities, plotted against the time spent in each. Each curve represents experiments with 100 nauplii.

effluent took a much longer time to recover this ability, and 10% did not recover at all whilst in the experimental solution. The effluent water at full strength also shows about 10% lower activity than that in full-strength sea water.

All the nauplii recovered swimming activity after 10 or 15 min in normal sea water.

The results show clearly that a major effect of the effluent is a result of its salinity, but that other effluent properties have a detectable effect.

(iii) Results of Layer Tests

In the field situation where refinery effluent is generally produced at variable salinities (but almost always lower than that of sea water, particularly after heavy rainfall), it might be expected that such water would remain at the surface of the sea due to its lower density. This is shown to be the case for this effluent, which remained on the surface even when only 3‰ salinity lower than the sea in Milford Haven. The results of an effluent distribution survey are given in the next section.

As nauplii swim towards the sea surface, they will encounter the effluent as a surface layer, and will not be forced to stay immersed in the solution as they are in the static tests. For this reason, the layer tests were carried out to simulate field conditions.

The same range of test solutions as in the static tests were layered on sea water. Nauplii were introduced and their initial reaction noted. Their distribution in the tubes after 5 min, 10 min and 1 h was recorded. The results are shown diagrammatically in Figs. 5 and 6. The reactions to effluent and low-salinity sea water were distinguishable by this test. The reactions to low-salinity sea water as an upper layer were as follows (Fig. 5). If the salinity of the surface layer was lower than 18‰ the nauplii sank out of this layer after ceasing to swim and recovered swimming ability immediately upon entering the normal sea water below. The attraction towards the light, however, made them swim up to the low-salinity layer again. They did not enter this layer, but remained swimming at the interface. This layer of nauplii immediately below the interface persisted for the whole experimental period. At 18‰ salinity, some of the stronger swimmers became acclimatised to the lower salinity at the interface after about 10 min, and slowly moved into the surface layer towards the light. The remainder stayed at the interface. Above 26‰ salinity of the surface layer the nauplii entered the surface layer without hesitation, but in the 23‰ tube, initial hesitation at the boundary

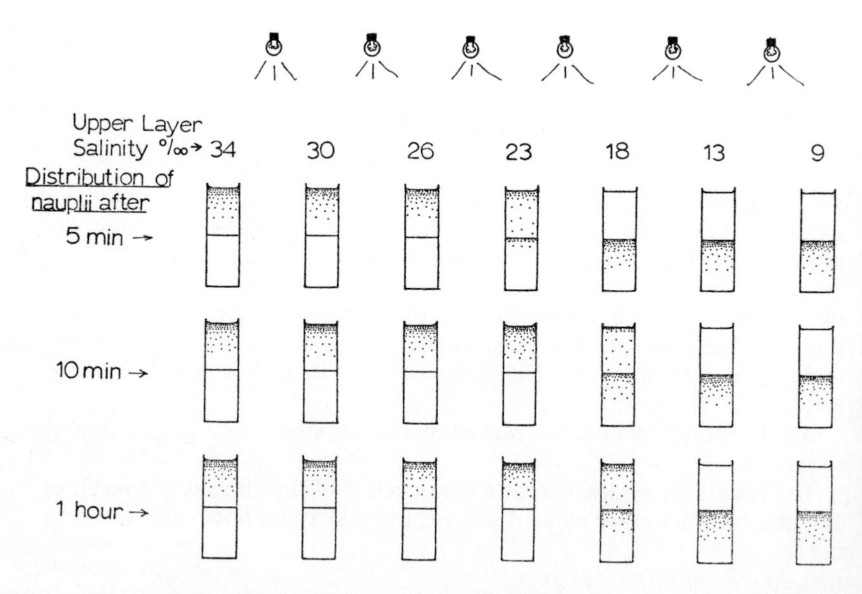

FIG. 5. The distribution of the nauplii of *Balanus balanoides* in test tubes of normal sea water, on the surface of which were layered sea waters of lower salinity. The shaded areas show the location of the larvae in the tubes after 5, 10 and 60 min in each. The tubes were illuminated from above.

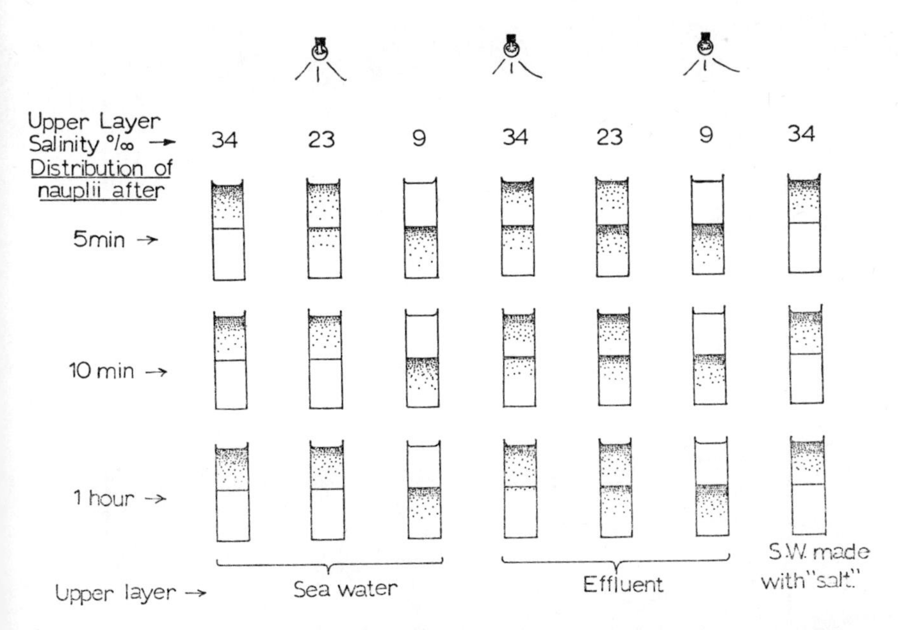

FIG. 6. The distribution of nauplii of *Balanus balanoides* in test tubes with refinery effluent and sea water of various salinities layered on normal sea water. The tubes were illuminated from above. The shaded areas show the location of the swimming larvae in the tubes after 5, 10 and 60 min in each.

was followed by entry of this layer and no detectable effect on the nauplii after 1 h. Whilst reactions to the effluent layer at various salinities was very similar to that of sea water at various salinities, some of the nauplii did not immediately enter the effluent layers of 23‰ and 34‰ salinity, but remained at the interface (Fig. 6). The majority of the nauplii eventually acclimatised to the 34‰ (after 1 h) and entered this layer, but they did not acclimatise at 23‰. These results agree well with those of the static tests, where a salinity effect and an effect of other effluent components was distinguishable.

Field Survey of Effluent Distribution and Salinity

The survey was carried out in June 1974, at the same time as the field survey of barnacle density. Slack water at the time of low tide was chosen for the measurements, and the salinity at the surface and at 2 m depth was taken at 22 sites around the outfall. The results are shown in Fig. 7. The effluent, though at only 3‰ lower salinity (as measured at the mouth of the outfall pipe) than the sea water in Milford Haven, clearly floats as a surface layer. The distribution of the effluent as a band along the shoreline moving up and down with the tide agrees well with the findings of Crapp (1971).

Conclusions

The results indicate that effluent has a dual influence on barnacle larvae, firstly because of reduced salinity, and secondly because of other effluent constituents which were not measured. Salinity of below 18‰ produces an

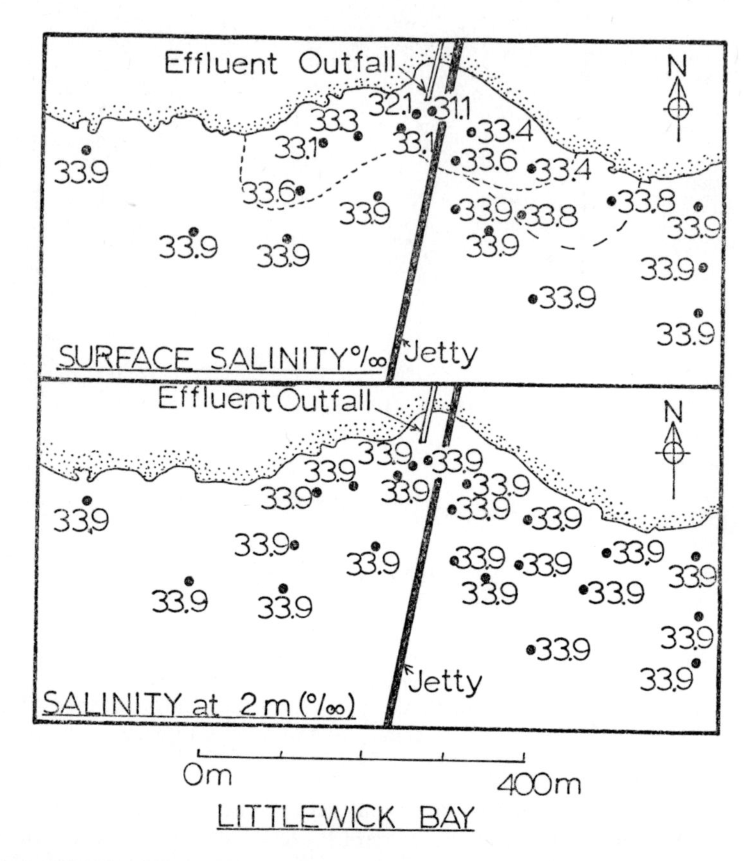

FIG. 7. The salinity of the sea at the surface and at a depth of 2 m at 22 sites in Little Wick Bay. The Little Wick refinery effluent was discharging at 31·1‰ salinity whilst the sea water salinity in Milford Haven was 33·9‰. The measurements were made at the period of slack water of a low tide.

immediate cessation of swimming activity in the nauplius, followed by sinking of the larva. In the field situation with low-salinity effluent, this results in the nauplius sinking out of the effluent (which floats on the surface) into normal sea water where activity recommences. As the effluent interferes with larval activity, it is reasonable to assume that the settling activity of the cyprid larva would also be interfered with by effluent. However, as the tides move the surface layer of effluent up and down the shore, there are long periods at high tide when effluent has no effect on the middle shore, during which time some cyprids will have the chance to settle. Hence settlement would only be seriously influenced where the effluent occurred continuously (very close to the outfall pipe) with a gradient of effect as distance from the outfall increased, produced by dispersion of the effluent. This fits well with the observed spatfall distribution.

The experiments show that the behavioural response of the larvae to the effluent (an immediate cessation in swimming) results in their avoidance of the effluent, and that this avoidance may well produce the observed ecological effect of lower barnacle density near the outfall, but without mortality of the larvae. The low salinity of the effluent appears to be at least as important in producing this effect as other characteristics of the effluent.

SOME RESPONSES OF THE LIMPET, *PATELLA VULGATA*, TO CRUDE OIL

The limpet, *Patella vulgata* L., has been shown to be of particular importance to the ecological balance of the rocky shore as the removal of this species produces far-reaching effects, especially in the balance between algal and animal abundance (Southward, 1964; Crapp, 1971). It is also very sensitive to oil and dispersants. For these two reasons a toxicity test was developed with this species which utilised criteria important to the limpet on the shore, and not mortality, as the end-point.

Though this work is described in detail elsewhere (Dicks, 1973), it is summarised here to illustrate three important points concerning responses to pollution. These points are (a) that sub-lethal doses of crude oil produce a response in the limpet which has severe ecological consequences; (b) that a diurnal rhythm in limpet activity considerably affects its susceptibility to pollution; and (c) that very short exposures to pollutants, such as can occur in one tide, can have considerable effects on the shore.

The limpet, which feeds by crawling over the rock surface, grazing off algal films and seaweed spores with its special tongue, has an interesting adaptation to the conditions of a rocky shore. After feeding, it always returns to the same point on the rock and settles down facing the same way. This results in the edge of the shell and the rock surface wearing where they come into contact, resulting over a period of time in a perfect fit. This provides protection for the animal against desiccation, rainfall and predators, which might be able to get through a small gap. The mark left on the rock at this site where the limpet fits exactly is called a home scar.

Methods

If this organism is collected from the shore and brought back to the laboratory, it no longer has its home scar available as a site of relative safety, which may well affect its susceptibility to pollution. To test the effect of the home scar, limpets were collected, along with their home scars, on stones from a boulder beach at Martinshaven, in Milford Haven. They were exposed to Kuwait crude oil firstly when all were on their home scars, and secondly when the majority were off their home scars feeding. An oiling time of 3 h was chosen as it was considered to be a time for which limpets in the middle shore were likely to be exposed to oil in the natural situation, oil normally being dumped on shore organisms only during low tide.

Limpets were tested experimentally in tanks containing 20 or 30 specimens on the rocks on which they were collected. Limpets of all sizes (from 8 to 50 mm shell length) were used experimentally.

Sea water for experimental use was collected from an adjacent beach on Pennar Point. The temperature was maintained at 12°C ± 2°C for the oiling experiments, a temperature close to normal sea temperatures at the time of the experiments. Limpets on rocks were oiled by drawing the rocks upwards through a thin film of oil on a tank of sea water. This is essentially similar to oiling on the shore, the tide depositing its surface film on the rocks as it recedes. Oil and other pollutants are normally floated off or otherwise dispersed by the incoming tide.

During oiling the animals were left in the air, as was an unoiled control. The oil was removed from the animals by washing twice with sea water and then re-submerging them in tanks of clean aerated sea water. The numbers of animals which detached were recorded before the oil was washed away at the end of the oiling period and again 3 h and 6 h after their return to clean sea water. Detachment was recorded in those cases where the animal had fallen from its rock or where very gentle mechanical disturbance was not followed by successful adhesion.

Results

The results are shown in Table I. It was noted that the oil caused a large number of active limpets to detach from the rocks, and a much smaller number to detach from their home scars. After detachment, the foot of those which detached was insensitive to mechanical stimulation and the animal was incapable of reattachment, but subsequent cleaning for 1 h to 3 h in fresh sea water gave 100% recovery and restored powers of attachment to the foot. The effect of the oil was presumably by anaesthetisation or irritation of the foot. All limpets which detached had oil under the foot, supporting this hypothesis. The higher detachment rate of active animals probably reflects the greater chance of oil getting under the foot.

Whilst detached animals were not killed by oil, it is likely that in the shore situation they would not have time to recover sufficiently to reattach before being eaten by gulls or other predators or washed into unsuitable environments by the incoming tide. This was supported by observation on the shore. Limpets were marked by a paint spot on the shell apex and oiled during low tide. Any which detached were searched for carefully during the next low tide, but very few were found. Hence the detachment response to irritation by oil, whilst not killing the limpets, ultimately results in their death on the shore.

Two other important points arise from these experiments. The first is that the very short times of exposure to oil (3 h) can cause as much as 64% detachment (and, therefore, probably mortality on the shore) whereas this sort of

TABLE I

Percentages of limpets which detached from the rocks after three hours exposure to fresh Kuwait crude oil. The results are from two experiments, one with 30 limpets, the other with 20 limpets.

Oiled for 3 h when—	*Actively feeding*	*Settled on home scars*
% detaching from the rock	60–64%	15–24%

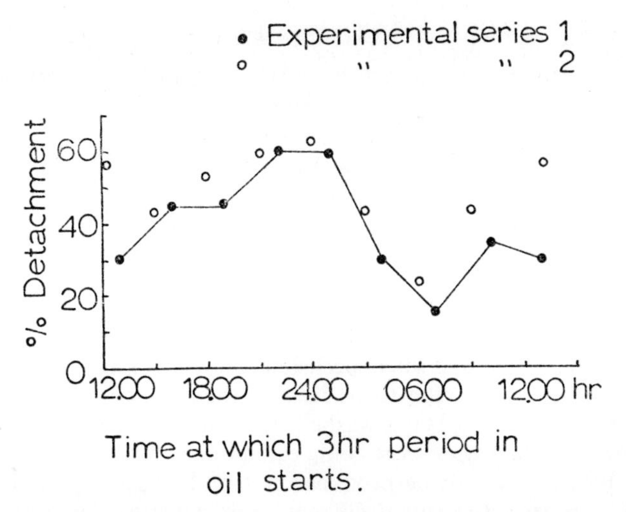

FIG. 8. The results of two experiments to test the detachment of *Patella vulgata* after a 3 h period in Kuwait crude oil at different times of day. The figure for percentage detachment is plotted at the time at which oiling commenced. In Experimental Series 1, each point represents an experiment with 20 animals, whilst in Series 2, each point represents an experiment with 30 animals.

exposure time does not produce direct mortality in the laboratory. The second point depends on the ecological importance of the limpet on the shore. If limpets are removed, grazing pressure on seaweeds is removed, which then grow abundantly causing changes in the abundance of many other species of animal by reducing space available for animals and upsetting the balance of food chains. Hence, what is apparently a sub-lethal pollution level to limpets results in wide changes in the shore ecology.

It has been shown that the limpet is more susceptible to crude oil pollution when active than when resting on its home scar. It can also be shown by respirometry measurement that this organism has a circadian rhythm in its activity, being most active during evening and night and least active in the early morning (Dicks, 1973). Oiling limpets experimentally as described above for 3 h periods starting at different times of day shows that three to four times higher detachment can be produced around midnight than in the early morning (Fig. 8). Repeating the experiment in the laboratory and on the shore when the animals were most active (around midnight) and least active (around 0700 hrs) produced the detachment figures shown in Table II. Hence the results of toxicity tests can vary substantially depending on the state of activity of the test organism, this being determined by its activity rhythm.

BEHAVIOURAL RESPONSES TO REFINERY EFFLUENT IN *NEREIS DIVERSICOLOR* AND *LITTORINA SAXATILIS*

Nereis diversicolor

Under certain conditions (usually during a power failure) a refinery effluent is discharged into a stream flowing to the sea through a saltmarsh at Martinshaven,

TABLE II

The percentage detachment of limpets in the laboratory and on the shore at midnight and 0700 h

Time at which oiling was started	*% detachment in the laboratory*	*% detachment on the shore*
0000 h	58·0	44·0
0700 h	19·0	9·5

Milford Haven. After one such occasion, large numbers of the ragworm, *Nereis diversicolor*, were seen to be leaving their burrows on the banks of the saltmarsh creeks and being washed down the stream across a mud flat towards the sea, where they were being eaten by sea birds. This response was shown by Baker (1975) to be produced not by low salinity of the effluent, but by some other unmeasured effluent characteristic. *Nereis* treated with effluent in the laboratory showed the same response, though the effluent did not kill the organisms.

Littorina saxatilis

Parsons (1972) describes a simple test which relies on a measure of activity as an indicator of pollution stress on the grooved periwinkle, *Littorina saxatilis*. This winkle, with its protective whorled shell, retracts into the shell under the influence of irritants or stress. Consequently the magnitude of the irritant or stress could be measured by recording the degree of retraction of the winkle. This can be done by using a scale dependent on activity. The scale used by Parsons recognised three states of activity: retracted in the shell, which scored 0; attached to the substrate, which scored 1; and crawling which scored 2. He was able to show that the activity scores of *Littorina* in refinery effluent were considerably lower than scores of those organisms in clean sea water (Fig. 9). It is possible to extend this activity scale to a wider number of categories or to fit organisms with different activity criteria.

DISCUSSION

The experimental work, both in the field and in the laboratory, which has been described in this paper shows that sub-lethal doses of pollutant can have varying ecological consequences, ranging from mortality of important organisms in a shore community (e.g. limpets) with accompanying ecological repercussion, to the avoidance of pollutants by larval stages (e.g. *B. balanoides*) resulting in localised depletion of barnacle populations. As Swedmark (1974) points out, many marine organisms, and shore organisms in particular, have a variety of protective mechanisms which are activated by irritation and can be used as indicators of irritation of the organisms by pollution. It is worth noting that one such reaction, the retraction response in *Littorina saxatilis*, may have both good and bad consequences for this organism in the field. Whilst retraction may allow weathering of a short spell of pollutant effect,

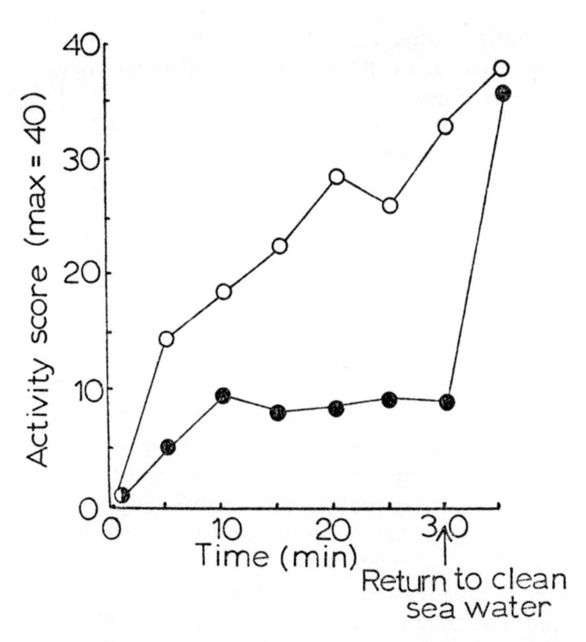

FIG. 9. A comparison of the mean recovery curves of *Littorina saxatilis* when induced to retract by physical disturbance and then placed either in clean sea water or refinery effluent. As 20 animals were used in each test (the results of which were averaged to give the curves shown), the maximum possible activity score was 40—see text. The characteristics of the effluent used in the experiments were:

	ppm		ppm
Sulphides	1	Ammonia	3
Oil	12	COD (Permanganate)	13
Total suspended solids	65	Total dissolved solids	17 000
Washed suspended solids	37	pH	7·2

Salinity 17%

The experiments were carried out at 15°C in a constant temperature room after 12 h acclimatisation. The data are reproduced from Parsons (1972).

waves or currents affecting the animal when retracted could wash it into a totally unsuitable environment where it will eventually die. The response of *Nereis* to pollutant irritation and its attempt to avoid it results in this instance in considerable mortality by predation.

In the cases described here, responses used in laboratory experiments have been very simple. It is, of course, possible to use much more sophisticated responses or measures of activity such as change in respiration rate (Hargrave and Newcombe, 1971) or change in respiratory movements, burrowing activity, and shell closure (Swedmark, 1974). However simple or complex, all fulfil an important criterion—that reactions or responses which may be of considerable importance to organisms in their environment are taken into account in laboratory experiments.

Baker and Crapp (1974) suggest that in assessing the impact of a pollutant on organisms and the environment, three experimental levels can be distinguished:

(a) straightforward bioassay techniques using mortality measurements;
(b) methods from which ecological predictions can be made; and
(c) full-scale field investigations.

The information presented here supports the distinction between levels (a) and (b), for ecological prediction from mortality tests is questionable if apparently sub-lethal levels of pollution produce high mortality in the field, as is the case here with the limpet, the ragworm and grooved periwinkle. However, these simple tests using behavioural responses fit into the second category, and it is hoped that they make a start in bridging the gap between laboratory mortality tests and time-consuming full-scale field investigation.

As a final point, the effects of refinery effluent on barnacle larvae indicate that salinity is in this case at least as important in producing the observed effects as other effluent characteristics, if not more important, but that both have an additive effect. There is a need for further work to determine the effluent components which have biologically and ecologically important effects and to determine the synergistic or antagonistc effects of the different components.

ACKNOWLEDGEMENTS

I am grateful to Texaco Petroleum Company Refinery, Milford Haven, for providing effluent for use in laboratory experiments, and to Roger Parsons for providing data on *Littorina* behaviour.

REFERENCES

Baker, J. M. (1973). Biological effects of refinery effluents, *Proceedings of EPA/API/USCG World Conference on the Prevention and Control of Oil Spills.*

Baker, J. M. (1975). Experimental investigation of refinery effluent (this volume).

Baker, J. M. and Crapp, G. (1974). Toxicity tests for predicting the ecological effects of oil and emulsifier pollution on littoral communities, in Beynon and Cowell (1974).

Beynon, L. R. and Cowell, E. B. (1974). *Ecological Aspects of Toxicity Testing of Oils and Dispersants*, Applied Science Publishers, for Institute of Petroleum, 149 pp.

Brown, F. A. J., Bennett, M. F., Webb, H. M. and Ralph, C. L. (1956). Persistent daily, monthly and 27 day cycles of activity in the oyster and quahog. *J. exp. Zool.*, **131**, 235–62.

Crapp, G. B. (1971). The biological effects of marine oil pollution and shore cleaning, Ph.D. thesis, University College of South Wales, Swansea.

Dicks, B. (1972). The occurrence of rhythm in the activity of the marine bivalve, *Mya arenaria* L., and some aspects of its importance and control, Ph.D. thesis, University of Reading.

Dicks, B. (1973). Some effects of Kuwait crude oil on the limpet, *Patella vulgata. Environ. Pollut.*, **5**, 219–29.

Hargrave, B. T. and Newcombe, C. P. (1973). Crawling and respiration as indices of sub-lethal effects of oil and a dispersant on an intertidal snail, *Littorina littorea. J. Fish. Res. Board Can.*, **30**, 1789–92.

Newell, G. E. and Newell, R. C. (1963). *Marine Plankton*, Hutchinson, London, 221 pp.

Parsons, R. (1972). Some sub-lethal effects of refinery effluent upon the winkle, *Littorina saxatilis*. Oil Pollution Research Unit, Field Studies Council, *Annual Report 1972*.

Portmann, J. E. and Wilson, K. (1971). The toxicity of 140 substances to the brown shrimp and other marine animals, Shellfish Information Leaflet No. 22, Ministry of Agriculture, Fisheries and Food.

Pyefinch, K. A. (1948). Methods of identification of the larvae of *Balanus balanoides* L., *B. crenatus* Brug and *Verruca stroemia* O. F. Müller. *J. mar. biol. Ass. U.K.*, **27**, 451–63.

Rao, K. P. (1954). Tidal rhythmicity of rate of water pumping in *Mytilus* and its modificability by transplanting. *Biol. Bull.*, **106**, 353–9.

Salanki, J. (1966). Comparative studies on the regulation of periodic activity in marine lamellibranchs. *Comp. Biochem. Physiol.*, **18**, 829–43.

Southward, A. J. (1964). Limpet grazing and the control of vegetation on rocky shores. *Symp. Br. Ecol. Soc.*, **4**, 265–73.

Swedmark, M. (1974). Toxicity testing at Kristineberg Zoological Station, in Beynon and Cowell (1974).

Wilson, K. (1974). Toxicity testing for ranking oils and oil dispersants, in Beynon and Cowell (1974).

Discussion

Dr R. H. Cook (Environmental Protection Service, Canada) referred to comments that indicated that the application of acute bioassay information on refinery effluents was not useful for ecological prediction purposes and that emphasis should be given to the increased application of field studies. He maintained that standardised acute toxicity evaluations of industrial effluents are complementary to the information derived from field studies conducted in the vicinity of the pollution source. Rather than measuring the many contaminants known to be contained in refinery effluents, in Canada a fish bioassay was used under specified conditions as a biological integrator. This was an effective interim pollution control device. Field studies should be designed to answer specific questions on whether imposed controls were sufficient.

Dr D. Straughan (University of Southern California) supported the approach taken at Orielton to the determination of toxicity, and commented that 96 h exposure tests are useless if the animal survives this but dies 5 days after a 6 h exposure period. **Dr I. C. White** (Ministry of Agriculture, Fisheries and Food) reminded participants that different types of tests are designed for different purposes and it is not a case of saying that one is better than another for all purposes. It was usually the interpretation of the results of such tests that were at fault rather than the tests themselves. **Mr W. G. Roberts** (British Petroleum Co. Ltd) was not happy about the use of biotests for routine control purposes unless they were short-term and very repeatable.

Offshore Monitoring and World Oil Spillages

17
Offshore Biological Monitoring

BRIAN DICKS
(*Oil Pollution Research Unit, Orielton Field Centre, Pembroke, Wales*)

SUMMARY

Little is known of the impact of oil developments on the offshore environment, though there are now a considerable number of these developments around the British Isles. At a more basic level, knowledge of the offshore biological communities of both the North Sea and the Celtic Sea is either limited or non-existent. This situation should be, and can be, rectified by establishing suitable offshore monitoring programmes.

The selection of suitable organisms for a monitoring programme, and the requirements and special problems of offshore monitoring design, are discussed. The aims of the programme should be carefully determined and will influence, in particular, the degree of quantification of sampling and levels of change which can be detected.

Data from a baseline survey in a North Sea oil field are used to illustrate problems of monitoring and the treatment of results. The importance of sample number to the quantification of data, and the use of diversity measurements, are illustrated and discussed.

There is a need for quantified data for future environmental decision-making offshore. However, the present lack of knowledge of the offshore environment suggests that such detail should be balanced by wide-ranging simple surveys capable of picking up, at the least, gross biological changes.

INTRODUCTION

The recent increasing development of offshore oil resources has not been accompanied by parallel growth and improvement of offshore biological monitoring designed to assess the present ecological status of development areas, and look at possible future impact of developments.

At present, knowledge of the biology of the offshore environment, though continually expanding, is still relatively limited. In particular, little is known of the impact of oil pollution offshore or, at a more basic level, of the complex interrelationships between organisms and communities in marine food chains (Steele, 1970; Thorson, 1971). The lack of detailed information on the effects of oil pollution on offshore communities and the now considerable number of developments in both the North and Celtic Seas suggest it should be regarded as a priority firstly to ascertain if ecological damage is being done, secondly to determine its extent, thirdly to watch its progress, and finally to estimate its importance.

Quantified data is the only basis for future environmental decision-making concerning the necessity for remedial action or otherwise. Such information can only be obtained by careful observation at one or more sites over a period of years—in other words by establishing monitoring programmes.

The initial biological survey in such a monitoring programme in the Ekofisk oil field, carried out in August 1973, is described herein and used to illustrate problems of design of offshore surveys and the interpretation of results.

APPLICABILITY OF MONITORING PROGRAMMES TO OFFSHORE POLLUTION SOURCES

Some sources of pollution offshore do not lend themselves to monitoring programmes. The effects of single oil spills and discharge of tanker washings or ballast water are not easy to monitor because areas which become affected are not predictable enough to allow siting of baseline surveys.

Monitoring techniques become useful where potential pollution comes from a static source, either intermittently or continuously over a period of time, and can be used to look at recovery of systems where damage has been known to occur. Oil installations offshore may produce repeated pollutions in a relatively small area, e.g. small spillages might occur around production platforms and single buoy moorings, toxic drilling muds (Falk and Lawrence, 1973) may be dumped in some areas, and the non-toxic but potentially smothering rock cuttings from the well are usually dispersed to the sea. All these sources produce continuous stress on certain parts of the marine ecosystem and can be monitored by a variety of methods.

THE OFFSHORE ENVIRONMENT AND THE SELECTION OF SUITABLE MONITORING ORGANISMS

The offshore marine ecosystem can be divided basically into four main interdependent categories:

(a) The plankton—minute animals and plants usually near the surface and generally drifting with currents. Certain components show a daily rhythm of vertical migration (Hardy, 1956; Newell and Newell, 1963).
(b) Pelagic fish—highly mobile, and many travel considerable distances.
(c) Epibenthic communities—live on the sea bed. Some are highly mobile, e.g. demersal fish; some are slow-moving, e.g. hermit crabs, gastropods; whilst others are static, e.g. oysters, bryozoans.
(d) Benthic infauna—organisms living in the sea bed. These are usually sedentary in habit or slow-moving.

The relationships between these groups can be represented diagrammatically as a basic feeding interconnection (Fig. 1).

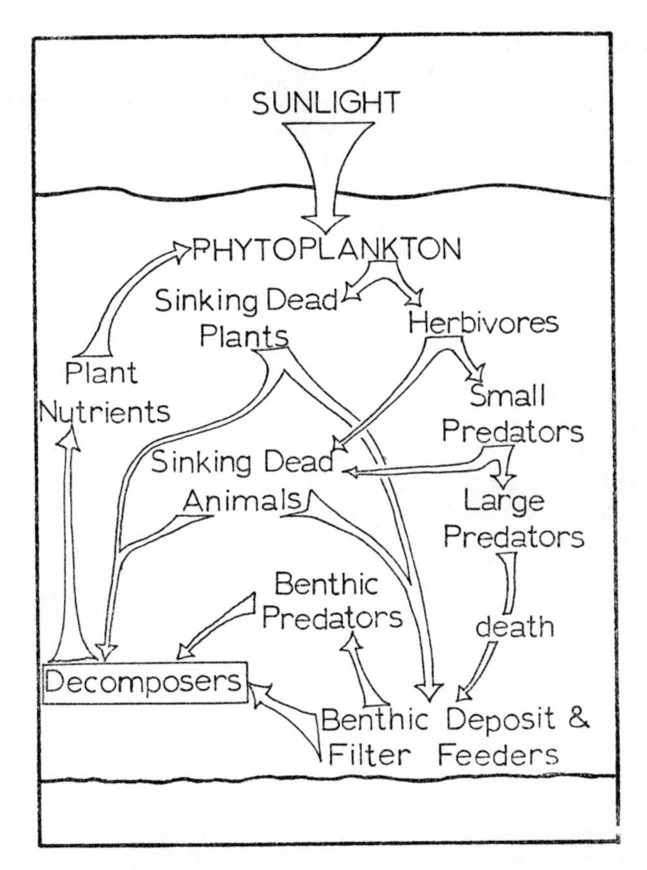

FIG. 1. Simplified food web for the marine ecosystem.

The first two categories, which live suspended in the water mass, present special problems to sampling. The pelagic fish are very active and can avoid both sampling devices and unfavourable conditions in their environment. This makes them difficult to sample quantitatively and unreliable as indicators of certain pollutants.

The plankton is essentially non-mobile relative to the water mass, but drifts with it, and is therefore susceptible to pollutants in the water. However, because the water masses move, the plankton is mobile relative to fixed points. This last factor, in combination with considerable patchiness of distribution (Cushing, 1953), daily fluctuations in depth of occurrence and considerable fluctuation in density, composition and distribution from season to season and year to year (Hardy, 1956; Newell and Newell, 1963), renders it extremely difficult to sample quantitatively for comparative purposes (see also Fraser, 1971). However, sampling the plankton in a semi-quantitative way can be a useful part of a monitoring programme for reasons given below.

A monitoring programme is most easily designed, and will produce comparative results, where sedentary communities of organisms which can be sampled quantitatively and repeatedly occur around a pollution source. It is in this context that slow-moving or sedentary epibenthic organisms and the infauna of the sea bed assume importance, as they can avoid neither pollutants nor samplers designed to assess their abundance and distribution.

Marine benthic communities have long been recognised as important integral parts of marine ecosystems (Petersen and Jensen, 1911; Thorson, 1957; Longhurst, 1964), and can therefore be expected to reflect changes occurring at any level in the marine environment. The feeding relationships illustrated in Fig. 1 emphasise the interdependence of the different communities.

Studies of the benthos are particularly useful where a foreign substance is dumped on or near the sea bed (e.g. drill cuttings, drilling muds, outfall of ballast water from oil storage installations), but may also reflect pollution effects where the contaminant stays on or near the surface. Oil can contaminate and kill planktonic organisms or adsorb on to particulate suspended matter which then 'rains' to the sea bed and is eaten by benthic organisms or incorporated into sediments (Nelson-Smith, 1972). Oil initially at the surface may also sink naturally, as a certain amount did after the Santa Barbara spill (Kolpack *et al.*, 1971) and the *Anne Mildred Brøvig* incident (Stehr, 1967), or may be sunk deliberately as part of a clean-up programme (Hofmann, 1949). In all cases, the benthos can be affected to a greater or lesser degree.

The processes by which oil can enter and affect the various parts of the marine ecosystem have been summarised by Nelson-Smith (1972) and are represented diagrammatically in Fig. 2. This figure correlates with Fig. 1 at planktonic and benthic levels.

The benthos may also reflect pollution effects where the contamination is restricted solely to surface waters by thermal or other stratification, for the following reason. The zooplankton of the surface waters is essentially composed of two types of organism:

(a) permanent members with their whole life-cycle spent in the plankton—the holoplankton;
(b) egg and larval stages of benthic organisms—the meroplankton.

If larval forms and eggs of benthic organisms are killed, then this will be reflected in the composition of benthic communities. It is in this context that qualitative or semi-quantitative studies of the plankton are useful correlates of benthic monitoring programmes.

A logical conclusion from the above evidence is that it is advisable to concentrate on studies of the benthos and plankton when designing offshore monitoring programmes.

Physical and chemical components of the marine environment which are responsible in part for determining the type and distribution of organisms present (Jones, 1950; Sanders, 1958, 1968; Lie, 1968; Thorson, 1966, 1971; Glemarec, 1973; Parker, 1975) should also be monitored to provide insight to reasons for natural fluctuations. These factors include temperature regimes, salinity, depth and sediment composition. It is also useful in terms of assessing oil pollution spread and effects to analyse sea water, sediment and organisms for hydrocarbon content.

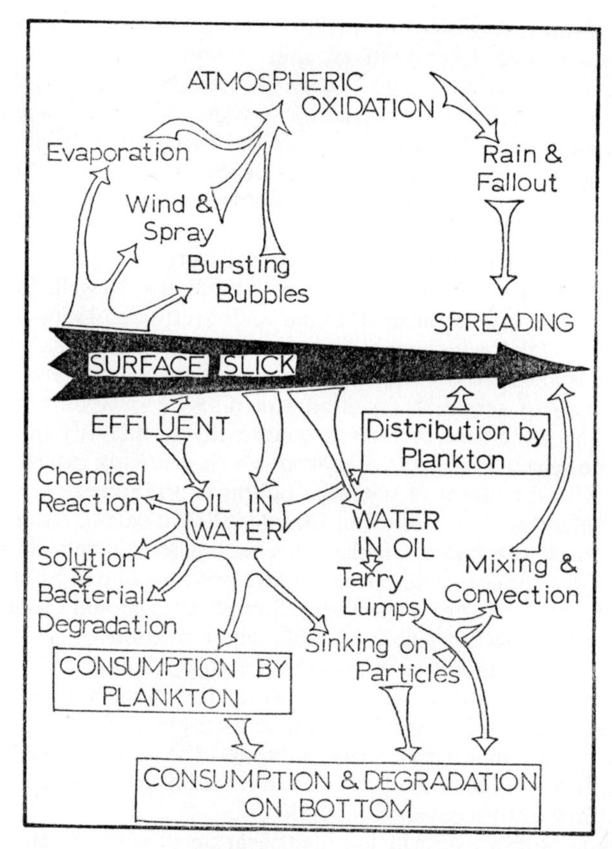

FIG. 2. A diagrammatic representation of the fate of oil on and in the sea (after Nelson-Smith, 1972).

FURTHER CONSIDERATIONS IN MONITORING PROGRAMME DESIGN

The usefulness of monitoring depends upon repeated sampling and the comparison of sample site results with each other and adequate control data from survey to survey in order to locate and assess the magnitude of ecological changes. Comparison is difficult and magnitude of effects are not measurable if data obtained are not quantitative. This presents problems in selection of sampling equipment and deciding on the number of samples to take. A full discussion of marine sampling equipment and its merits or otherwise can be found in Holme and McIntyre (1971). Recent improvement in grab sampler design (e.g. Day modification of Smith–McIntyre) and knowledge of operation of dredges on the bottom (Gage, 1972) have improved the situation. The influence of sample number on quantification of data is illustrated and discussed in a later section.

Accurate sample site location is important to repetitive sampling because, in some situations, considerable variation in population density and community type can occur over relatively short distances (Addy, 1975). There is evidence, however, that communities of waters where environmental stresses are less exacting are more uniform (Sanders, 1968; Glemarec, 1973). Location of sampling sites in oil fields around Britain is possible to within a few metres using a combination of the Decca Navigator or Decca Hi-fix system, and radar and horizontal sextent angles of fixed installations (Holme and McIntyre, 1971).

Seasonal and annual fluctuations occur in density and species composition of planktonic communities (Hardy, 1956; Newell and Newell, 1963; Fraser, 1971) and benthic communities (Holme and McIntyre, 1956; Feder *et al.* 1973; Parker, 1975). Such changes can mask other pollution-related effects but can be compensated for by adequate control data from nearby unaffected areas. Problems of seasonal variation can also be overcome by sampling annually at the same time of year or continually to quantify the changes.

Sample site selection and the area over which sampling extends should be compatible with the extent of potential pollution and time and money available. It is preferable that a series of samples should extend from affected to unaffected areas, the latter samples then acting as controls. If this is not possible, control data can come from surveys previous to pollution or samples from similar unaffected areas elsewhere. Sample sites should be close enough to each other to show a gradient of effect where possible.

The duration of a monitoring programme will be influenced by the particular problem under study, as well as with matters such as availability of finance. Any effects, unless obvious and catastrophic, will not be evident or provable without a number of years' data, and the same applies to proof of absence of effect. If effects are observed and remedial action taken, monitoring should continue until recovery is complete.

Problems may also occur in the identification of sample contents due to lack of suitable keys or available specialist knowledge. Numbers and types of different organisms are obviously important when comparing data but, in the strictest sense, monitoring does not require identification of every organism. It is only necessary to be able to recognise each type and know whether it is changing in abundance or distribution. This can be achieved by establishing a collection of 'type' specimens, suitably coded, until identification becomes possible.

OFFSHORE MONITORING AROUND THE BRITISH ISLES

Developments of the offshore oil industry around Britain provide us with two different situations with regard to offshore monitoring.

The Celtic Sea

This area is, as yet, relatively undeveloped and provides us with an opportunity to acquire a considerable amount of 'before' data to be followed up by intensive surveys around points of development. At present, knowledge of the marine communities of the Celtic Sea is limited to a small amount of benthic data off the Brittany coast and outside the present development area

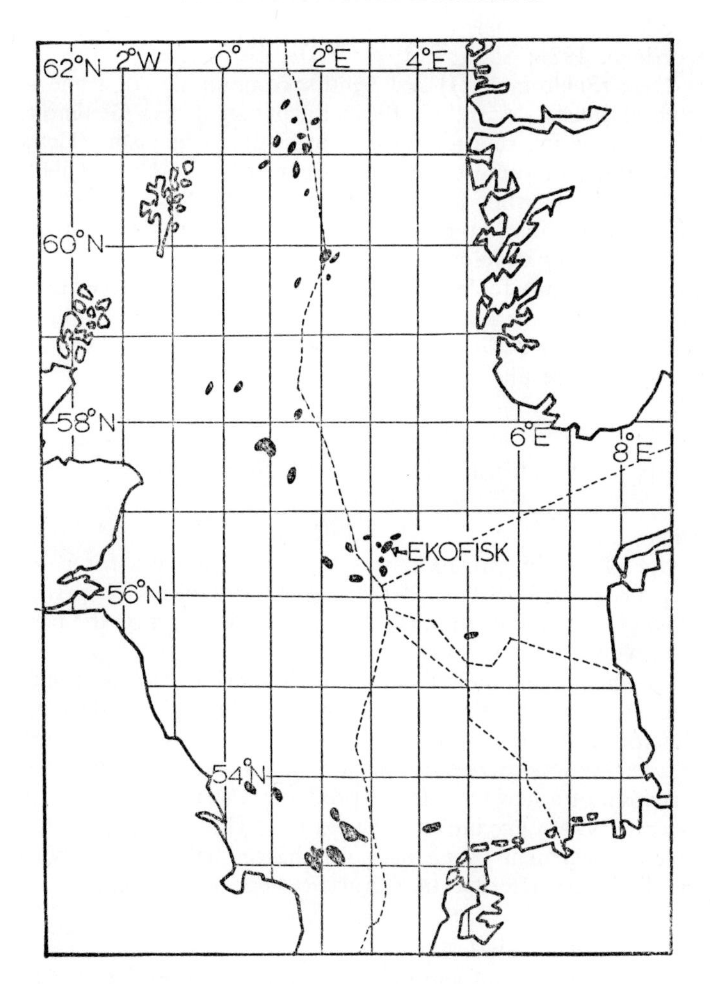

FIG. 3. The distribution of oil and gas fields in the North Sea.

(Glemarec, 1973), and some planktonic data (Southward, 1962). There is little published information on the benthos over most of the area. Plans are already formulated for a survey in the Celtic Sea to determine the distribution of the main community types, and samples have been taken from some areas already. This information will become available in the near future.

The North Sea

The North Sea presents a different face entirely. There is considerable development already (Fig. 3) and, in this case, some knowledge of the biology of the area, but little or no detailed 'before' information on biological aspects at the development sites is available.

Biological information is available on the plankton (Fraser, 1971; Edinburgh Oceanographic Soc., 1973) and provides useful material for comparison of planktonic samples. Data are also available for fisheries (Cole and Holden,

1971; Korringa, 1971; Shelton, 1971; FAO, 1972), the benthos of certain dumping areas (Shelton, 1971) and benthic communities of some areas off the north-east coasts (McIntyre, 1961; Buchanan, 1963; Glemarec, 1973). There is also some information on certain benthic organisms (Ursin, 1960; Kirkegaard, 1969) of some of the areas in which oil and gas fields occur. Little work on the benthos has been done recently, however. There is considerable scope for use of biological monitoring programmes associated with new and existing developments.

A unique opportunity to design and carry out a baseline survey upon which a monitoring programme could be based was presented by the Phillips Norway Group in their Ekofisk Field, and will be used to illustrate offshore monitoring and the analysis of results.

EKOFISK SURVEY, AUGUST 1973

Introduction

The purpose of the investigation was to provide a basis for a long-term monitoring programme by (a) identifying the organisms occurring in the benthos and plankton and (b) obtaining quantitative data on the benthos for comparison with future surveys.

Ekofisk, as the first big oil find in the North Sea (Fig. 3), has been functioning for longer than any other field. Though no biological information was obtained before operations started or during the first few years of operation, the present survey was carried out prior to major extensions which included the construction of a new complex of production structures, and a 1 million barrel concrete oil storage tank on the sea bed. This storage tank, operating on a water displacement by oil basis, discharges ballast water when being filled. The discharge, after treatment, occurs near the sea bed.

The Area

The sea bed is of firm sand and has an average depth of 220 ft (69 m) with maximum depth in the survey area of 231 ft (70·6 m) and minimum depth of 219 ft (66·9 m). Details of depth at each sample site are given in Table 1E, Appendix A.

Ekofisk lies approximately in the centre of an extensive area of medium and fine sandy deposits about 80 miles north to south by 60 miles east to west (*Oxford Atlas of Britain and Northern Ireland*, 1963). Currents and tidal flows in the area (Fig. 4) are weak, and these and data on water masses of the North Sea are summarised by Hill (1971) and Eisma (1971). Thermal stratification of the water mass occurs in the summer months (Fig. 9).

Sampling Site Selection

The chosen distribution pattern for sampling sites was a series of five transect lines radiating from a near-central point in the oil field, the new production complex (56°32′ N, 03°13′ E). The transects were between 6000 and 6500 m in length with five sample sites along each line except for one line with four sample sites, a total of 24 sample sites. They were positioned at approximately

FIG. 4. Surface drift currents in the North Sea in August (after Böhnecke, 1922).

250–500, 1000, 2000, 4000 and 6000 m along each line. Sample site positions and installations in the area are shown in Fig. 5. The positioning of sample sites along the transects and the total length of the lines were selected to provide sites near enough to the production complex to show up any gradient of effect from it (on the assumption that pollution potential of the production complex could be equivalent to a small refinery effluent), and to have sites far enough away to be considered as controls to provide comparable material. The homogeneity of the sea bed in terms of sediment type and animal community removed many of the problems of comparison of sites which could have been encountered if a number of different types of substrate and animal community had been found (Feder *et al.*, 1973).

Site Location

Each site was located by a series of correlated measurements. These were Decca Navigator co-ordinates (North British Chain), and bearings and distances of fixed installations using radar, horizontal and vertical sextant angles. Complementary information of depth, length of anchor cable out, wind, sea and swell conditions were also recorded to assist in re-positioning a vessel for future re-sampling. Some of this information is summarised in Table IE, Appendix A.

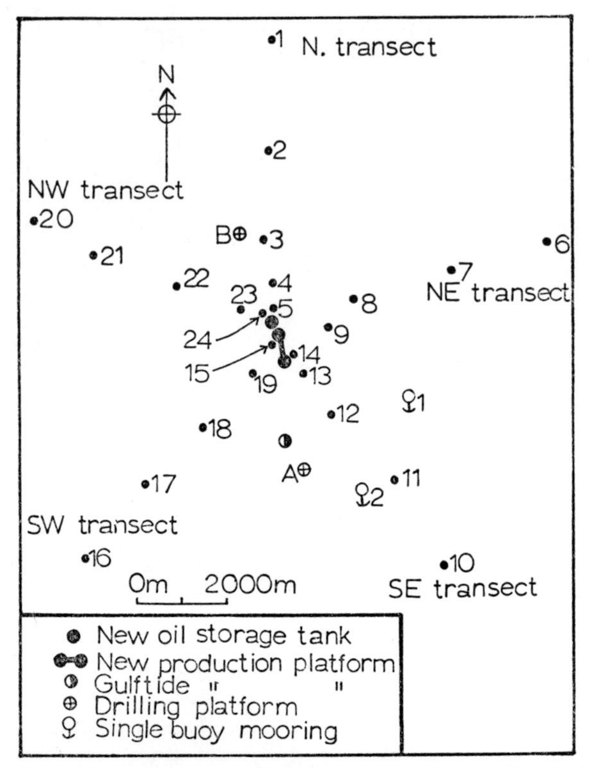

FIG. 5. Map of Ekofisk Field showing the position of sample sites (numbered 1–24), and the location of installations at the time of the survey (installations not to scale). The five transect lines radiating from the central complex (56°32′ N, 03°13′ E) are named by compass direction.

Sampling Methods

Sampling methods used during the Ekofisk survey are described fully in Appendix A, and summarised briefly here.

Benthos

Samples were taken using a 0·1 m^2 Day grab weighted with 100 lb lead. This grab samples to approximately 10 cm depth in the substrate. Eleven grabs were taken per site, ten for analysis of the benthos and one to provide a sediment sample for particle size analysis, organic carbon analysis and, at some sites, hydrocarbon analysis. Movements of the vessel at each site effectively 'randomise' the samples at each site. Samples were sieved on board through a 1 mm mesh sieve and the organisms preserved in 4% formalin containing a general dye, eosin. This stained the organisms, simplifying subsequent sorting. Sites were sampled at anchor where possible, or, if not, with the vessel held stationary using its engines.

Plankton

Standard Plymouth MBA nets for phyto- and zooplankton were used. 100 m hauls were made at various times of day, and samples preserved in the same manner as described for benthic organisms.

Water Temperature
A water temperature profile was taken using a portable NIO temperature probe.

Water Sampling for Hydrocarbon Analysis
Water samples were taken from the surface and 200 ft using a stainless-steel-lined 7·0 litre NIO bottle. Immediately upon receipt of the sample on board, subsamples were solvent extracted with cyclohexane for UV analysis and *n*-heptane for GLC analysis. Extracts were stored in a deep freeze until analysed.

Sediment Sampling for Hydrocarbon Analysis
Subsamples of benthic sediment from a Day grab were solvent extracted for UV and GLC hydrocarbon analysis in the same manner as the water samples, and extracts stored in a deep freeze until analysed. Subsequent analysis was carried out by the Microbiology Department, University College of South Wales, Cardiff.

Whilst it is realised that considerable problems arise in obtaining uncontaminated and representative samples of both sea water and sediment for analysis of hydrocarbon content, and that methods used herein are subject to all the usual limitations, the techniques are simple and repeatable and capable of quantifying changes in measured hydrocarbons of the order of 1 or 2 ppm.

Laboratory Analysis and Presentation of Results
Fully tabulated data and method description are presented in Appendix A. These are diagrammatically summarised here using selected examples. Two methods of diagrammatic data presentation are used. As the sample stations are arranged in five transect lines radiating from the centre of the area, data are presented as a series of five graphs, each graph representing a transect line (Fig. 6). This method is useful for the detection of gradients along the sample lines. The second method is based upon the sample distribution map (Fig. 5) with data for each site being superimposed in the site location. This method is useful in assessing distribution throughout the field.

Sediment Particle-Size Analysis
Sediment samples were dried at 100°C and passed through a series of Endecott test sieves. Cumulative oversize percentages were calculated and plotted against sieve mesh size to give a soil composition skew diagram, and to interpolate median particle size for each site. The results are summarised in Figs. 7 and 8. The sediment was very uniform in composition throughout the sample area, and was a fine sand with median particle size averaging 176 μm. Quartile deviation and skewness were calculated using the method of Morgans (1956) to give an idea of the efficiency with which the samples were sorted (Appendix A). Sorting appears to be fairly good with very even sorting between small and large particles.

Thermal Stratification
A water temperature profile for station 1 in August 1973 is plotted in Fig. 9. Thermal stratification with a thermocline between 115 and 130 ft was found. Temperatures in the surface 115 ft were around 14°C with the water from 130 ft to the bottom at 7°C.

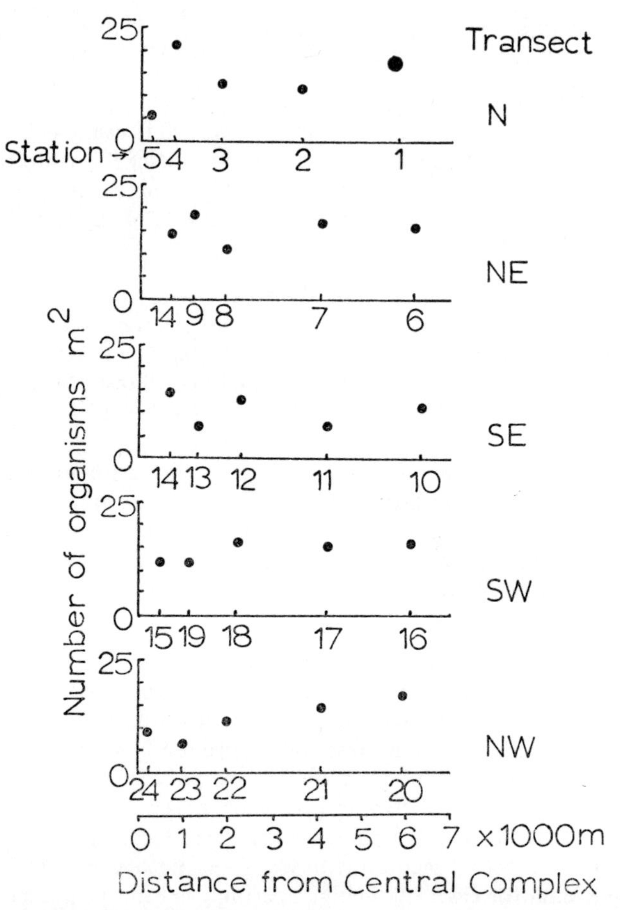

FIG. 6. The presentation of sample data as a series of five histograms showing number of organisms per m^2 plotted against distance from the central complex. Each graph represents one transect in the direction marked. Station position and number (from which the data were collected) are marked along each horizontal axis (compare with Figs. 13 and 14). The density of *Nephtys incisa* at each sample site is used to illustrate the method.

Organic Content of the Sediment

Sediment samples were analysed immediately on return to the laboratory by a potassium dichromate/ferrous sulphate titration method described in Leeper (1948). The results are summarised in Fig. 10. The results were not corrected to 'available' organic carbon (Morgans, 1956), but are in a suitable form for comparison with future samples. The results show similar values of organic carbon in mg/g of sediment throughout the sample area.

Zooplankton

The samples were subsampled in the laboratory and organisms identified to group or, on some cases, species. Using certain approximations (Appendix

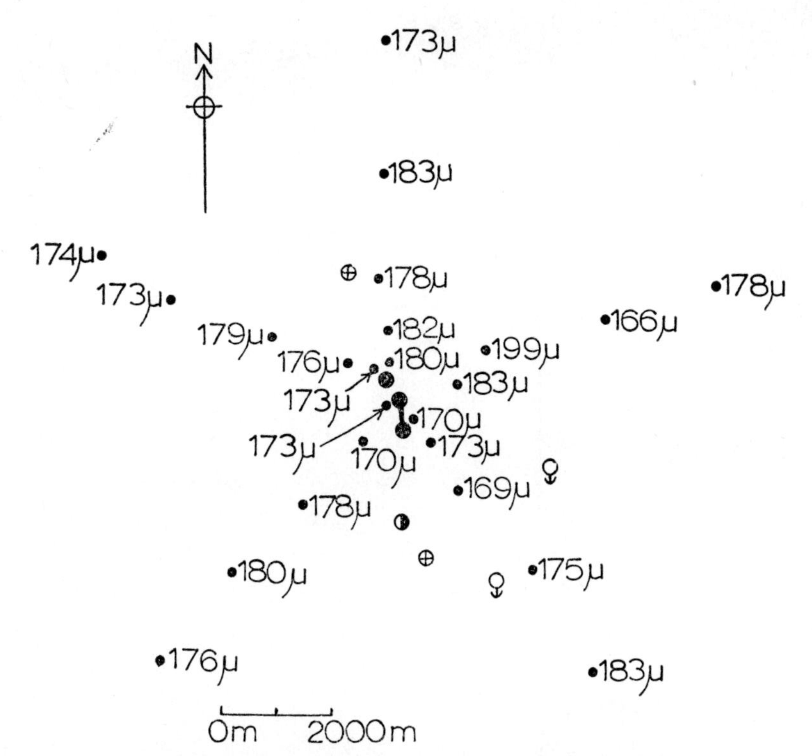

FIG. 7. Median particle diameter of the sediment at each sample station in the Ekofisk Field. For key to installations and site numbers see Fig. 5.

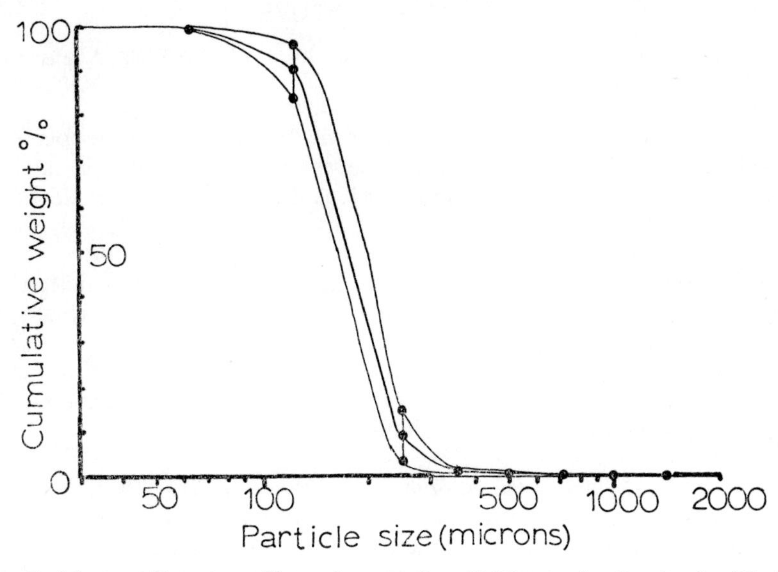

FIG. 8. Mean sediment profile and range for all 24 sample sites in the Ekofisk Field. Variation in cumulative percent weight for each particle size is very small.

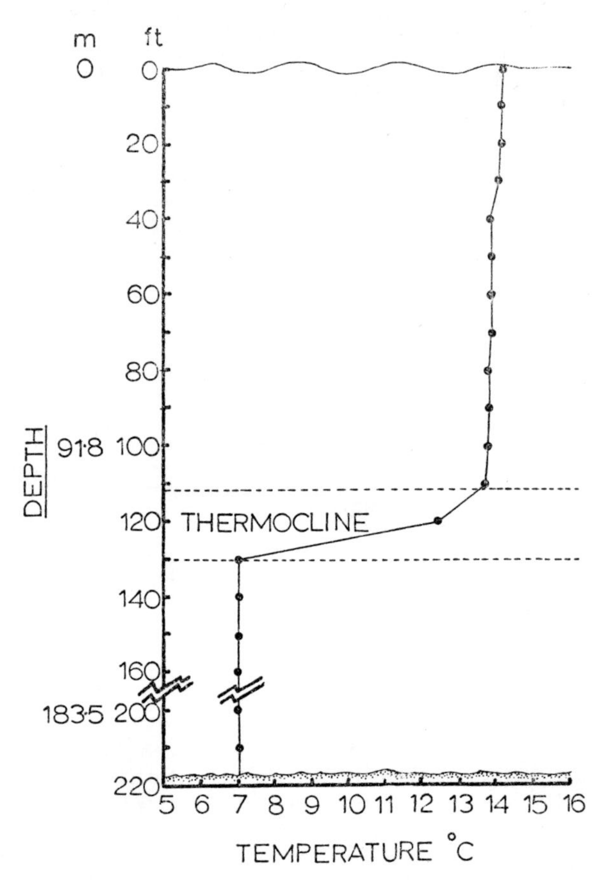

FIG. 9. Water temperature profile at sample station 1, Ekofisk Field, August 1973.

A), results were quantified to numbers per m^3. Variations in the zooplankton abundance with time of day were noted and a typical variation (in calanoid copepod numbers) is illustrated in Fig. 11. A diurnal rhythm is evident.

Phytoplankton

Quantification was not possible in the same way as the zooplankton (see Appendix A). Samples were sorted and species identified. The results are shown in Table III, Appendix A.

The results of both plankton sampling will be compared with Continuous Plankton Recorder records of the North Sea (Edinburgh Oceanographic Soc., 1973) when these become available.

Analysis of Water and Sediment Samples for Hydrocarbons

GLC analysis was used to detect and quantify *n*-paraffin hydrocarbons of C No. 24–33 in the *n*-heptane extracts, and ultra-violet absorption analysis to detect aromatic ring structures in the cyclohexane extracts. Quantification of ultra-violet results was arbitrary as no standard samples are available for

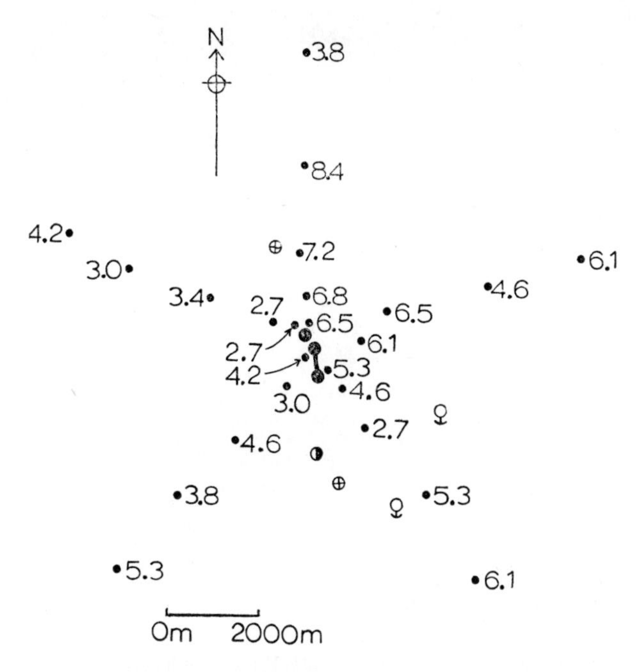

FIG. 10. Organic carbon content of the sediment in mg/g at each sample station in the Ekofisk Field. For the key to sample station numbers and installations see Fig. 5.

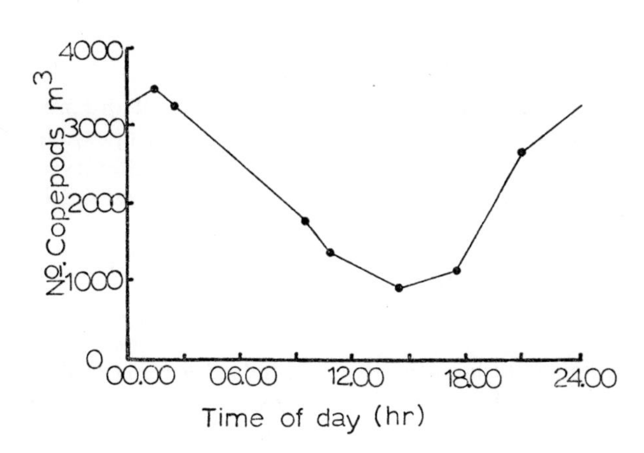

FIG. 11. The change in abundance (numbers of individuals per m^3 sea water) of planktonic calanoid copepods with time of day. The samples were taken from the surface 1 m.

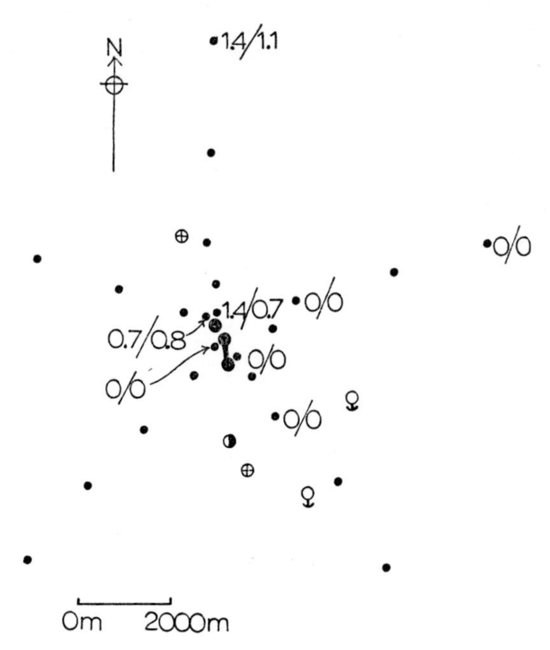

FIG. 12. Results of GLC analysis for *n*-paraffins at selected sample sites in the Ekofisk Field. First figure is for surface, second figure for 200 ft. All figures are in ppm × 10^{-2}. For the key to sample station numbers and installations see Fig. 5.

the complex admixture of hydrocarbons in sea water and sediment. For convenience a Kuwait crude oil standard was used. The results, though they are not absolute measures of aromatic ring compounds in the environment, are comparative if the same technique is used each survey.

Results are summarised in Fig. 12. Amounts of hydrocarbons found were low.

Benthos

The ten samples per station were sorted and different organisms identified and counted. In most cases, identification to species was possible. Where not, the organism was identified as far as possible then preserved and coded. Preservation of sorted samples was in hexamine buffered 4% formalin. The results are fully tabulated in Table IA, Appendix A.

Statistical Treatment of Results

Several statistical tests were applied to the counts of benthic organisms:

(a) *Standard error and 95% confidence limits of the mean.* Where organisms occurred in sufficiently large numbers, standard error and 95% confidence limits of the mean number per station were calculated. This will allow accurate comparison of any future results and calculation of the significance of differences or otherwise. This test was applied to total density of organisms per m^2, numbers of *Myriochele heeri* per m^2 and numbers of Pleurogonid ascidians (a mixture of *Eugyra arenosa* and *Polycarpa fibrosa*) per m^2.

Total counts per m^2 and numbers of *Myriochele heeri* per m^2 for each sample station are presented as transect line graphs with 95% confidence limits (Figs. 13 and 14) and as distribution maps (Figs. 15 and 16). Calculations were based on:

$$\text{Standard error } (\sigma_n) = \left(\frac{\sum (x^2) - (\sum x)^2/n}{n(n-1)}\right)^{1/2}$$

and 95% limit (0·05 probability limits)

$$= \bar{x} \pm t\sigma_n$$

where t = probability constant dependent on $n - 1$, n = number of samples, x = each sample result and $\bar{x}$ = sample mean.

(b) *Optimum sample number* was tested by three methods: (i) serial calculation of 95% limits from $n_2{}^{10}$; (ii) cumulative mean method; and (iii) occurrence of new species with increasing sample number.

(i) 95% limits of total number of organisms per m^2 were calculated at station 1 for a series of sample numbers from 2 to the maximum of 10 to assess how limits changed with sample number. The results are plotted in Fig. 17.
(ii) The cumulative mean number of organisms per m^2 was calculated for station 1 by finding the mean for sample 1, sample $1 + 2$, sample $1 + 2 + 3$, and so on to $1 + 2 + \cdots + 10$. Considerable variation in the mean occurs initially, but as sample number increases, fluctuation decreases. Optimum sample number occurs where further samples do not significantly influence the cumulative mean. The results of cumulative mean plotted against sample number for station 1 are shown in Fig. 17.
(iii) The numbers of new species encountered in each successive sample from 1 to 10 at station 1 was plotted against number of samples. Optimum sample number occurs at the point where small numbers of new species are encountered with increasing number of samples. The results of this test at station 1 are shown in Fig. 18.

The reason for calculating optimum sample number is to determine the minimum number of samples which provide a reasonable picture of the environment. This is compatible with minimum cost of survey and obtaining results for useful comparison of one survey with another.

Diversity Measurement

The benthic data were further described by the use of diversity measurements. Whilst the above statistical tests are designed for numerical quantification and comparison of organism density either at a community or specific level, a diversity index provides a single figure describing the community in terms of both density and the number of different types of organism. The usefulness of diversity measurement stems from the supposition that changes in the environment result in change in the species found and the population density (Pearson *et al.* 1967). Such a measurement can have value in a monitoring

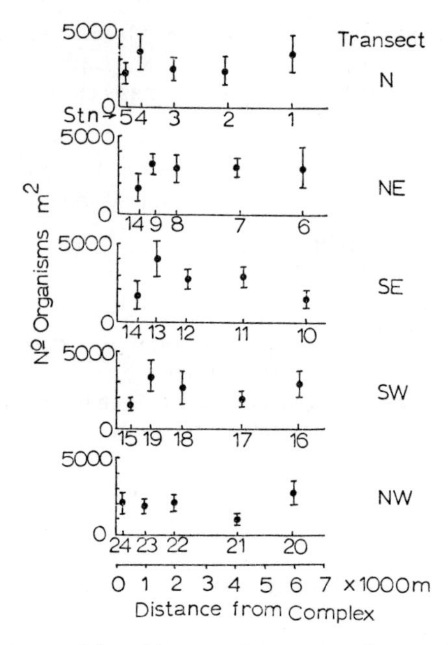

FIG. 13. Total numbers of benthic organisms per m² at each sample site. 95% confidence limits are marked.

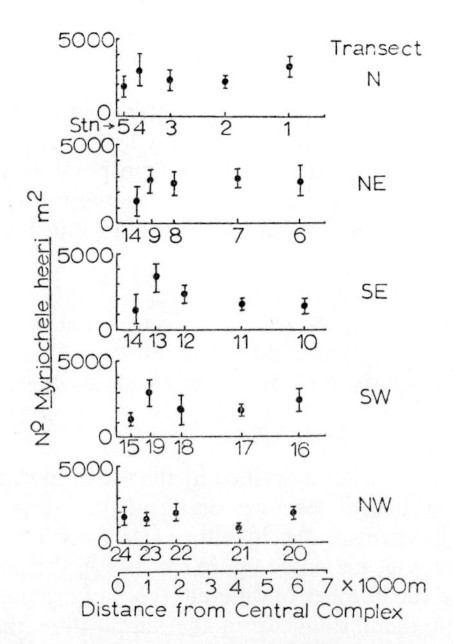

FIG. 14. Numbers of *Myriochele heeri* per m² at each sample station with 95% confidence limits marked.

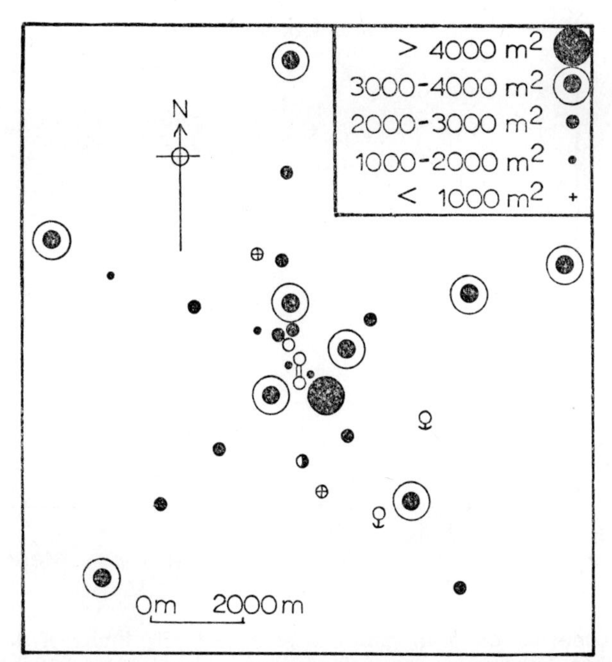

FIG. 15. Numerical distribution of individual organisms of the benthic community in the Ekofisk Field. The key to sample station numbers and installations can be found in Fig. 5. To avoid confusion, the central complex structures have not been shaded.

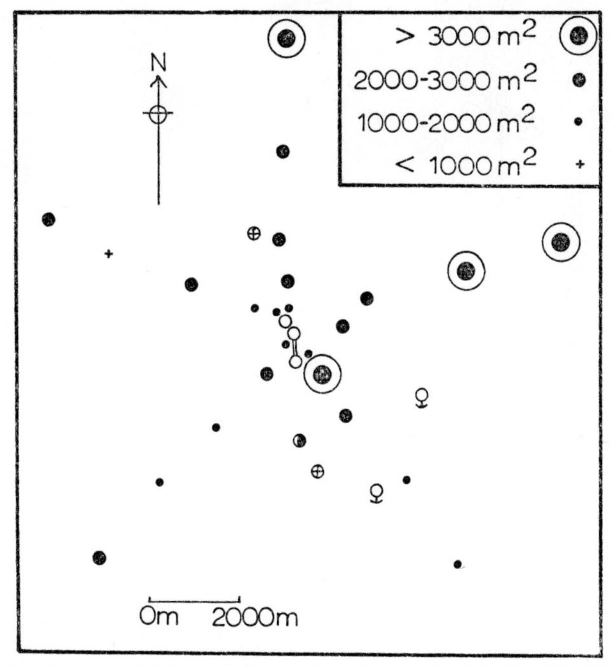

FIG. 16. Distribution of *Myriochele heeri* on the sea bed in the Ekofisk Field. For the key to sample station numbers and installations see Fig. 5. To avoid confusion, the central complex structures are not shaded.

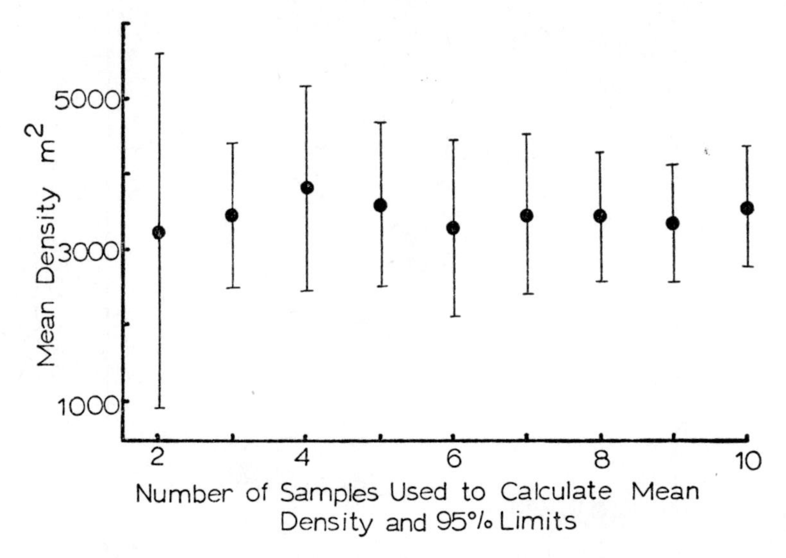

FIG. 17. Dependence of mean sample density and 95% limits of the mean on sample number. The benthic data from sample station 1 of the Ekofisk Field are used to calculate the mean density per m² and its 95% limits when sample number increases from 2 to 10.

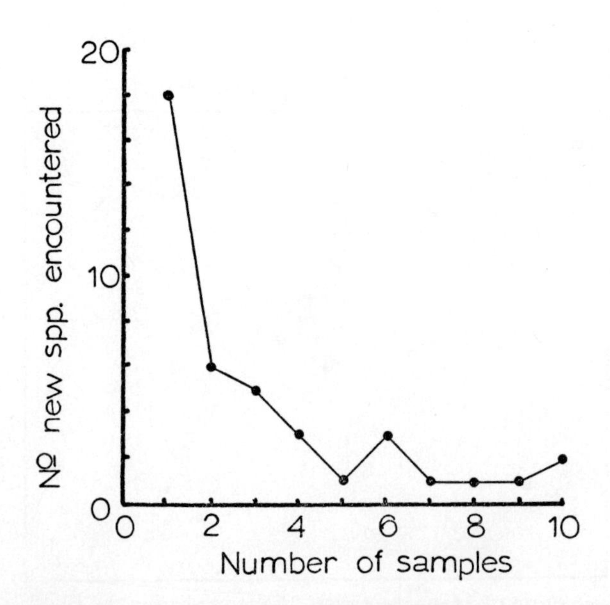

FIG. 18. Number of new species encountered with increasing number of samples, benthic data, sample station 1, Ekofisk Field.

programme when looking for change in an ecosystem. Three different indices were calculated to provide full description of each sample site, and to compare the relative sensitivity of each index to species change and numerical change at the densities and species numbers found in the Ekofisk survey.

The indices calculated were:

(a) *The Shannon–Wiener information function* (from Margalef, 1968):

$$H_{(s)} = -\sum p_i \log_2 p_i$$

where $H_{(s)}$ = diversity, p_i = probability of occurrence of each species calculated from n_i/N where n_i = number of that species and N = total number of individuals in all species.

(b) *Gleason index* (from Margalef, 1968):

$$D' = \frac{s-1}{\log_e N}$$

where s = number of species, N = number of individuals and D' = diversity.

(c) *Menhinick index* (from Odum, 1959):

$$D = \frac{s}{(N)^{1/2}}$$

where s = number of species, N = number of individuals and D = diversity.

The results are tabulated in Table I, and the Shannon–Wiener index results are also summarised diagrammatically in Fig. 19. Each diversity index was

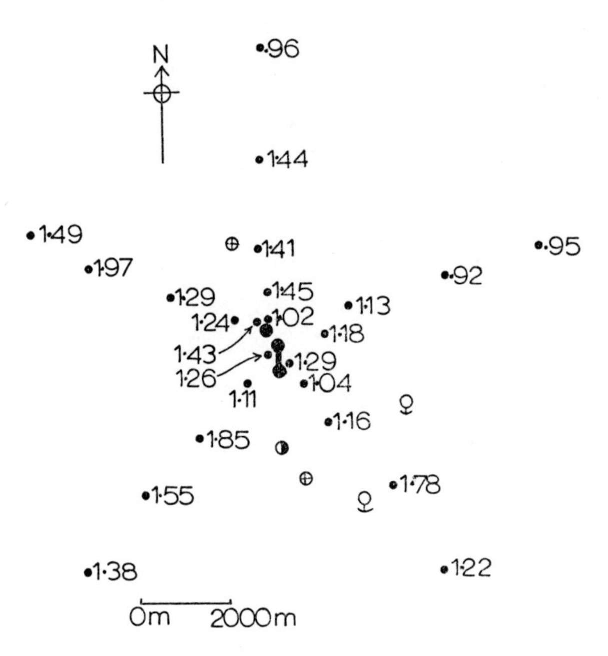

FIG. 19. Shannon–Wiener index at each sample station. For key to sample station numbers and installations see Fig. 5.

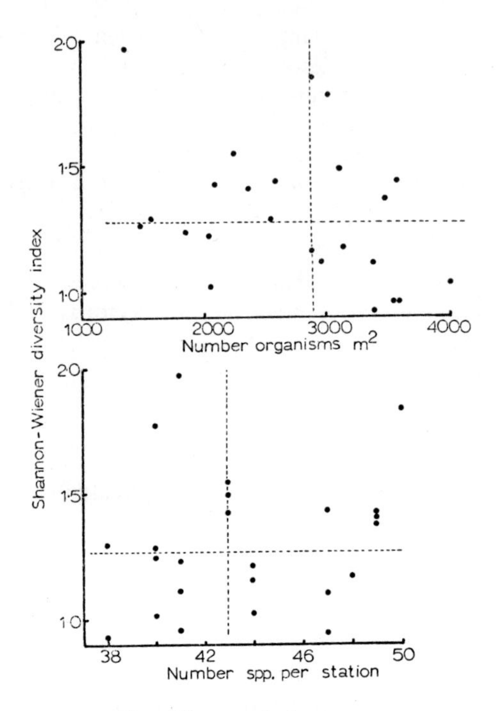

FIG. 20—*see facing page.*

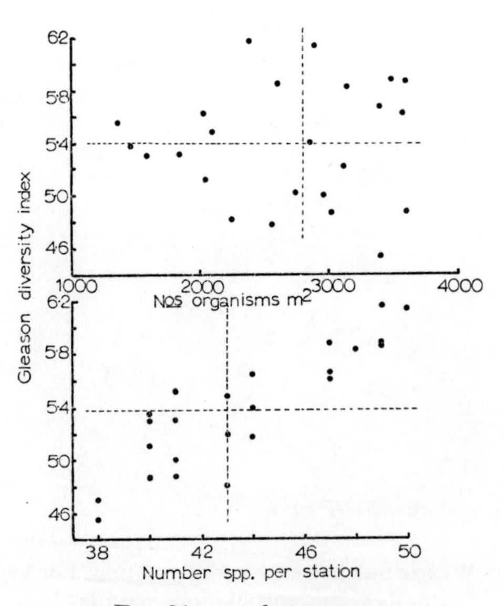

FIG. 21—*see facing page.*

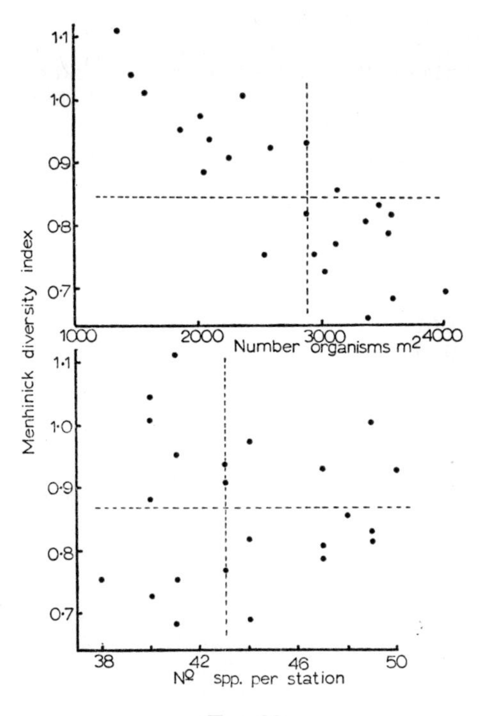

FIG. 22

FIGS. 20–22. Plots of the Shannon-Wiener (Fig. 20), Gleason (Fig. 21) and Menhinick (Fig. 22) indices against number of species and number of individuals for the 24 sample stations to reveal bias of the indices to either number of individuals or species. To find significant trends in the scatter of points, the quick regression test of Lewis and Taylor (1967) was applied. The quadrants for this test are marked by broken lines. The vertical broken line divides the scatter of points equally right and left, and the horizontal broken line divides them equally above and below. To show a significant trend in the scatter, the number of points in the quadrant(s) containing the smallest number must be equal to or less than 2, and in the quadrant(s) with the greatest number equal to or more than 9 for a scatter of 24 points. Points lying on the quadrant lines are ignored.

There is no significant trend in either scatter of Fig. 20, but Fig. 21 shows a significant trend in number of species plotted against Gleason's index, whilst Fig. 22 shows a significant trend with Menhinick's index plotted against sample size (number of organisms per m^2).

regression tested for weighting towards exaggeration of changes either in the number of species or number of individuals. This was achieved by plotting the calculated diversity indices against the numbers of species and numbers of individuals represented by each index. The quick graphical regression test of Lewis and Taylor (1966) for significant correlation was then applied to detect a trend in the scatter diagrams. The results are shown in Figs. 20–22. The diversity indices for the 24 sample sites were also correlated with each other using the Spearman rank correlation test to 0·5 probability as described in Snedecor and Cochran (1967).

TABLE I

The values of the Shannon–Wiener, Gleason and Menhinick diversity indices for each of the 24 sample stations, Ekofisk Field, August 1973

Station	*Shannon–Wiener index*	*Gleason index*	*Menhinick index*
1	0·96	4·89	0·68
2	1·44	5·85	0·92
3	1·41	6·18	1·01
4	1·45	5·86	0·82
5	1·02	5·11	0·88
6	0·95	5·62	0·79
7	0·92	4·55	0·65
8	1·13	5·01	0·75
9	1·18	5·84	0·86
10	1·22	5·64	0·97
11	1·78	4·87	0·73
12	1·16	5·39	0·82
13	1·04	5·18	0·69
14	1·29	5·30	1·01
15	1·26	5·35	1·04
16	1·38	5·89	0·83
17	1·55	4·82	0·91
18	1·85	6·15	0·93
19	1·11	5·66	0·81
20	1·49	5·22	0·77
21	1·96	5·54	1·11
22	1·29	4·72	0·75
23	1·24	5·32	0·95
24	1·43	5·49	0·94

(d) *The Spearman rank coefficient*:

$$r_s = 1 - \frac{6 \sum d^2}{n(n^2 - 1)}$$

where d = the difference between rank 1 and 2 and n = the number of ranks.

Rank correlating the Shannon–Wiener index with the Gleason index results in a value for r_s of 1·3726, a value which falls well short of the lower limit of 2·074 for 0·05 probability positive correlation. Rank correlating the Shannon–Wiener index with the Menhinick index gives a value for r_s of 2·497, which is greater than the required value of 2·074 for positive correlation at 0·05 probablity. The Mehinick index therefore correlates well in result with the Shannon–Wiener index, whereas the Gleason index does not.

THE EKOFISK BENTHIC COMMUNITY

The data for bottom sediments particle size and organic carbon content in association with evenness of depth and thermal stratification indicate considerable uniformity of the sea bed. This is borne out by the similarity between communities of organisms found at each sample site, and their densities, some of which are illustrated in Figs. 13–16. Of the 92 species isolated and four groups of more than one species (Appendix A, Table IA), 15 were found at 95% or more of sites, 28 at 75% or more of sites, and 41 at 50% or more of sites. Feder *et al.* (1973) classified all species in this last category as 'biologically important species' and considered these the most important to look at in the monitoring context. The occurrence of species at each sample station is given in Table II. The biologically important species are marked. Those species which occur at 75% or more sites are used here to define the overall community type of the sample area. They are:

Amphipod crustaceans: *Ampelisca gibba, Hippomedon denticulatus, Tryphosites longipes.*
Bivalve molluscs: *Cyprina islandica, Cultellus pellucidus.*
Gastropod molluscs: *Siphonodentalium lofotense, Dentalium entalis, Philine* sp. *Acteon tornatilis, Chaetoderma nitidulum.*
Echinoderms: *Echinocardium cordatum, Amphiura filiformis.*
Ascidians: The Pleurogonid ascidians, *Eugyra arenosa, Polycarpa fibrosa.*
Polychaetes: *Myriochele heeri, Owenia fusiformis, Stenhelais limicola, Phyllodoce lamelligera, Nephtys incisa, Glycera alba, Goniada maculata, Lumbriconereis fragilis, Scoloplos armiger, Euclymene lumbricoides, Pectinaria korenii, Potamilla torelli.*
Others: *Cerianthus* sp.

Difficulties occur in fitting the Ekofisk benthic community into any of the North Sea communities summarised by Glemarec (1973), but it fits well to the boreal offshore mud association of Jones (1950). Assessment of the community type is of value in that it allows a certain degree of extrapolation of the physical environmental parameters which determined the community (Jones, 1950). There is also a fair degree of agreement in the species of polychaetes found with those found by Kirkegaard (1969) at his nearest site (13 miles to the west of Ekofisk).

OPTIMUM SAMPLE NUMBER

Figure 17 and 18 illustrate the effects of increasing sample number at station 1 on 95% confidence limits, numbers of new species, and the variation of the sample mean. These graphs make several points clear.

(i) The greatest reduction in 95% limits occurs between sample numbers 2 and 6. After 8 samples, the reduction in the 95% limits is small (Fig. 17). In order to improve 95% limits to appreciably less than those calculated for the 10 samples obtained will require a considerable increase in number of samples. 10 samples per site has produced 95% confidence limits for the 24 sample sites which average 25% of the mean. These figures suggest that between 8 and 10 samples per station adequately describes the majority of the benthos.

(ii) Sizeable fluctuation in the cumulative mean of successive samples at station 1 occur until 7 samples have been summed. The curve then begins to level out at between 3400 and 3600 organisms per m^2 (Fig. 17). However, fluctuations are still occurring after 10 samples.

TABLE II

The occurrence of benthic species at each of the 24 sample stations in the Ekofisk field, August 1973. Biologically important species (those occurring at 50% or more stations) are marked ●.

		Station																							
		1	2	3	4	5	6	7	8	9	10	11	12	13	14	15	16	17	18	19	20	21	22	23	24
Ampelisca gibba	●	×	×	×	×	×	×	×	×	×	×	×		×	×	×	×	×	×		×	×	×	×	×
Hippomedon denticulatus	●		×	×	×	×	×	×	×	×	×	×	×	×	×	×	×	×	×	×	×	×	×	×	×
Tryphosites longipes	●	×	×	×	×	×	×	×	×	×	×	×	×	×	×	×	×	×	×	×	×	×	×	×	×
Cumaceans			×	×	×		×	×		×			×	×	×					×			×		
Macropippus sp.				×		×							×	×	×	×	×	×	×						×
Pagarus sp.			×	×													×							×	
Corophium bonelli																	×				×		×		
Podoceros odontyx																×									
Corystes cassivelaunus																	×								
Cyprina islandica	●	×	×	×	×	×	×	×		×		×	×	×		×	×		×	×	×	×		×	×
Cultellus pellucidus	●	×	×	×	×	×	×	×	×	×	×	×	×	×	×	×	×	×	×	×	×	×	×	×	×
Mysella bidentata	●	×	×	×	×		×	×	×				×	×	×	×			×	×	×			×	
Cardium echinatum	●	×			×	×	×	×	×	×			×	×		×		×	×	×		×			×
Montacuta ferruginosa					×						×	×					×	×				×			
Mya arenaria		×	×	×							×		×					×		×					
Mya truncata		×									×						×		×						
Nucula hanleyi			×	×				×			×	×	×	×											
Astarte borealis				×											×		×		×		×				
Gari fervensis		×			×												×					×			
Lucinoma borealis		×																							
Venus casina	●			×	×	×	×				×	×	×	×					×	×		×		×	

Venus striatula					×	×																			
Lutraria lutraria								×								×					×		×	×	
Spisula elliptica							×					×					×		×						
Musculus niger																							×		
Thyasira flexuosa											×	×								×					
Siphonodentalium lofotense	●	×	×	×	×	×	×	×	×		×		×		×	×	×		×	×	×	×	×	×	×
Dentalium entalis	●		×	×	×	×	×		×	×	×		×	×	×	×	×	×		×	×	×	×	×	
Philine sp.	●	×	×	×	×	×	×	×	×	×	×	×	×	×	×	×	×	×	×	×	×	×	×	×	
Acteon tornatilis	●	×	×	×	×	×		×	×	×	×	×	×	×		×		×	×	×	×		×	×	×
Colus gracilis				×			×					×		×					×			×			×
Clathrus clathrus					×					×															
Natica alderi																×	×	×	×					×	×
Natica catena							×		×																
Nudibranch sp. 41.41					×																				
Chaetoderma nitidulum	●	×	×	×	×	×	×	×	×	×	×	×	×	×	×	×	×	×	×	×	×		×	×	×
Echinocardium flavescens		×	×	×		×				×	×		×					×	×		×			×	
Echinocardium cordatum	●	×	×	×	×	×	×	×	×	×	×	×	×	×	×		×	×	×	×		×	×	×	×
Spatangus purpureus																	×								
Amphiura filiformis	●	×	×	×	×	×	×	×	×	×	×	×	×	×	×	×	×	×	×	×	×	×	×	×	×
Amphiura chiajei		×		×			×		×																
Ophiuroid juveniles	●	×	×	×	×	×	×	×	×	×	×	×	×	×	×	×	×	×	×		×	×	×	×	×
Ophiura texturata				×																×					
Asteropecten irregularis	●	×	×	×	×	×	×	×		×	×	×			×		×				×	×	×	×	×
Pleurogonid spp.	●		×	×	×	×	×	×	×	×	×	×	×	×	×	×	×	×	×	×	×	×	×	×	×
Ciona intestinalis										×						×									
Ascidian sp. 55.98										×								×						×	

TABLE II—*contd.*

		Station																							
		1	2	3	4	5	6	7	8	9	10	11	12	13	14	15	16	17	18	19	20	21	22	23	24
Nemertine sp.	●				×	×	×		×	×	×		×	×	×	×		×	×	×	×			×	×
Lineus ruber																				×	×			×	×
Pennatulid sp.	●	×	×	×	×	×		×	×		×	×	×	×	×			×	×	×		×			×
Flatworm sp.		×	×		×					×	×		×					×	×	×	×				×
Sponge 52.95										×	×			×					×		×			×	×
Anemone sp. 16.27			×										×		×	×	×	×	×	×				×	×
Anemone sp. 17.21			×							×					×		×			×	×				
Cerianthus sp.	●	×	×	×	×	×	×	×	×	×	×	×	×	×	×	×	×	×	×	×	×	×	×	×	×
Flustra foliacea								×																	
Flustrella hispida																						×			
Phascolion strombi	●	×					×					×	×		×	×		×	×	×	×		×	×	×
Crystallogobius linearis			×																						
Myriochele heeri	●	×	×	×	×	×	×	×	×	×	×	×	×	×	×	×	×	×	×	×	×	×	×	×	×
Owenia fusiformis	●	×	×	×	×	×	×	×	×	×	×	×	×	×	×	×	×	×	×	×	×	×	×	×	×
Stenhelais limicola	●	×	×	×	×	×	×	×	×	×	×	×	×	×	×	×	×	×	×	×	×	×	×	×	×
Aphrodita aculeata							×	×			×			×					×						×
Lepidonotus squamatus			×	×	×		×			×	×	×					×			×					
Harmothoe sp.										×			×												
Phyllodoce lamelligera	●	×	×	×	×	×	×		×	×	×		×	×	×	×	×	×	×	×	×	×	×	×	

Anaitides sp.(juv.)			×																						
Ophiodromus flexuosus														×		×	×	×				×			
Nephtys incisa	●	×	×	×	×	×	×	×	×	×	×	×	×	×	×	×	×	×	×	×	×	×	×	×	×
Nephtys cirrosa	●		×	×	×	×		×	×	×	×	×		×	×	×	×	×	×			×	×	×	
Glycera alba	●	×	×	×	×	×	×	×	×		×	×	×	×	×	×	×	×	×	×		×	×	×	×
Goniada maculata	●	×	×	×	×	×	×	×	×	×	×	×	×	×	×	×	×	×	×	×	×	×	×	×	×
Lumbrineris fragilis	●	×	×	×		×	×		×	×	×	×	×	×	×	×	×	×	×	×	×	×	×	×	×
Lumbrineris impatiens				×	×	×	×	×		×															
Scoloplos armiger	●	×	×	×	×	×	×	×	×	×	×	×	×	×	×	×	×	×	×	×	×	×	×	×	×
Spiophanes bombyx	●	×	×	×	×		×		×			×		×	×				×	×	×	×	×		×
Spio filicornis														×											
Poecilochaetus serpens																			×						
Chaetozone setosa	●	×		×	×	×	×		×	×	×	×		×	×	×		×	×	×	×		×		×
Cirratulid sp. 33.39	●		×	×	×	×		×	×	×	×		×	×	×	×	×			×	×	×	×		×
Diplocirrus glaucus	●	×		×		×	×		×	×		×	×				×	×		×	×	×			×
Caulleriella sp.								×		×															×
Ophelina aulogaster					×																×				
Ophelia limacina		×	×		×		×		×	×			×			×	×					×			×
Euclymene lumbricoides	●	×	×		×		×	×	×	×	×	×	×	×	×	×	×	×	×	×	×	×	×	×	×
Pectinaria koreni	●	×	×	×	×	×	×	×	×	×	×	×	×	×	×	×	×	×	×	×	×	×	×	×	×
Sosane gracilis																			×	×					
Ampharetid sp. (juv.)	●	×	×	×	×	×	×	×	×	×		×		×			×	×	×	×	×				
Ampharetid sp. 46.610	●			×	×	×	×		×	×	×	×	×	×		×	×	×	×	×	×	×	×	×	
Lanice conchilega																						×			
Thelepus setosus														×											
Lysilla loveni			×																						
Bispira volutacornis		×					×		×	×	×														
Potamilla torelli	●	×	×	×	×		×	×	×	×	×	×	×	×		×	×		×	×	×	×	×		×
Potamilla reniformis																		×				×			
Jasmineira sp. cf. *caudata*																	×								

(iii) The numbers of previously unencountered species occurring in successive samples at station 1, plotted against the number of samples taken, reveals new species being encountered at the rate of about 1 per new sample after 10 samples (Fig. 18). The curve for new encounters starts to level at between 7 and 10 samples, suggesting an optimum sample number which will describe the majority of the species between these limits.

The results of these tests of optimum sample number suggest that between 7 and 10 samples is near the optimum number in terms of return for effort. The disagreement between the cumulative mean method and the other two tests of sample number substantiates the suggestions of Longhurst (1959) and Lie (1968) that cumulative mean may be an unreliable estimate of minimum sample number. Fewer than 5 samples per site in this survey would mean missing approximately 20% of the species encountered in 10 grabs, and 95% confidence between 1·2 and 2·9 times larger than those for 10 samples. A larger number of samples than 10 would not greatly reduce 95% limits or substantially increase the number of new species found.

Feder *et al.* (1973) found that between 68% and 72% of species found in 8 grabs were encountered in the first 3 grabs of the 8, and considered 3 grabs sufficient to quantify the abundant species. Longhurst (1959) suggests that $5 \times 0{\cdot}1\ m^2$ grabs are necessary to quantify abundant species. At station 1, 70·7% of species are encountered after 3 grabs, which agrees well with the results of Feder *et al.* (1973). After the first 3 grabs, the less abundant of the species (marginal species) are encountered. However, knowledge of the marginal species is valuable to a monitoring programme as these are likely to be the first to suffer under an environmental stress (Feder *et al.*, 1973). Their disappearance will considerably affect sample diversity (see next section) and they are thus a potential 'first line' pollution indicator. For this reason, taking more than 5 grabs per site may be very important to a monitoring programme. It is realised that although abundant species can be quantified using between 3 and 5 grabs per site, 10 may not be adequate to quantify marginal species due to inherent difficulties in adequately sampling widely distributed organisms (Longhurst, 1959). At the least, taking 10 grabs allows encountering the majority of the marginal species, and quantification of the more common of these. For this reason, and the improvement of 95% confidence limits of density measurement, 10 grabs per site should be regarded as minimal in a monitoring programme where detailed study of change is required and where large numbers of different species are likely to be encountered at each site. Smaller numbers of grabs may be useful where fewer species occur, though adequate field tests should be carried out to determine the useful number.

Ultimately, it is only feasible to make suggestions about necessary sample number above the minimum to quantify abundant organisms. A decision must be made at the start of the programme as to the sensitivity required. In the case of the Ekofisk survey, high sensitivity was required in order to detect changes which may only be small, or indeed to show that no changes occur. For this purpose, a sample number of 10 is necessary. Where lower sensitivity is needed, sample number should be chosen to fit.

Further non-environmental considerations which influence sample number are availability of finance and the time and cost of sample analysis. Where both are limited and prevent detailed sampling, the taking of even a small number of samples will allow at least the detection of catastrophic change.

DIVERSITY

Regression testing of the Shannon–Wiener index for bias towards either exaggeration of species number or changes in the number of individuals (scatter diagrams, Fig. 20) shows no bias to either at the densities found in Ekofisk. The Shannon–Wiener index, therefore, does not favour change in number of species over number of individuals or vice versa, but is sensitive to change in both. This agrees well with the evidence presented by Sanders (1968) and amplified by Fager (1972), who suggested in addition that the Shannon–Wiener is more a measure of intra-interspecific relationships than evenness of distribution.

Testing the Gleason (Fig. 21) and Menhinick (Fig. 22) indices in the same manner showed the Gleason index to be more sensitive to species changes and the Menhinick index more sensitive to numerical change at the species number and animal densities encountered in the Ekofisk field. Lie (1968), however, found the Gleason index more sensitive to numerical change and Fager (1972) dismisses this index as 'erratically variable'.

Calculation of more than one index for a set of data provides a more useful description of the diversity spectrum of a community than any one individually (Margalef, 1968; Feder *et al.* 1973), and the three indices selected here provide a full description of the Ekofisk community. There is a tendency to utilise indices such as the Gleason index instead of the more reliable Shannon–Wiener index because of their considerably easier computation. Such short-cutting is acceptable if good correlation can be obtained between the simple index and a reliable index such as the Shannon–Wiener, and should even then be limited to use only in one type of environment, for the variability in these indices, e.g. Gleason's (Fager, 1972; Lie, 1968), makes comparison of different communities unreliable. The Spearman rank correlation test applied between the Gleason index and Shannon–Wiener index showed no positive correlation between the two for the Ekofisk data, but the same test of the Menhinick index with the Shannon–Wiener index showed positive correlation of 0·05 probability for these data. It is feasible, therefore, to use the Menhinick index as a short-cut in this case, but the dependence of this index on numerical change suggests that it should not be used alone as a complete description of the data, but in conjunction with the information provided by the other two indices.

Whilst it is easy to stress the usefulness of diversity measurements, comments should be made on limitations of the indices. Two of the main limitations of diversity indices when comparing environments are that (a) one numerical value of an index can represent entirely different sets of information from completely different environments (Margalef, 1968) and (b) that the input data cannot be extracted from the index value.

These limitations, along with variability in index value with differing numbers of species and individuals, suggest that the prime use of diversity should be as a means of expression of the information content of a set of data (Margalef, 1968), and should not be used independently of these data. It follows from these points that they are most reliable when describing changes in one type of community at one site over a period of time, and are thus particularly applicable to monitoring studies.

'ARTIFICIAL' MONITORING METHODS

In addition to sampling the composition of the benthos and plankton by the use of grabs and plankton nets, other techniques are available to obtain estimates of the condition of the marine environment. Two such methods are the use of settlement plates, and the making of transplants.

Settlement plates

These are usually squares of asbestos or sand-blasted formica, and act as a clean area on which planktonic larvae can settle and grow. Regular harvesting and replacing of the plates provides a measure of the viable larval forms in the plankton. The disappearance or significant decrease in abundance of certain organisms on these plates may be used as an indicator of pollution. The advantage of such plates is that they can be attached easily to offshore installations or buoys, and their distribution can be selected to suit the problem under study. It is, of course, also possible to look at the organisms growing on pilings or rig legs in a monitoring programme, for these are, as far as many marine organisms are concerned, no more than gigantic settlement plates. The use of different settling media is also possible, e.g. baskets containing stones or rocks as a settlement substrate. These have the disadvantage of being more complex to sort and handle, but provide a wider variety of microhabitat.

Transplants

Organisms known to be sensitive to pollutants can be placed in suitable cages, transplanted to the test areas and attached to fixed installations or buoys. This technique suffers from the disadvantages that cages can be unwieldy, and adequate controls are needed to ascertain whether the transplanting has an effect on the organisms. The technique may be useful where a plentiful supply of transplant organisms is available.

CONCLUSION

The data obtained in this survey in the Ekofisk field are of a baseline nature, and their usefulness lies in comparison with data from future surveys of the same area. They have fulfilled their initial objective of defining and quantifying the existing benthic community and its distribution, obtaining related information on the plankton, and recording some of the physical and chemical factors on which the present community is dependent. This will allow determination of any future change in the communities, and assist in ascribing any such changes to natural environmental fluctuations or other factors such as the activities of the oil industry. At present, the distribution of organisms is relatively uniform throughout the area and there is no evidence of ecological damage to the benthos associated with the existing oil installations.

The detailed nature of the Ekofisk survey is an integral part of its objectives of accuracy and the potential of detecting and quantifying small changes or trends over a period of time, or, indeed, showing that no significant changes occur. The lack of knowledge of the impact of offshore developments, however, suggests that some effort should also be made towards a simple and wider survey of a larger number of oil fields to at least be able to detect ecological damage of large dimensions. Extensive surveys, such as the one described herein, then assume increased importance in the provision of detailed correlative information for such a wider survey, and for making decisions as to type and extent of remedial action where damage occurs.

The present trend suggested and exemplified by Parker (1975) of increase in the use of sophisticated, complex and costly computerised techniques for analysis of offshore environmental data should be balanced by wide-ranging simplified surveys which allow, as a minimum, the detection of environmental changes of substantial proportions. At present we are unable to achieve even this minimum.

ACKNOWLEDGEMENTS

I am grateful to the following people and organisations: to the Phillips Norway Group for sponsoring the Ekofisk survey and the opportunity to present the data; to Messrs J. Addy, D. Levell and I. Tew for their help in sorting the Ekofisk samples: to NERC Research Vessel Base, Barry Dock, MAFF Laboratory, Burnham-on-Crouch, and British Petroleum Company Ltd, Sunbury-on-Thames, for the loan of sampling equipment; to Dr M. Rolfe, MAFF Laboratories, Burnham-on-Crouch, and Dr P. Gibbs, Plymouth MBA Laboratories, for their help in the identification of difficult polychaetes and sipunculids and comments on the manuscript; and finally to Dr J. Baker and my wife for their constructive criticism on the manuscript.

The Phillips Norway Group consists of the following companies:

Phillips Petroleum Company—Operator
American Petrofina Exploration Company of Norway
Norsk Agip A/S
Elf Aquitaine
Norsk Hydro A/S
Total Marine Norsk A/S
Eurofrep Norge A/S
Comparex Norge A/S
Cofranord A/S

REFERENCES

Addy, J. (1975). The sublittoral fauna of Milford Haven (this volume).

Böhnecke, G. (1922). Salzgehalt und Strömungen der Nordsee. *Veröff. Inst. Meeresk. Univ. Berl. Neue Folge A., Geog-naturwiss. Reihe*, **10**, 1–34.

Buchanan, J. B. (1963). The bottom fauna communities and their sediment relationships off the coast of Northumberland. *Oikos*, **14** (2), 154–75.

Cole, H. A. and Holden, M. J. (1971). History of the North Sea Fisheries 1950–1969, in: *North Sea Science*, ed. E. D. Goldberg, MIT Press, Cambridge, Mass., pp. 337–69.

Cushing, D. H. (1953). Studies on plankton populations. *J. Cons. Perm. Intern. Explor. Mer.*, **19**, 3–22.

Edinburgh Oceanographic Society (1973). Continuous Plankton Records: a plankton atlas of the North Atlantic and North Sea. *Bull. Mar. Ecol.*, **7**, 1–174.

Eisma, D. (1971). Sediment distribution in the North Sea in relation to marine pollution, in: *North Sea Science*, ed. E. D. Goldberg, MIT Press, Cambridge, Mass., pp. 131–52.

Fager, E. W. (1972). Diversity: a sampling study. *Amer. Naturalist*, **106**, 293–310.

Falk, M. R. and Lawrence, M. J. (1973). Acute toxicity of petrochemical drilling fluid components and wastes to fish.

FAO (1972). *Atlas of the Living Resources of the Seas*, 3rd ed., FAO, Rome.

Feder, H. M., Mueller, G. J., Dick, M. H. and Hawkins, D. B. (1973). Preliminary benthos study, in: *Environmental Studies of Port Valdez*, ed. D. W. Hood, W. E. Shiels and E. J. Kelly, Institute of Marine Science, Univ. of Alaska, Occasional Publication No. 3.

Fraser, J. H. (1971). Zooplankton of the North Sea, in: *North Sea Science*, ed. E. D. Goldberg, MIT Press, Cambridge, Mass., 267–92.

Gage, J. (1972). A preliminary survey of the benthic macrofauna and sediment in Lochs Etive and Creran, sea-lochs on the west coast of Scotland. *J. mar. biol. Ass. U.K.*, **52**, 237–76.

Glemarec, M. (1973). The benthic communities of the European North Atlantic continental shelf. *Oceanogr. Mar. Biol. Ann. Rev.*, **11**, 263–89.

Hardy, A. C. (1956). The world of the plankton, in: *The Open Sea: Its Natural History*, New Naturalist series, Collins, London, 336 pp.

Hill, H. W. (1971). Currents and water masses, in: *North Sea Science*, ed. E. D. Goldberg, MIT Press, Cambridge, Mass., pp. 17–42.

Hofmann, R. E. (1949). Control of oil pollution. *Publ. Works, N.Y.*, **80**, 26–7.

Holme, N. A. and McIntyre, A. D. (1971). *Methods for Study of the Marine Benthos*, IBP Handbook No. 16, Blackwell Scientific Pub., Oxford and Edinburgh, 334 pp.

Jones, N. S. (1950). Marine bottom communities. *Ext. Biol. Rev.*, **25**, 283–313.

Kirkegaard, J. B. (1969). A quantitative investigation of the central North Sea polychaetes. *Spolia zool. Mus. haun.*, **29**.

Kolpack, R. L., Mattson, J. S., Mark, H. D. and Ta-Ching, Y. (1971). Hydrocarbon content of Santa Barbara Channel sediments, in: *Biological and Oceanographic Survey of the Santa Barbara Channel Oil Spill, 1969–1970*, vol. II: *Physical, Chemical and Geological Studies*, Allan Hancock Foundation, Univ. of Southern California, Los Angeles.

Korringa, P. (1971). The edge of the North Sea as a nursery ground and shellfish area, in: *North Sea Science*, ed. E. D. Goldberg, MIT Press, Cambridge, Mass., pp. 361–82.

Leeper, G. W. (1948). *Introduction to Soil Science*, Melbourne Univ. Press, Melbourne, 222 pp.

Lewis, T. and Taylor, L. R. (1966). *Introduction to Experimental Ecology*, Academic Press, London and New York, 401 pp.

Lie, U. (1968). A quantitative study of benthic infauna in Puget Sound, Washington, USA, in 1963–1964. *Fisk. Dir. Skr.* (*Ser Havundas*), **14**, 223–356.

Longhurst, A. R. (1964). A review of the present situation in benthic synecology. *Bull. Inst. Oceanogr. Monaco*, **63**, 1–54.

Margalef, R. (1968). *Perspectives in Ecological Theory*, Univ. of Chicago Press, Chicago, 111 pp.
McIntyre, A. D. (1961). Quantitative differences in the fauna of boreal mud associations. *J. mar. biol. Ass. U.K.*, **41**, 599–616.
Morgans, J. F. C. (1956). Note on the analysis of shallow-water soft substrata. *J. Anim. Ecol.*, **25**, 367–87.
Nelson-Smith, A. (1972). *Oil Pollution and Marine Ecology*, Paul Elek (Scientific Books), Ltd, London, 260 pp.
Newell, G. E. and Newell, R. C. (1963). *Marine Plankton*, Hutchinson, London, 221 pp.
Odum, E. P. (1959). *Fundamentals of Ecology*, 4th ed., Saunders, Philadelphia and London.
Oxford Atlas of Britain and Northern Ireland (1963). Clarendon Press, Oxford.
Parker, R. H. (1975). *The Study of Benthic Communities*, Elsevier Oceanography series, Elsevier Pub. Co., Amsterdam, Oxford, New York, 279 pp.
Pearson, A. J., Storrs, P. N. and Selleck, R. E. (1967). Some physical parameters and their significance in marine waste disposal, in: *Pollution and Marine Ecology*, ed. T. A. Olson and F. J. Burgess, Interscience, New York, 364 pp.
Petersen, C. G. J. and Jensen, P. B. (1911). Valuation of the Sea. I. Animal life of the sea bottom, its food and quantity. *Rept. Danish Biol. Sta.*, **20**, 1–76.
Sanders, H. L. (1958). Benthic studies in Buzzards Bay. I. Animal–sediment relationships. *Limnol. Oceanogr.*, **3**, 245–58.
Shelton, R. G. J. (1971). Some effects of dumped, solid wastes on marine life and fisheries, in: *North Sea Science*, ed. E. D. Goldberg, MIT Press, Cambridge, Mass., pp. 415–36.
Snedecor, G. W. and Cochran, W. G. (1967). *Statistical Methods*, 6th ed., Iowa State Univ. Press, Ames, Iowa, 593 pp.
Southward, A. J. (1962). The distribution of some plankton animals in the English Channel and approaches. II. Surveys with the Gulf high-speed sampler, 1958–1960. *J. mar. biol. Ass. U.K.*, **42**, 275–375.
Steele, J. H. (ed.) (1970). *Marine Food Chains*, University of California Press, Berkeley and Los Angeles, 552 pp.
Stehr, E. (1967). Über Ölverschmutzung durch Tankerunfäller auf hoher See. *Gas Wasserfach.*, **108**, 53–4.
Thorson, G. (1966). Some factors influencing the recruitment and establishment of marine benthic communities. *Netherlands J. Sea Res.*, **3** (2), 267–93.
Thorson, G. (1971). *Life in the Sea*, World University Library, New York, 256 pp.
Ursin, E. (1960). A quantitative investigation of the echinoderm fauna of the central North Sea. *Medd. Danmarks Fisk. Havundersøg. N.S.Z.*, no. 24.

Appendix A

OFFSHORE BIOLOGICAL MONITORING: SAMPLING METHODS AND TABULATED DATA

SAMPLING METHODS

Benthos

The benthos was sampled using two types of standard grab, the Smith–McIntyre grab, and a recent modification (and improvement) of this grab—the Day grab. Sample site locations are shown in Fig. A.2. Working details of the former grab and information on the methods of benthic sampling can be found in Holme and McIntyre (1971). The Day grab proved more efficient and robust than the Smith–McIntyre, giving 95% sampling success when weighted with 100 lb of lead, and hence was used for all sampling after initial tests. It samples an area of 0·1 m^2 and to a depth of about 10 cm when properly filled. If the grab was not full, the sample was discarded and another taken. Eleven grabs were taken at each site, ten for analysis for benthic organisms and one to provide sediment sample for organic carbon analysis and, at some sites, hydrocarbon analysis. Where possible, the sites were sampled at anchor, but if not, the ship was held in position using the engines. Movements of the vessel at each site effectively 'randomised' the sampling.

When the samples had been brought on board they were sieved through a 1 mm nylon mesh and any large stones or dead shells with no animals attached were removed. The remainder was then placed in a storage jar and preserved in 4% formalin with a general dye (eosin) added. This served to stain the animals, making subsequent sorting much easier. Every sample was double coded, once by writing on the jar and a second time with a tag inside the jar.

Plankton

The plankton was sampled using standard Plymouth MBA nets, one for zooplankton (mesh size 170 μm) and one for phytoplankton (mesh size 61 μm). Hauls were made by slowly hand-lining the nets for 100 m. Samples were preserved in 4% formalin with added eosin.

To quantify the zooplankton samples roughly, it was assumed firstly that the planton was homogeneous, secondly that all water ahead of the net passed through, thirdly that everything larger than the mesh size was caught, and fourthly that no organisms swam out of the net or its path. These conditions are rarely fulfilled together. The fineness of mesh of the phytoplankton net had sufficient water resistance to prevent a great deal of water in the path of the net from going through, so these results were not quantified. Samples were taken at various times during night and day, when different organisms were encountered.

The samples provide material for comparison with similar hauls in future surveys and with the records produced by the Edinburgh Oceanographic Society (1973).

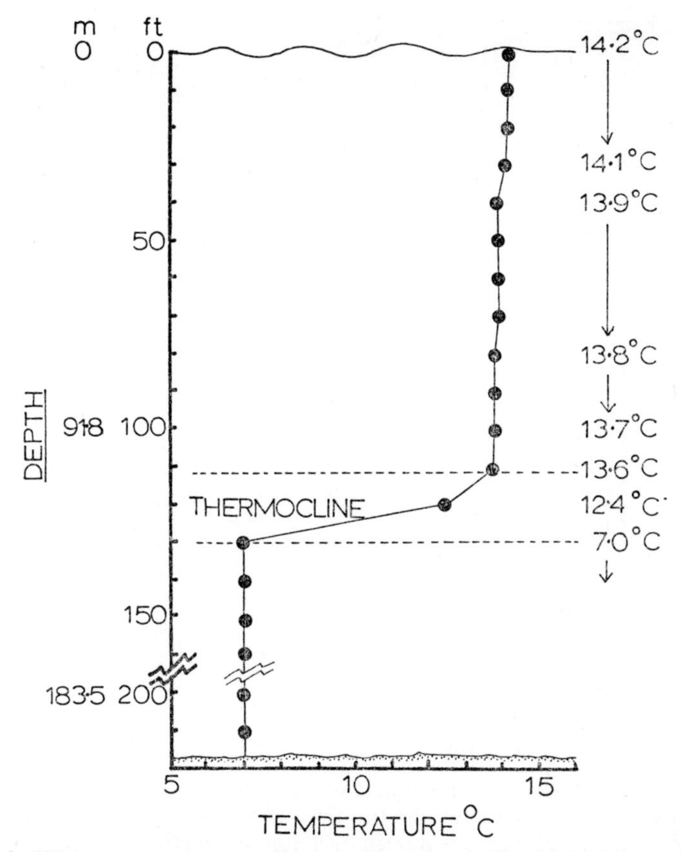

FIG. A.1. Water temperature profile, sample station 1, Ekofisk Field, August 1973.

Temperature Profile

This was taken using a portable NIO temperature recorder. The results are shown in Fig. A.1.

Water Sampling and Solvent Extraction for Hydrocarbon Analysis

Stainless-steel-lined 7 litre National Institute of Oceanography water sampling bottles were used to obtain water samples for analysis. Samples were taken at sites marked in Fig. A.2. Surface samples were taken from the surface 20 cm by closing the bottom stopper and the bottle and allowing it to fill from the top. The deep-water samples were taken by closing both stoppers with a messenger at the required depth. Solvent extraction of the samples was carried out by the following methods immediately after the sample was received on board the vessel:

(i) *Extraction for analysis by ultra-violet absorption spectrophotometry.* 2 litres of sea-water was shaken vigorously with two aliquots of 40 ml of cyclohexane for 2 min each in a separating funnel. This gave a total extract volume of 80 ml.

(ii) *Extraction for gas-liquid chromatograph analysis.* 5 litres of sea-water was extracted three times with 100 ml aliquots of *n*-heptane (IP specification

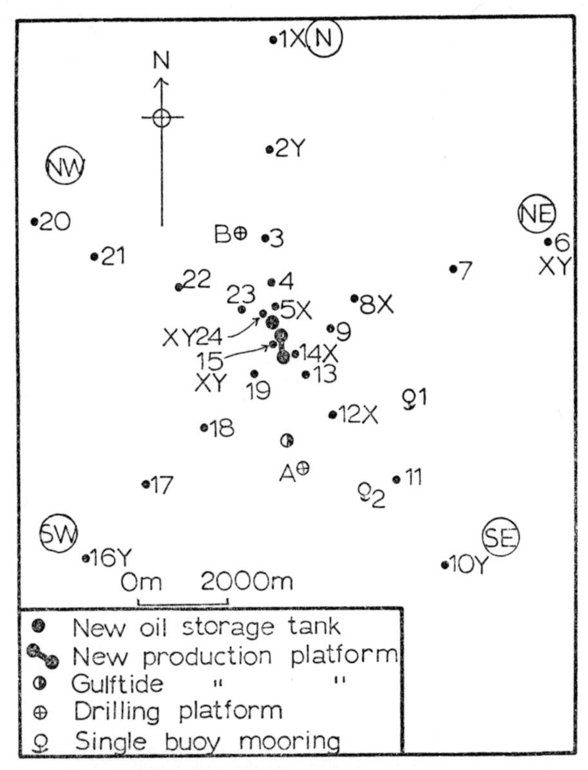

FIG. A.2. Map of the Ekofisk Field showing the position of sample stations (1–24) and the location of installations at the time of the survey (installations not to scale). The five transect lines are named by compass direction. Water samples and sediment samples for hydrocarbon analysis were taken at stations marked *x* and *y* respectively. The production complex is at 56°32′ N 03°13′ E.

for *n*-heptane) by shaking for 2 min each in a separating funnel. The final extract volume was therefore 300 ml.

The extracts were stored in a deep-freeze until subsequent analysis by the Department of Microbiology, University College, Cardiff. Sites from which the samples were taken are illustrated in Fig. A.2.

Sediment Sampling and Solvent Extraction for Hydrocarbon Analysis

Samples of benthic sediment were taken at selected sites (Fig. A.2) from the Day grab during the benthos sampling programme. These were solvent extracted on board the vessel as follows.

(i) *For GLC analysis.* 200 g of sediment was shaken for several minutes with 6 litres of distilled water, and the resultant mixture was extracted with two aliquots of 500 ml of *n*-heptane to give a final extract volume of 1 litre.

(ii) *For UV analysis.* 100 g of sediment was shaken for several minutes with 2 litres of distilled water and then extracted twice with 300 ml aliquots of cyclohexane to give an extract volume of 600 ml. Half of the extract was shaken with 5 g Florisil (30/60 mesh) to remove polar substances and was then filtered. This was done because naturally occurring organic compounds tend to be polar whilst the constituents of mineral oils tend to be non-polar. The halves were analysed separately and the results compared.

The extracts were stored in a deep-freeze until analysed by the Department of Microbiology, University College, Cardiff.

The considerable and well-known problems associated with obtaining uncontaminated and representative samples and the subsequent problems of laboratory analysis of sea water and sediments for hydrocarbon content make it clear that the techniques and results presented here are of a comparatively crude nature, and are capable of quantifying changes of the order of 1 or 2 ppm.

ANALYSIS OF SAMPLES IN THE LABORATORY AND RESULTS

Benthos

All the organisms in each sample were identified where possible and counted. Where specific identification of an organism was not possible, it was identified to group or genus and given a code number. Sorted samples were preserved in hexamine buffered 4% formalin. The following keys were used to identify organisms: Barret and Yonge (1958); Bell (1853); Chevreux and Page (1925); Clark (1960); Day (1967); Forbes and Hanley (1855); Fauvel (1923, 1927); Graham (1971); MacMillan (1968); Millar (1970); Tebble (1966); Sars (1895); and Southward (1972). The nomenclature of these authors is used throughout. Assistance with difficulties in identification of polychaetes was received from Dr M. Rolfe (MAFF) and Dr P. Gibbs (MBA). The results for each sample, totals per site, and organisms per m^2 are tabulated in Table IA. Numbers of selected species and main groupings of organisms are given in Table IB. Standard error and 95% confidence limits of the means at each site were calculated for total number of organisms per m^2, numbers of *Myriochele heeri* per m^2 and numbers of pleurogonid ascidians per m^2. These results are tabulated in Table IC.

Standard error and 95% confidence limits of the mean were calculated as follows:

$$\text{Standard error } (\sigma_n) = \left(\frac{\sum (x^2) - (\sum x)^2/n}{n(n-1)}\right)^{1/2}$$

and 0·05 limits $= \bar{x} \pm t\sigma_n$
where t = probability constant dependent on $n - 1$, n = number of samples, x = each sample result and $\bar{x}$ = sample size.

Diversity indices were calculated by three methods: (i) the Shannon–Wiener information function; (ii) Gleason's index; and (iii) Menhinick's index. The results are tabulated in Table ID.

The indices were calculated as follows:

(i) *Shannon–Wiener* (taken from Sanders, 1968).

$$H_{(s)} = -\sum p_i \log_2 p_i$$

where $H_{(s)}$ = diversity, p_i = probability of occurrence of each species calculated from n_i/N, where n_i = number of that species, N = total number of individuals in all species.

(ii) *Gleason index* (taken from Margalef, 1968):

$$D' = \frac{s - 1}{\log_e N}$$

where s = number of species, N = number of individuals and D' = diversity.

(iii) *Menhinick index* (from Odum, 1959):

$$D = \frac{s}{(N)^{1/2}}$$

where s = number of species, N = number of individuals and D = diversity.

Information on the location of sample sites for correlation with benthic data is given in Tables IE and IF.

Zooplankton

Each sample was thoroughly mixed and three subsamples of $\frac{1}{25}$ of the volume taken. All the organisms in the subsamples were counted by a petri dish/grid method. It was considered unnecessarily time-consuming to identify everything in the samples to species. The organisms were either counted under a group heading, or identified to species if common. (Useful guides to identifying main types of organisms were found in Wimpenny, 1966, and Newell and Newell, 1963.) The number in the whole sample was then calculated.

By calculating the area of the net mouth and knowing that the net was towed 100 m, the density of organisms per m^3 was calculated. The results are shown in Table II.

Phytoplankton

These organisms were identified to species and recorded as present or absent at each of the sample sites. The results are shown in Table III. The major work used for identification was Lebour (1930).

Organic Content of the Sand

The following method was used to determine the organic carbon content of the sand (Leeper, 1948). 10 ml of N-potassium dichromate and 20 ml of concentrated sulphuric acid were added to a 1·0 g (dry weight) sample of the sand substrate dried at 60°C. Oxidation of organic matter took place in the flask and excess dichromate was back-titrated with standard ferrous sulphate. With this method, 1 ml of normal oxidising agent is equivalent to 3·8 mg of organic carbon. The figures obtained have not been corrected to 'available' organic carbon (Morgans, 1956), but are in suitable form for comparison with future surveys. The results are shown in Table IV.

Sediment Particle Size Analysis

The sediment samples were dried at 100°C and large aggregations gently broken up. Samples were then shaken through a series of Endecott test sieves and the cumulative oversize weight percentages of various particle sizes calculated and plotted against sieve mesh size. The results and median particle size for each sample site are given in Tables V and VI. The mean and range of cumulative oversize percentage weight per mesh size for all stations are tabulated in Table VII.

Quartile duration ($Qd\phi = \frac{1}{2}[Q_3\phi - Q_1\phi]$: lower value = better sorting) and skewness ($SkQ\phi = \frac{1}{2}[Q_3\phi + Q_1\phi - 2Md\phi]$: positive = particles larger than the mean are better sorted; negative = particles smaller than the mean are better sorted) were calculated (Morgans, 1956) to give an idea of the efficiency of the samples.

Analysis of Extracts of Sediment and Water Samples for Hydrocarbon Content

(i) *GLC analysis.* Subsamples of the extracts were run through various temperature programmes in the GLC after a pristane marker had been added to allow quantification of results. Using this method, it was possible to quantify the occurrence of *n*-paraffins from carbon numbers 24 to 33. No other hydrocarbons could be adequately identified or quantified. The results of this analysis in ppm are shown in Table VIII. No measurable quantities of the *n*-paraffins were found in the sediment extracts.

(ii) *UV analysis.* Subsamples of the extracts were placed in a UV spectrophotometer and absorption of radiation of wavelength 260 nm was measured. Absorption of radiation at this wavelength is due to the presence of aromatic ring structures and other conjugated bonds in the sample. When the type or types of oil extracted from a sample are unknown, as is the case here, the choice of a suitable standard for quantification of the results presents a considerable problem and has to be arbitrary. The chosen standard in this case was Kuwait crude as a matter of convenience. This means, of course, that the results in terms of the concentration of oil are not absolute and can only be used comparatively from year to year using the same method.

The results for the water samples are shown in Table IX, and for the sediments in Table X.

Results

Due to considerable difficulties concerning detection and identification of hydrocarbons by any methods at present available, it is impossible to do a complete quantitative analysis of the amount of oil and mud and water samples containing unknown mixtures of many types. One must therefore be selective about what is measured and use the chosen indicator as a point of comparison from one survey to the next. In this context, the results do not need to be absolute.

In general, the amounts of oil measured by the above methods were very low at all sampling sites (Tables VIII–X).

TABLE IA

Lists of Benthic Species and Their Numbers in Each of the 10 Grabs at 24 Sample Sites in the Ekofisk Field, August 1973

Notes on Table IA:
Counting *Myriochele heeri:* During the sieving process, even when great care was taken, the tubes of *Myriochele heeri* were somewhat broken up. Counting was therefore difficult, but accuracy was achieved by counting only tubes with a worm inside.
Pleurogonid ascidians: The group Pleurogonid ascidians consists of two species of ascidian, *Eugyra arenosa* and *Polycarpa fibrosa.* These were difficult to differentiate during sorting, and were consequently grouped together.

Polychaete nomenclature:

Myriochele heeri Malmgren
Owenia fusiformis delle Chiaje
Stenhelais limicola Ehlers
Aphrodite aculeata (Linnaeus)
Lepidonotus squamatus (Linnaeus)
Harmothoe sp.
Phyllodoce lamelligera (Linnaeus)
Anaitides sp. (juv.)
Ophiodromus flexuosus (delle Chiaje)
Nephtys incisa Malmgren
Nephtys cirrosa Ehlers
Glycera alba (Müller)
Goniada maculata Oersted
Lumbrineris fragilis (Müller)
Lumbrineris impatiens (Claparède)
Scoloplos armiger (Müller)
Spiophanes bombyx (Claparède)
Spio filicornis (Müller)
Poecilochaetus serpens Allen
Chaetozone setosa Malmgren
Cirratulid sp. 33.39
Diplocirrus glaucus (Malmgren)
Caulleriella sp.
Ophelina aulogaster (Rathke)
Ophelia limacina (Rathke)
Euclymene lombricoides (Quatrefages)
Pectinaria (Lagis) *koreni* Malmgren
Sosane gracilis (Malmgren)
Ampharetid sp. (juv.)
Ampharetid sp. 46.610
Lanice conchilega (Pallas)
Thelepus setosus (Quatrefages)
Lysilla loveni Malmgren
Bispira volutacornis (Montagu)
Potamilla torelli (Malmgren)
Potamilla reniformis (Müller)
Jasmineira sp. cf. *caudata* Langerhans

The above nomenclature is based on:

Hartman, O. (1959). Catalogue of the polychaetous annelids of the world. *Allan Hancock Foundation Occasional Papers,* **23**, 1–628.
Hartman O. (1965). Ibid., Supplement 1960–1965 and Index. Ibid., 1–197.

STATION 1

	Sample number										*Total per station*
	1	*10*	*3*	*8*	*2*	*7*	*6*	*5*	*4*	*9*	
Crustaceans:											
Ampelisca gibba		1	2		4	1	1	2	3		14
Hippomedon denticulatus											
Tryphosites longipes	2	3			1				1		7

TABLE IA—*contd.*

STATION 1

	Sample number										*Total per station*
	1	*10*	*3*	*8*	*2*	*7*	*6*	*5*	*4*	*9*	
Cumaceans											
Macropippus sp.											
Pagurus sp.											
Corophium bonelli											
Podoceros odontyx											
Corystes cassivelaunus											
Molluscs:											
Cyprina islandica		3		3	1	3	2	3	5	3	23
Cultellus pellucidus	1	1								1	3
Mysella bidentata	1	3				1			2	3	10
Cardium echinatum								1			1
Montacuta ferruginosa											
Mya arenaria					1			1			2
Mya truncata									1		1
Nucula hanleyi											
Astarte borealis											
Gari fervensis	1	1									2
Lucinoma borealis										1	1
Venus casina											
Venus striatula											
Lutraria lutraria											
Spisula elliptica											
Musculus niger											
Thyasira flexuosa											
Siphonodentalium lofotense	1	1		1	1	1	2	2		2	11
Dentalium entalis											
Philine sp.			2								2
Acteon tornatilis					1	2					3
Colus gracilis											
Clathrus clathrus											
Natica alderi											
Natica catena											
Nudibranch 41.41											
Chaetoderma nitidulum		1								1	2
Echinoderms:											
Echinocardium flavescens				1							1
Echinocardium cordatum						1					1
Spatangus purpureus											
Amphiura filiformis	3	9	2	2		6	2	1	11	4	40
Amphiura chiajei			1			2	1				4
Ophiuroid juveniles	1		6				3	2			12
Ophiura texturata											
Asteropecten irregularis		1									1

TABLE IA—*contd.*

STATION 1

	Sample number										*Total per station*
	1	*10*	*3*	*8*	*2*	*7*	*6*	*5*	*4*	*9*	
Ascidians:											
Pleurogonid sp.											
Ciona intestinalis											
Ascidian sp. 55.98											
Other groups:											
Nemertine spp.											
Lineus ruber											
Pennatulid sp.			2					1	1	1	5
Flatworm sp.									1		1
Sponge 52.95											
Sea anemone sp. 16.27											
Sea anemone sp. 17.21											
Cerianthus sp.	1	1		1	1	4		1	2	1	12
Flustra foliacea											
Flustrella hispida											
Phascolion strombi							1				1
Crystallogobius linearis											
Polychaetes:											
Myriochele heeri	282	500	350	300	306	424	148	250	453	259	3272
Owenia fusiformis	1	1	3	4	16	5		4	5	4	43
Stenhelais limicola			3	1			1	1	2	2	10
Aphrodite aculeata											
Lepidonotus squamatus											
Harmothoe sp.											
Phyllodoce lamelligera	1	2	1		1	2			1		8
Anaitides sp. (juv.)											
Ophiodromus flexuosus											
Nephtys incisa	3	1	1	2	3	1	1	1	2	2	17
Nephtys cirrosa											
Glycera alba	1	3	3	1	2	1	1	2	1	2	17
Goniada maculata						7	3				10
Lumbrineris fragilis				1		1	1				3
Lumbrineris impatiens											
Scoloplos armiger		1	2	2	1	1	1	3	2	3	16
Spiophanes bombyx	2	1	1								4
Spio filicornis											
Poecilochaetus serpens											
Chaetozone setosa				1	1						2
Cirratulid sp. 33.39											
Diplocirrus glaucus									1		1
Caulleriella sp.											
Ophelina aulogaster											
Ophelia limacina	1		1		1	1		1	1		6

TABLE IA—*contd.*

STATION 1

	Sample number										Total per station
	1	*10*	*3*	*8*	*2*	*7*	*9*	*5*	*4*	*6*	
Euclymene lumbricoides	1		1	1					1		4
Pectinaria koreni	2		3				1		3	2	11
Sosane gracilis											
Ampharetid sp. (juv.)					2				3		5
Ampharetid sp. 46.610											
Lanice conchilega											
Thelepus setosus											
Lysilla loveni											
Bispira volutacornis		1									1
Potamilla torelli	1	1						1			3
Potamilla reniformis											
Jasmineira sp. cf. *caudata*											
Number spp. per sample	18	20	17	14	16	17	16	17	21	16	42
Number individuals/sample	306	536	384	321	343	464	169	277	502	291	3593

STATION 2

	Sample number										Total per station
	1	*5*	*2*	*6*	*9*	*3*	*10*	*7*	*4*	*8*	
Crustaceans:											
Ampelisca gibba	1	1	4		1		2	2			11
Hippomedon denticulatus	1	1	1	2	1						6
Tryphosites longipes		1		1		3	1		2		8
Cumaceans					2						2
Macropippus sp.											
Pagurus sp.		1									1
Corophium bonelli											
Podoceros odontyx											
Corystes cassivelaunus											
Molluscs:											
Cyprina islandica		4				3				1	8
Cultellus pellucidus				1	1						2
Mysella bidentata	3	1		6							10
Cardium echinatum											
Montacuta ferruginosa											
Mya arenaria	1			3							4
Mya truncata											
Nucula hanleyi		1									1

TABLE IA—*contd.*

STATION 2

	Sample number										*Total*
	1	*5*	*2*	*6*	*9*	*3*	*10*	*7*	*4*	*8*	*per station*
Astarte borealis											
Gari fervensis											
Lucinoma borealis											
Venus casina											
Venus striatula											
Lutraria lutraria											
Spisula elliptica											
Musculus niger											
Thyasira flexuosa											
Siphonodentalium lofotense			1		1						2
Dentalium entalis				1							1
Philine sp.					1					1	2
Acteon tornatilis			1		1		2				4
Colus gracilis											
Clathrus clathrus											
Natica alderi											
Natica catena											
Nudibranch 41.41											
Chaetoderma nitidulum	1		1						1		3
Echinoderms:											
Echinocardium flavescens										1	1
Echinocardium cordatum	1	1	1	1		1		2			7
Spatangus purpureus											
Amphiura filiformis	10	10	6	3	3		2	2	6		42
Amphiura chiajei											
Ophiuroid juveniles	1				2				1		4
Ophiura texturata											
Asteropecten irregularis					1					1	2
Ascidians:											
Pleurogonid sp.		30	25	60	33	7	8	16			179
Ciona intestinalis											
Ascidian sp. 55.98											
Other groups:											
Nemertine spp.											
Lineus ruber											
Pennatulid sp.					1					1	2
Flatworm sp.				1	1						2
Sponge 52.95											
Sea anemone sp. 16.27								1			1
Sea anemone sp. 17.21	1										1
Cerianthus sp.			2				1	1	1		5
Flustra foliacea											
Flustrella hispida											

TABLE IA—*contd.*

STATION 2

	Sample number										*Total per station*
	1	*5*	*2*	*6*	*9*	*3*	*10*	*7*	*4*	*8*	
Phascolion strombi											
Crystallogobius linearis	1										1
Polychaetes:											
Myriochele heeri	349	242	304	209	195	170	188	156	113	168	2094
Owenia fusiformis	7	6	5	2	6	4	4	10	5	4	53
Stenhelais limicola		1	1	4	2		2			2	12
Aphrodite aculeata											
Lepidonotus squamatus		1									1
Harmothoe sp.											
Phyllodoce lamelligera		1			2						3
Anaitides sp. (juv.)					1						1
Ophiodromus flexuosus											
Nephtys incisa	1	2	2	1	1				3	1	11
Nephtys cirrosa	1		1		1	2		2			7
Glycera alba	1	1	3	2		2	5	4	2	1	21
Goniada maculata	4	2		5	2					2	15
Lumbrineris fragilis								1	3	4	8
Lumbrineris impatiens											
Scoloplos armiger	3	4	3	2	1			1	2	2	18
Spiophanes bombyx		1									1
Spio filicornis											
Poecilochaetus serpens											
Chaetozone setosa											
Cirratulid sp. 33.39				1							1
Diplocirrus glaucus											
Caulleriella sp.											
Ophelina aulogaster											
Ophelia limacina							1			1	2
Euclymene lumbricoides		1			1		1				3
Pectinaria koreni	1	1	3		3		3	1	2	1	15
Sosane gracilis											
Ampharetid sp. (juv.)	2	1		4	3				1		11
Ampharetid sp. 46.610											
Lanice conchilega											
Thelepus setosus											
Lysilla loveni										1	1
Bispira volutacornis											
Potamilla torelli				1	1						2
Potamilla reniformis											
Jasmineira sp. cf. *caudata*											
Number spp. per sample	19	23	17	20	26	8	13	13	13	16	47
Number individuals/sample	390	315	364	310	268	192	220	200	142	192	2593

TABLE IA—*contd.*

STATION 3

	Sample number										Total per station
	1	*2*	*3*	*5*	*4*	*8*	*6*	*7*	*10*	*9*	
Crustaceans:											
Ampelisca gibba	1			3	6		3				13
Hippomedon denticulatus	3		1				2	1		3	10
Tryphosites longipes		2	2	2	1	1	2			2	12
Cumaceans			1								1
Macropippus sp.		1								1	2
Pagurus sp.										1	1
Corophium bonelli											
Podoceros odontyx											
Corystes cassivelaunus											
Molluscs:											
Cyprina islandica	3		2	1	4			4		5	19
Cultellus pellucidus								2			2
Mysella bidentata	1	1	1								3
Cardium echinatum											
Montacuta ferruginosa											
Mya arenaria										1	1
Mya truncata											
Nucula hanleyi	1	1									2
Astarte borealis	1										1
Gari fervensis											
Lucinoma borealis											
Venus casina				1							1
Venus striatula											
Lutraria lutraria											
Spisula elliptica											
Musculus niger											
Thyasira flexuosa											
Siphonodentalium lofotense								1			1
Dentalium entalis			2								2
Philine sp.				2							2
Acteon tornatilis	1										1
Colus gracilis										1	1
Clathrus clathrus											
Natica alderi											
Natica catena											
Nudibranch 41.41											
Chaetoderma nitidulum	1			1	2			1		1	6
Echinoderms:											
Echinocardium flavescens								1			1
Echinocardium cordatum	1			2						1	4
Spatangus purpureus											
Amphiura filiformis	1	3		5	4	1	1	1		7	23
Amphiura chiajei			1		1						2

TABLE IA—*contd.*

STATION 3

	Sample number										Total per station
	1	*2*	*3*	*5*	*4*	*8*	*6*	*7*	*10*	*9*	
Ophiuroid juveniles			1		1	5	2				9
Ophiura texturata				1							1
Asteropecten irregularis		1					1				2
Ascidians:											
Pleurogonid sp.	4	26	45	6	51	6	31	4		3	176
Ciona intestinalis											
Ascidian sp. 55.98											
Other groups:											
Nemertine spp.											
Lineus ruber											
Pennatulid sp.										1	1
Flatworm sp.											
Sponge 52.95											
Sea anemone sp. 16.27											
Sea anemone sp. 17.21											
Cerianthus sp.	1			1						1	3
Flustra foliacea											
Flustrella hispida											
Phascolion strombi											
Crystallogobius linearis											
Polychaetes:											
Myriochele heeri	371	293	280	206	198	148	72	22		139	1928
Owenia fusiformis	5	1	1	4		2	3	2		11	29
Stenhelais limicola	2	1	2	1		3	2	6			17
Aphrodite aculeata											
Lepidonotus squamatus			3		2						5
Harmothoe sp.											
Phyllodoce lamelligera						2		3		2	7
Anaitides sp. (juv.)											
Ophiodromus flexuosus											
Nephtys incisa	1	2				2	1	3		3	12
Nephtys cirrosa					1						1
Glycera alba			2		1						3
Goniada maculata	3	5	3	1	1	3	2	5		2	25
Lumbrineris fragilis						1					1
Lumbrineris impatiens			1								1
Scoloplos armiger	3	2			1	1	1			1	9
Spiophanes bombyx		2		1		1	1			1	6
Spio filicornis											
Poecilochaetus serpens											
Chaetozone setosa	2							1			3
Cirratulid sp. 33.39										1	1
Diplocirrus glaucus		1	2	1	1			1			6
Caulleriella sp.											

TABLE 1A—*contd.*

STATION 3

	Sample number										*Total per station*
	1	*2*	*3*	*5*	*4*	*8*	*6*	*7*	*10*	*9*	
Ophelina aulogaster											
Ophelia limacina											
Euclymene lumbricoides											
Pectinaria koreni		3	2	1			2	3		1	12
Sosane gracilis											
Ampharetid sp. (juv.)		1									1
Ampharetid sp. 46.610								1		1	2
Lanice conchilega											
Thelepus setosus											
Lysilla loveni											
Bispira volutacornis											
Potamilla torelli					1						1
Potamilla reniformis											
Jasmineira sp. cf. *caudata*											
Number spp. per sample	19	17	18	18	16	13	15	18		23	49
Number individuals/sample	406	346	352	240	276	176	126	261		190	2373

STATION 4

	Sample number										*Total per station*
	2	*5*	*8*	*7*	*1*	*6*	*3*	*4*	*10*	*9*	
Crustaceans:											
Ampelisca gibba			1				1		1		3
Hippomedon denticulatus	1	1	1	2	1	2	1				9
Tryphosites longipes	1			1	1				1		4
Cumaceans					2						2
Macropippus sp.											
Pagurus sp.											
Corophium bonelli											
Podoceros odontyx											
Corystes cassivelaunus											
Molluscs:											
Cyprina islandica	1		2	1	1	2	3	3		2	14
Cultellus pellucidus	1		1		1		1		1	1	6
Mysella bidentata	1	4						2		2	9
Cardium echinatum			1								1
Montacuta ferruginosa		1	1								2
Mya arenaria											
Mya truncata											

TABLE IA—*contd.*

STATION 4

	Sample number										*Total per station*
	2	*5*	*8*	*7*	*1*	*6*	*3*	*4*	*10*	*9*	
Nucula hanleyi											
Astarte borealis											
Gari fervensis			1		1					3	5
Lucinoma borealis											
Venus casina		1									1
Venus striatula										1	1
Lutraria lutraria											
Spisula elliptica											
Musculus niger											
Thyasira flexuosa											
Siphonodentalium lofotense						1			1	2	4
Dentalium entalis		1							1		2
Philine sp.	1			2						1	4
Acteon tornatilis				1							1
Colus gracilis											
Clathrus clathrus								1			1
Natica alderi											
Natica catena											
Nudibranch 41.41					1						1
Chaetoderma nitidulum	1		1		1	1					4
Echinoderms:											
Echinocardium flavescens											
Echinocardium cordatum		1	1	1			1	1	1		6
Spatangus purpureus											
Amphiura filiformis	1	4		11		6	6	5	1	4	38
Amphiura chiajei											
Ophiuroid juveniles	1	1			2	3	1	2		1	11
Ophiura texturata											
Asteropecten irregularis				1	1		1			1	4
Ascidians:											
Pleurogonid sp.	29	18	37	38	35	82	42	13	22	81	397
Ciona intestinalis											
Ascidian sp. 55.98											
Other groups:											
Nemertine spp.			2	2			1				5
Lineus ruber											
Pennatulid sp.					1						1
Flatworm sp.		2									2
Sponge 52.95											
Sea anemone sp. 16.27											
Sea anemone sp. 17.21											
Cerianthus sp.	2		1	4						3	10
Flustra foliacea											

TABLE IA—*contd.*

STATION 4

	Sample number										*Total per station*
	2	*5*	*8*	*7*	*1*	*6*	*3*	*4*	*10*	*9*	
Flustrella hispida											
Phascolion strombi											
Crystallogobius linearis											
Polychaetes:											
Myriochele heeri	294	34	165	486	340	51	416	378	249	405	2818
Owenia fusiformis	4	13	14	8	5	5	3	7	1	7	67
Stenhelais limicola	2	1	3	1		2		2	2	1	14
Aphrodite aculeata											
Lepidonotus squamatus								1			1
Harmothoe sp.											
Phyllodoce lamelligera	1		1	1	2	1		1			7
Anaitides sp. (juv.)											
Ophiodromus flexuosus											
Nephtys incisa	6			1	2	2	3	3		5	22
Nephtys cirrosa						1					1
Glycera alba	1	2					2	1			6
Goniada maculata	2	1	1	2	6	4	1	6		4	27
Lumbrineris fragilis											
Lumbrineris impatiens	2	1									3
Scoloplos armiger	4	2	2	3	2	2	3	1		1	20
Spiophanes bombyx				1	1			2		1	5
Spio filicornis											
Poecilochaetus serpens											
Chaetozone setosa	3	1	2					1		1	8
Cirratulid sp. 33.39					1						1
Diplocirrus glaucus											
Caulleriella sp.											
Ophelina aulogaster									1		1
Ophelia limacina				1							1
Euclymene lumbricoides	1		1					1			3
Pectinaria koreni	3	2		5	1		1	3		2	17
Sosane gracilis											
Ampharetid sp. (juv.)			1		4			2		1	8
Ampharetid sp. 46.610	1	1					1			3	6
Lanice conchilega											
Thelepus setosus											
Lysilla loveni											
Bispira volutacornis											
Potamilla torelli		1				1			1		3
Potamilla reniformis											
Jasmineira sp. cf. *caudata*											
Number spp. per sample	24	21	21	21	22	16	18	21	13	23	49
Number individuals/sample	364	93	240	573	412	166	488	435	283	533	3587

TABLE IA—*contd.*

STATION 5

	Sample number										*Total per station*
	7	*6*	*9*	*1*	*5*	*10*	*8*	*3*	*2*	*4*	
Crustaceans:											
Ampelisca gibba				1		1			1	1	4
Hippomedon denticulatus			1	1			1	1		2	6
Tryphosites longipes	1	2		1					2	1	7
Cumaceans											
Macropippus sp.						1					1
Pagurus sp.											
Corophium bonelli											
Podoceros odontyx											
Corystes cassivelaunus											
Molluscs:											
Cyprina islandica				2					1		3
Cultellus pellucidus			1		1	1	1		1		5
Mysella bidentata											
Cardium echinatum	1			1		1					3
Montacuta ferruginosa											
Mya arenaria											
Mya truncata											
Nucula hanleyi											
Astarte borealis											
Gari fervensis											
Lucinoma borealis											
Venus casina					1						1
Venus striatula				1							1
Lutraria lutraria											
Spisula elliptica											
Musculus niger											
Thyasira flexuosa											
Siphonodentalium lofotense					1						1
Dentalium entalis	1			1			1		1	1	5
Philine sp.		1	2						3		6
Acteon tornatilis										1	1
Colus gracilis											
Clathrus clathrus											
Natica alderi											
Natica catena											
Nudibranch 41.41											
Chaetoderma nitidulum							1				1
Echinoderms:											
Echinocardium flavescens						1					1
Echinocardium cordatum			1								1
Spatangus purpureus											
Amphiura filiformis	8		1	1	1	3	5	3	1	1	24

TABLE IA—*contd.*

STATION 5

	Sample number										*Total*
	7	*6*	*9*	*1*	*5*	*10*	*8*	*2*	*3*	*4*	*per station*
Amphiura chiajei											
Ophiuroid juveniles	2	1	1	1			1	1		1	8
Ophiura texturata											
Asteropecten irregularis									1		1
Ascidians:											
Pleurogonid sp.	10	8		17	10	1	20	1		7	74
Ciona intestinalis											
Ascidian sp. 55.98											
Other groups:											
Nemertine spp.						1					1
Lineus ruber											
Pennatulid sp.				1							1
Flatworm sp.											
Sponge 52.95											
Sea anemone sp. 16.27											
Sea anemone sp. 17.21											
Cerianthus sp.				2	1			2			5
Flustra foliacea											
Flustrella hispida											
Phascolion strombi			1								1
Crystallogobius linearis											
Polychaetes:											
Myriochele heeri	208	122	169	299	234	65	268	116	43	278	1802
Owenia fusiformis	3	1			1		1	1		1	8
Stenhelais limicola	1	2	1	2				3	3	1	13
Aphrodite aculeata											
Lepidonotus squamatus											
Harmothoe sp.											
Phyllodoce lamelligera							1	1			2
Anaitides sp. (juv.)											
Ophiodromus flexuosus											
Nephtys incisa				2	1			1	1	1	6
Nephtys cirrosa							1				1
Glycera alba			1	1	2					1	5
Goniada maculata				1	1		1		1	2	6
Lumbrineris fragilis			1		1		1				3
Lumbrineris impatiens		1				1					2
Scoloplos armiger	1	2	1		1	5	1			1	12
Spiophanes bombyx											
Spio filicornis											
Poecilochaetus serpens											
Chaetozone setosa				2			1				3
Cirratulid sp. 33.39				1							1

TABLE IA—*contd.*

STATION 5

	Sample number										Total per station
	7	*6*	*9*	*1*	*5*	*10*	*8*	*2*	*3*	*4*	
Diplocirrus glaucus						2					2
Caulleriella sp.											
Ophelina aulogaster											
Ophelia limacina											
Euclymene lumbricoides											
Pectinaria koreni	2		1	11	1	3			1		19
Sosane gracilis											
Ampharetid sp. (juv.)					1	1					2
Ampharetid sp. 46.610			1								1
Lanice conchilega											
Thelepus setosus											
Lysilla loveni											
Bispira volutacornis											
Potamilla torelli											
Potamilla reniformis											
Jasmineira sp. cf. *caudata*											
Number spp. per sample	11	9	14	20	15	14	15	10	13	15	40
Number individuals/sample	238	140	183	349	258	87	305	130	60	300	2050

STATION 6

	Sample number										Total per station
	1	*5*	*9*	*6*	*4*	*2*	*10*	*7*	*3*	*8*	
Crustaceans:											
Ampelisca gibba	4	1	2	4	2	4	2	4	5	1	29
Hippomedon denticulatus	3					1			2		6
Tryphosites longipes						2				1	3
Cumaceans	1										1
Macropippus sp.											
Pagurus sp.											
Corophium bonelli											
Podoceros odontyx											
Corystes cassivelaunus											
Molluscs:											
Cyprina islandica		1		1							2
Cultellus pellucidus	2		3	1	1	2		2			11
Mysella bidentata					1						1
Cardium echinatum	1		1			1					3
Montacuta ferruginosa											
Mya arenaria											

TABLE IA—*contd.*

STATION 6

	Sample number										*Total per station*
	1	*5*	*9*	*6*	*4*	*2*	*10*	*7*	*3*	*8*	
Mya truncata											
Nucula hanleyi											
Astarte borealis											
Gari fervensis											
Lucinoma borealis											
Venus casina						1					1
Venus striatula											
Lutraria lutraria											
Spisula elliptica								1			1
Musculus niger											
Thyasira flexuosa											
Siphonodentalium lofotense	1			2		1					4
Dentalium entalis	1		1								2
Philine sp.			1							1	2
Acteon tornatilis											
Colus gracilis						1					1
Clathrus clathrus											
Natica alderi											
Natica catena						1					1
Nudibranch 41.41											
Chaetoderma nitidulum		2			1						3
Echinoderms:											
Echinocardium flavescens											
Echinocardium cordatum		1				1			1		3
Spatangus purpureus											
Amphiura filiformis	2	5	3	3	2		6	3	5	2	31
Amphiura chiajei	1										1
Ophiuroid juveniles	2		1						1	2	6
Ophiura texturata											
Asteropecten irregularis						1		1	1		3
Ascidians:											
Pleurogonid sp.	41	13	8	6	8	34	3	37	8	20	178
Ciona intestinalis											
Ascidian sp. 55.98											
Other groups:											
Nemertine spp.				2			1				3
Lineus ruber											
Pennatulid sp.											
Flatworm sp.											
Sponge 52.95											
Sea anemone sp. 16.27											
Sea anemone sp. 17.21											
Cerianthus sp.	2	4		1		1			4	3	15
Flustra foliacea											

TABLE IA—*contd.*

STATION 6

	Sample number										*Total per station*
	1	*5*	*9*	*6*	*4*	*2*	*10*	*7*	*3*	*8*	
Flustrella hispida											
Phascolion strombi	1										1
Crystallogobius linearis											
Polychaetes:											
Myriochele heeri	443	365	329	405	240	242	502	329	88	209	3152
Owenia fusiformis		1	1			1	2	1		1	7
Stenhelais limicola	2	1	1	2	2	5	3	1	2	1	20
Aphrodite aculeata						1	1				2
Lepidonotus squamatus	1										1
Harmothoe sp.											
Phyllodoce lamelligera					1					1	2
Anaitides sp. (juv.)											
Ophiodromus flexuosus											
Nephtys incisa	4	1	3	1	2	2		1		2	16
Nephtys cirrosa											
Glycera alba							2			1	3
Goniada maculata		3	2	2	1	1	1			1	11
Lumbrineris fragilis									1		1
Lumbrineris impatiens										2	2
Scoloplos armiger	1	2	2	1			1	3	3	2	15
Spiophanes bombyx							1				1
Spio filicornis											
Poecilochaetus serpens											
Chaetozone setosa				1				1			2
Cirratulid sp. 33.39											
Diplocirrus glaucus									1		1
Caulleriella sp.											
Ophelina aulogaster											
Ophelia limacina							1				1
Euclymene lumbricoides				1	1	1		1			4
Pectinaria koreni	1	1	2	1			2		1	2	10
Sosane gracilis											
Ampharetid sp. (juv.)						1		1	1	1	4
Ampharetid sp. 46.610		1		1		1					3
Lanice conchilega											
Thelepus setosus											
Lysilla loveni											
Bispira volutacornis				1							1
Potamilla torelli				1	1						2
Potamilla reniformis											
Jasmineira sp. cf. *caudata*											
Number spp. per sample	19	15	15	19	13	22	14	14	15	18	47
Number individuals/sample	514	402	360	437	263	306	528	386	124	253	3573

TABLE IA—*contd.*

STATION 7

	Sample number										*Total per station*
	7	*6*	*9*	*10*	*4*	*1*	*3*	*5*	*2*	*8*	
Crustaceans:											
Ampelisca gibba		1					1	2	1	2	7
Hippomedon denticulatus	2			1	1	1				2	7
Tryphosites longipes										1	1
Cumaceans					1						1
Macropippus sp.											
Pagurus sp.											
Corophium bonelli											
Podoceros odontyx											
Corystes cassivelaunus											
Molluscs:											
Cyprina islandica	1	1		1			1				4
Cultellus pellucidus	2	1	1								4
Mysella bidentata					1				2		3
Cardium echinatum			1			1					2
Montacuta ferruginosa											
Mya arenaria											
Mya truncata											
Nucula hanleyi	1										1
Astarte borealis											
Gari fervensis											
Lucinoma borealis											
Venus casina											
Venus striatula											
Lutraria lutraria							1				1
Spisula elliptica											
Musculus niger											
Thyasira flexuosa											
Siphonodentalium lofotense			1		1					1	3
Dentalium entalis											
Philine sp.			1	1				1			3
Acteon tornatilis				1		1					2
Colus gracilis											
Clathrus clathrus											
Natica alderi											
Natica catena											
Nudibranch 41.41											
Chaetoderma nitidulum		1			1			1			3
Echinoderms:											
Echinocardium flavescens											
Echinocardium cordatum				1	1		2				4
Spatangus purpureus											
Amphiura filiformis	6	11	2	7	11	12	5	7	9	11	81

TABLE IA—*contd.*

STATION 7

	Sample number										*Total per station*
	7	*6*	*9*	*10*	*4*	*1*	*3*	*5*	*2*	*8*	
Amphiura chiajei											
Ophiuroid juveniles		1	1	2		1	3		1		9
Ophiura texturata											
Asteropecten irregularis					1					1	2
Ascidians:											
Pleurogonid sp.	19	5	12	12	13	7	16	4	22	7	117
Ciona intestinalis											
Ascidian sp. 55.98											
Other groups:											
Nemertine spp.											
Lineus ruber											
Pennatulid sp.		1		1							2
Flatworm sp.											
Sponge 52.95											
Sea anemone sp. 16.27											
Sea anemone sp. 17.21											
Cerianthus sp.		2	1				1	1		1	6
Flustra foliacea		1									1
Flustrella hispida											
Phascolion strombi											
Crystallogobius linearis											
Polychaetes:											
Myriochele heeri	221	281	242	230	247	331	325	315	392	431	3015
Owenia fusiformis	2	2	2		1	3	1	2	1	5	19
Stenhelais limicola	2		1	1	1	1	2	1		1	10
Aphrodite aculeata				1							1
Lepidonotus squamatus											
Harmothoe sp.											
Phyllodoce lamelligera											
Anaitides sp. (juv.)											
Ophiodromus flexuosus											
Nephtys incisa	1	1	2	3	1	1	1	1	4	2	17
Nephtys cirrosa										1	1
Glycera alba						1			1		2
Goniada maculata	1	2	2	3	3	1	1	2	4	1	20
Lumbrineris fragilis											
Lumbrineris impatiens										1	1
Scoloplos armiger	1	3		1	1	8	3	2	1	2	22
Spiophanes bombyx											
Spio filicornis											
Poecilochaetus serpens											
Chaetozone setosa											
Cirratulid sp. 33.39		1		1							2
Diplocirrus glaucus											

TABLE IA—*contd.*

STATION 7

	Sample number										*Total per station*
	7	*6*	*9*	*10*	*4*	*1*	*3*	*5*	*2*	*8*	
Caulleriella sp.										1	1
Ophelina aulogaster											
Ophelia limacina											
Euclymene lumbricoides				1		1					2
Pectinaria koreni	1	1	3	1	1	7		1	1	1	17
Sosane gracilis											
Ampharetid sp. (juv.)	1				1		1		1		4
Ampharetid sp. 46.610											
Lanice conchilega											
Thelepus setosus											
Lysilla loveni											
Bispira volutacornis											
Potamilla torelli				1					1		2
Potamilla reniformis											
Jasmineira sp. cf. *caudata*											
Number spp. per sample	14	17	14	19	17	15	15	13	14	18	38
Number individuals/sample	261	316	272	270	287	377	364	334	437	472	3390

STATION 8

	Sample number										*Total per station*
	1	*3*	*2*	*4*	*10*	*7*	*5*	*6*	*8*	*9*	
Crustaceans:											
Ampelisca gibba		2			1						3
Hippomedon denticulatus		2	1		5						8
Tryphosites longipes	1			1							2
Cumaceans											
Macropippus sp.											
Pagurus sp.											
Corophium bonelli											
Podoceros odontyx											
Corystes cassivelaunus											
Molluscs:											
Cyprina islandica											
Cultellus pellucidus		1		1				2			4
Mysella bidentata				2							2
Cardium echinatum		1	1			1					3
Montacuta ferruginosa											
Mya arenaria											

TABLE IA—*contd.*

STATION 8

	Sample number										Total per station
	1	*3*	*2*	*4*	*10*	*7*	*5*	*6*	*8*	*9*	
Mya truncata											
Nucula hanleyi											
Astarte borealis											
Gari fervensis											
Lucinoma borealis											
Venus casina											
Venus striatula											
Lutraria lutraria											
Spisula elliptica											
Musculus niger											
Thyasira flexuosa											
Siphonodentalium lofotense	1					1					2
Dentalium entalis		1						2		1	4
Philine sp.		1	1								2
Acteon tornatilis									1	2	3
Colus gracilis											
Clathrus clathrus											
Natica alderi											
Natica catena				1							1
Nudibranch 41.41											
Chaetoderma nitidulum				1		1		1	2		5
Echinoderms:											
Echinocardium flavescens											
Echinocardium cordatum	2	3	1				1			1	8
Spatangus purpureus											
Amphiura filiformis	5	7	8	16	3	7	16	7	3	1	73
Amphiura chiajei		1									1
Ophiuroid juveniles	1	1	1	2			3				8
Ophiura texturata											
Asteropecten irregularis											
Ascidians:											
Pleurogonid sp.	2	11	5	41	3		14	32			108
Ciona intestinalis											
Ascidian sp. 55.98											
Other groups:											
Nemertine spp.		1				1					2
Lineus ruber											
Pennatulid sp.					1						1
Flatworm sp.											
Sponge 52.95											
Sea anemone sp. 16.27											
Sea anemone sp. 17.21											
Cerianthus sp.	1	3	1	1	1		1	1			9
Flustra foliacea											

TABLE IA—*contd.*

STATION 8

	Sample number										*Total per station*
	1	*3*	*2*	*4*	*10*	*7*	*5*	*6*	*8*	*9*	
Flustrella hispida											
Phascolion strombi											
Crystallogobius linearis											
Polychaetes:											
Myriochele heeri	116	418	263	281	195	264	393	249	227	141	2547
Owenia fusiformis	2		2	1	1		2	1	3	5	17
Stenhelais limicola	1	4	1	2	2	1	2	2		1	16
Aphrodite aculeata											
Lepidonotus squamatus											
Harmothoe sp.											
Phyllodoce lamelligera	1		1	1		1	1				5
Anaitides sp. (juv.)											
Ophiodromus flexuosus											
Nephtys incisa				3	1	2	3		1	1	11
Nephtys cirrosa		1									1
Glycera alba	1				1				2		4
Goniada maculata	1	4	1	3	2	2	4			1	18
Lumbrineris fragilis									1	1	2
Lumbrineris impatiens											
Scoloplos armiger	1	2	4	5	5	1	3	4	3		28
Spiophanes bombyx					1	1				1	3
Spio filicornis											
Poecilochaetus serpens											
Chaetozone setosa					2				1		3
Cirratulid sp. 33.39					2						2
Diplocirrus glaucus						1					1
Caulleriella sp.											
Ophelina aulogaster											
Ophelia limacina	1										1
Euclymene lumbricoides		1	1					1	1		4
Pectinaria koreni	2	5	4	4			1			6	22
Sosane gracilis											
Ampharetid sp. (juv.)		2			1	5	2		3	1	14
Ampharetid sp. 46.610	1				1						2
Lanice conchilega											
Thelepus setosus											
Lysilla loveni											
Bispira volutacornis				1				1			2
Potamilla torelli		1				1		1			3
Potamilla reniformis											
Jasmineira sp. cf. *caudata*											
Number spp. per sample	18	22	16	18	18	15	14	13	12	13	41
Number individuals/sample	140	473	296	367	228	290	446	304	248	163	2955

TABLE IA—*contd.*

STATION 9

	Sample number										*Total per station*
	1	*2*	*10*	*3*	*6*	*7*	*5*	*4*	*9*	*8*	
Crustaceans:											
Ampelisca gibba				1				1			2
Hippomedon denticulatus	1	1	2		1	1			1		7
Tryphosites longipes				1				1			2
Cumaceans			1					1			2
Macropippus sp.											
Pagurus sp.											
Corophium bonelli											
Podoceros odontyx											
Corystes cassivelaunus											
Molluscs:											
Cyprina islandica	1	1	1						1		4
Cultellus pellucidus	1	1	1	2	1	2	1	1	1	1	12
Mysella bidentata											
Cardium echinatum		1									1
Montacuta ferruginosa											
Mya arenaria											
Mya truncata											
Nucula hanleyi											
Astarte borealis											
Gari fervensis											
Lucinoma borealis											
Venus casina											
Venus striatula											
Lutraria lutraria											
Spisula elliptica											
Musculus niger											
Thyasira flexuosa											
Siphonodentalium lofotense											
Dentalium entalis	2	1	1		1				1	3	9
Philine sp.		1						3			4
Acteon tornatilis		1	1					1		1	4
Colus gracilis											
Clathrus clathrus		2									2
Natica alderi											
Natica catena											
Nudibranch 41.41											
Chaetoderma nitidulum	1					1					2
Echinoderms:											
Echinocardium flavescens								1			1
Echinocardium cordatum							1				1
Spatangus purpureus											
Amphiura filiformis	8	9	5	6	8	13	10	4	7	9	79

TABLE IA—*contd.*

STATION 9

	Sample number										*Total per station*
	1	*2*	*10*	*3*	*6*	*7*	*5*	*4*	*9*	*8*	
Amphiura chiajei											
Ophiuroid juveniles	1		2	1	2	1				2	9
Ophiura texturata											
Asteropecten irregularis	1		1	1		1					4
Ascidians:											
Pleurogonid sp.	9	11	30	57	18	9	9	66	10	10	229
Ciona intestinalis		1									1
Ascidian sp. 55.98										1	1
Other groups:											
Nemertine spp.									1		1
Lineus ruber											
Pennatulid sp.											
Flatworm sp.						1				1	2
Sponge 52.95							1				1
Sea anemone sp. 16.27											
Sea anemone sp. 17.21	1	1									2
Cerianthus sp.	1				2			1	1		5
Flustra foliacea											
Flustrella hispida											
Phascolion strombi											
Crystallogobius linearis											
Polychaetes:											
Myriochele heeri	225	118	439	252	290	367	209	297	245	186	2628
Owenia fusiformis	3	2	2	3		1	7	1	3	4	26
Stenhelais limicola	1									1	2
Aphrodite aculeata											
Lepidonotus squamatus	1										1
Harmothoe sp.	1										1
Phyllodoce lamelligera							1				1
Anaitides sp. (juv.)											
Ophiodromus flexuosus											
Nephtys incisa	2	1	4	1	2	1	1	1	2	3	18
Nephtys cirrosa			1			1	1			1	4
Glycera alba											
Goniada maculata	3	1	2	3	2			3	5	4	23
Lumbrineris fragilis		1						1			2
Lumbrineris impatiens									1		1
Scoloplos armiger	4	1	1	2	1	1	2	1	2	4	19
Spiophanes bombyx											
Spio filicornis											
Poecilochaetus serpens											
Chaetozone setosa				2							2
Cirratulid sp. 33.39		1									1

TABLE IA—*contd.*

STATION 9

	Sample number										Total per station
	1	*2*	*10*	*3*	*6*	*7*	*5*	*4*	*9*	*8*	
Diplocirrus glaucus			1								1
Caulleriella sp.										1	1
Ophelina aulogaster											
Ophelia limacina										1	1
Euclymene lumbricoides		2							1	1	4
Pectinaria koreni	1	1		1	1	1	1		1	1	8
Sosane gracilis											
Ampharetid sp. (juv.)			1		1			1			3
Ampharetid sp. 46.610	1						1				2
Lanice conchilega											
Thelepus setosus											
Lysilla loveni											
Bispira volutacornis	1										1
Potamilla torelli	1				1					1	3
Potamilla reniformis											
Jasmineira sp. cf. *caudata*											
Number spp. per sample	23	21	18	14	14	14	13	17	16	20	48
Number individuals/sample	271	159	496	333	331	401	245	385	283	236	3140

STATION 10

	Sample number										Total per station
	8	*10*	*5*	*7*	*9*	*4*	*1*	*3*	*2*	*6*	
Crustaceans:											
Ampelisca gibba						3		2	1	1	7
Hippomedon denticulatus			1			2	1	3	1	1	9
Tryphosites longipes								1			1
Cumaceans											
Macropippus sp.											
Pagurus sp.											
Corophium bonelli											
Podoceros odontyx											
Corystes cassivelaunus											
Molluscs:											
Cyprina islandica											
Cultellus pellucidus		1	1			2			2		6
Mysella bidentata											
Cardium echinatum											
Montacuta ferruginosa						1					1
Mya arenaria									1		1

TABLE IA—*contd.*

STATION 10

	Sample number										Total per station
	8	*10*	*5*	*7*	*9*	*4*	*1*	*3*	*2*	*6*	
Mya truncata						1					1
Nucula hanleyi									1		1
Astarte borealis											
Gari fervensis											
Lucinoma borealis											
Venus casina			1								1
Venus striatula											
Lutraria lutraria											
Spisula elliptica											
Musculus niger											
Thyasira flexuosa											
Siphonodentalium lofotense						1			1		2
Dentalium entalis			1		1			1			3
Philine sp.									1		1
Acteon tornatilis						1	1		2		4
Colus gracilis											
Clathrus clathrus											
Natica alderi											
Natica catena											
Nudibranch 41.41											
Chaetoderma nitidulum		1		1		1				1	4
Echinoderms:											
Echinocardium flavescens			1					1			2
Echinocardium cordatum					1	1		1			3
Spatangus purpureus											
Amphiura filiformis	8	1	2	6	3	12	9	6	7	2	56
Amphiura chiajei											
Ophiuroid juveniles							1	1			2
Ophiura texturata											
Asteropecten irregularis					1						1
Ascidians:											
Pleurogonid sp.	1	6	4		2	8	1	5	21	3	51
Ciona intestinalis											
Ascidian sp. 55.98											
Other groups:											
Nemertine spp.					1			1		1	3
Lineus ruber											
Pennatulid sp.		1		1							2
Flatworm sp.	1						1	1			3
Sponge 52.95								1			1
Sea anemone sp. 16.27											
Sea anemone sp. 17.21											
Cerianthus sp.	1		2	1	1	3	1	3	2	2	16

TABLE IA—*contd.*

STATION 10

	Sample number										*Total per station*
	8	*10*	*5*	*7*	*9*	*4*	*1*	*3*	*2*	*6*	
Flustra foliacea											
Flustrella hispida											
Phascolion strombi											
Crystallogobius linearis											
Polychaetes:											
Myriochele heeri	134	240	75	309	179	98	126	229	138	212	1740
Owenia fusiformis	1	3	2	2	1	1		3	1	7	21
Stenhelais limicola	3		1	1	2	1		3	2	2	15
Aphrodite aculeata						1					1
Lepidonotus squamatus							1				1
Harmothoe sp.											
Phyllodoce lamelligera	1			1			1				3
Anaitides sp. (juv.)											
Ophiodromus flexuosus											
Nephtys incisa		2			1	2	2	4	1		12
Nephtys cirrosa						1					1
Glycera alba	1							1			2
Goniada maculata		3	2	1		2	3	1	2	3	17
Lumbrineris fragilis		1	1	1	1						4
Lumbrineris impatiens											
Scoloplos armiger	1	1	1	2	1	2	2		1	5	16
Spiophanes bombyx											
Spio filicornis											
Poecilochaetus serpens											
Chaetozone setosa				2							2
Cirratulid sp. 33.39			1						1		2
Diplocirrus glaucus											
Caulleriella sp.											
Ophelina aulogaster											
Ophelia limacina											
Euclymene lumbricoides					1			1			2
Pectinaria koreni		2	1			3	2	3	1	1	13
Sosane gracilis											
Ampharetid sp. (juv.)											
Ampharetid sp. 46.610							2				2
Lanice conchilega											
Thelepus setosus											
Lysilla loveni											
Bispira volutacornis			1								1
Potamilla torelli										1	1
Potamilla reniformis											
Jasmineira sp. cf. *caudata*											
Number spp. per sample	10	12	17	12	14	21	15	21	19	14	44
Number individuals/sample	152	262	98	328	196	147	154	272	187	242	2038

TABLE IA—*contd.*

STATION 11

	Sample number										Total
	4	*1*	*3*	*6*	*8*	*10*	*5*	*9*	*7*	*2*	*per station*
Crustaceans:											
Ampelisca gibba			3				1			1	5
Hippomedon denticulatus	1	2		1	2		1		3	2	12
Tryphosites longipes			3			1	2		3		9
Cumaceans											
Macropippus sp.											
Pagurus sp.											
Corophium bonelli											
Podoceros odontyx											
Corystes cassivelaunus											
Molluscs:											
Cyprina islandica						1		1			2
Cultellus pellucidus	2	1				1					4
Mysella bidentata											
Cardium echinatum											
Montacuta ferruginosa		1		1							2
Mya arenaria											
Mya truncata											
Nucula hanleyi		1									1
Astarte borealis											
Gari fervensis											
Lucinoma borealis											
Venus casina							1		1		2
Venus striatula											
Lutraria lutraria											
Spisula elliptica		1									1
Musculus niger											
Thyasira flexuosa									1		1
Siphonodentalium lofotense											
Dentalium entalis											
Philine sp.				1	1				1	1	4
Acteon tornatilis	2		5								7
Colus gracilis							1	1			2
Clathrus clathrus											
Natica alderi											
Natica catena											
Nudibranch 41.41											
Chaetoderma nitidulum	1	1	1			1					4
Echinoderms:											
Echinocardium flavescens											
Echinocardium cordatum		1		2	1		2				6
Spatangus purpureus											
Amphiura filiformis	8	7	9	8	9	7	6	7	17	12	90

TABLE IA—*contd.*

STATION 11

	Sample number										*Total per station*
	4	*1*	*3*	*6*	*8*	*10*	*5*	*9*	*7*	*2*	
Amphiura chiajei											
Ophiuroid juveniles	5	1	1		4	2	2	4	1	1	21
Ophiura texturata											
Asteropecten irregularis					1	1					2
Ascidians:											
Pleurogonid sp.	172	28	44	14	184	73	88	11	116	128	858
Ciona intestinalis											
Ascidian sp. 55.98											
Other groups:											
Nemertine spp.											
Lineus ruber											
Pennatulid sp.				1			1				2
Flatworm sp.											
Sponge 52.95											
Sea anemone sp. 16.27											
Sea anemone sp. 17.21											
Cerianthus sp.		1	1				1	1	3		7
Flustra foliacea											
Flustrella hispida											
Phascolion strombi							1				1
Crystallogobius linearis											
Polychaetes:											
Myriochele heeri	159	178	269	152	226	140	221	190	112	173	1820
Owenia fusiformis	7	2	2	1	4	3	2	1	2	5	29
Stenhelais limicola	1	3	1	2	1	1	1	2	3	3	18
Aphrodite aculeata											
Lepidonotus squamatus								1			1
Harmothoe sp.											
Phyllodoce lamelligera											
Anaitides sp. (juv.)											
Ophiodromus flexuosus											
Nephtys incisa	1			1	1	2	1			1	7
Nephtys cirrosa			2						1		3
Glycera alba		1	1	1	1			1			5
Goniada maculata	2	5	2	3	4		3	2	1	5	27
Lumbrineris fragilis	1		1	1		1		1			5
Lumbrineris impatiens											
Scoloplos armiger	4	1	4	1	1	1	3	2	3	1	21
Spiophanes bombyx				1							1
Spio filicornis											
Poecilochaetus serpens											
Chaetozone setosa		1			1			1	1		4
Cirratulid sp. 33.39											

TABLE IA—*contd.*

STATION 11

	Sample number										*Total per station*
	4	*1*	*3*	*6*	*8*	*10*	*5*	*9*	*7*	*2*	
Diplocirrus glaucus									1		1
Caulleriella sp.											
Ophelina aulogaster											
Ophelia limacina											
Euclymene lumbricoides					1	1	1	1	2		6
Pectinaria koreni	5	1		2	1	1	1	1	5	4	21
Sosane gracilis											
Ampharetid sp. (juv.)						1				1	2
Ampharetid sp. 46.610						1			2	1	4
Lanice conchilega											
Thelepus setosus											
Lysilla loveni											
Bispira volutacornis											
Potamilla torelli			1	1						1	3
Potamilla reniformis											
Jasmineira sp. cf. *caudata*											
Number spp. per sample	15	19	17	18	17	18	20	17	20	16	40
Number individuals/sample	371	237	350	194	443	239	340	228	279	340	3021

STATION 12

	Sample number										*Total per station*
	4	*6*	*8*	*10*	*5*	*3*	*7*	*2*	*9*	*1*	
Crustaceans:											
Ampelisca gibba											
Hippomedon denticulatus	3	1	1				1	2	2	1	11
Tryphosites longipes			1		2			1		1	5
Cumaceans				1							1
Macropippus sp.	1										1
Pagurus sp.											
Corophium bonelli											
Podoceros odontyx											
Corystes cassivelaunus											
Molluscs:											
Cyprina islandica								1			1
Cultellus pellucidus		1				1	2		1	1	6
Mysella bidentata		1									1
Cardium echinatum		1									1
Montacuta ferruginosa											
Mya arenaria										1	1

TABLE IA—*contd.*

STATION 12

	Sample number										*Total*
	4	*6*	*8*	*10*	*5*	*3*	*7*	*2*	*9*	*1*	*per station*
Mya truncata											
Nucula hanleyi		1									1
Astarte borealis					1		1				2
Gari fervensis											
Lucinoma borealis											
Venus casina		1									1
Venus striatula											
Lutraria lutraria											
Spisula elliptica											
Musculus niger											
Thyasira flexuosa											
Siphonodentalium lofotense					1				2		3
Dentalium entalis			3			1	1	2			7
Philine sp.	1									1	2
Acteon tornatilis	1			1	1			2			5
Colus gracilis											
Clathrus clathrus											
Natica alderi											
Natica catena											
Nudibranch 41.41											
Chaetoderma nitidulum	4										4
Echinoderms:											
Echinocardium flavescens						1					1
Echinocardium cordatum				1							1
Spatangus purpureus											
Amphiura filiformis	8	7	8	5	3	2	6	6	7	7	59
Amphiura chiajei											
Ophiuroid juveniles		4		1	2	3			3		13
Ophiura texturata											
Asteropecten irregularis											
Ascidians:											
Pleurogonid sp.	3	18	19	8	43	4	10	16	26	30	177
Ciona intestinalis											
Ascidian sp. 55.98											
Other groups:											
Nemertine spp.							1				1
Lineus ruber											
Pennatulid sp.									2		2
Flatworm sp.					1						1
Sponge 52.95											
Sea anemone sp. 16.27							1				1
Sea anemone sp. 17.21											
Cerianthus sp.				1	2			1			4
Flustra foliacea											

TABLE IA—*contd.*

STATION 12

	Sample number										*Total per station*
	4	*6*	*8*	*10*	*5*	*3*	*7*	*2*	*9*	*1*	
Flustrella hispida											
Phascolion strombi					1						1
Crystallogobius linearis											
Polychaetes:											
Myriochele heeri	308	265	244	308	142	196	375	122	253	222	2435
Owenia fusiformis	5	1	5	5	1		7	3	1	1	29
Stenhelais limicola	2		1		1		2	1		1	8
Aphrodite aculeata											
Lepidonotus squamatus											
Harmothoe sp.						1					1
Phyllodoce lamelligera		1	1			1			1		4
Anaitides sp. (juv.)											
Ophiodromus flexuosus											
Nephtys incisa	2		1	1	1	2	1		3	2	13
Nephtys cirrosa											
Glycera alba			1	1							2
Goniada maculata	2			3	1		1	2	4	4	17
Lumbrineris fragilis							2		2		4
Lumbrineris impatiens											
Scoloplos armiger	1		3	4	1	1	6	3	2	2	23
Spiophanes bombyx											
Spio filicornis											
Poecilochaetus serpens											
Chaetozone setosa											
Cirratulid sp. 33.39									1		1
Diplocirrus glaucus	1										1
Caulleriella sp.											
Ophelina aulogaster											
Ophelia limacina					1						1
Euclymene lumbricoides	1					1				2	4
Pectinaria koreni		2	1	7			1	2	6	1	20
Sosane gracilis											
Ampharetid sp. (juv.)											
Ampharetid sp. 46.610	1						1				2
Lanice conchilega											
Thelepus setosus											
Lysilla loveni											
Bispira volutacornis											
Potamilla torelli					1						1
Potamilla reniformis											
Jasmineira sp. cf. *caudata*											
Number spp. per sample	16	13	13	14	18	12	17	14	16	15	45
Number individuals/sample	344	304	289	347	206	214	419	164	316	277	2880

TABLE IA—*contd.*

STATION 13

	Sample number										Total per station
	3	*9*	*5*	*2*	*6*	*8*	*7*	*10*	*1*	*4*	
Crustaceans:											
Ampelisca gibba		1		2		1	1	1			6
Hippomedon denticulatus			2					2		2	6
Tryphosites longipes			2			1			2		5
Cumaceans		1									1
Macropippus sp.						1			1		2
Pagurus sp.											
Corophium bonelli											
Podoceros odontyx											
Corystes cassivelaunus											
Molluscs:											
Cyprina islandica	1					1					2
Cultellus pellucidus	1			1		3	4		1	2	12
Mysella bidentata						2	1				3
Cardium echinatum	1			1			1				3
Montacuta ferruginosa											
Mya arenaria											
Mya truncata											
Nucula hanleyi			1								1
Astarte borealis											
Gari fervensis											
Lucinoma borealis											
Venus casina		1	1								2
Venus striatula											
Lutraria lutraria											
Spisula elliptica											
Musculus niger											
Thyasira flexuosa											
Siphonodentalium lofotense											
Dentalium entalis	1		1	1							3
Philine sp.											
Acteon tornatilis		1				1	1	2			5
Colus gracilis							1				1
Clathrus clathrus											
Natica alderi											
Natica catena											
Nudibranch 41.41											
Chaetoderma nitidulum		1	1				1				3
Echinoderms:											
Echinocardium flavescens											
Echinocardium cordatum		1				1			1	1	4
Spatangus purpureus											
Amphiura filiformis	8	4	9	13	3	9	10	9	10	11	86
Amphiura chiajei											

TABLE IA—*contd.*

STATION 13

	Sample number										Total per station
	3	*9*	*5*	*2*	*6*	*8*	*7*	*10*	*1*	*4*	
Ophiuroid juveniles		1	5	4			1	1		2	14
Ophiura texturata											
Asteropecten irregularis											
Ascidians:											
Pleurogonid sp.	16	16	31	35	16	49	10	9	34	14	230
Ciona intestinalis											
Ascidian sp. 55.98											
Other groups:											
Nemertine spp.	1						1	2			4
Lineaus ruber											
Pennatulid sp.							1			1	2
Flatworm sp.											
Sponge 52.95					1						1
Sea anemone sp. 16.27											
Sea anemone sp. 17.21											
Cerianthus sp.		1		1	1			3	1		7
Flustra foliacea											
Flustrella hispida											
Phascolion strombi											
Crystallogobius linearis											
Polychaetes:											
Myriochele heeri	200	244	306	506	177	527	501	149	415	428	3453
Owenia fusiformis	1	1	6	6	1	3	5	2	2	2	29
Stenhelais limicola			3				1		1	1	6
Aphrodite aculeata				1					1		2
Lepidonotus squamatus											
Harmothoe sp.											
Phyllodoce lamelligera		1					1		1	1	4
Anaitides sp. (juv.)											
Ophiodromus flexuosus											
Nephtys incisa			1			2			3	1	7
Nephtys cirrosa	1			2							3
Glycera alba					2	2			1	1	6
Goniada maculata		1	7	1	2	3	1	1	3		19
Lumbrineris fragilis	2			1		2	1	1			7
Lumbrineris impatiens											
Scoloplos armiger		4		5		3	2	3	2	5	24
Spiophanes bombyx			1								1
Spio filicornis						1					1
Poecilochaetus serpens											
Chaetozone setosa						1					1
Cirratulid sp. 33.39							1				1
Diplocirrus glaucus											

TABLE IA—*contd.*

STATION 13

	Sample number										*Total per station*
	3	*9*	*5*	*2*	*6*	*8*	*7*	*10*	*1*	*4*	
Caulleriella sp.											
Ophelina aulogaster											
Ophelia limacina											
Euclymene lumbricoides			1								1
Pectinaria koreni	4			3	3		4	5	2	3	24
Sosane gracilis											
Ampharetid sp. (juv.)			1								1
Ampharetid sp. 46.610	1	1	1						1		4
Lanice conchilega											
Thelepus setosus						1					1
Lysilla loveni											
Bispira volutacornis											
Potamilla torelli		1		1							2
Potamilla reniformis											
Jasmineira sp. cf. *caudata*											
Number spp. per sample	13	17	18	17	9	20	21	14	18	15	44
Number individuals/sample	238	281	380	584	206	618	550	190	488	475	4010

STATION 14

	Sample number										*Total per station*
	7	*9*	*6*	*8*	*4*	*5*	*10*	*2*	*1*	*3*	
Crustaceans:											
Ampelisca gibba									1		1
Hippomedon denticulatus	2	1	2		2	4	1	1	2	1	16
Tryphosites longipes	3	1	1		1			1	1		8
Cumaceans				1							1
Macropippus sp.				1		3					4
Pagurus sp.											
Corophium bonelli											
Podoceros odontyx											
Corystes cassivelaunus											
Molluscs:											
Cyprina islandica											
Cultellus pellucidus	1							1	2	2	6
Mysella bidentata					1	1					2
Cardium echinatum			1								1
Montacuta ferruginosa											
Mya arenaria											

TABLE IA—*contd.*

STATION 14

	Sample number										*Total per station*
	7	*9*	*6*	*8*	*4*	*5*	*10*	*2*	*1*	*3*	
Mya truncata											
Nucula hanleyi											
Astarte borealis			1								1
Gari fervensis											
Lucinoma borealis											
Venus casina											
Venus striatula											
Lutraria lutraria											
Spisula elliptica											
Musculus niger											
Thyasira flexuosa											
Siphonodentalium lofotense					1						1
Dentalium entalis	1							1		1	3
Philine sp.	1			1							2
Acteon tornatilis											
Colus gracilis											
Clathrus clathrus											
Natica alderi											
Natica catena											
Nudibranch 41.41											
Chaetoderma nitidulum	1										1
Echinoderms:											
Echinocardium flavescens											
Echincardium cordatum						1					1
Spatangus purpureus											
Amphiura filiformis			6		4		10	5	7	2	34
Amphiura chiajei											
Ophiuroid juveniles		1						2	1		4
Ophiura texturata											
Asteropecten irregularis				1							1
Ascidians:											
Pleurogonid sp.	4	1	11	1			3		13		33
Ciona intestinalis											
Ascidian sp. 55.98											
Other groups:											
Nemertine spp.		1			1					1	3
Lineus ruber											
Pennatulid sp.									1		1
Flatworm sp.											
Sponge 52.95											
Sea anemone sp. 16.27					1						1
Sea anemone sp. 17.21						1					1
Cerianthus sp.				2	2	1					5
Flustra foliacea											

TABLE IA—*contd.*

STATION 14

	Sample number										*Total per station*
	7	*9*	*6*	*8*	*4*	*5*	*10*	*2*	*1*	*3*	
Flustrella hispida											
Phascolion strombi		1									1
Crystallogobius linearis											
Polychaetes:											
Myriochele heeri	119	238	103	75	18	9	426	98	229	14	1329
Owenia fusiformis	1	1	1	1			2	1	3		10
Stenhelais limicola	1		2								3
Aphrodite aculeata											
Lepidonotus squamatus											
Harmothoe sp.											
Phyllodoce lamelligera			1	1							2
Anaitides sp. (juv.)											
Ophiodromus flexuosus				1							1
Nephtys incisa	1	3			1	1	2	1	4	1	14
Nephtys cirrosa								1			1
Glycera alba			1	1			1				3
Goniada maculata	1	1		1			4	2	3		12
Lumbrineris fragilis	1		1		1		2	1	1		7
Lumbrineris impatiens											
Scoloplos armiger	6	2	2	2	1		1	1	2		17
Spiophanes bombyx			1								1
Spio filicornis											
Poecilochaetus serpens											
Chaetozone setosa										1	1
Cirratulid sp. 33.39							1				1
Diplocirrus glaucus											
Caulleriella sp.											
Ophelina aulogaster											
Ophelia limacina											
Euclymene lumbricoides							1				1
Pectinaria koreni		5	7	5	3		3	1	3	6	33
Sosane gracilis											
Ampharetid sp. (juv.)											
Ampharetid sp. 46.610											
Lanice conchilega											
Thelepus setosus											
Lysilla loveni											
Bispira volutacornis											
Potamilla torelli											
Potamilla reniformis											
Jasmineira sp. cf. *caudata*											
Number spp. per sample	14	12	15	14	13	8	13	14	15	9	40
Number individuals/sample	143	255	141	94	37	21	457	117	273	29	1567

TABLE IA—*contd.*

STATION 15

	Sample number										*Total*
	1	*3*	*4*	*8*	*5*	*10*	*6*	*9*	*7*	*2*	*per station*
Crustaceans:											
Ampelisca gibba							1				1
Hippomedon denticulatus			1	1		1		1	1	2	7
Tryphosites longipes				2	2			1			5
Cumaceans											
Macropippus sp.	1						1				2
Pagurus sp.											
Corophium bonelli											
Podoceros odontyx								1			1
Corystes cassivelaunus											
Molluscs:											
Cyprina islandica				1	1			1			3
Cultellus pellucidus	1	1				2				1	5
Mysella bidentata		1									1
Cardium echinatum		1									1
Montacuta ferruginosa											
Mya arenaria											
Mya truncata											
Nucula hanleyi											
Astarte borealis											
Gari fervensis											
Lucinoma borealis											
Venus casina											
Venus striatula											
Lutraria lutraria		1									1
Spisula elliptica											
Musculus niger											
Thyasira flexuosa											
Siphonodentalium lofotense		1				1		1			3
Dentalium entalis	1		1						1		3
Philine sp.							1		1		2
Acteon tornatilis			1				1			2	4
Colus gracilis											
Clathrus clathrus											
Natica alderi			1								1
Natica catena											
Nudibranch 41.41											
Chaetoderma nitidulum		1							1		2
Echinoderms:											
Echinocardium flavescens											
Echinocardium cordatum											
Spatangus purpureus											
Amphiura filiformis	5	7	1	5	2	3	3	2	2	4	34
Amphiura chiajei											

TABLE IA—*contd.*

STATION 15

	Sample number										*Total per station*
	1	*3*	*4*	*8*	*5*	*10*	*6*	*9*	*7*	*2*	
Ophiuroid juveniles	4		1			1		1			7
Ophiura texturata											
Asteropecten irregularis											
Ascidians:											
Pleurogonid sp.	6	5	6	2	5	4	1			1	30
Ciona intestinalis				1							1
Ascidian sp. 55.98											
Other groups:											
Nemertine spp.		1						1		1	3
Lineus ruber											
Pennatulid sp.											
Flatworm sp.											
Sponge 52.95											
Sea anemone sp. 16.27			1								1
Sea anemone sp. 17.21											
Cerianthus sp.				2	1						3
Flustra foliacea											
Flustrella hispida											
Phascolion strombi			1								1
Crystallogobius linearis											
Polychaetes:											
Myriochele heeri	255	175	90	65	169	51	128	92	134	94	1253
Owenia fusiformis	3	2	2	1	1	1			1	1	12
Stenhelais limicola		1		1							2
Aphrodite aculeata											
Lepidonotus squamatus											
Harmothoe sp.											
Phyllodoce lamelligera		1				2		1	1		5
Anaitides sp. (juv.)											
Ophiodromus flexuosus											
Nephtys incisa	3	1		2	1	1	2		1	1	12
Nephtys cirrosa					2						2
Glycera alba	1		1	1	2						5
Goniada maculata		1		3	3	1	1		3	2	14
Lumbrineris fragilis		4						1	1	1	7
Lumbrineris impatiens											
Scoloplos armiger	3	3		2	1	2		1		1	13
Spiophanes bombyx											
Spio filicornis											
Poecilochaetus serpens											
Chaetozone setosa								1			1
Cirratulid sp. 33.39					1						1
Diplocirrus glaucus											

TABLE IA—*contd.*

STATION 15

	Sample number										Total per station
	1	*3*	*4*	*8*	*5*	*10*	*6*	*9*	*7*	*2*	
Caulleriella sp.											
Ophelina aulogaster											
Ophelia limacina	1										1
Euclymene lumbricoides											
Pectinaria koreni	1		2	2	2			3		1	11
Sosane gracilis											
Ampharetid sp. (juv.)											
Ampharetid sp. 46.610	2									1	3
Lanice conchilega											
Thelepus setosus											
Lysilla loveni											
Bispira volutacornis											
Potamilla torelli		2							1		3
Potamilla reniformis											
Jasmineira sp. cf. *caudata*											
Number spp. per sample	14	18	13	15	14	12	9	14	12	14	40
Number individuals/sample	287	209	109	89	194	71	139	108	148	113	1467

STATION 16

	Sample number										Total per station
	3	*9*	*4*	*7*	*10*	*1*	*2*	*6*	*8*	*5*	
Crustaceans:											
Ampelisca gibba				1		1					2
Hippomedon denticulatus	1	2				3			1		7
Tryphosites longipes	1	3				2			1	1	8
Cumaceans											
Macropippus sp.					1						1
Pagurus sp.				1							1
Corophium bonelli									1		1
Podoceros odontyx											
Corystes cassivelaunus				1				1			2
Molluscs:											
Cyprina islandica								1			1
Cultellus pellucidus	1	2	1		2	1				3	10
Mysella bidentata											
Cardium echinatum											
Montacuta ferruginosa								2			2
Mya arenaria											
Mya truncata		1									1

TABLE IA—*contd.*

STATION 16

	Sample number										*Total*
	3	*9*	*4*	*7*	*10*	*1*	*2*	*6*	*8*	*5*	*per station*
Nucula hanleyi											
Astarte borealis							1				1
Gari fervensis							1				1
Lucinoma borealis											
Venus casina											
Venus striatula											
Lutraria lutraria											
Spisula elliptica					1						1
Musculus niger											
Thyasira flexuosa											
Siphonodentalium lofotense	1			1							2
Dentalium entalis				1	1			1	2	1	6
Philine sp.						1			1	2	4
Acteon tornatilis											
Colus gracilis											
Clathrus clathrus											
Natica alderi									1	2	3
Natica catena											
Nudibranch 41.41											
Chaetoderma nitidulum				1							1
Echinoderms:											
Echinocardium flavescens											
Echinocardium cordatum								1			1
Spatangus purpureus								1			1
Amphiura filiformis	5	6	4	10	5	3	11	4	8		56
Amphiura chiajei											
Ophiuroid juveniles				1	1				3		5
Ophiura texturata											
Asteropecten irregularis		1									1
Ascidians:											
Pleurogonid sp.	59	65	11	68	73	73	10	26	165	73	623
Ciona intestinalis											
Ascidian sp. 55.98											
Other groups:											
Nemertine spp.											
Lineus ruber											
Pennatulid sp.											
Flatworm sp.											
Sponge 52.95											
Sea anemone sp. 16.27										1	1
Sea anemone sp. 17.21				1							1
Cerianthus sp.				2				2		1	5
Flustra foliacea											

TABLE IA—*contd.*

STATION 16

	Sample number										*Total per station*
	3	*9*	*4*	*7*	*10*	*1*	*2*	*6*	*8*	*5*	
Flustrella hispida											
Phascolion strombi											
Crystallogobius linearis											
Polychaetes:											
Myriochele heeri	316	140	158	259	159	440	222	383	283	244	2594
Owenia fusiformis	1	1	1	4	4	2	3	1		1	18
Stenhelais limicola	1	2	1	1	1	2	1	2	2	2	15
Aphrodite aculeata											
Lepidonotus squamatus				1							1
Harmothoe sp.											
Phyllodoce lamelligera								1			1
Anaitides sp. (juv.)											
Ophiodromus flexuosus								2			2
Nephtys incisa	3	1	2	1	3		2	1		3	16
Nephtys cirrosa								1		1	2
Glycera alba				1		1		1		1	4
Goniada maculata	1	4		1	5	4	5	1	6	2	29
Lumbrineris fragilis				1				2			3
Lumbrineris impatiens											
Scoloplos armiger	2	3	1	1			1	2	3	1	14
Spiophanes bombyx											
Spio filicornis											
Poecilochaetus serpens											
Chaetozone setosa											
Cirratulid sp. 33.39		1									1
Diplocirrus glaucus	1			1							2
Caulleriella sp.											
Ophelina aulogaster											
Ophelia limacina					1						1
Euclymene lumbricoides										1	1
Pectinaria koreni	1	2	1			3	3	3	2	3	18
Sosane gracilis											
Ampharetid sp. (juv.)			2		1						3
Ampharetid sp. 46.610			1						1		2
Lanice conchilega											
Thelepus setosus											
Lysilla loveni											
Bispira volutacornis											
Potamilla torelli			1						1	2	4
Potamilla reniformis											
Jasmineira sp. cf. *caudata*									1		1
Number spp. per sample	15	15	12	21	14	13	11	21	17	19	49
Number individuals/sample	394	234	184	349	258	536	261	438	482	344	3480

TABLE Ia—*contd.*

STATION 17

	Sample number										*Total*
	5	*10*	*6*	*2*	*7*	*8*	*4*	*1*	*9*	*3*	*per station*
Crustaceans:											
Ampelisca gibba		1	1							1	3
Hippomedon denticulatus	1		1	1		2	2		7		14
Tryphosites longipes						1	2				3
Cumaceans											
Macropippus sp.							1				1
Pagurus sp.											
Corophium bonelli											
Podoceros odontyx											
Corystes cassivelaunus											
Molluscs:											
Cyprina islandica											
Cultellus pellucidus	1	1	2	1			2		1	1	9
Mysella bidentata											
Cardium echinatum			1								1
Montacuta ferruginosa		1									1
Mya arenaria					2						2
Mya truncata											
Nucula hanleyi											
Astarte borealis											
Gari fervensis											
Lucinoma borealis											
Venus casina											
Venus striatula											
Lutraria lutraria											
Spisula elliptica											
Musculus niger											
Thyasira flexuosa											
Siphonodentalium lofotense											
Dentalium entalis		1		1	1			2			5
Philine sp.									1		1
Acteon tornatilis	1			1				1			3
Colus gracilis											
Clathrus clathrus											
Natica alderi			1								1
Natica catena											
Nudibranch 41.41											
Chaetoderma nitidulum					1					2	3
Echinoderms:											
Echinocardium flavescens									1		1
Echinocardium cordatum	1	1	1			2	1				6
Spatangus purpureus											
Amphiura filiformis	3	15	9	8	8	9	12	4	3	11	82

TABLE IA—*contd.*

STATION 17

	Sample number										*Total per station*
	5	*10*	*6*	*2*	*7*	*8*	*4*	*1*	*9*	*3*	
Amphiura chiajei											
Ophiuroid juveniles		1							1	1	3
Ophiura texturata											
Asteropecten irregularis											
Ascidians:											
Pleurogonid sp.	4	30	18		31	16	15	72	20	19	225
Ciona intestinalis											
Ascidian sp. 55.98							1				1
Other groups:											
Nemertine spp.				1		1	1			1	4
Lineus ruber											
Pennatulid sp.										1	1
Flatworm sp.	1	1			1						3
Sponge 52.95											
Sea anemone sp. 16.27			1								1
Sea anemone sp. 17.21											
Cerianthus sp.	2						4		2	2	10
Flustra foliacea											
Flustrella hispida											
Phascolion strombi								1			1
Crystallogobius linearis											
Polychaetes:											
Myriochele heeri	174	174	107	138	158	303	144	230	154	142	1724
Owenia fusiformis	1	1	2			2	3	1	1		11
Stenhelais limicola		1	4	1	2		1	3	2	1	15
Aphrodite aculeata											
Lepidonotus squamatus											
Harmothoe sp.											
Phyllodoce lamelligera	1	1		1							3
Anaitides sp. (juv.)											
Ophiodromus flexuosus						1					1
Nephtys incisa	3		1	3	3	2		1	2		15
Nephtys cirrosa				1							1
Glycera alba	1	1	1								3
Goniada maculata	3	5	3	2		3	3	5	4	3	31
Lumbrineris fragilis		1			1						2
Lumbrineris impatiens											
Scoloplos armiger	1	2	3	2	1	2	2	4	1		18
Spiophanes bombyx											
Spio filicornis											
Poecilochaetus serpens											
Chaetozone setosa					2						2
Cirratulid sp. 33.39											

TABLE IA—*contd.*

STATION 17

	Sample number										*Total per station*
	5	*10*	*6*	*2*	*7*	*8*	*4*	*1*	*9*	*3*	
Diplocirrus glaucus				1							1
Caulleriella sp.											
Ophelina aulogaster											
Ophelia limacina											
Euclymene lumbricoides				1	1		1				3
Pectinaria koreni	3	2	1	2	2	2	3	5	2	4	26
Sosane gracilis											
Ampharetid sp. (juv.)		1					1				2
Ampharetid sp. 46.610		1							1		2
Lanice conchilega											
Thelepus setosus											
Lysilla loveni											
Bispira volutacornis											
Potamilla torelli											
Potamilla reniformis							1				1
Jasmineira sp. cf. *caudata*											
Number spp. per sample	16	20	17	16	14	13	19	12	16	13	43
Number individuals/sample	201	242	157	165	214	346	200	330	203	188	2246

STATION 18

	Sample number										*Total per station*
	1	*2*	*10*	*5*	*6*	*8*	*9*	*7*	*4*	*3*	
Crustaceans:											
Ampelisca gibba		1							2		3
Hippomedon denticulatus			1	1		3	1	2		1	9
Tryphosites longipes						4		1			5
Cumaceans											
Macropippus sp.	1										1
Pagurus sp.											
Corophium bonelli											
Podoceros odontyx											
Corystes cassivelaunus											
Molluscs:											
Cyprina islandica			1								1
Cultellus pellucidus					1	1			1	2	5
Mysella bidentata										1	1
Cardium echinatum		1		1							2
Montacuta ferruginosa											

TABLE IA—*contd.*

STATION 18

	Sample number										Total per station
	1	*2*	*10*	*5*	*6*	*8*	*9*	*7*	*4*	*3*	
Mya arenaria											
Mya truncata			1								1
Nucula hanleyi											
Astarte borealis	1										1
Gari fervensis											
Lucinoma borealis											
Venus casina			2					2			4
Venus striatula											
Lutraria lutraria											
Spisula elliptica	1										1
Musculus niger											
Thyasira flexuosa											
Siphonodentalium lofotense	1		1		1	1	1				5
Dentalium entalis											
Philine sp.					1						1
Acteon tornatilis		2						2			4
Colus gracilis							1				1
Clathrus clathrus											
Natica alderi						1					1
Natica catena											
Nudibranch 41.41											
Chaetoderma nitidulum				1	1					1	3
Echinoderms:											
Echinocardium flavescens	2				1	1					4
Echinocardium cordatum	1	2		1			2	1	1	2	10
Spatangus purpureus											
Amphiura filiformis	3	6	8	9	5	3		6	8	4	52
Amphiura chiajei											
Ophiuroid juveniles		3	1	1	1	1		2			9
Ophiura texturata											
Asteropecten irregularis											
Ascidians:											
Pleurogonid sp.	61	68	66	70	107	62	67	65	78	74	718
Ciona intestinalis											
Ascidian sp. 55.98											
Other groups:											
Nemertine spp.	1		1				1	1		1	5
Lineus ruber											
Pennatulid sp.	1				1						2
Flatworm sp.			1								1
Sponge 52.95								1			1
Sea anemone sp. 16.27								1			1
Sea anemone sp. 17.21											
Cerianthus sp.			1			2		1	1		5

TABLE IA—*contd.*

STATION 18

	Sample number										*Total*
	1	*2*	*10*	*5*	*6*	*8*	*9*	*7*	*4*	*3*	*per station*
Flustra foliacea											
Flustrella hispida											
Phascolion strombi	1			1				1			3
Crystallogobius linearis											
Polychaetes:											
Myriochele heeri	139	120	109	71	540	243	142	103	102	256	1825
Owenia fusiformis	6	3	4	1	2	5	6	4	4	3	38
Stenhelais limicola						1	1		1		3
Aphrodite aculeata			1			1				1	3
Lepidonotus squamatus											
Harmothoe sp.											
Phyllodoce lamelligera		1	2					1		2	6
Anaitides sp. (juv.)											
Ophiodromus flexuosus	1										1
Nephtys incisa	1		2	2	1	3	1	3	1	2	16
Nephtys cirrosa					1						1
Glycera alba	2		1		1	1	2	1		1	9
Goniada maculata			3	1	4	3	3	3	1	4	22
Lumbrineris fragilis					1	1			1	2	5
Lumbrineris impatiens											
Scoloplos armiger	1	1	3	2	5	2	5	1	8	2	30
Spiophanes bombyx			2				1				3
Spio filicornis											
Poecilochaetus serpens										1	1
Chaetozone setosa							1	2	1	1	5
Cirratulid sp. 33.39											
Diplocirrus glaucus											
Caulleriella sp.											
Ophelina aulogaster											
Ophelia limacina											
Euclymene lumbricoides					1		1	1		1	4
Pectinaria koreni		2	2	4	4	9	3	1	2	3	30
Sosane gracilis							1				1
Ampharetid sp. (juv.)			1		4	3	5	1		1	15
Ampharetid sp. 46.610	1		2		2		1				6
Lanice conchilega											
Thelepus setosus											
Lysilla loveni											
Bispira volutacornis											
Potamilla torelli		2			2	1	2		2		9
Potamilla reniformis											
Jasmineira sp. cf. *caudata*											
Number spp. per sample	18	13	23	14	22	22	21	24	16	22	51
Number individuals/sample	225	212	216	166	687	352	248	207	214	364	2891

TABLE Ia—*contd.*

STATION 19

	Sample number										Total per station
	2	*7*	*6*	*8*	*3*	*4*	*5*	*9*	*10*	*1*	
Crustaceans:											
Ampelisca gibba											
Hippomedon denticulatus	1		1		2	1		4		2	11
Tryphosites longipes	1	1	1					1	1	2	7
Cumaceans		1	2								3
Macropippus sp.											
Pagurus sp.											
Corophium bonelli											
Podoceros odontyx											
Corystes cassivelaunus											
Molluscs:											
Cyprina islandica							1				1
Cultellus pellucidus		1	1				2		1		5
Mysella bidentata	1										1
Cardium echinatum	1										1
Montacuta ferruginosa											
Mya arenaria	1										1
Mya truncata											
Nucula hanleyi											
Astarte borealis											
Gari fervensis											
Lucinoma borealis											
Venus casina						1	1				2
Venus striatula											
Lutraria lutraria											
Spisula elliptica											
Musculus niger											
Thyasira flexuosa										1	1
Siphonodentalium lofotense								1			1
Dentalium entalis			1		1				1		3
Philine sp.					1						1
Acteon tornatilis		2	1				1			1	5
Colus gracilis											
Clathrus clathrus											
Natica alderi											
Natica catena											
Nudibranch 41.41											
Chaetoderma nitidulum		1		1			1				3
Echinoderms:											
Echinocardium flavescens											
Echinocardium cordatum				2						1	3
Spatangus purpureus											
Amphiura filiformis	5	4	7	4	8	5	12	8	5	14	72
Amphiura chiajei											

TABLE IA—*contd.*

STATION 19

	Sample number										*Total*
	2	*7*	*6*	*8*	*3*	*4*	*5*	*9*	*10*	*1*	*per station*
Ophiuroid juveniles											
Ophiura texturata			1		1		1	1		4	8
Asteropecten irregularis											
Ascidians:											
Pleurogonid sp.	13	3	4		15		12	24	16	8	95
Ciona intestinalis											
Ascidian sp. 55.98											
Other groups:											
Nemertine spp.		1	1		1			1	2	2	8
Lineus ruber									1	1	2
Pennatulid sp.		1					1				2
Flatworm sp.					1			1			2
Sponge 52.95											
Sea anemone sp. 16.27					1						1
Sea anemone sp. 17.21						1	2	1		1	5
Cerianthus sp.	3	2	1	2		1	1			4	14
Flustra foliacea											
Flustrella hispida											
Phascolion strombi			1				1				2
Crystallogobius linearis											
Polychaetes:											
Myriochele heeri	231	251	281	226	514	266	287	393	405	92	2946
Owenia fusiformis	3	6	1	2	4	4	3	2	1	4	30
Stenhelais limicola			2			1			1		4
Aphrodite aculeata											
Lepidonotus squamatus							1				1
Harmothoe sp.											
Phyllodoce lamelligera		1			1		1			1	4
Anaitides sp. (juv.)											
Ophiodromus flexuosus											
Nephtys incisa	1		1		4	1	2	1	1	1	12
Nephtys cirrosa											
Glycera alba		1			1		1	1		2	6
Goniada maculata	6	3	4	2		2		7	2	3	29
Lumbrineris fragilis	1		1		1		2		1	1	7
Lumbrineris impatiens											
Scoloplos armiger	2	5	2		3		5	3	1	5	26
Spiophanes bombyx	1					1		1		1	4
Spio filicornis											
Poecilochaetus serpens											
Chaetozone setosa			2				1				3
Cirratulid sp. 33.39					1		2	1			4
Diplocirrus glaucus		1	1								2

TABLE IA—*contd.*

STATION 19

	Sample number										*Total*
	2	*7*	*6*	*8*	*3*	*4*	*5*	*9*	*10*	*1*	*per station*
Caulleriella sp.											
Ophelina aulogaster											
Ophelia limacina											
Euclymene lumbricoides			1	2	1	1	2				7
Pectinaria koreni	3	3		3	7	1	4			3	24
Sosane gracilis	1										1
Ampharetid sp. (juv.)				2					1		3
Ampharetid sp. 46.610	1	1			1		3			1	7
Lanice conchilega											
Thelepus setosus											
Lysilla loveni											
Bispira volutacornis											
Potamilla torelli	2		1		1		1	1		1	7
Potamilla reniformis											
Jasmineira sp. cf. *caudata*											
Number spp. per sample	19	19	23	10	21	13	26	18	15	24	47
Number individuals/sample	278	289	319	246	570	286	351	452	440	156	3387

STATION 20

	Sample number										*Total*
	1	*3*	*5*	*9*	*6*	*2*	*7*	*10*	*4*	*8*	*per station*
Crustaceans:											
Ampelisca gibba	3			3					1		7
Hippomedon denticulatus	1	3		1			1				6
Tryphosites longipes	2			1					1		4
Cumaceans											
Macropippus sp.											
Pagurus sp.											
Corophium bonelli		1									1
Podoceros odontyx											
Corystes cassivelaunus											
Molluscs											
Cyprina islandica	2	1				1	1				5
Cultellus pellucidus		1		2				1			4
Mysella bidentata	1				1	1					3
Cardium echinatum											
Montacuta ferruginosa											
Mya arenaria											

TABLE IA—*contd.*

STATION 20

	Sample number										*Total per station*
	1	*3*	*5*	*9*	*6*	*2*	*7*	*10*	*4*	*8*	
Mya truncata											
Nucula hanleyi											
Astarte borealis		1				1					2
Gari fervensis											
Lucinoma borealis											
Venus casina											
Venus striatula											
Lutraria lutraria					4						4
Spisula elliptica											
Musculus niger											
Thyasira flexuosa											
Siphonodentalium lofotense	1	1	1								3
Dentalium entalis		1					1				2
Philine sp.											
Acteon tornatilis										1	1
Colus gracilis											
Clathrus clathrus											
Natica alderi											
Natica catena											
Nudibranch 41.41											
Chaetoderma nitidulum	1	1						1		1	4
Echinoderms:											
Echinocardium flavescens	1										1
Echinocardium cordatum											
Spatangus purpureus											
Amphiura filiformis	9	10	10	11	7	6	8	5	6	4	76
Amphiura chiajei											
Ophiuroid juveniles	1	1									2
Ophiura texturata											
Asteropecten irregularis				3							3
Ascidians:											
Pleurogonid sp.	17	88	179	23	11	63	34		82	24	521
Ciona intestinalis											
Ascidian sp. 55.98											
Other groups:											
Nemertine spp.	1			1					2		4
Lineus ruber									1		1
Pennatulid sp.											
Flatworm sp.							1				1
Sponge 52.95									1		1
Sea anemone sp. 16.27											
Sea anemone sp. 17.21				1					1		2
Cerianthus sp.	1		2	1	1	2	3	1	1		12
Flustra foliacea											

TABLE IA—*contd.*

STATION 20

	Sample number										*Total per station*
	1	*3*	*5*	*9*	*6*	*2*	*7*	*10*	*4*	*8*	
Flustrella hispida											
Phascolion strombi	1					1	1	1	1		5
Crystallogobius linearis											
Polychaetes:											
Myriochele heeri	333	141	275	193	307	235	287	168	148	207	2294
Owenia fusiformis	6	3	2	1	4	5	1	2	2	3	29
Stenhelais limicola	2				2	2	1	1		1	9
Aphrodite aculeata											
Lepidonotus squamatus											
Harmothoe sp.											
Phyllodoce lamelligera			1								1
Anaitides sp. (juv.)											
Ophiodromus flexuosus											
Nephtys incisa			1	2	2	4	2	2		4	17
Nephtys cirrosa											
Glycera alba											
Goniada maculata	5	2	3	2	3	6	4	5	2	4	36
Lumbrineris fragilis				1			1			1	3
Lumbrineris impatiens											
Scoloplos armiger		3	3	2	1	2	1	1	1	1	15
Spiophanes bombyx					1	1		1			4
Spio filicornis											
Poecilochaetus serpens											
Chaetozone setosa		1						1			2
Cirratulid sp. 33.39		1								1	2
Diplocirrus glaucus							1	1			2
Caulleriella sp.											
Ophelina aulogaster		1									1
Ophelia limacina											
Euclymene lumbricoides					2			1			3
Pectinaria koreni	2	1	2	2	1	2	1			2	13
Sosane gracilis											
Ampharetid sp. (juv.)						2	1		1	1	5
Ampharetid sp. 46.610		1				1		1	1		4
Lanice conchilega											
Thelepus setosus											
Lysilla loveni											
Bispira volutacornis											
Potamilla torelli						1	1			1	3
Potamilla reniformis											
Jasmineira sp. cf. *caudata*											
Number spp. per sample	19	20	11	17	14	18	19	16	16	14	43
Number individuals/sample	390	263	479	250	343	340	351	193	252	256	3117

TABLE IA—*contd.*

STATION 21

	Sample number										Total per station
	1	*8*	*7*	*6*	*9*	*3*	*5*	*4*	*10*	*2*	
Crustaceans:											
Ampelisca gibba			1			2		1		1	5
Hippomedon denticulatus				1			1	1			3
Tryphosites longipes		1		1							2
Cumaceans											
Macropippus sp.											
Pagurus sp.											
Corophium bonelli											
Podoceros odontyx											
Corystes cassivelaunus											
Molluscs:											
Cyprina islandica		3		3	2	2		2	2		14
Cultellus pellucidus			2		3		1	1	1		8
Mysella bidentata											
Cardium echinatum				1							1
Montacuta ferruginosa					2						2
Mya arenaria											
Mya truncata											
Nucula hanleyi											
Astarte borealis											
Gari fervensis	1										1
Lucinoma borealis											
Venus casina							1				1
Venus striatula											
Lutraria lutraria											
Spisula elliptica											
Musculus niger											
Thyasira flexuosa											
Siphonodentalium lofotense	1		1								2
Dentalium entalis	1		2		1	1					5
Philine sp.									1		1
Acteon tornatilis											
Colus gracilis								1			1
Clathrus clathrus											
Natica alderi											
Natica catena											
Nudibranch 41.41											
Chaetoderma nitidulum											
Echinoderms:											
Echinocardium flavescens											
Echinocardium cordatum	1				1			1			3
Spatangus purpureus											
Amphiura filiformis	5	7	9	3	8	5	8	7	12	2	66

TABLE IA—*contd.*

STATION 21

	Sample number										Total per station
	1	*8*	*7*	*6*	*9*	*3*	*5*	*4*	*10*	*2*	
Amphiura chiajei											
Ophiuroid juveniles						1			1		2
Ophiura texturata											
Asteropecten irregularis								1			1
Ascidians:											
Pleurogonid sp.	7	9	10		23	29	2	33	2	20	135
Ciona intestinalis											
Ascidian sp. 55.98											
Other groups:											
Nemertine spp.											
Lineus ruber											
Pennatulid sp.							2				2
Flatworm sp.											
Sponge 52.95											
Sea anemone sp. 16.27											
Sea anemone sp. 17.21											
Cerianthus sp.		1	1	2	2		1		1	2	10
Flustra foliacea											
Flustrella hispida	1										1
Phascolion strombi											
Crystallogobius linearis											
Polychaetes:											
Myriochele heeri	92	111	119	67	89	174	69	127	47	58	953
Owenia fusiformis	2	1	1	3		1		1	1	1	11
Stenhelais limicola	2	1	3	1	2	3		1	1		14
Aphrodite aculeata											
Lepidonotus squamatus											
Harmothoe sp.											
Phyllodoce lamelligera				1							1
Anaitides sp. (juv.)											
Ophiodromus flexuosus											
Nephtys incisa	2	3	3	1		2	2		1		14
Nephtys cirrosa			1		1	1	1			1	5
Glycera alba	1			1		1		1	1		5
Goniada maculata	2	7	2	4	4	6	4	1	6	1	37
Lumbrineris fragilis	1			1							2
Lumbrineris impatiens											
Scoloplos armiger	2	3		4	1	2	1	3	2	1	19
Spiophanes bombyx						1					1
Spio filicornis											
Poecilochaetus serpens											
Chaetozone setosa											
Cirratulid sp. 33.39	1										1

TABLE IA—*contd.*

STATION 21

	Sample number										*Total per station*
	1	*8*	*7*	*6*	*9*	*3*	*5*	*4*	*10*	*2*	
Diplocirrus glaucus						1					1
Caulleriella sp.											
Ophelina aulogaster											
Ophelia limacina	1								1		2
Euclymene lumbricoides	1	1							2		4
Pectinaria koreni			2		2	5	2	1	1	2	15
Sosane gracilis											
Ampharetid sp. (juv.)											
Ampharetid sp. 46.610					1		1	1			3
Lanice conchilega						1					1
Thelepus setosus											
Lysilla loveni											
Bispira volutacornis											
Potamilla torelli						1			1		2
Potamilla reniformis	1										1
Jasmineira sp. cf. *caudata*											
Number spp. per sample	19	12	14	15	15	19	14	17	18	10	41
Number individuals/sample	125	148	157	94	142	239	96	185	84	89	1359

STATION 22

	Sample number										*Total per station*
	1	*10*	*4*	*5*	*8*	*3*	*9*	*2*	*6*	*7*	
Crustaceans:											
Ampelisca gibba			1	1	1			1			4
Hippomedon denticulatus	1	1	2			2		1		1	8
Tryphosites longipes		1	1	4	2				2		10
Cumaceans						1					1
Macropippus sp.											
Pagurus sp.											
Corophium bonelli								1			1
Podoceros odontyx											
Corystes cassivelaunus											
Molluscs:											
Cyprina islandica											
Cultellus pellucidus		2	3			4	1		1	1	12
Mysella bidentata											
Cardium echinatum											
Montacuta ferruginosa											

TABLE IA—*contd.*

STATION 22

	Sample number										*Total*
	1	*10*	*4*	*5*	*8*	*3*	*9*	*2*	*6*	*7*	*per station*
Mya arenaria											
Mya truncata											
Nucula hanleyi											
Astarte borealis											
Gari fervensis											
Lucinoma borealis											
Venus casina											
Venus striatula											
Lutraria lutraria								1			1
Spisula elliptica											
Musculus niger		1									1
Thyasira flexuosa											
Siphonodentalium lofotense								1	2		3
Dentalium entalis			1		1	1		1			4
Philine sp.									1		1
Acteon tornatilis		1			1	1					3
Colus gracilis											
Clathrus clathrus											
Natica alderi											
Natica catena											
Nudibranch 41.41											
Chaetoderma nitidulum	1			1							2
Echinoderms:											
Echinocardium flavescens											
Echinocardium cordatum		1				1	1	2			5
Spatangus purpureus											
Amphiura filiformis	10	9	5	7	6	3	5	17	10	7	79
Amphiura chiajei											
Ophiuroid juveniles				2	1	6	2	1	2		14
Ophiura texturata											
Asteropecten irregularis				1							1
Ascidians:											
Pleurogonid sp.	7	3	18	10	5	12	4	33	7	11	120
Ciona intestinalis											
Ascidian sp. 55.98											
Other groups:											
Nemertine spp.											
Lineus ruber											
Pennatulid sp.											
Flatworm sp.											
Sponge 52.95											
Sea anemone sp. 16.27											
Sea anemone sp. 17.21											
Cerianthus sp.	1	2	2	2			1		1		9

TABLE IA—*contd.*

STATION 22

	Sample number										*Total per station*
	1	*10*	*4*	*5*	*8*	*3*	*9*	*2*	*6*	*7*	
Flustra foliacea											
Flustrella hispida											
Phascolion strombi				1							1
Crystallogobius linearis											
Polychaetes:											
Myriochele heeri	195	152	306	160	229	283	285	144	105	263	2122
Owenia fusiformis	3	2	1	3	1	4	4	2	3	2	25
Stenhelais limicola	1	2	1	4	3	2	3	2	1	1	20
Aphrodite aculeata											
Lepidonotus squamatus											
Harmothoe sp.											
Phyllodoce lamelligera				1		2		1			4
Anaitides sp. (juv.)											
Ophiodromus flexuosus				1							1
Nephtys incisa	3		1	2	1	2	1		1		11
Nephtys cirrosa		2							1		3
Glycera alba			1				1				2
Goniada maculata	4	2	1	5	4	2	2	4	4	5	33
Lumbrineris fragilis						1			1		2
Lumbrineris impatiens											
Scoloplos armiger	2		2	1	3		2	2		2	14
Spiophanes bombyx		1	1						1		3
Spio filicornis											
Poecilochaetus serpens											
Chaetozone setosa			1			1					2
Cirratulid sp. 33.39								1	1		2
Diplocirrus glaucus											
Caulleriella sp.											
Ophelina aulogaster											
Ophelia limacina											
Euclymene lumbricoides		1					1	1	1		4
Pectinaria koreni	3	2	2	2	1	2	2		1	1	16
Sosane gracilis											
Ampharetid sp. (juv.)											
Ampharetid sp. 46.610						1	1	1			3
Lanice conchilega											
Thelepus setosus											
Lysilla loveni											
Bispira volutacornis											
Potamilla torelli		2				2					4
Potamilla reniformis											
Jasmineira sp. cf. *caudata*											
Number spp. per sample	12	18	18	18	14	20	16	19	19	10	38
Number individuals/sample	231	187	350	208	259	333	316	217	146	294	2541

TABLE IA—*contd.*

STATION 23

	Sample number										*Total per station*
	1	*2*	*3*	*4*	*7*	*10*	*6*	*9*	*5*	*8*	
Crustaceans:											
Ampelisca gibba									1	1	2
Hippomedon denticulatus		1			1	1	1		2	1	7
Tryphosites longipes	2		2			1	1				6
Cumaceans											
Macropippus sp.											
Pagurus sp.										1	1
Corophium bonelli											
Podoceros odontyx											
Corystes cassivelaunus											
Molluscs:											
Cyprina islandica						1	1		2		4
Cultellus pellucidus	1		1			2		1			5
Mysella bidentata					1						1
Cardium echinatum											
Montacuta ferruginosa											
Mya arenaria											
Mya truncata											
Nucula hanleyi											
Astarte borealis											
Gari fervensis											
Lucinoma borealis											
Venus casina					1						1
Venus striatula											
Lutraria lutraria	1										1
Spisula elliptica											
Musculus niger											
Thyasira flexuosa											
Siphonodentalium lofotense	2						1	1			4
Dentalium entalis	2				1				1		4
Philine sp.								1	1		2
Acteon tornatilis		1							1		2
Colus gracilis											
Clathrus clathrus											
Natica alderi		1	1								2
Natica catena											
Nudibranch 41.41											
Chaetoderma nitidulum		1				1			1		3
Echinoderms:											
Echinocardium flavescens							1				1
Echinocardium cordatum		1	1	2	1				2		7
Spatangus purpureus											
Amphiura filiformis	5	9	6	13	5	6	8	7	4	1	64
Amphiura chiajei											

TABLE IA—*contd.*

STATION 23

	Sample number										Total per station
	1	*2*	*3*	*4*	*7*	*10*	*6*	*9*	*5*	*8*	
Ophiuroid juveniles				1					1		2
Ophiura texturata											
Asteropecten irregularis								1			1
Ascidians:											
Pleurogonid sp.		5	15	1	4	6		6	1	1	39
Ciona intestinalis											
Ascidian sp. 55.98						1					1
Other groups:											
Nemertine spp.				1	1				1		3
Lineus ruber				1							1
Pennatulid sp.											
Flatworm sp.											
Sponge 52.95			1			1					2
Sea anemone sp. 16.27		1									1
Sea anemone sp. 17.21											
Cerianthus sp.	2		2				1	1		1	7
Flustra foliacea											
Flustrella hispida											
Phascolion strombi			1								1
Crystallogobius linearis											
Polychaetes:											
Myriochele heeri	141	192	44	179	137	94	113	143	287	242	1572
Owenia fusiformis	1	4	3		3	3		2	1		17
Stenhelais limicola		1	1			2	1	2	1	1	9
Aphrodite aculeata											
Lepidonotus squamatus											
Harmothoe sp.											
Phyllodoce lamelligera									1		1
Anaitides sp. (juv.)											
Ophiodromus flexuosus											
Nephtys incisa	1			1	1	1	1	1			6
Nephtys cirrosa	1										1
Glycera alba						1				1	2
Goniada maculata		1	2	6	2	3		4	3		21
Lumbrineris fragilis	1			1	1						3
Lumbrineris impatiens											
Scoloplos armiger	1	1	2	4	2	3		3	1	2	19
Spiophanes bombyx											
Spio filicornis											
Poecilochaetus serpens											
Chaetozone setosa											
Cirratulid sp. 33.39											
Diplocirrus glaucus											

TABLE IA—*contd.*

STATION 23

	Sample number										*Total per station*
	1	*2*	*3*	*4*	*7*	*10*	*6*	*9*	*5*	*8*	
Caulleriella sp.											
Ophelina aulogaster											
Ophelia limacina											
Euclymene lumbricoides		1		1	1			1		1	5
Pectinaria koreni	2		2	2	2	3		2	2		15
Sosane gracilis											
Ampharetid sp. (juv.)											
Ampharetid sp. 46.610					2	1		1			4
Lanice conchilega											
Thelepus setosus											
Lysilla loveni											
Bispira volutacornis											
Potamilla torelli											
Potamilla reniformis											
Jasmineira sp. cf. *caudata*											
Number spp. per sample	14	14	15	13	17	18	10	16	19	11	41
Number individuals/sample	163	220	83	213	166	132	129	177	314	253	1850

STATION 24

	Sample number										*Total per station*
	5	*4*	*3*	*6*	*7*	*8*	*1*	*10*	*2*	*9*	
Crustaceans:											
Ampelisca gibba					1					1	2
Hippomedon denticulatus				1		1	1	1		3	7
Tryphosites longipes			1	1	1		1			1	5
Cumaceans											
Macropippus sp.		1									1
Pagurus sp.											
Corophium bonelli											
Podoceros odontyx											
Corystes cassivelaunus											
Molluscs:											
Cyprina islandica										1	1
Cultellus pellucidus					1						1
Mysella bidentata											
Cardium echinatum						1					1
Montacuta ferruginosa											
Mya arenaria											

TABLE IA—*contd.*

STATION 24

	Sample number										*Total per station*
	5	*4*	*3*	*6*	*7*	*8*	*1*	*10*	*2*	*9*	
Mya truncata											
Nucula hanleyi											
Astarte borealis											
Gari fervensis											
Lucinoma borealis											
Venus casina											
Venus striatula											
Lutraria lutraria											
Spisula elliptica											
Musculus niger											
Thyasira flexuosa											
Siphonodentalium lofotense		1									1
Dentalium entalis											
Philine sp.											
Acteon tornatilis			1			1		1	1		4
Colus gracilis						1					1
Clathrus clathrus											
Natica alderi					1	1					2
Natica catena											
Nudibranch 41.41											
Chaetoderma nitidulum						1		1	1		3
Echinoderms:											
Echinocardium flavescens											
Echinocardium cordatum		3		3	2		2				10
Spatangus purpureus											
Amphiura filiformis	3	7			4	8	7	4	9	6	48
Amphiura chiajei											
Ophiuroid juveniles	1	1	1				1	1		1	6
Ophiura texturata											
Asteropecten irregularis	1				1			1			3
Ascidians:											
Pleurogonid sp.	36	30	16	19	10	16	18	9	18	26	198
Ciona intestinalis											
Ascidian sp. 55.98											
Other groups:											
Nemertine spp.	1										1
Lineus ruber	1										1
Pennatulid sp.									1		1
Flatworm sp.			1								1
Sponge 52.95		1									1
Sea anemone sp. 16.27		1									1
Sea anemone sp. 17.21											
Cerianthus sp.				1		1		2	1	2	7
Flustra foliacea											

TABLE IA—*contd.*

STATION 24

	Sample number										*Total per station*
	5	*4*	*3*	*6*	*7*	*8*	*1*	*10*	*2*	*9*	
Flustrella hispida											
Phascolion strombi		1				1					2
Crystallogobius linearis											
Polychaetes:											
Myriochele heeri	229	262	114	74	317	94	99	120	110	241	1660
Owenia fusiformis	2	9	1	1	11	1	4	3	1		33
Stenhelais limicola		2	1	1	1	2	1			3	11
Aphrodite aculeata					1						1
Lepidonotus squamatus											
Harmothoe sp.											
Phyllodoce lamelligera											
Anaitides sp. (juv.)											
Ophiodromus flexuosus											
Nephtys incisa	2		1	2			1	2	1		9
Nephtys cirrosa											
Glycera alba				1		1		1	1		4
Goniada maculata	4	5	2	1	3	2	5	4	4		30
Lumbrineris fragilis		2		1	1						4
Lumbrineris impatiens											
Scoloplos armiger	2	4		1		1	2	2	4		16
Spiophanes bombyx					1						1
Spio filicornis											
Poecilochaetus serpens											
Chaetozone setosa					1						1
Cirratulid sp. 33.39		1		1							2
Diplocirrus glaucus						1					1
Caulleriella sp.									1		1
Ophelina aulogaster											
Ophelia limacina	1										1
Euclymene lumbricoides		1									1
Pectinaria koreni	1	1		1	2	4	2	2			13
Sosane gracilis											
Ampharetid sp. (juv.)											
Ampharetid sp. 46.610											
Lanice conchilega											
Thelepus setosus											
Lysilla loveni											
Bispira volutacornis											
Potamilla torelli										1	1
Potamilla reniformis											
Jasmineira sp. cf. *caudata*											
Number spp. per sample	13	18	10	15	17	18	13	15	13	11	43
Number individuals/sample	284	233	139	109	359	138	144	154	153	286	2099

TABLE IB

Numbers of selected species and groups of organisms per m², at all sampling stations

Station number	*Crustaceans*	*Bivalves*	Amphiura filiformis	*Other echinoderms*	*Prosobranchs*	Owenia fusiformis	Nephtys incisa	Goniada maculata	Scoloplos armiger	Pectinaria korenii	*Other polychaetes*	*Pleurogonid ascidians*	*All other species*
1	19	42	40	19	17	43	17	10	16	11	76	77	8
2	28	25	42	14	9	53	11	15	18	15	69	181	8
3	39	29	23	19	7	29	12	25	9	12	57	176	7
4	18	39	38	21	13	67	22	27	20	17	77	397	12
5	18	13	24	13	14	8	6	6	13	19	35	74	3
6	39	19	31	13	11	7	16	11	15	10	65	178	6
7	16	15	81	15	8	19	17	20	22	17	32	117	6
8	13	9	73	16	12	17	11	18	28	22	72	108	8
9	13	17	79	15	19	26	18	23	19	8	35	229	10
10	17	11	56	8	10	21	12	17	16	13	53	51	13
11	26	13	90	29	14	29	7	27	21	21	60	858	6
12	18	14	59	15	17	29	13	17	23	20	33	177	10
13	20	23	86	8	9	29	7	19	24	24	48	230	10
14	30	10	34	6	6	10	14	12	17	33	26	33	7
15	16	11	34	7	13	12	12	14	13	11	33	30	8
16	22	17	56	8	15	18	16	29	14	18	48	623	2
17	21	13	82	10	10	11	15	3	18	26	46	225	14
18	18	16	52	23	12	38	16	22	30	30	77	718	16
19	21	12	72	11	10	30	12	29	26	24	74	95	19
20	18	18	76	6	6	29	17	36	15	13	52	521	16
21	10	27	66	6	9	11	14	37	19	15	53	135	3
22	24	14	79	20	13	25	11	33	14	16	59	120	3
23	16	12	64	11	14	17	6	21	19	15	32	39	12
24	15	3	48	19	8	33	9	30	16	13	36	198	1

TABLE Ic

Standard error and 95% confidence limits of means at each site for total and selected organisms per m^2

Station number	*Total population*	Myriochele	*Pleurogonid ascidians*
1	3593 ± 809	3272 ± 759	
2	2593 ± 830	2094 ± 405	179 ± 139
3	2373 ± 610	2142 ± 698	196 ± 147
4	3587 ± 1112	2818 ± 1109	397 ± 171
5	2050 ± 710	1802 ± 651	74 ± 51
6	3573 ± 877	3152 ± 878	178 ± 103
7	3390 ± 460	3015 ± 506	117 ± 43
8	2955 ± 870	2547 ± 687	108 ± 104
9	3140 ± 680	2628 ± 654	229 ± 154
10	2038 ± 496	1740 ± 521	51 ± 44
11	3021 ± 562	1820 ± 331	858 ± 450
12	2880 ± 547	2435 ± 550	177 ± 90
13	4010 ± 1170	3453 ± 1023	230 ± 95
14	1567 ± 935	1329 ± 936	33 ± 34
15	1467 ± 470	1253 ± 438	30 ± 55
16	3480 ± 820	2594 ± 707	623 ± 318
17	2246 ± 460	1724 ± 399	225 ± 142
18	2891 ± 1100	1825 ± 996	718 ± 96
19	3387 ± 850	2946 ± 835	95 ± 56
20	3117 ± 607	2294 ± 489	521 ± 385
21	1359 ± 347	953 ± 274	135 ± 85
22	2541 ± 476	2122 ± 504	120 ± 64
23	1850 ± 475	1572 ± 507	39 ± 33
24	2099 ± 663	1660 ± 619	198 ± 61

TABLE ID

The values of three diversity indices (Shannon–Wiener, Gleason and Menhinick) at each sample station, Ekofisk Field, August 1973

Station number	*Shannon–Wiener index*	*Gleason index*	*Menhinick index*
1	0·96	4·89	0·68
2	1·44	5·85	0·92
3	1·41	6·18	1·01
4	1·45	5·86	0·82
5	1·02	5·11	0·88
6	0·95	5·62	0·79
7	0·92	4·55	0·65
8	1·13	5·01	0·75
9	1·18	5·84	0·86
10	1·22	5·64	0·97
11	1·78	4·87	0·73
12	1·16	5·39	0·82
13	1·04	5·18	0·69
14	1·29	5·30	1·01
15	1·26	5·35	1·04
16	1·38	5·89	0·83
17	1·55	4·82	0·91
18	1·85	6·15	0·93
19	1·11	5·66	0·81
20	1·49	5·22	0·77
21	1·96	5·54	1·11
22	1·29	4·72	0·75
23	1·24	5·32	0·95
24	1·43	5·49	0·94

TABLE IE

Conditions at the sample stations during sampling, Ekofisk Field, August 1973

Station number	*Date*	*Time*	*Wind direction and Force*		*Sea*	*Swell*	*Vessel*	*Depth* (fath)
1	12/8	0745	235°	3	3	Slight	Steady	36·5
2	12/8	1110	230°	3	3	Slight	Steady	37·5
3	12/8	1315	210°	3	3	Slight	Steady	37
4	12/8	1505	220°	3	3	Slight	Steady	38·5
5	12/8	1637	200°	3	3	Slight	Steady	38
6	12/8	1930	200°	3	3	Slight	Steady	37·5
7	12/8	2200	200°	3	3	Slight	Steady	37·5
8	12/8	2345	180°	3–4	3	Slight	Steady	38
9	13/8	0300	145°	3	3	Slight	Steady	37·5
10	13/8	0820	180°	3–4	3	Slight	Steady	37·5
11	13/8	1020	180°	3	3	Slight	Steady	37·5
12	13/8	1200	180°	3	3	Low	Steady	37·5
13	13/8	1325	150°	4	4	Low	Steady	37·5
14	13/8	1440	150°	4	4	Low	Steady	37·5
15	13/8	1620	170°	3–4	3	Low	Steady	38
16	13/8	1820	170°	3–4	3	Low	Steady	37·5
17	14/8	0115	160°	3	3	Low	Steady	38
18	14/8	0320	150°	3	3	Low	Steady	38·5
19	14/8	0615	110°	3	3	Low	Steady	38·5
20	14/8	0900	100°	3	3	Low	Steady	38
21	14/8	1145	090°	3–4	3	Low	Steady	37·5
22	14/8	1430	110°	3–4	3	Low	Steady	37·5
23	14/8	1545	130°	3	3	Slight	Steady	37·5
24	14/8	1655	130°	3	3	Slight	Steady	—

TABLE IF

Distance of sample stations from the central complex in nautical miles and metres

Station number	*Nautical miles*	*Metres*	*Station number*	*Nautical miles*	*Metres*
1	3·05	5654	13	0·70	1298
2	1·93	3578	14	0·40	742
3	0·93	1724	15	0·23	426
4	0·40	741	16	3·30	6118
5	0·13	241	17	2·25	4171
6	3·33	6173	18	1·16	2151
7	2·16	4004	19	0·63	1168
8	1·00	1853	20	3·15	5840
9	0·66	1223	21	2·20	4079
10	3·40	6303	22	1·06	1965
11	2·30	4264	23	0·53	983
12	1·23	2280	24	0·12	229

TABLE II

Numbers of zooplanktonic organisms per m^3 at various times of day, within 6000 m of the Ekofisk Central Complex

	Time of Day						
Species	0130	0230	0930	1045	1430	1730	2100
Copepods	3482	3234	1766	1395	903	1133	2687
Eggs	82	171	87	80	88	72	55
Oikopleura dioica	38	42	77	45	22	16	19
Nauplii	38	55	43	16	19	58	88
Gastropod veligers	88	64	10	13	2	19	126
Bivalve larvae	9	5	1	11	0	6	3
Echinoderm larvae	6	8	3	0	0	3	6
Trochophore larvae	0	17	0	5	0	5	116
Sagitta setosa	3	25	10	13	2	11	13
Tomopteris sp.	0	3	0	0	0	2	0
Cladocerans	6	9	3	3	14	0	3
Cyphonauts	3	3	0	0	0	0	3
Anthomedusae	6	3	0	0	0	0	2
Amphipods	0	3	0	0	0	0	0

TABLE III

The occurrence of phytoplanktonic organisms at selected stations in the Ekofisk Field, August 1973

Species	*Station number* 3	6	9	10	15	17	20
Coscinodiscus excentricus	+	+	+	+	+	+	+
Coscinodiscus lineatus	+	+		+	+	+	+
Navicula sp.	+						+
Rhizosolenia styliformis	+	+	+	+	+	+	+
Bacillaria paradoxa					+		
Biddulphia alternans					+		
Pleurosigma sp.		+	+			+	+
Chaetoceros teres	+	+	+	+	+	+	
Chaetoceros densus	+	+					
Chaetoceros decipiens		+	+	+	+	+	
Diplosalis acuta		+	+	+	+		
Peridinium sp.			+	+		+	+
Ceratium longipes	+	+	+	+	+	+	+
Ceratium extensum		+		+	+		+
Ceratium macroceros		+		+	+	+	+
Ceratium lineatum	+	+	+	+	+	+	+
Ceratium tripos		+					
Ceratium candelabrum				+			

TABLE IV

The organic carbon content of sediments at each sample station in the Ekofisk Field, August 1973

Station number	*Organic carbon*	*Station number*	*Organic carbon*	*Station number*	*Organic carbon*
1	3·8	9	6·1	17	3·8
2	8·4	10	6·1	18	4·6
3	7·2	11	5·3	19	3·0
4	6·8	12	2·7	20	4·2
5	6·5	13	4·6	21	3·0
6	6·1	14	5·3	22	3·4
7	4·6	15	4·2	23	2·7
8	6·5	16	5·3	24	2·7

TABLE V

Results of particle size analysis of sediments at the 24 sample sites, Ekofisk Field, August 1973

Sieve aperture (μm)	*Weight* (g)	*Cumulative %*	*Weight* (g)	*Cumulative %*
	Station 1		*Station 2*	
> 1400	0·515	0·3447	0·010	0·0077
1000	0·025	0·3612	0·016	0·0201
710	0·980	1·0167	0·105	0·1013
500	0·153	1·1904	0·078	0·1617
355	2·319	2·6702	0·312	0·4030
250	8·245	8·1851	4·452	3·8470
125	118·474	87·4303	111·398	90·0210
63	17·740	99·2963	12·120	99·3970
< 63	1·052	100·0000	0·780	100·0000
	Station 3		*Station 4*	
> 1400	0·045	0·0287	0·010	0·0053
1000	0·089	0·0854	0·038	0·0255
710	0·085	0·1396	0·150	0·1051
500	0·925	0·7295	0·475	0·3570
355	1·915	1·9510	3·370	2·1450
250	9·371	7·9260	14·899	10·0510
125	131·595	91·8380	161·452	95·7240
63	11·220	98·9930	6·980	99·4280
< 63	1·580	100·0000	1·078	100·0000
	Station 5		*Station 6*	
> 1400	0·000	0·0000	0·023	0·0160
1000	0·012	0·0069	0·014	0·0250
710	0·064	0·0441	0·068	0·0720
500	0·280	0·2070	0·279	0·2640
355	1·243	0·9280	0·799	0·8100
250	13·783	8·9270	13·703	10·1890
125	147·405	94·4720	117·270	90·4530
63	8·510	99·4720	13·368	99·6030
< 63	1·015	100·0000	0·580	100·0000
	Station 7		*Station 8*	
> 1400	0·055	0·0270	0·237	0·1480
1000	0·029	0·0410	0·060	0·1860
710	0·082	0·0810	0·201	0·3120
500	0·149	0·1680	0·530	0·6440
355	2·940	1·5810	2·655	2·3070
250	19·075	10·8460	19·884	14·7620
125	169·252	93·0530	121·941	91·1460
63	13·741	99·7280	12·965	99·2670
< 63	0·561	100·0000	1·170	100·0000

TABLE V—*contd.*

Sieve aperture (μm)	*Weight* (g)	*Cumulative %*	*Weight* (g)	*Cumulative %*
	Station 9		*Station 10*	
> 1400	0·015	0·008	0·100	0·078
1000	0·034	0·027	0·020	0·093
710	0·117	0·092	0·061	0·141
500	0·317	0·269	0·125	0·238
355	2·822	1·841	3·052	2·611
250	24·905	15·712	14·110	13·582
125	138·648	92·933	105·135	95·329
63	11·969	99·599	5·578	99·666
< 63	0·720	100·000	0·429	100·000
	Station 11		*Station 12*	
> 1400	0·967	0·386	0·265	0·099
1000	0·160	0·450	0·199	0·174
710	0·246	0·549	0·258	0·271
500	0·338	0·684	0·338	0·397
355	0·899	1·043	1·065	0·796
250	21·675	9·707	21·658	8·914
125	198·873	89·203	204·988	85·743
63	25·509	99·399	33·942	98·464
< 63	1·502	100·000	4·099	100·000
	Station 13		*Station 14*	
> 1400	0·100	0·052	0·050	0·023
1000	0·155	0·132	0·045	0·044
710	0·306	0·290	0·082	0·083
500	0·378	0·486	0·130	0·143
355	1·088	1·049	2·040	1·097
250	19·855	11·327	15·730	8·446
125	148·211	88·045	160·442	83·412
63	21·965	99·415	33·438	99·036
< 63	1·130	100·000	2·063	100·000
	Station 15		*Station 16*	
< 1400	0·214	0·092	0·748	0·386
1000	0·015	0·098	0·106	0·409
710	0·063	0·125	0·242	0·565
500	0·118	0·176	0·364	0·754
355	2·660	1·318	1·088	1·315
250	17·108	8·660	19·855	11·565
125	181·420	86·524	148·211	88·077
63	29·820	99·322	21·965	99·416
< 63	1·579	100·000	1·130	100·000

TABLE V—*contd.*

Sieve aperture (μm)	*Weight* (g)	*Cumulative %*	*Weight* (g)	*Cumulative %*
	Station 17		*Station 18*	
> 1400	0·043	0·017	0·208	0·075
1000	0·070	0·044	0·089	0·107
710	0·063	0·068	0·233	0·191
500	0·300	0·195	0·451	0·353
355	4·335	1·881	1·427	0·867
250	18·950	9·259	21·385	8·571
125	217·879	94·087	232·670	92·383
63	14·130	99·589	19·323	99·344
< 63	1·056	100·000	1·822	100·000
	Station 19		*Station 20*	
> 1400	0·100	0·030	0·382	0·176
1000	0·222	0·098	0·148	0·244
710	0·204	0·159	0·090	0·285
500	0·668	0·363	0·237	0·394
355	5·267	1·965	2·310	1·455
250	16.335	6·932	15·177	8·429
125	260·295	86·080	172·482	87·686
63	43·545	99·321	25·723	99·506
< 63	2·233	100·000	1·075	100·000
	Station 21		*Station 22*	
> 1400	0·232	0·082	0·240	0·086
1000	0·025	0·091	0·095	0·120
710	0·093	0·124	0·118	0·163
500	0·417	0·272	0·370	0·297
355	2·402	1·123	5·056	2·114
250	15·857	6·747	24·020	10·753
125	231·393	88·731	223·878	91·268
63	30·471	99·527	22·420	99·367
< 63	1·334	100·000	1·760	100·000
	Station 23		*Station 24*	
> 1400	0·603	0·201	0·105	0·045
1000	0·298	0·300	0·182	0·122
710	0·488	0·463	0·399	0·292
500	0·705	0·698	0·595	0·546
355	1·614	1·236	1·230	1·069
250	19·758	7·830	11·868	6·123
125	248·712	90·699	192·592	88·141
63	26·643	99·577	26·560	99·515
< 63	1·269	100·000	1·288	100·000

TABLE V—*contd.*

Skewness of sediment particle size analysis curves

1	−0·005	13	0·010
2	+0·070	14	0·000
3	0·000	15	0·010
4	−0·050	16	0·000
5	−0·005	17	0·000
6	0·000	18	0·000
7	−0·005	19	−0·010
8	0·040	20	0·015
9	−0·010	21	−0·015
10	−0·015	22	−0·010
11	0·000	23	0·000
12	0·000	24	0·010

Quartile deviation of sediment particle size analysis curves

1	0·315	13	0·300
2	0·290	14	0·320
3	0·300	15	0·320
4	0·220	16	0·320
5	0·295	17	0·290
6	0·300	18	0·300
7	0·295	19	0·320
8	0·270	20	0·325
9	0·340	21	0·325
10	0·315	22	0·310
11	0·300	23	0·300
12	0·300	24	0·300

TABLE VI

Median particle diameter of the sediment at each of the 24 sample stations, Ekofisk Field, August 1973

Station number	*Median particle diameter* (μm)	*Station number*	*Median particle diameter* (μm)
1	173	13	173
2	183	14	170
3	178	15	173
4	182	16	176
5	180	17	180
6	178	18	178
7	166	19	170
8	199	20	174
9	183	21	173
10	183	22	179
11	175	23	176
12	169	24	173

TABLE VII

The mean and range of cumulative oversize percentages for all sediment fractions at the 24 sample stations combined, Ekofisk Field, August 1973

Sieve mesh size (μm)	*Cumulative % mean*	*Maximum*	*Minimum*
>1400	0·1007	0·386	0·000
1000	0·1449	0·450	0·007
710	0·2388	1·017	0·044
500	0·4126	1·119	0·143
355	1·4823	2·670	0·403
250	9·4693	15·712	3·847
125	90·1031	95·724	83·412
63	99·3820	99·728	98·464
<63	100·0000	100·000	100·000

TABLE VIII

The results of GLC analysis of water samples at 8 sites in the Ekofisk field. Samples at site 6, 8, 12, 14 and 15 were from the surface and 200 ft at each. The others are as indicated.

Stn	*Depth*	*C No.*	*Conc.* (ppm)	*Total** (ppm)	*Stn*	*Depth*	*C No.*	*Conc.* (ppm)	*Total** (ppm)
5	Surface	24	0·002 75		5	200 ft	24	0·000 13	
		25	0·003 91				25	0·000 51	
		26	0·005 76				26	0·002 00	
		27	0·006 16				27	0·000 36	
		28	0·006 28				28	0·000 58	
		29	0·004 78				29	0·000 88	
		30	0·004 78				30	0·000 86	
		31	0·002 90				31	0·000 82	
		32	0·001 99	0·0143			32	0·000 63	
							33	0·000 44	0·0072
24	Surface	24	0·000 52		24	200 ft			
		25	0·000 84						
		26	0·000 95						
		27	0·001 17				27	0·001 11	
		28	0·001 45				28	0·001 50	
		29	0·001 26				29	0·000 25	
		30	0·001 26				30	0·000 87	
		31	0·000 84				31	0·001 67	
		32	0·000 63	0·0074			32	0·001 17	
							33	0·001 33	0·0079
1	Surface				1	200 ft	23	0·000 51	
		24	0·000 44				24	0·001 16	
		25	0·000 71				25	0·001 35	
		26	0·001 27				26	0·000 96	
		27	0·001 72				27	0·001 74	
		28	0·002 01				28	0·001 16	
		29	0·002 24				29	0·002 35	
		30	0·002 01				30	0·001 62	
		31	0·001 84				31	0·001 09	
		32	0·001 29				32	0·001 09	
		33	0·001 04	0·0146			33	0·001 22	0·0111

Note: Samples were also taken at stations 6, 8, 12, 14 and 15 but no measurable amounts of *n*-paraffins were found.

* Total concentration within the given range of carbon numbers.

TABLE IX

The results of UV analysis of water samples taken in the Ekofisk field. The results are from a comparison with a Kuwait crude oil standard.

Station	*Depth*	*Concentration* (ppm)
1	Surface	0·038
1	200 ft	0·035
5	Surface	0·041
5	200 ft	0·012
6	Surface	0·033
6	200 ft	0·009
8	Surface	0·033
8	200 ft	0·009
12	Surface	0·013
12	200 ft	0·009
14	Surface	0·674
14	200 ft	0·009
15	Surface	0·035
15	200 ft	0·015
24	Surface	0·013
24	200 ft	0·009

TABLE X

Oil in the mud samples by UV analysis. The right-hand column of figures are the results after treatment with Florisil. The results are from a comparison with a Kuwait crude oil standard.

Station	*Total conc. in extract* (ppm)	*Conc. of mineral oil in extract* (ppm)
2	7	0·458
6	15	15
10	15	15
15	7	7
16	1·04	0·916
24	30	22·5

REFERENCES

Barret, J. and Yonge, C. M. (1958). *Collins Pocket Guide to the Sea Shore*, Collins, London, 272 pp.

Bell, T. (1853). *British Stalk-eyed Crustacea*, Woodfall & Kinder, London.

Chevreux, E. and Fage, L. (1925). *Faune de France*, Vol. **9**, *Amphipods*, Lechevalier, Paris, 488 pp.

Clark, R. B. (1960). *The Fauna of the Clyde Sea Area: Polychaeta*, Bell & Bain, Glasgow, for Scottish MBA ,Millport.
Day, J. H. (1967). *The Polychaeta of Southern Africa, I and II*, Brit. Mus. (Nat. Hist.).
Edinburgh Oceanographic Society (1973). Continuous Plankton Records: a plankton atlas of the North Atlantic and North Sea. *Bull. Mar. Ecol.*, **7**, 1–174.
Fauvel, P. (1923). *Faune de France*, Vol. **5**, *Polychaetes errantes*. Lechevalier, Paris, 488 pp.
Fauvel, P. (1927). *Faune de France*, Vol. **16**, *Polychaetes sédentaires*, Lechevalier, Paris, 494 pp.
Forbes, E. and Hanley, S. (1853). *A History of British Mollusca, and Their Shells*, 4 vols, Van Voorst, London.
Graham, A. (1971). *British Prosobranchs*, Synopses of the British Fauna, **2**, Academic Press, London.
Holme, N. A. and McIntyre, A. D. (1971). *Methods for Study of the Marine Benthos*, IBP Handbook 16, Blackwell, Oxford, 334 pp.
Lebour, M. V. (1930). *The Planktonic Diatoms of Northern Seas*, Ray Soc. Monograph, London, 224 pp.
Leeper, G. W. (1948). *Introduction to Soil Science*, Melbourne Univ. Press, Melbourne, 222 pp.
MacMillan, N. F. (1968). *British Shells*, F. Warne, London.
Margalef, R. H. (1968). *Perspectives in Ecological Theory*, Univ. of Chicago Press, Chicago, 111 pp.
Millar, R. H. (1970). *British Ascidians*, Synopses of the British Fauna, **1**, Academic Press, London, 92 pp.
Morgans, J. F. C. (1956). Note on the analysis of shallow-water soft substrata. *J. Anim. Ecol.*, **25**, 367–87.
Newell, G. E. and Newell, R. C. (1963). *Marine Plankton*, Hutchinson, London, 221 pp.
Odum, E. P. (1959). *Fundamentals of Ecology*, 4th ed., Saunders, London.
Sanders, H. L. (1968). Marine benthic diversity: a comparative study. *Amer. Naturalist*, **102**, 243–82.
Sars, G. O. (1895). *An Account of the Crustacea of Norway*, Vol. 1, Mallingske Boktrykken, Christiana.
Southward, E. C. (1972). Keys for the identification of Echinodermata of the British Isles (unpublished manuscript).
Tebble, N. (1966). *British Bivalve Seashells*, Brit. Mus., London, 212 pp.
Wimpenny, R. S. (1966). *The Plankton of the Sea*, Faber & Faber, London.

18

An Oil Spill in the Straits of Magellan

JENIFER M. BAKER
(Oil Pollution Research Unit),

ITALO CAMPODONICO, LEONARDO GUZMAN, JEAN JORY TEXERA, BILL TEXERA, CLAUDIO VENEGAS
(Instituto de la Patagonia)

and ALFREDO SANHUEZA
(Instituto de Fomento Pesquero)

SUMMARY

On 9 August 1974 the Metula *grounded at Satellite Patch in the eastern Straits of Magellan, Chile. About 47 000 tons of light Arabian crude oil and 3000–4000 tons of Bunker 'C' are estimated to have been lost between this date and the successful flotation on 25 September.*

As a result of the spill, large volumes of 'chocolate mousse' were produced, and most of this landed on Tierra del Fuego beaches between Punta Piedra and Punta Anegada. Calculations based upon measurements made by the Chilean Marines indicate that a little over 45 000 m^3 of mousse landed on this 77 km stretch of shore. An undetermined amount of mousse also landed on some beaches of the mainland and at other localities on Tierra del Fuego. Estimates of the volume of oil contained in this mousse, and of the proportions of the spilt oil that evaporated, dispersed, dissolved or absorbed on to sediments, vary, and there are insufficient data for accurate calculations.

The affected shores were relatively poor in macroflora and macrofauna, especially from the mid-intertidal zone to the supralittoral zone. However, towards the sublittoral zone there is a much richer flora and fauna and there some mussel beds and algae were contaminated.

The main bird species affected were cormorants and the Magellan penguin. 1500–2000 individuals are estimated to have been killed up to the beginning of October 1974; since that time there is evidence of continued oiling and the February 1975 estimate is a total of 3000–4000 individuals killed. It does not seem likely, however, that the losses will have a serious effect on the total populations of these species.

Limited surveys in the kelp beds during September 1974 showed that kelp near polluted shores still supported a good number and variety of animals. Apart from one catch of tainted fish, no reports have been received of damage to commercial fisheries, most of which are well to the south of the affected area.

INTRODUCTION

On 9 August 1974 the *Metula* grounded at Satellite Patch in the eastern Straits of Magellan, Chile (Figs. 1 and 2), whilst carrying approximately 190 000 tons of light Arabian crude oil from the Persian Gulf to Chile. Initial spillage of Bunker 'C' and crude oil from the forward part was about 6000 tons. A few days later the tidal current swung the ship's stern on to the bank and further spillage occurred. The total lost between grounding and the successful flotation on 25 September is estimated as 50 000–52 000 tons.

It has been calculated that a little over 45 000 m^3 of 'chocolate mousse' were deposited on Tierra del Fuego beaches, mainly between Punta Piedra and Punta Anegada. An undetermined amount of mousse was deposited on Punta Catalina and in two inlets at Punta Espora (Tierra del Fuego) and between Posesión and Punta Dungeness on the mainland. The loading on the mainland was smaller than the loading on Tierra del Fuego. Oil not incorporated in the mousse could have evaporated, dispersed towards and into the Atlantic Ocean, or become absorbed on to sediments.

This paper is based on a number of surveys carried out during August, September and October 1974 and January and February 1975. It describes the short-term effects of the oil spill and discusses possible longer-term effects and other problems in the eastern Straits of Magellan.

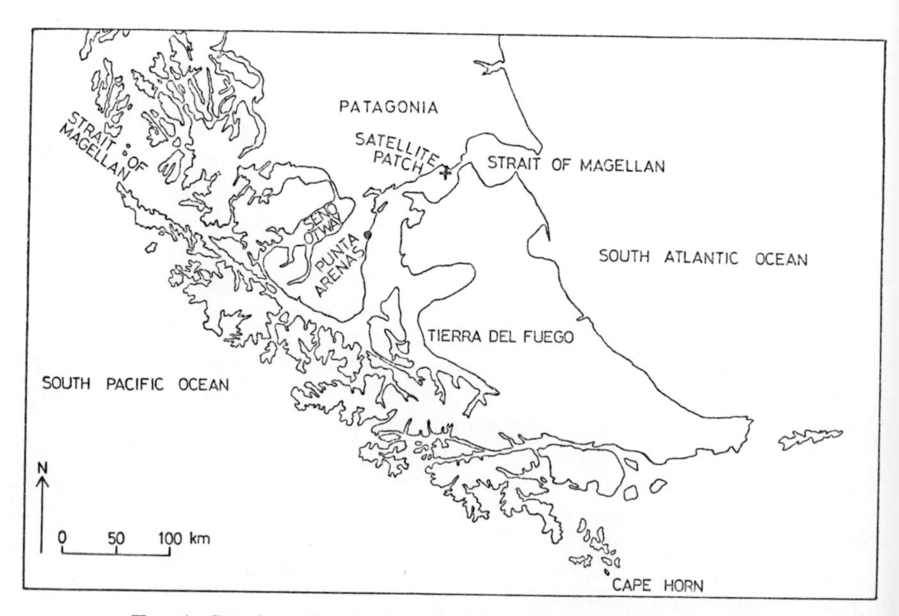

FIG. 1. Southern South America, showing Straits of Magellan.

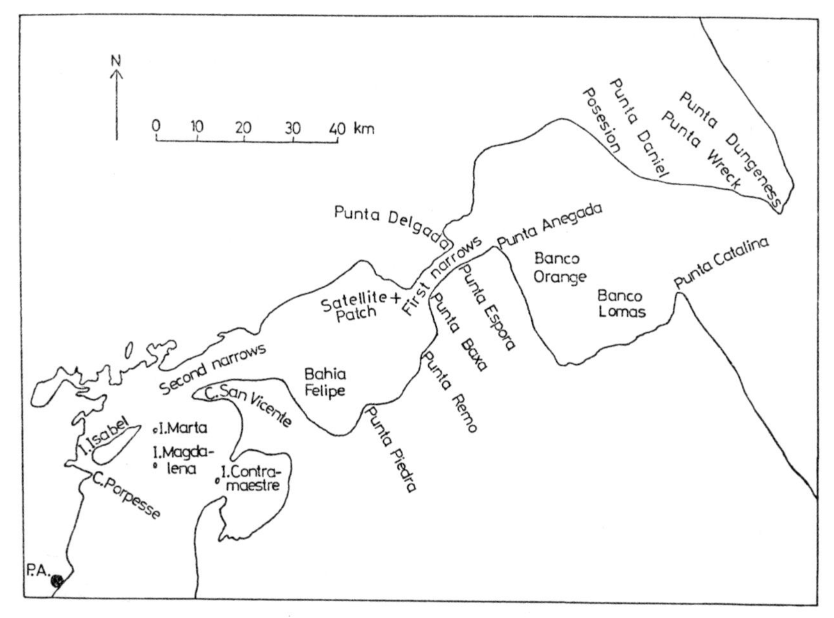

FIG. 2. Eastern Straits of Magellan, showing site of tanker grounding (Satellite Patch). (Based upon British Admiralty Charts with the permission of the Controller of HM Stationery Office and of the Hydrographer of the Navy.)

The climate of the north-east sector of the Straits is that corresponding to Cold Steppe (BSK), following Koppen's classification (Fuenzalida, 1967). Topographically both coasts of this sector are flat or slightly undulating, with fluvial glacial origins. The vegetation is that of Patagonian Steppe which has been characterised by Pisano (1971). According to Jerez and Arancibia (1972) this area is included between the isohyets of 250 to 300 mm per year. Winds are strong and sustained, predominantly from west and south-west directions, with the force and frequency increasing in spring and summer. There are sometimes squalls of more than 120 km/h.

The intertidal zone consists largely of shingle, gravel and sand, with mudflats and saltmarshes in places, but in some localities there are also shores with boulders and large stones. Rocky shores are very rare north of Punta Arenas, but there are occasional areas of hard clay. Kelp beds are common offshore nearly everywhere in the Straits, and are shown on the Admiralty and Chilean large-scale charts. All over the region there is a semidiurnal tide pattern with a periodicity of approximately 6 h 15 min between each high and low tide (Alveal *et al.*, 1973). As Atlantic and Pacific tidal characteristics differ considerably, and the Straits form a narrow channel linking the two, tidal streams are strong and reach a rate of 8–9 knots in the narrows at spring tides. Accurate predictions of tidal times and ranges are not possible from the tide tables as there is marked variability over short distances. In some areas the tidal range may reach approximately 10 m or more.

According to Fuenzalida (1967), Punta Arenas has a transandine climate with Steppe Degeneration (Cfc) following Koppen's classification. The mean annual temperature is 6·7°C, the coldest month being July with a mean temperature of 2·5°C and the warmest month January with a mean of 11·7°C. The yearly rainfall pattern is rather uniform with values ranging from 350 to 450 mm per year (Jerez and Arancibia, 1972).

Very little modern literature on the biology of the area is available, and the taxonomy of many groups of organisms is in a primitive state.

MOVEMENTS AND FATE OF *METULA* OIL

Observations from a number of aerial surveys (Fig. 3) give a general idea of oil movements during the three weeks following the grounding. Initial deposits of mousse were on the northern shore of Tierra del Fuego, mainly

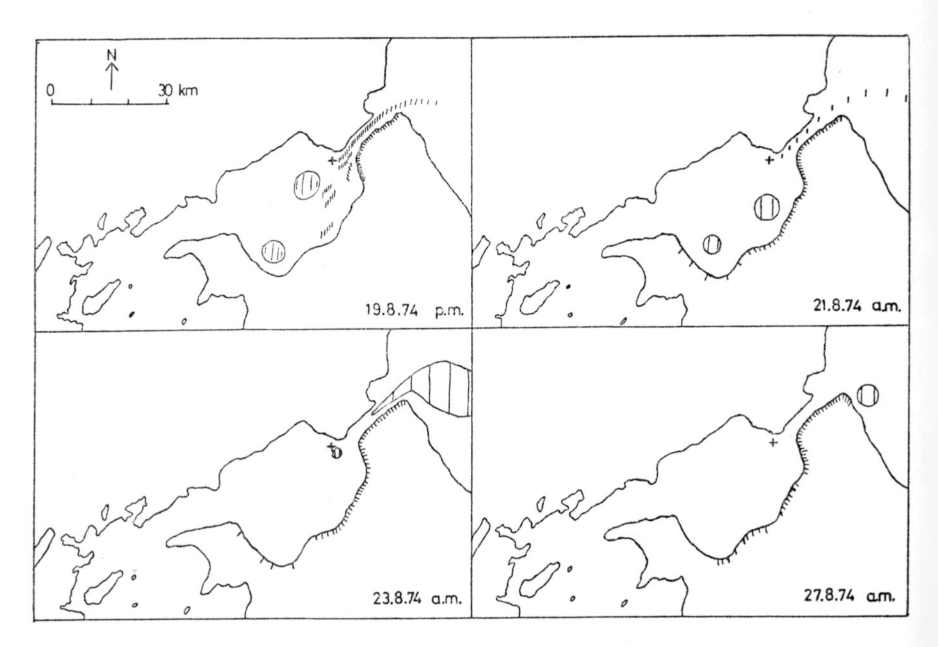

FIG. 3. Aerial survey observations, showing oil movements. Thin oil films are indicated by widely spaced lines, thicker oil films by closely spaced lines; patches of sheen are circled. (From information supplied by J. Wardley Smith.)

between Punta Anegada and Punta Piedra. The survey on 11 September showed new deposits on the mainland between Posesión and Punta Dungeness; these probably resulted from new spills and/or redistribution during a period of strong south-west winds the preceding week.

During some of the surveys, oil was seen moving towards the Atlantic past Punta Anegada, thinning out quickly. The limit of visibility of patches of sheen was about 50 km outside the eastern Straits entrance (J. Butt, pers. com., 1975).

As much as possible of the intertidal zone was examined on foot during August and September 1974 and January and February 1975, a total of about 100 km being covered altogether. This included a number of unpolluted sites as indicated by the aerial survey, and the following information was obtained which the aerial surveys did not give:

(a) Mousse present on Banco Orange and Punta Catalina saltmarshes.
(b) Mousse buried or covered with blown sand in several places.
(c) Non-*Metula* oil present in many places in comparatively very small quantities, usually in the form of tarballs (2–10 cm in diameter), or weathered patches on stones.

The distribution of surface mousse deposits is shown in Fig. 4.

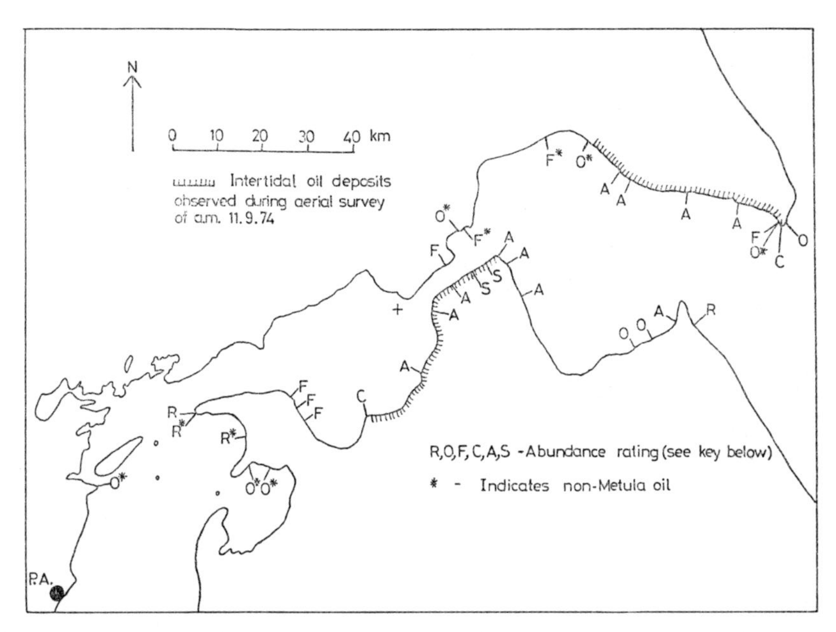

FIG. 4. Intertidal surface oil deposits observed during aerial survey of 11 September 1974 and from the ground subsequently. Abundance scale: R, Rare: one or two small spots. O, Occasional: widely scattered small spots. F, Frequent: numerous small spots or scattered larger patches. C, Common: numerous large patches, sometimes thick. A, Abundant: continuous belts of mousse, often thick. S, Superabundant: large accumulations of mousse in inlets. (Based upon British Admiralty Charts with the permission of the Controller of HM Stationery Office and of the Hydrographer of the Navy.)

Although the mousse layers were unstable, measurements of the width and thickness of the mousse on Tierra del Fuego beaches (Table I) were obtained every 1 km by Chilean Marines, during low tides (14–27 September). The thickness was measured at three points across the band of mousse (top, middle and bottom).

TABLE I
Width and Thickness of Mousse on Most Affected Sector of Tierra del Fuego Shoreline

Date	*Distance* (km)	*Width of band of mousse* (m)	*Thickness of band of mousse*		
			Top	*Middle* (cm)	*Bottom*
24 Nov. 1974	0	—	—	—	—
Punta Piedra	1	2	0·5	0·5	0·5
Punta Remo	2	2	0·5	0·5	0·5
	3	3	0·5	0·5	0·5
	4	2	0·5	0·5	0·5
25 Nov. 1974	5	2	0·5	0·5	0·5
	6	3	1	1	1
	7	3·5	1	1	1
	8	3·5	1	1	1
	9	4	1	1	1
	10	4	1·5	1·5	1·5
	11	4	1·5	1·5	1·5
	12	5	2	2	2
	13	5	2	2	2
26 Nov. 1974	14	6	2	2	2
	15	8	2	2	2
	16	9	3	3	3
	17	9	3	3	3
	18	8	3	3	3
	19	7	3	3	3
	20	7	3	3	3
	21	8	3	3	3
	22	8	3	3	3
27 Nov. 1974	0	3	6	4	1
Punta Remo	1	4	4	1	1
Punta Baja	2	2	3	2	1
	3	5	3	3	2
	4	2	5	2	2
	5	4	2	3	1
	6	4	1·5	1	1
	7	3	2	1·5	2
	8	2	1	1·5	1·5
	9	5	2	2	1
	10	2	2	1	1
	11	4	1	1·5	1
	12	8	2	1	2
	13	8	3	2	2
	14	12	1	2	1
	15	12	14	11	5
	16	8	12	9	2
	17	10	9	6	2
	18	6	5	3	1
	19	4	5	2	2
	20	2	6	1	0·5
	21	2·5	4	2	1

TABLE I—*contd.*

Date	*Distance* (km)	*Width of band of mousse* (m)	*Thickness of band of mousse* *Top*	*Middle* (cm)	*Bottom*
27 Nov. 1974	22	1·5	2	1	0·5
Punta Remo	23	2	3	2	0·5
Punta Baja—*contd*	24	1	1	1	0·5
17 Nov. 1974	0	15	5	2	1·5
Punta Baja	1	16	2	5	2
Punta Mendez	2	7	0·5	5	2
	3	9	1	2	2
	4	10	1	1·5	3
	5	2	2	1·5	1
	6	3	2	1	1
17 Nov. 1974	0	3	2	1	1
Punta Mendez	1	10	2	3	3
Punta Espora	2	12	8	20	2·5
	3	7	2	3	1·5
	4	8	2	2·5	3
	5	3	1	1	2
	6	1·5	3	1	3
17 Nov. 1974	0	25	15	5	3
Punta Espora	1	50	30	15	5
Punta Anegada	2	7	12	5	3
	3	6	15	7	5
	4	20	10	8	4
	5	14	30	23	18
	6	20	10	5	3
	7	8	15	8	5
18 Nov. 1974	8	15	10	8	6
	9	6	8	4	3
	10	4	6	2	1
	11	3	6	3	1
	12	3	5	2	1
	13	1	2	2	1
	14	1	1	1	0·5
	15	0·5	1	0·5	0·5
24 Nov. 1974	0	1	5	5	3
Punta Anegada,	1	2	5	3	1
11 km towards	2	1·5	4	2	2
Banco Orange	3	1	2	2	0·5
	4	0·5	2	1	0·5
	5	Drops	1	0·5	0·25
	6	Drops	2	0·5	0·25
25 Nov. 1974	7	Drops	2	0·5	0·25
	8	Drops	2	1	0·5
	9	Drops	1	1	0·5
	10	—	—	—	—
	11	—	—	—	—

The width of the belt of mousse varied from 2 to 50 m, often from 2 to 8 m, and the thickness varied from 0·5 to 15 cm, often between 0·5 and 5 cm, the middle level being the thickest. Buried mousse was observed as follows:

(a) Large patches of mousse, 1–10 cm thick, buried under coarse sand and gravel to a depth of 10–15 cm (usually) and up to 30 cm (exceptionally), giving a 'quicksand' effect. Common between Punta Piedra and Punta Anegada, and near Posesión.
(b) Mousse absorbed into sand, i.e. mousse no longer apparent as discrete patches. Observed at intervals between Punta Piedra and Punta Anegada, near Posesión.
(c) Mousse covered with thin layers of blown sand. Observed near Punta Catalina and Cabo Orange.

The mousse appeared as two different layers, 'dark mousse' and 'fresh mousse', the first being much drier and located at the top of the high-tide mark. The fresh mousse was a light-brown semi-solid.

Mousse on and in shingle, gravel and coarse sand was being churned and mixed with these substrates, with re-flotation of small lumps at intervals. Run-off from affected shores produced patchy sheen and occasional mousse lumps in near-shore waters. Sheen was also observed coming from the kelp beds near affected shores, and occasional small flecks of mousse were caught in the wrinkles on the kelp fronds.

A calculation of the total volume of mousse on the shores was very difficult, as the deposits were unstable and sometimes patchy or partly buried. Based upon the information obtained by the Chilean Marines, the following calculation (Table II) was made of the total volume of mousse which landed on 77 km of Tierra del Fuego shores from Punta Piedra to 4 km past Punta Anegada. No measurements were made of the mousse in the inlets near Punta Espora, near Punta Catalina, or on the mainland. It should be emphasised that the mousse measured by the Chilean Marines represents only that portion of the oil that landed on the beaches of the more affected area.

TABLE II
Volume of Mousse (in m^3) in each Coastal Area of Tierra del Fuego

	Distance patrolled[a] (km)	*Top*	*Middle*	*Bottom*	*Total*
Punta Piedra–Punta Remo	22	1 613	1 613	1 613	4 839
Punta Remo–Punta Baja	24	3 842	3 133	1 435	8 410
Punta Baja–Punta Mendez	6	688	1 278	812	2 778
Punta Mendez–Punta Espora	6	835	1 973	687	3 495
Punta Espora–Punta Anegada	15	7 370	11 206	6 113	24 689
Punta Anegada	4	800	72	50	922
Total	77	15 148	19 275	10 710	45 133

[a] Distance was measured walking along the shore.

Hann (1974), from samples analysed by the US Coast Guard, assumed a 5% and 25% water content in the dark mousse and fresh mousse respectively. He estimated that each cubic foot of mousse will contain 50 lb of oil, which is equivalent to nearly 0·8 tons of oil per cubic metre of mousse. Using this estimate and the data of the Chilean Marines, not less than 70% of the spilt oil landed as mousse. This appears to be rather a high estimate, bearing in mind possible evaporative and other losses.

Original estimates of up to 20% given by Baker (1974) differ from the above mainly in the assumptions made about water content of the mousse. The work of Berridge *et al.* (1968) defined 'chocolate mousse' as emulsions of from 50% to 80% water content, and their experimental work and *Torrey Canyon* analyses showed mousse water contents to be often up to 80%. A fresh mousse sample separated in Punta Arenas contained roughly one-third oil, two-thirds water plus organic debris. It is clear that the mousse varied considerably in its properties both from place to place and with time. Also, it was continuously moving on the beaches, incorporating varying amounts of sand, gravel and organic debris, so accurate calculations as to the volume of oil involved are impossible.

Taking Hann's water content figures into account, 20% appears to be rather a low estimate. A reasonable figure between 20% and 70% is difficult to deduce and, in the absence of large numbers of mousse analyses, must be a matter of opinion.

If 70% of the spilt oil was stranded on the shore, this would leave 30% to be lost either by evaporation, sinking to the bottom, solution or dispersion. Hann estimates 15–20% lost by evaporation (pers. com., 1975). The estimate of Baker (1974) for evaporation and dispersed oil included an allowance of 30–35% evaporative loss. This was based on the experimental work and *Torrey Canyon* analyses described by Brunnock *et al.* (1968).

The relatively low temperatures in the Straits at the time of the spill would reduce evaporation rate, but on the other hand the high wind speeds and choppy seas would increase the rate and favour the 'aerosol effect' described by Nelson-Smith (1972). The *Torrey Canyon* samples analysed by Brunnock *et al.* showed an evaporative loss of about one-third, and the *Metula* oil was a lighter crude than that from the *Torrey Canyon*, as shown by the following distillation data (Table III). Evaporative losses are normally limited to materials boiling below 300°C (Brunnock *et al.*, 1968).

No measurements were made concerning dissolved or sedimented oil, or oil dispersing as droplets, mousse flecks, slicks or sheen. Some aerial observations showed oil moving from the eastern end of the First Narrows and dispersing in the Atlantic, being no longer visible 50 km from the Straits entrance.

TABLE III

	Kuwait crude (*Torrey Canyon*)	*Light Arabian crude* (*Metula*)
Total distillate to 149°C (% wt)	15·3	17·3
Total distillate to 232°C (% wt)	27·6	31·5
Total distillate to 343°C (% wt)	44·6	51·7

By early November 1974 the Chilean Navy reported that in all areas where there was sand or coarse gravel, the mousse had been covered. There was at this time still a band of weathered dried-out mousse at high-tide mark. Mousse was still visible on rocks, though in lesser quantities. Later observations made in January and February 1975 showed that along the worst-affected stretches of beaches, the sand covered a thick layer of mousse that was still fresh and only partially mixed with sand in many places. There has been a general easterly drift of mousse and deposits have been building up in protected areas of little wave action, notably the two inlets at Punta Espora, the area of greatest biological damage, which are likely to remain oily for a long time.

CLEANING

Dispersants and spraying equipment were moved to the area as quickly as possible after the spill. However, the use of these to treat oil on the water was not practical because of the weather conditions and the mousse formation.

Mousse on the shore would have been difficult to treat either with dispersants or by physical removal, because of the large area involved and its isolation. The value of treatment in this case is open to question. From the ecological point of view the massive use of dispersants in areas of gravel, sand, mud and saltmarsh is likely to create further problems. Physical removal of mousse from the biologically richer areas of mussel beds and saltmarsh would have been difficult to achieve without further damage to the organisms involved. However, mousse on the beaches and especially in the inlets at Punta Espora constitutes a source of chronic pollution which is likely to remain for many years.

From the amenity point of view, physical removal of mousse from the upper parts of gravel and sand shores would improve the appearance, but few people live along most of the coastline and access for heavy machinery is very limited.

The two ferry landing slips at Punta Espora were in the end the only areas that were treated—with grit and gravel so that vehicles could continue to use them.

OTHER OIL

It was clear from the shore surveys (Fig. 4) that some minor localised pollution results from activities such as deballasting, well drilling and testing, and drilling mud disposal.

CONDITION OF FLORA AND FAUNA

Intertidal and Splash Zone

Mobile Shingle, Gravel and Coarse Sand

These shores are so unstable that there is little permanent macroflora and macrofauna, especially from the mid-intertidal zone to the supralittoral zone.

In places, masses of rotting washed-up kelp (*Macrocystis*) support temporary insect and amphipod populations, which in turn are fed upon by strandline birds. In general terms at least 80% of the upper and middle shores of the eastern Straits of Magellan are of this type, and most of the mousse has landed on such shores.

Stable Shingle, Hard Clay and Rock

In places, the mobile deposits of the upper shore are above a more stable type of lower shore, consisting of larger pieces of rock, stones embedded in mud, or outcrops of hard clay. Such places are comparatively rich in plants and animals. Completely rocky shores were not found at all in the eastern Straits north of Punta Arenas, the nearest one examined being Punta Santa Ana (approximately 60 km south of Punta Arenas, Fig. 1). The sites visited are shown on Fig. 5, and a summary of observations given in Table IV.

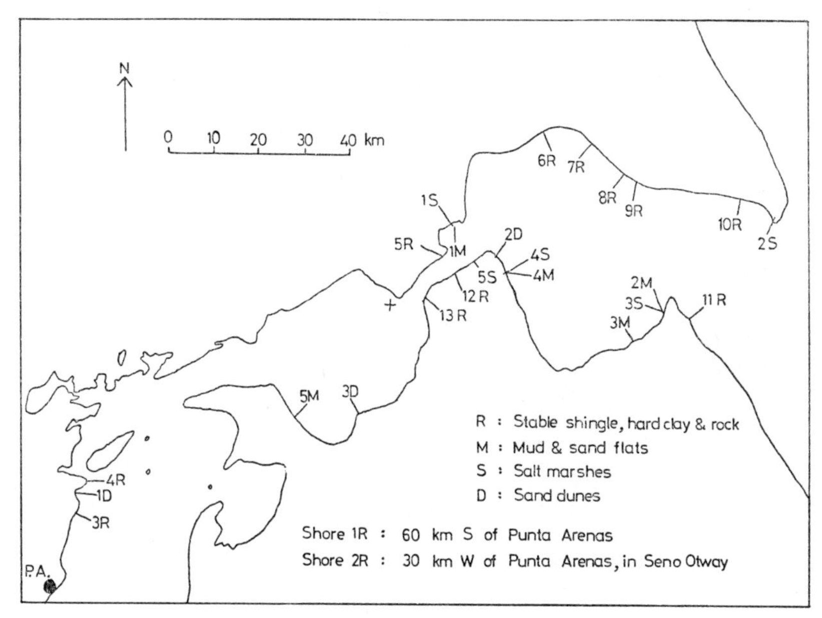

FIG. 5. Locations of shores visited. (Based upon British Admiralty Charts with the permission of the Controller of HM Stationery Office and of the Hydrographer of the Navy.)

Characteristically, this type of site is dominated by mussels, the common species being *Mytilus chilensis*. This is sometimes mixed with *Perumytilus purpuratus* and *Aulacomya ater*. Associated with the mussel beds are the algae *Porphyra* sp., *Ulva* spp., *Adenocystis* sp., *Enteromorpha* sp., *Iridaea* sp. and a variety of filamentous and leafy red algae, with the limpet *Nacella magellanica*, large chitons, and the red anemone *Corynactis*.

Where mousse landed on these sites, it was usually caught between the mussels and was patchy, only occasionally being thick enough to cover them.

TABLE IV
Summary of Observations on Stable Shingle, Hard Clay and Rocky Shores, 13–21 September 1974

Site	*Shore type*	*Flora, main types*	*Fauna, main types*	*Oil and mousse*	*Effects of oil and mousse*
1R	Bedrock, with pools	*Porphyra*, and a mixed algal turf (A–S)	*Mytilus* (A–S), *Chthamalus*, with *Nacella* (C), chitons (O) and *Corynactis* (C) in pools	None	None
2R	Stable shingle with pools	*Porphyra*	*Mytilus* (A–S), *Chthamalus*, with *Nacella* (C), chitons (O) and *Corynactis* (C) in pools	None	None
3R	Shingle in mud with pools	*Enteromorpha* (A), filamentous brown algae, diatoms	*Mytilus* (A) and small limpets (F), polychaetes, amphipods	None	None
4R	Shingle and boulders	*Enteromorpha* (A), *Porphyra* (F)	Mytilus (C) and small limpets (F)	Occ. non-Met.	None
5R	Shingle and boulders	Filamentous red algae (C), leafy red algae (C), *Ulva*, *Codium* and *Adenocystis* (all O)	*Mytilus* (S), *Nacella* (F), *Corynactis* (O)	Occ. mousse	Thin oil films on algae but plants appeared healthy
6R	Shingle with pools	Filamentous and leafy red algae (F), *Lithothamnion* (F)	*Mytilus* (A), *Nacella* (F), *Corynactis* (F), plus *Trophon* (O), other gastropods (O) and bivalves (F)	Occ. non-Met.	None
7R	Shingle with pools	Filamentous and leafy red algae (F), *Lithothamnion* (F)	,,	Occ. non-Met.	None
8R	Shingle with pools	Filamentous and leafy red algae (F), *Lithothamnion* (F)	*Mytilus* (O), *Nacella* (O)	Lower shore clean, upper shore with mousse	None

TABLE IV—*contd.*

Site	*Shore type*	*Flora, main types*	*Fauna, main types*	*Oil and mousse*	*Effects of oil and mousse*
9R	Shingle with large pools	Filamentous and leafy red algae (F), *Lithothamnion* (F) plus *Macrocystis* (A) in pools	*Mytilus* (A–S), *Nacella* (C–A), *Corynactis* (O)	Mousse patchy on mussels	Most mussels still alive
10R	Shingle, boulders and hard clay, with pools	Filamentous red algae (C) and *Lithothamnion*	*Mytilus* (A–S), *Nacella* (C–A), *Corynactis* (O) plus *Balanus* (C), various gastropods (O–F) and asteroids	Mousse patchy on mussels	Most mussels still alive
11R	Hard clay with pools	*Enteromorpha* (C), filamentous brown algae (A), *Porphyra* (F)	*Mytilus* (A), *Nacella* (O), boring bivalves (O), nemertines	None	None
12R	Shingle in mud with pools	Filamentous red and brown algae (O), *Ulva* (O) *Porphyra* (O)	*Mytilus* (C), *Nacella* (F), chitons (O), asteroids, small polychaetes (more than 100 per m^2), *Corynactis* (F)	Mousse common on mussels, etc.	*Porphyra* and other algae sick. Occ. dropped-off limpets. Most mussels still alive
13R	Shingle in mud with pools	Filamentous red and brown algae (O), *Ulva* (O) *Porphyra* (O)	*Mytilus* (A), *Nacella* (C), *Corynactis* (C)	Mousse common on mussels, etc.	*Porphyra* and other algae sick. Occ. dropped-off limpets. Most mussels still alive

Notes:
1. The letters O–S refer to abundance ratings (see abundance scales given in Paper 5, this volume).
2. Only the commoner organisms are listed in this summary.

TABLE V

Observations on Mud and Sand Flats, 13–21 September

Site	*Observations*
1M	Mud with 30 cm ridges, and pools. Diatoms abundant on mud surface. Large (10 cm) polychaetes, family Nereidae, common. Flatworms (Platyhelminthes) common in pools. No oil.
2M	Fine dark sand, in places hard and layered. Old worm burrows in places, filled with sand. Patches of mousse, some covered with blown sand.
3M	Fine dark sand with occasional polychaetes. Spots of mousse rare.
4M	Sandy mud, polychaetes generally rare. Polychaetes (family Nereidae) abundant, however, in beds of some of the saltmarsh creeks. Oil not present on mud as it is trapped on saltmarsh plants at a higher level.
5M	Sandy mud with abundant sedentary polychaetes in sand-grain tubes (family Sabellidae). Gulls and oystercatchers feeding. Mousse rare on mud, worms healthy and spawning.

TABLE VI

Observations on Saltmarshes, 13–21 September

Site	*Observations*
1S	Dominant *Salicornia ambigua*, *Festuca* and *Armeria* at higher levels. *Monostroma* and amphipods in pans. A broad marsh with numerous creeks and pans, occasional chunks of drilling mud. East end of marsh has patches of stones with *Verrucaria* and some hard old oil patches.
2S	Marsh fringing Shark Creek. Dominant *Salicornia ambigua*. Mousse common along strandline for 800 m along creek, but not thick.
3S	Patch of saltmarsh with one main creek. Dominant *Salicornia ambigua*. Abundant mousse in creek, patches on vegetation.
4S	Saltmarsh 200–300 m wide, with numerous creeks and pans. Two oiled zones each up to 50 m wide, mousse abundant in these zones but not thick. Dominant *Salicornia ambigua*, underground systems of these perennial plants not affected by oil.
5S	Dominant *Salicornia ambigua*, with *Senecio patagonicus* at higher levels and *Lepidophyllum cupressiforme* on the edges of the marsh. Much *Salicornia* covered with mousse, *Lepidophyllum* affected to a minor extent. When revisited in January 1975, plants were re-sprouting in the more thinly covered areas. The woodier *Senecio* and *Lepidophyllum* were sprouting less than the *Salicornia*. This saltmarsh is in an area where mousse deposits have been accumulating as a result of a general easterly drift of deposits.

TABLE VII

Observations on Sand Dunes, 13–21 September

Site	*Observations*
1D	Pioneer *Senecio candidans*. Also present *Elymus arenarius*. No oil.
2D	Mainly *Elymus arenarius*. Patches of mousse blown into grass for up to 10 m. Underground systems in good condition. Post-oiling growth indicated by green shoots 10–15 cm long, with oil on the tips only.
3D	Mainly *Elymus arenarius*. Patches of mousse blown into grass for up to 5 m. Underground systems in good condition.

Very few dead mussels were observed in September, and as the beds were usually still crowded with mussels it may be assumed that not many had washed away. In February 1975 mussel beds were comparatively more affected (in terms of dead individuals) and were still covered by a thin layer of oil as a result of the run-off. Mousse-coated *Porphyra* on the mussel beds was frequently sick or dead, as were other species of algae, especially *Ulva* spp.

Pools on the lower levels of badly oiled beaches generally had oily sheen on them, but pool inhabitants usually appeared healthy. These pools are flushed out daily by the tides. Occasional sick and dead dropped-off limpets were observed at Punta Espora, but in September attached limpets were still frequent at this site.

Stable Fine Mud and Sand Flats, Saltmarshes and Sand Dunes

Summaries of the data for these three types of shore are given in Tables V to VII, and the locations of sites visited are shown in Fig. 5.

Giant Kelp Beds

The giant kelp *Macrocystis pyrifera* is common and widely distributed in the subtidal levels of the Straits of Magellan in depths up to 10–15 m. Although no studies are available on the organisms associated with the giant kelp beds of the Straits, some observations have demonstrated that a great variety of plants and animals largely depend on these beds, either living attached to the fronds, on the surface and/or inside the complicated structure of the holdfast, or simply being protected by the dense populations of the alga. No information is available on kelp bed predators in this area, but the beds appear to be favoured feeding areas of many birds and dolphins.

TABLE VIII
Macrocystis **Net Samples: Organisms found 29 September 1974**

Punta Espora site 1: Net breaking surface during tow. Nothing caught.

Punta Espora site 2:

	No. of individuals	*No. of species*
Platynereis magalhaensis Kinberg	2	1
Other Polychaeta	1	1
Isopoda	48	1
Amphipoda	14	3
Munida gregaria	1	1
Total	66	7

Punta Delgada:

	No. of individuals	*No. of species*
Isopoda	112 (many small specimens)	1
Amphipoda	58	3
Total	170	4

TABLE IX
Macrocystis **Holdfast Samples: Organisms found**
Punta Espora: Large holdfast collected 29 Sept. 1974. Size: *c*. 40 × 50 cm

	No. of individuals	*No. of species*
Filamentous red algae	Numerous tufts	2+
Lithothamnion spp.	Numerous patches	1+
Bryozoa	Numerous patches	3+
Potamilla sp.	3	1
Harmothoë sp.	2	1
Platynereis magalhaensis Kinberg	30	1
Thelepus sp. cf. *plagiostoma* Schmarda	1	1
Polynoinae sp.	10	1
Lumbrinerinae sp.	1	1
Golfingia margaritacea (Sars)	1	1
Holothuroidea white sp.	109	1
Holothuroidea dark-brown sp.	2	1
Halicarcinus planatus	5	1
Isopoda	4	2
Amphipoda	2	2
Chitonida	1	1
Fissurella sp.	8	1
Octopus sp.	1	1
Asteroidea	14 (includes 1 10 cm *Calvasterias* sp. brooding numerous uncounted young ones)	3
Unidentified eggs	31	1
Total	228+	27+

Bahia Felipe: Large holdfast collected Aug. 1974 after *Metula* stranding.
Size: *c*. 40 × 50 cm

Filamentous red algae	Numerous tufts	1+
Lithothamnion	Numerous patches, especially on *Aulacomya*	1+
Bryozoa	Numerous patches	2+
Potamilla sp.	8	1
Spirorbis	Frequent	1+
Other polychaetes	16	4
Halicarcinus planatus	2	1
Balanus sp.	3	1
Calyptraeidae	168	1
Aulacomya ater	24	1
Calvasterias sp.	2	1
Holothuroidea	1	1
Total	228+	16+

A mollusc dominated holdfast cf. Punta Espora sample.

TABLE IX—*contd.*

Punta Delgada: 2 small holdfasts collected 29 Sept. 1974. Size: 20–25 cm

	No. of individuals	*No. of species*
Potamilla sp.	1	1
Thelepus sp. cf. *plagiostoma* Schmarda	2	1
Platynereis magalhaensis Kinberg	3	1
Polynoinae sp.	1	1
Holothuroidea white sp.	2	1
Halicarcinus planatus	1	1
Eurypodius latrellei	1	1
Calvasterias sp.	1	1
Total	12	8

As pointed out by Braud *et al.* (1974), beds of *M. pyrifera* constitute an extremely favourable biotope for the development and reproduction of certain crustaceans and fishes. Algae, bryozoans, molluscs, copepods and fishes have been observed to be associated with the kelp beds in the Argentinian Patagonian region (Kuhnemann, 1970). Because of bad weather, only three sites were sampled: two near Punta Espora were areas subjected to constant run-off from the nearby abundant upper shore mousse deposits; one near Punta Delgada was clean. In addition, a preserved holdfast taken from Bahia Felipe after the *Metula* grounding was examined.

The Punta Espora and Punta Delgada sites were sampled from the lowered ramp of a landing craft. *Macrocystis* fronds were grappled aboard and examined, holdfasts were brought up on the ship's anchor, and net samples were taken over approximately 250 m tows through the upper fronds, using a net of mouth diameter 1 m, mesh 6 mm. The results are shown in Tables VIII and IX.

At no site did the *Macrocystis* itself appear unhealthy. From the very limited sample numbers it is, of course, impossible to quantify any oil effects, but the following points emerge: (a) at the time of the observations, the *Macrocystis* fronds and holdfast from the most oiled kelp beds supported a large number of healthy individual animals of a wide range of species; (b) holdfast size and age, and bottom characteristics in the area where it came from, appear to have a great effect on the fauna. However, a detailed study on the possible longer-term effects of oil run-off would be desirable.

Birds

Beached Bird Counts

At the end of August, approximately 250 dead oiled birds were counted along the stretch of coast between Punta Remo and Punta Anegada. Most were cormorants; there were also some penguins, gulls and other species (see Table X).

TABLE X
Bird Species Affected by Oil

Species seen killed by oil:	
Phalacrocorax atriceps	Blue-eyed Cormorant
Phalacrocorax albiventer	King Cormorant
Phalacrocorax magellanicus	Rock Cormorant
Spheniscus magellanicus	Magellanic Penguin
Eudyptes crestatus	Rockhopper Penguin
Larus dominicanus	Kelp Gull
Larus maculipennis	Brown-hooded Gull
Fulmarus glacialoides	Southern Fulmar
Oceanites oceanicus	Wilson's Storm Petrel
Polyborus plancus	Crested Caracara
Diomedea melanophris	Black-browed Albatross
Lophonetta specularioides	Crested Duck
Pelecanoides magellanicus	Magellanic Diving Petrel
Rollandia rolland	White-tufted Grebe
Species seen oiled but not dead:	
Zonibyx modestus	Winter Plover
Numenius phaeopus	Whimbrel
Chloëphaga picta	Upland Goose

Between 13 September and 2 October, birds were counted at the sites shown in Fig. 6. The results are given in Table XI. Dying birds were killed when appropriate. The main species affected were the cormorants *Phalacrocorax albiventer*, *P. atriceps* and *P. magellanicus*, the Magellan Penguin *Spheniscus magellanicus* and the Kelp Gull *Larus dominicanus*.

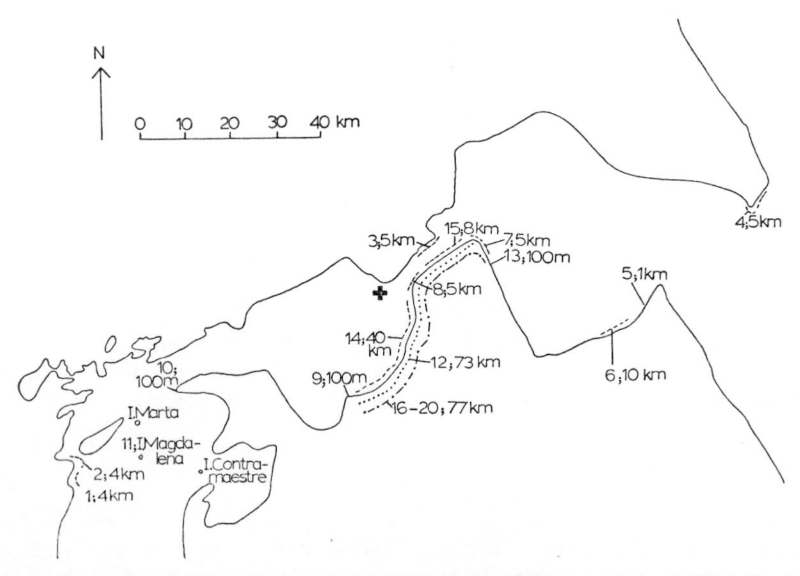

FIG. 6. Sites where birds were counted, with lengths of shore examined. (Based upon British Admiralty Charts with the permission of the Controller of HM Stationery Office and of the Hydrographer of the Navy.)

TABLE XI
Beached Bird Counts

Site	*Observers and date*	*Dead, oiled*	*Dead, not oiled*	*Live, oiled (usually sick)*	*Other comments*
1	J. Baker 14 Sept. 1974	1 cormorant	—	—	—
2	J. Baker 14 Sept. 1974	—	4 cormorants 5 gulls	2 cormorants	—
3	J. Baker J. Butt 15 Sept. 1974	*c*. 30 cormorants 4 penguins	—	2 cormorants 1 penguin	200–300 clean healthy cormorants
4a	J. Texera C. Venegas 16 Sept. 1974	42 penguins	—	10 penguins	A few hundred clean healthy penguins
4b	J. Baker J. Butt 2 Oct. 1974	13 cormorants 71 penguins 2 rockhopper penguins	8 cormorants 10 penguins 1 rockhopper penguin	13 penguins	About 20 000 clean healthy penguins. 1 sick unoiled cormorant. Oiled bird count probably includes a good proportion of those listed in 4a
5	J. Baker J. Butt 13 Sept. 1974	1 ? shearwater	—	—	—
6	J. Baker and Chilean Marines 20 Sept. 1974	—	—	—	—
7	J. Baker and Chilean Marines 20 Sept. 1974	136 cormorants 15 penguins 22 gulls	4 gulls	9 cormorants 6 penguins	—
8	J. Baker and Chilean Marines 19 Sept. 1974	—	—	6 cormorants 4 penguins	Only recently killed birds were counted
9	21 Sept. 1974	—	—	6 cormorants	About 20 clean cormorants
10	21 Sept. 1974	1 penguin	—	—	—
11	J. Texera 17 Sept. 1974	—	—	20 penguins	About 3000 clean healthy penguins

TABLE XI—*contd.*

Site	*Observers and date*	*Dead, oiled*	*Dead, not oiled*	*Live, oiled (usually sick)*	*Other comments*
12	J. Texera C. Venegas B. Texera R. Hann J. Wonham 1–2 Sept. 1974	149 cormorants 35 penguins 17 gulls 4 rockhopper penguins 7 other birds (see Table X)	—	7 cormorants 9 penguins 83 gulls 1 plover 1 whimbrel	
13	K. Raedeke 9 Sept. 1974	—	—	30 upland geese	—
14	J. Texera C. Venegas R. Hann C. Gunnerson K. Adams 12–13 Jan. 1975	22 cormorants 5 penguins 2 diving petrels	—	—	Only recently killed birds were counted
15	J. Texera C. Venegas R. Hann C. Gunnerson 14 Jan. 1975	23 cormorants	—	—	Recently killed birds
16	Chilean Marines 24–26 Sept. 1974	1 gull 4 cormorants 2 rockhopper penguins	—	—	Could include many of the birds counted in 12 (sites 16–20)
17	27 Sept. 1974	42 cormorants 8 penguins 14 rockhopper penguins	—	—	
18	Chilean Marines 17 Sept. 1974	17 cormorants 4 rockhopper penguins	—	—	
19	17 Sept. 1974	106 cormorants 2 penguins			
20	17–18 Sept. 1974	78 gulls 4 ducks 229 cormorants 30 penguins	—	6 penguins 13 cormorants	

The number of oiled birds (including live oiled birds) counted up to the beginning of October was 1320, but a minor percentage (perhaps 5–10%) could have been counted twice. During the reconnaissance of the area made during January 1975 it was possible to see comparatively recently killed oiled birds, primarily cormorants but including penguins and other birds. Almost 50 birds were found over the stretch of beach from Punta Piedra to Punta

Anegada, with the highest density being found nearer the last point. The cause of these new deaths can be attributed to the refloating of oil and mousse from the beaches and to large accumulations of still unweathered mousse such as are found at the large saltmarsh about 1 km to the east of the ferry landing at Punta Espora. Even if the numbers of newly killed birds are low, the fact in itself probably means that oiling of birds has been continuing since the last census was made in 1974.

When estimating from these observations, the following points were considered:

1. The coastal zone covered by observers, while it includes most of the badly affected areas, does not include all areas contaminated by oil.
2. Large areas of bushes (*Lepidophyllum cupressiforme*) where many dead cormorants and penguins were found could not be examined completely owing to the large area involved.
3. Dead cormorants were discovered up to 1 km inland, which increases the possible area included in their dispersion from the coast, and which certainly includes uncensused areas.
4. Some birds that were relatively little oiled were able to cover considerable distances either flying or swimming before dying. In this way it was possible to find live or dead oiled birds in areas that were not directly affected by the spill.
5. In certain spots the layer of mousse was so thick that it was very difficult to recognise dead birds at all. In many cases only a vague outline of a wing or head showed and it is certain that many birds were overlooked in this way.
6. It was not possible in areas of heavy oiling to distinguish between birds which had been killed by the oil, and birds which died for other reasons and were subsequently coated with oil. On the almost clean beach between the Punta Dungeness lighthouse and the nearby penguin colony (4a and b, Table XI), 105 dead birds were counted, of which 19 were clean. Though this is not necessarily typical, it does demonstrate that a small proportion of the dead birds on badly oiled beaches may have died before oiling.
7. Some oiled birds may never have been stranded within the Straits, but may have been washed out to sea.
8. Some oiled birds may have been washed off the beaches and been carried elsewhere by the action of tides and currents.
9. Some oil escaping from the eastern mouth of the Straits may have affected birds in the Atlantic.

Bearing all these points in mind, it is believed that a reasonable estimate of oil-killed birds up to February 1975 is 3000–4000 individuals. These figures probably do not substantially affect the total penguin and cormorant populations. Accurate population counts and knowledge of natural yearly fluctuations in numbers would be necessary to quantify the effects properly, but such data do not exist at present. It is expected that some oiling will continue, but probably with a gradual decrease in the number of dead birds.

Bird Colonies

The positions of the main breeding colonies are shown in Fig. 6. The islands of Marta and Magdalena form a National Park 'Los Pinguinos' (Decree No. 207 of 22 August 1966) and are described by Pisano (1971).

Punta Dungeness. This colony of the penguin *Spheniscus magellanicus* located on the Argentine side of the border was visited on 16 September and 2 October. On the earlier date a few hundred clean healthy penguins were observed, but most burrows were unoccupied. On the later date as many as 20 000 clean healthy penguins were seen, mainly paired and occupying burrows under bushes of *Lepidophyllum cupressiforme* near the shore. When the colony is full it probably contains about 30 000 birds. 105 oiled birds of various species were observed (see Table XI, site 4b).

Isla Marta. This island was observed from the sea on 17 September. Several hundred *Spheniscus* and many thousands of cormorants *Phalacrocorax albiventer* and *P. atriceps* were seen, clean and healthy.

Isla Magdalena. This was observed from the sea on 17 September. Owing to the earliness of the season, only about 1500 pairs of clean healthy *Spheniscus* were seen, occupying about 10% of the available burrows. The full colony, therefore, may be as many as 30 000—cf. Olrog (1948) who estimated 50 000 and Scott (1954) who estimated 30 000 (15 000 pairs). About 20 live oiled birds were observed (see Table XI, site 11). A return trip to Isla Magdalena on 19 January showed that virtually all burrows were occupied and the birds healthy.

Isla Contramaestre. On the basis of aerial observation the full colony has been estimated at about 3000 pairs of *Spheniscus* (J. Butt, pers. com.). A few penguins are resident throughout the year in the Straits, but some winter in the channels regions and some on the Atlantic coasts of Argentina and Tierra del Fuego, migrating during September to the breeding colonies. One migration route is through the eastern Straits entrance, but there is also a route from the south-west. From the Punta Dungeness visits it appears that a large migration took place between 16 September and 2 October. This was a time when the water surface was almost entirely clean.

Comparing numbers of dead oiled cormorants and penguins with numbers of clean healthy birds, it seems unlikely that the breeding populations of these species will be much affected.

Offshore Birds

From the comments of Bourne (pers. com., 1975) and the paper of Jehl (1974), consideration should be given to possible effects on offshore birds over the continental shelf of Argentina. Jehl's winter observations showed that the seabirds were concentrated in areas of upwelling or strong tidal currents where vertical mixing could enrich surface waters. The highest concentrations were at the edge of the continental shelf, especially in the northern half (though there appears to be no consistent relationship between latitude and seabird abundance).

Other relevant observations are:

1. Oil seen moving towards the Atlantic was thinning rapidly and no sheen was observed further than 50 km outside the Straits entrance.
2. Magellan penguins successfully migrated from their wintering areas to breeding colonies in the Straits. Virtually all burrows on Isla Magdalena were occupied on 19 January 1975.
3. Mousse was occasional or rare on the Atlantic sides of Punta Dungeness and Punta Catalina.

These observations suggest that oil that escaped into the Atlantic was not in the form of slicks that would be a serious hazard to offshore birds. Some may have been affected, but data are virtually impossible to collect.

Mammals

About 200 sea lions (*Otaria byronia*) and fur seals (*Arctocephalus australis*) were seen on Isla Marta on 17 September. No dead or sick animals have been reported. Grazing sheep are in most places prevented from wandering on the shore by wire fences. On Punta Catalina (west side), a creek near Punta Espora, and possibly at a few other sites, there is a chance of some sheep and cattle straying on to oiled patches of saltmarsh.

Commercial Fisheries and Shellfisheries

Commercial fisheries and shellfisheries are based on the species shown in Table XII and the main fishing grounds are shown in Figs. 7–10. Most of the important fishing grounds are located south of the area affected by the *Metula*, the exceptions being areas which account for about 80% of the catches of silverside (pejerrey) and sea perch (robalo). These are shown on Fig. 10.

Fisheries statistics provided by the Servicio Agricola y Ganadero are shown in Tables XIII and XIV. Most of the fisheries and shellfisheries are seasonal (see Table XIV), especially in the case of silverside (pejerrey), mackerel (sierra) and king crab (centolla).

At the time of the oil spill only a few fishermen were fishing sea perch (robalo) in the affected area. To the south of this area, mussels (cholga and chorito) were being caught and the king crab (centolla) season was starting.

The only complaint reaching the Instituto de la Patagonia and the Instituto de Fomento Pesquero concerns tainted silverside and sea perch caught in Bahia Felipe on 7 September. The catch was for supplying a local canteen of the Empresa Nacional del Petroleo. Some specimens from this catch were examined and no gross traces of oil were found. Recent information from the Servicio Nacional de Salud of Punta Arenas indicates that fishes from the north-eastern Straits of Magellan have been found to be palatable and fit for human consumption, though no tissue analyses were made of these samples.

Possible longer-term effects of oil on fish and shellfish might be detected by examining statistics from 1975 onwards. For fish, statistics from 1970 to 1974 have been relatively constant, so a large drop in catches could be detected. The shellfisheries statistics are relatively constant up to 1973, when the reduction registered was caused by a red tide phenomenon. The effects of this have been analysed by Guzman and Campodonico (1975).

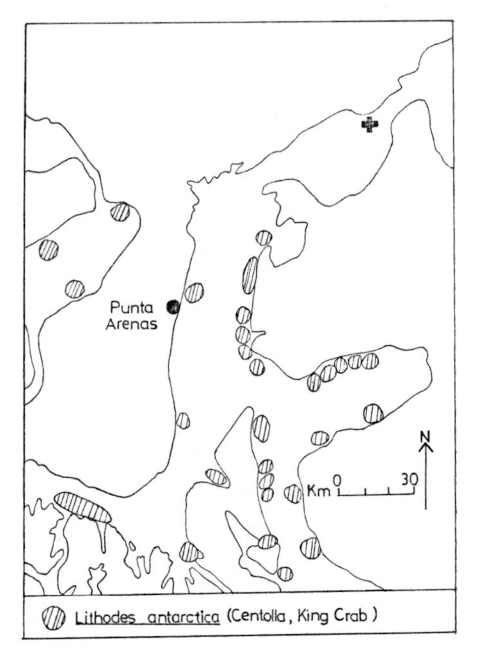

FIG. 7. Fishing areas for king crab.

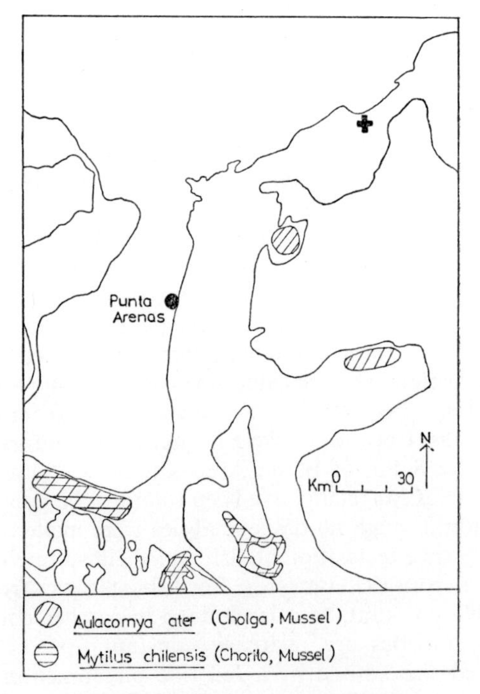

FIG. 8. Fishing areas for mussels.

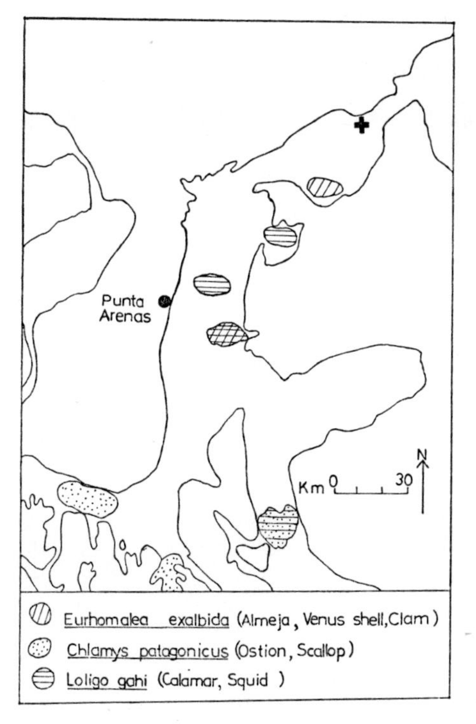

FIG. 9. Fishing areas for Venus shells, scallops and squid.

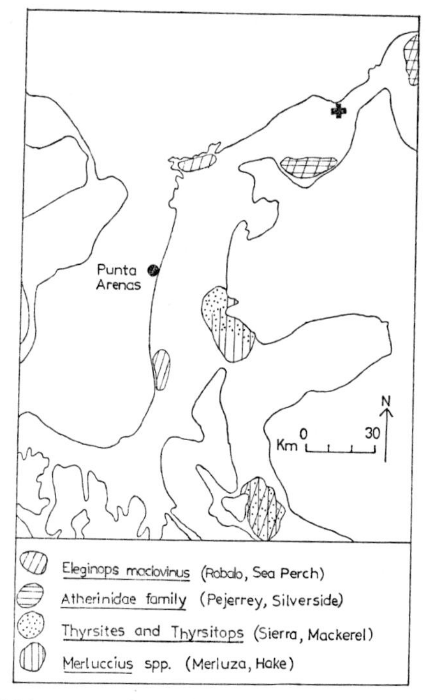

FIG. 10. Fishing areas for robalo, pejerrey, sierra and merluza.

TABLE XII
Commercial and Potentially Commercial Species in the Straits of Magellan

English name	Chilean name	Scientific name
I. Commercial species:		
Bivalves:		
Mussel	Cholga	*Aulacomya ater*
Mussel	Chorito	*Mytilus chilensis*
Clam	Almeja	*Eurhomalea exalbida*
Scallop	Ostion	*Chlamys patagonicus*
Other invertebrates:		
Squid	Calamar	*Loligo gahi*
King crab	Centolla	*Lithodes antarctica*
Sea urchin	Erizo	*Loxechinus albus*
Fish:		
Sea perches, glassfishes	Robalo	*Eleginops maclovinus*
Silversides, hardyheads	Pejerrey	Fam. Atherinidae
Mackerel-type fish	Sierra	*Thyrsites atun* and *Thyrsitops lepidopodes*
Hake	Merluza	*Merluccius* spp.
II. Potential commercial species:		
Fish:		
Conger	Congrio	*Genypterus blacodes*
Long-tailed hake	Merluza de cola	*Macruronus magellanicus*
Hake	Brotula	*Salilota australis*
Falkland herring, sardine	Sardina	*Clupea* spp.
Invertebrates:		
Lobster-krill	Langostino	*Munida gregaria*
Lobster-krill	Langostino	*Munida subrugosa*
Seaweeds:		
Giant kelp	Huiro	*Macrocystis pyrifera*

TABLE XIII
Straits of Magellan: Annual Tonnages by Species from 1970 to 1974

	1970	*1971*	*1972*	*1973*	*1974*
Silverside (Pejerrey)	89	81	134	110	119
Sea perch (Robalo)	42	43	29	42	66
Mackerel (Sierra)	1	8	5	4	14
Total fish	132	132	168	156	199
King crab (Centolla)	407	347	340	346	517
Mussel (Cholga)	3009	3573	3764	10	1998
Mussel (Chorito)	123	175	179	58	721
Scallop (Ostion)				129	45
Total shellfish	3539	4095	4283	543	3281
General total	3671	4227	4451	699	3480

Source: Servicio Agricola y Ganadero.

TABLE XIV
Straits of Magellan: Mean Monthly Tonnage by Species from 1970 to 1973 for Fresh and Industrial Use

	J	*F*	*M*	*A*	*M*	*J*	*J*	*A*	*S*	*O*	*N*	*D*	*Total*
Silverside (Pejerrey)	18	32	35	19	7	5	1	2	1	5	5	3	133
Sea perch (Robalo)	6	2	7	3	5	3	2	2	1	5	7	7	50
Mackerel (Sierra)	5	6	1									1	13
Total fish	29	40	43	22	12	8	3	4	2	10	12	11	196
King crab (Centolla)	31						1	5	9	38	136	154	374
Mussel (Cholga)	3	132	314	446	436	537	525	393	278	10	5	3	3082
Mussel (Chorito)	45	8	35	22	12	21	17	23	32	13	3	7	238
Scallop (Ostion)						26	35	24	2	11	13	9	120
Total shellfish	79	140	349	468	448	584	578	445	321	72	157	173	3814
General total	108	180	392	490	460	592	581	449	323	82	169	184	4010

Source: Servicio Agricola y Ganadero.

It would be difficult to prove or disprove the oil pollution as being the causative factor of any future changes in catches.

According to the Instituto de Fomento Pesquero, the principal potential resources existing in the Straits of Magellan are those shown in Table XII. On the other hand, the *Atlas of the Living Resources of the Sea* (FAO, 1972) shows distribution and potential for hake and sprats in the Atlantic between the eastern mouth of the Straits and the Falkland Islands. Further information is given by Gulland (1971). Certainly an evaluation of the possible effects on these resources would be even more difficult.

Plankton

Very little is known about the plankton of the eastern Straits. In the FAO *Atlas of the Living Resources of the Sea* the area falls into the following zones:

Phytoplankton production: 150–250 $mgC/m^2/d$
Zooplankton abundance: 51–200 mg/m^3

No information on the effects of the oil was collected. Data from other spills, however, suggest that different species vary in their sensitivity to crude oil, that growth and reproduction are sometimes affected, but that effects are localised and short-term. The literature is reviewed in the IMCO Report of Study No. VI (1973).

DISCUSSION AND CONCLUSIONS CONCERNING DAMAGE

1. The *Metula* mousse landed largely on shingle, gravel and coarse sand upper and mid-shore areas where macroflora and macrofauna are relatively scarce.

2. It is being mixed into these shores, with re-flotation of small lumps and sheen at intervals.

3. Up to 2 October 1974 the only oil observed to have passed in a south-westerly direction through the Second Narrows was in the form of three small spots of mousse on Cabo San Vicente.

4. It was not possible with the facilities available to make any measurement of microbial degradation rates on the shore. These rates are determined by factors such as type and surface area of oil, availability of oxygen and nutrients such as nitrates and phosphates, presence of appropriate microbial species, and temperature.

In the case of most of the Straits' shores, the coarse grinding substrates result in good penetration of oxygen and gradual increase of oil surface area, but the temperatures are low. In February 1975, six months after the *Metula* grounding, most of the mousse was still on the shores under a sand layer.

5. At extreme high water mark and in the splash zone, oil deposits are far less likely to be mixed and refloated, and are darkening and hardening to form probably long-lived asphaltic deposits.

6. In the lower intertidal areas, where generally much less mousse has been deposited, several large patches of mussels (*Mytilus*) have been contaminated but not extensively killed, and associated plants of laver-bread (*Porphyra*) and other species have been damaged. Limpets (*Nacella*), which occur locally, have been affected in places, but frequent apparently healthy individuals remain in the rock pools.

Mytilus and *Porphyra* are both edible species, and in contaminated areas are no longer fit for consumption. However, few people live along most of the affected coastline.

It is difficult to say at this stage if there will be any long-term effect on the mussels. Chan (1973) reports a case which is comparable in that there was no immediate large kill of mussels. He studied extensive beds of *Mytilus californianus* which were covered with Bunker 'C' oil following the January 1971 San Francisco spill. He found the death rate was about 1% and that the beds were in good health in December 1971.

There is always the possibility of unpredictable effects, and the Instituto de la Patagonia plan to sample the mussel beds at intervals to determine if there is any long-term effect on health and growth.

7. In places, large patches of saltmarsh and narrow belts of sand-dune plants (*Elymus*) have mousse on them. The plants affected are perennials, and as the accident happened towards the end of winter, the parts contaminated are usually the old shoots from the previous summer. The underground systems of these perennials have not usually been touched by the mousse, and it was expected that as the spring proceeded, new shoots would grow up through the old oily ones. These predictions are based upon field experiments with similar types of plant (Baker, 1971). Observations made in January 1975 show this to be true in places, although not where plants have been smothered by thick and accumulating mousse near Punta Espora.

8. Sampling of the *Macrocystis* beds and their fauna was very limited because of bad weather. A bed near Punta Espora had been subjected to oily run-off for several weeks when sampled, but still supported large numbers of healthy individuals of many species.

Everywhere the *Macrocystis* itself appeared in good condition, though small flecks of mousse were occasionally caught in wrinkles in the fronds, and the water surface in the beds had patches of sheen.

Effects of crude oils on *M. pyrifera* are poorly documented, though it is known from laboratory studies (Clendenning and North, 1960) that photosynthesis is reduced by cresols, phenols and boiler fuel oil. Studies following the wreck of the *Tampico Maru* (North *et al.*, 1964) showed that it was affected by diesel oil, photosynthesis of young blades being severely reduced by diesel oil films of initial thickness 0·02 mm or emulsions of 0·01% diesel oil in sea water, after three days exposure or more. These results cannot be directly applied to the very different circumstances in the eastern Straits, but indicate that any kelp situated in continual oily run-off may suffer some reduction of photosynthesis.

9. The estimate of birds killed in August and September 1974 is 1500–2000. This estimate is based on extensive shore walking. In particular, the whole of the badly affected shore between Punta Piedra and Cabo Orange was covered on foot. Later observations indicate continued mortality of birds and, bearing this and other factors in mind, the estimate of birds killed up to February 1975 is 3000–4000. In estimating from the actual counts, additions have been made for shores not visited and birds washed out to sea, dispersion by flying and swimming, and the fact that many birds were completely covered by the oil and therefore either barely visible or completely invisible. Subtractions then have to be made to allow for the fact that some birds die from natural mortality, even though they may be counted as dead oiled birds on the beach. (Counts of dead unoiled birds on some beaches (see Table XI) give a rough idea of the numbers involved.)

The main species affected are the Magellan penguin *Spheniscus* and the cormorants *Phalacrocorax*. Many thousands of clean healthy birds of these species successfully reached their breeding colonies, and it seems unlikely that the populations as a whole will be much affected. An accurate population count and knowledge of yearly fluctuations in numbers would be necessary to quantify effects of an oil spill on these bird colonies, but such data do not exist at present. A project of this kind could usefully be undertaken by the Instituto de la Patagonia.

10. There has been one complaint concerning commercial fish species. It seems unlikely that oil will move southwards to the main commercial fishing grounds. With oily run-off from the shore to be expected for a long time in Bahia Felipe, there is the possibility of near-shore sea perch and silverside in this area being tainted or moving elsewhere. Information collected during January and February 1975 indicated that fishes from the north-eastern Straits were palatable at this time.

11. It was impossible with the facilities available to do any work on immediate or possible long-term effects of dissolved or finely dispersed oil on behaviour, growth, reproduction and survival of plankton and young of commercial and other species. This sort of work would be worthwhile but requires time and a lot of pioneering biological study. It could best be carried out locally by the Instituto de la Patagonia. A possible monitoring scheme (both for *Metula* oil run-off and release from beaches, and other oil sources and effluents) involves settlement plates fixed to buoys. Some laboratory

experimental work could be useful in interpreting results, but it should be related to field conditions and knowledge of the natural behaviour of the species studied.

12. Without previous ecological studies of the communities of the affected areas, the immediate effects of this spill cannot be properly quantified. From an economic point of view, the immediate effects of this spill appear to be of relatively small importance. As movement of the oil seems to be restricted mainly to the presently affected area, long-term effects of refloated oil are unlikely to the south of this area. Nevertheless, the moving mousse has accumulated in and has greatly affected the two inlets at Punta Espora, which are likely to remain oily for a long time. According to Guzman and Campodonico (1973), the sea perch, *Eleginops maclovinus*, a nothotenid fish, enters this type of inlet during certain periods of its life-cycle and a part of the population may have already been affected by the oil.

13. There were three fortunate factors that mitigated the immediate effects of the oil spill:

(a) The location of the grounding was located far away from the most important fisheries areas. They were favoured in addition by the meteorological and hydrographic conditions, which prevented the oil from passing through the Second Narrows to these areas.
(b) Most of the mousse has landed on beaches which were naturally impoverished in terms of macroflora and macrofauna. However, mousse may remain for several years, so recovery of these shores could be very slow.
(c) The accident happened before the main migration time of the penguins.

ACKNOWLEDGEMENTS

This work would not have been possible without the help and co-operation of a great many people. We should like to thank in particular Rear-Admiral Eduardo Allen and the Chilean Navy; the Chilean Marines on Tierra del Fuego and their commanding officer Jaime Weidenslaufer; the director of the Instituto de la Patagonia, Mateo Martinic; the Empresa Nacional del Petroleo; John Butt of Shell International Marine Ltd; Shell agents Peter Fussell of Ultramar and Dick Gibbons; Maria Beros, Jorge Canas, Dr P. Gibbs, J. Wardley Smith, Dr J. Wonham and Dr Roy Hann, Jr.

REFERENCES

Alveal, K., Romo, H. and Valenzuela, J. (1973). Consideraciones ecologicas de las regiones de Valparaiso y Magallanes, *Rev. Biol. Mar., Valparaiso*, **15**(1), 1–29.

Baker, J. (1971). Studies on saltmarsh communities, in *The Ecological Effects of Oil Pollution on Littoral Communities* (Proceedings of a Symposium held at the Zoological Society of London, 30 Nov.–1 Dec. 1970), Applied Science Publishers, London, pp. 16–101.

Baker, J. (1974). Grounding of *Metula*, Magellan Straits Ecological Survey (immediate report, MS).

Berridge, S. A., Thew, M. T. and Loriston-Clarke, A. G. (1968). The formation and stability of emulsions of water in crude petroleum and similar stocks, in *Scientific Aspects of Pollution of the Sea by Oil*, Institute of Petroleum, London.

Braud, J.-P., Etcheverry, H. and Perez, R. (1974). Développement de l'algue *M. pyrifera* (L.) Ag. sur les côtes Bretonnes. *Science et Pêche, Bull. Inst. Pêches marit.*, **233**, 1–15.

Brunnock, J. V., Duckworth, D. F. and Stephens, G. C. (1968). Analysis of beach pollutants, in *Scientific Aspects of Pollution of the Sea by Oil*, Institute of Petroleum, London.

Chan, G. (1973). A study of the effects of the San Francisco oil spill on marine organisms, in *Prevention and Control of Oil Spills*, Proceedings of an API/EPA/USCG Symposium held in Washington, 13–15 Mar. 1973, pp. 741–79.

Clendenning, K. and North, W. (1960). Effects of waste on the giant Kelp, *Macrocystis pyrifera*, in Pearson, E. A. (ed.), *Proc. 1st Int. Conf. Waste Disposal in the Marine Environment*, Pergamon Press, New York.

FAO (1972). *Atlas of the Living Resources of the Sea*, Department of Fisheries, FAO, Rome.

Fuenzalida, H. (1967). Clima, in *Geografia Economica de Chile*, Texto Refundido, CORFO, Fundación Pedro Aguirre Cerda, Santiago.

Gulland, J. A. (1971). *The Fish Resources of the Ocean*, FAO, Rome.

Guzman, L. and Campodonico, I. (1973). Algunos aspectos de la biologia de *Eleginops maclovinus* (Cuv. y Val.) 1830, con especial referencia a su morfometria, caracteres meristicos y alimentación. *Ans. Inst. Pat.*, Punta Arenas (Chile), **4**(1–3), 343–71.

Guzman, L. and Campodonico, I. (1975). *Marea Roja en la Région de Magallanes*, Publicaciones del Instituto de la Patagonia, serie Monografias No. 9, Punta Arenas (Chile), 44 pp.

Hann, R., Jr. (1974). Regarding oil pollution from the Tanker *Metula*: Report to US Coast Guard, Research and Development Program, Texas A and M University, Environmental Engineering Division, Civil Engineering Department.

Intergovernmental Maritime Consultative Organisation (1973). Report of Study No. VI, *The Environmental and Financial Consequences of Oil Pollution from Ships*, Programmes Analysis Unit, Chilton, Didcot, Berks.

Jehl, J. R. (1974). The distribution and ecology of marine birds over the continental shelf of Argentina in winter, *Trans. San Diego Soc. Nat. Hist.*, **17**(16), 217–34.

Jerez, M. and Arancibia, M. (1972). *Trazado de Isoyetas del Sector Centro-Oriental de la Provincia de Magallanes*, Publicaciones del Instituto de la Patagonia, serie Monografias No. 4, Punta Arenas (Chile), 32 pp.

Kuhnemann, O. (1970). Algunas consideraciones sobre los bosques de *Macrocystis pyrifera*, *Physis*, **29**(79), 273–96.

Nelson-Smith, A. (1972). *Oil Pollution and Marine Ecology*, Elek Science, London, pp. 82–3.

North, W. J., Neushul, M. and Clendenning, K. A. (1964). Successive biological changes observed in a marine cove exposed to a large spillage of mineral oil, *Symp. Poll. mar. Micro-org. Prod. Petrol., Monaco*, pp. 335–54.

Olrog, O. C. (1948). Observaciones sobre la avifauna de Tierra del Fuego y Chile, *Acta Zoologica Lilloana*, **5**, 437–531.

Pisano, E. (1971). Estudio ecologico preliminar del Parque Nacional 'Los Pinguinos' (Estrecho de Magallanes), *Ans. Inst. Pat.*, Punta Arenas (Chile), **2**(1–2), 76–92.

Scott, P. (1954). South America 1953, in *Annual Report of the Wildfowl Trust*, **6**, 54–69.

Servicio Agricola y Ganadero, División de Pesca y Caza. Estadisticas Pesqueras de Magallanes, anos 1970–1974.

19
Oil Spills from Tankers

J. WARDLEY SMITH
(International Tanker Owners Pollution Federation Ltd)

The review of world oil spillages from 1960 to 1975 circulated at this Conference by Sheila van Gelder-Ottway and Marion Knight (see Paper 20) gives a rather horrifying picture of the very large number of spills which have occurred during the period under review. Even so, the statistics are probably incomplete and may not be entirely accurate. This is no reflection on the compilers, but rather upon the large number of interests concerned with oil and its spillage and the highly subjective nature of some of the reports. Over the past two and a half years the International Tanker Owners Pollution Federation has been collecting details of oil spills from tankers. These have largely been received from the various organisations who insure the clean-up costs and/or the legal liability for oil pollution. It should perhaps be explained that 99·8% of the world's tankers have signed a voluntary agreement to pay for expenditure incurred in avoiding or mitigating damage to coastlines from pollution produced by crude oil and its residuals such as asphalt, bitumen, fuel oil, heavy diesel oil or lubricating oil. The tanker causing the discharge is presumed to be negligent unless the owner can establish that it occurred without the tanker's fault. These costs are, of course, covered by insurance, and the removal costs have a maximum of US$100 per gross registered ton of the tanker in question, or US$10 million, whichever is the less. These payments for clean-up are made either to the national Government whose coastline is polluted or threatened, or to the tanker owner who himself carries out the clean-up.

Within the last few months the Federation has entered the data into a computer so that the results can be analysed more rapidly and accurately, and also so that specialised reports can be prepared to encourage the various insuring agencies to be more zealous in making their initial reports. The Federation is, of course, aware of and in touch with the world's tanker owners and it might be argued that it would be easier and more satisfactory were spillage reports made directly by them. However, it should be appreciated that there are more than 6800 tankers owned by over 3000 firms (and this excludes tankers below 3000 tons), so that to deal with such a large number would be very difficult, whereas there are only some 30 groups dealing with the insurance side.

Before discussing the amount of oil spilled from tankers in operation or accident, it is desirable to consider all the various ways in which oil enters the marine environment. Crude oil is a natural product. It seeps from the earth both on the ground and below sea level, and this must have been going on for many thousands of years. Marco Polo comments on oil being carried away from seeps in the Middle East, and there are other early historical references. Recently a very careful study has been made of the amount of oil emerging from these seeps (Wilson, 1973). This long and detailed paper finally concludes that the probable range of seepage into the marine environment is 0·2–6·0 million metric tons per year. Within this range the best estimate for present marine seepage, worldwide, is given as 0·6 million metric tons per year. It also points out that this seepage is not uniformly distributed around the world but that certain high seepage areas contribute 45% of the total.

The total amount of oil entering the marine environment from various sources has been the subject of many estimates. The latest and perhaps one of the most accurate was prepared by members of the US Coast Guard (Charter *et al.*, 1973) in a paper given at the same conference. Their basic results are given in the second column of Table I. The figure for tanker operations is

TABLE I
Sources of Oil in the Marine Environment

	Jeffery's estimate	*Charter et al.*	*Author's estimate*
(a) Marine operational loss:			
Tankers using load-on-top	100 000	1 069 763	1 000 000
Tankers *not* using load-on-top	600 000		
Bilge discharges, etc.	50 000	301 536	300 000
(b) Accidental discharges:			
From all sources	200 000	346 086	350 000
(c) Offshore production		118 026	150 000
and oil seepage			600 000
Total	150 000		
(d) Land-based discharges:			
Refineries and petrochemical plant	300 000		
Waste oils, etc.	500 000		
Total		1 974 032	1 300 000
Total	1 900 000	3 809 443	3 700 000

divided into operating losses and bilge discharges, etc. It is interesting to compare these figures with the figures prepared by Jeffery (1971), so these are given in the first column of the table. It will be seen that Charter's figure was 3·8 million tons, whereas Jeffery's was 1·9 million, neglecting the fact that Jeffery had included 150 000 tons for offshore production and oil seepage. On the other hand, Charter's figures for waste oil were based on measurements made in various US rivers, and as the disposal of waste oil in America is rather different than in many other countries, on a global basis this figure may well be too high. The third column in this table is a further estimate. This is substantially the same as Charter's estimate for tanker losses but is an increase on Jeffery's to make up for the fact that 'load-on-top' has not developed as well as one hoped, and that oil carried by sea has rather increased. The figure for bilge discharges, etc., is taken as Charter's figure, as is the figure for accidental discharge for all sources. Offshore production is taken at the higher figure of Jeffery's at 150 000 and, in addition, the seepage figure given by Wilson of 600 000 is also added. Under item (d), land-based discharges, a figure of 1·3 million has been chosen, considering that Jeffery's figure is probably too low and, as explained above, Charter's is too high. This gives a total of 3·7 million tons a year for the global estimate, but this high figure does not take account of the potential hydrocarbon in the ocean produced by exhaust gases from automobiles, stationary engines, and from the flue discharges from oil power stations and central heating plants. Some authorities believe that this pollution in the air is washed from the atmosphere by rain and so must enter the marine environment. This would, however, appear to be a more speculative approach and has been omitted.

It will be seen that vessel accidents of all kinds account for about 10% of this grand total and yet it is these vessel accidents which produce such alarm and despondency in the general public and which encourage a vociferous group of self-styled environmentalists to argue against the use of large tankers anywhere, and in particular the approach of such tankers to their own shores.

The reason for this is fairly clear. Before the Second World War, oil pollution of any kind on beaches or in recreational areas of the sea was unusual, largely because the world use of oil was only perhaps 10% of the present figure and even less was carried by sea. The sinking of tankers caused by the hostilities of the Second World War brought oil pollution to many places in Western Europe and the Pacific which had never seen it before. At the end of the war the increasing use of oil and the increasing need to transport it by sea produced a great increase in oil pollution caused by tank washings and the dumping of slops. This was, of course, particularly prevalent close to the major oil ports of Europe and America. This led directly to the 1954 International Conference, the setting up of IMCO and the various International Conventions which have resulted. The implementation of these Conventions, and the anxiety of the major tanker owners to reduce pollution which resulted in the introduction of load-on-top operation, have probably reduced the total pollution from the maximum which was achieved in the mid-1950s, but nevertheless, oil in the form of 'tarry lumps' can be found on most of the sandy beaches of the world. This worldwide beach pollution has coincided with the immense expansion of leisure activities on or by the sea so that more and more people in more and more countries are aware of the annoyance, discomfort and in some cases actual financial loss due to pollution of this kind. The reasoning of the 'anti-tankers' is as follows: 'Oil comes from tankers, tankers have accidents, the bigger tankers must have the bigger accidents, therefore keep the big tankers away from the shore and oil pollution will decrease.'

This argument on tanker-derived oil pollution always fails to take account of the other sources of oil pollution which, as shown above, are considerably larger than the oil discharged by tankers in total and even more greatly exceed that lost by tankers having accidents or making accidental discharges. The effects on the biology of the ocean of such discharges have been dealt with in other papers at this Conference. It is an interesting fact that, despite the anxious search by many scientists all over the world for the effect of the chronic pollution of the world's oceans, little has been found except where the chronic pollution is highly localised. Even a recent report on the state of Lake Maracaibo (Templeton *et al.*, 1975), where oil has been produced for the past 60 years and has been discharged from pipeline breaks and other incidents over the whole period, suggests that the potential impact of non-petroleum waste from increase of industrial and domestic waste waters discharged to the lake, is likely to contribute more to the degradation of water quality than the petroleum which has apparently had little or no effect on the ecology of the lake over the whole period.

The tanker industry is acutely conscious of this public feeling and consequently is anxious to do whatever it can to reduce discharges to the ocean still further. If, and when, the 1973 International Convention on Marine Pollution is ratified, then the operational discharges will be still further reduced and it is believed that the deposits of 'tarry lumps' on beaches will virtually cease. Even so, however good the operating procedures, accidents, caused by failures of equipment or by human error, will still occur. In general terms, these are most likely to occur where the tanker is at its terminal or on the final approach to the terminal. Consequently, the Federation's collection of statistics is aimed at determining the reason for these accidental spillages occurring so that remedial action can be taken (Wardley Smith, 1973).

The past year has seen a number of unfortunate accidents to large tankers. One of these, the stranding of the *Metula* in the Straits of Magellan, has already been discussed. Another stranding, that of the *Showa Maru* close to Singapore, spilled some 4000 tons out of a cargo of 230 000 tons, and though coastal pollution and environmental damage was very small, it has nevertheless re-emphasised the problems of large tankers navigating through this difficult channel. Nearer home, the spill of the *Universe Leader* in Bantry Bay was widely reported. But another incident, that of the *Jakob Maersk*, which may well have been the second largest loss of oil to the marine environment since the *Torrey Canyon*, was but little reported. This tanker, carrying 83 000 tons of light Iranian crude, grounded while attempting to enter the harbour of Oporto, Portugal, had an engine-room explosion resulting in the deaths of seven crew members and caught alight. The fire burned with great vigour for over 48 h and then went out. It is believed that all but one tank, containing 17 000 tons, was consumed. The heavy swell rapidly broke up the fore part of the vessel and the oil in this last tank was released and fortunately carried away into the Atlantic by an offshore wind. The wind, while the tanker was burning, was onshore, yet surprisingly the amount of pollution was very limited and was restricted to some 2·5 km of the shore immediately adjacent to the stranding. Ecological damage was quite severe in this area but very limited elsewhere, as was the pollution. One can only assume that the type of oil, the manner in which it burned in the ship, and the intensity of the conflagration, resulted in an almost complete consumption of the oil rather than leaving a great amount of heavy residues, as is the more general experience.

The pollution insurers, known as P & I Clubs, have been asked to inform the Federation when an oil spill occurs. The information (a copy of the form used is reproduced opposite), including the name of the ship and the date,

ADVICE OF OIL SPILLAGE

PART 1

I D A

Parts 2 and 3 of this document should be completed and returned to us as further information becomes available.

As soon as a spill is reported complete Section A and as much of Section B and C as is possible and return Part 1 of this set to us as soon as you can.

Part 4 forms your office copy.

N.B.: 1. "Not known" should only be used if it will never be known.
2. Only one letter or figure in each box.

SECTION A (Basic Data) Complete items 2 to 6

1. Advice Number

2. Date of Incident — DAY ☐☐ MONTH ☐☐ YEAR ☐☐

3. Local time ☐☐ hours (24 hr. clock)

4. Name of Tanker ☐☐☐☐☐☐☐☐☐☐☐☐☐☐☐☐☐☐☐☐☐☐☐☐☐☐☐☐☐☐

5. Port/Position Incident Occurred .... ☐☐☐ ☐☐☐ **Leave blank**

6a. If buoy mooring spill put X in box [M]

6b. If NO oil entered water put X in box [D]

SECTION B (Spill Data)

7. **Operation in Progress/Circumstances** (put X in one box only)

Loading	A	Bunkering	B	Deballasting	C
Discharging	D	Ballasting	E	Pumping Bilges	F
Cleaning Tanks	G	Internal Transfers	H	Intentional discharge	I
Stranding/grounding	J	Collision	K		
Not known	L	Other, specify	M	....	

8. **Type of Oil** (put X in one box only)

Crude	A	Bunker	B	Bilges	C
Fuel (cargo)	D	White product	E	Tank washings	F
Lube oil	G	Bitumen	H		
Not known	I	Other, specify	J	....	

9. **Quantity Spilt** (put X in one box only)

Trace	A	Less than ½ bl.	B	½-5 bls.	C
5-50 bls.	D	50-5,000 bls.	E	Over 5,000 bls.	F
Not known	G				

10. **Reason for Spill** (put X in one box only)

Hull failure	A	Equipment/ material failure	B	Human error	C
Hull defect	D	Incident denied	E	Shore fault	F
Not known	G	Other, specify	H	....	

11. **Cause—Equipment/Material element** (put X in one box only)

Defective pipeline	A	Hose failure	B	Loading arm failure	C
Open valve	D	Leaking valve	E	Manifold failure	F
Sea suction	G	None	H		
Not known	I	Other, specify	J	....	

12. **Cause—Human element** (put X in one box only)

Improper supervision	A	Improper procedure	B	Inattention	C
Lack of Communication	D	None	E		
Not known	F	Other, specify	G	....	

SECTION C (Cost Data)

13. **Costs (in U.S. Dollars)**
If actual costs known enter in appropriate box/s
If not known leave box/s blank
If zero cost enter 0 in box/s
If estimate available enter amount and put X in E box

Clean Up Cost ☐☐☐☐☐☐☐☐ [E]

Third Party Cost ☐☐☐☐☐☐☐☐ [E]

Fine Cost ☐☐☐☐☐☐☐☐ [E]

time and place of the accident, also has data on the circumstances under which the spill occurred; thus details are requested of the operation in progress under 13 headings, including such things as loading, discharging, ballasting, bilge pumping, etc. The type of oil or oily material discharged is also to be reported under ten headings, including the heading 'unknown' as well as 'tank washings'. The quantity spilled is of considerable importance, but this is very difficult to obtain. As is well known, oil rapidly spreads when spilled on water so that the half-barrel of a light oil in daylight can look much more than several barrels of heavy fuel oil at dusk. The amount reported spilled is broken down into six groups with a seventh for 'unknown', and ranges from a trace to over 5000 barrels. The reasons for spills can be hull failure or defect, an equipment or material failure, a human error or a mistake made by the terminal, while the cause, if of equipment or material, can be failures of pipelines or valves or hoses and smaller equipment, or human error, or, as quite frequently reported, 'unknown'. If human error is the reason, then this can have arisen owing to improper supervision or an improper procedure being carried out or, of course, inattention to the task in hand. The cost of the clean-up is an item of considerable interest as in third-party claims for consequential damage which can be claimed in some areas, or if the ship or its Master is fined, all these are recorded on the form.

These data are being entered into a computer which has two main stores. The first contains details of all the tankers entered with the Federation, including owner, flag, size, date of construction and type of cargo, i.e. a black oil carrier, a white oil carrier or an oil or bulk cargo carrier. The other store contains the details extracted from the spillage advice forms. This operation has only recently started, but in the period under review nearly 900 reports have been received. From this information a number of tables have been prepared.

Table II gives the operation in progress when the spill occurred. Discharging is the largest item with 33%. The form has an interesting element in that there are three sections which deal in part with the reason or cause of the spill. Section 10 asks for the reason for the spill and among the items is 'human error'. Section 11 discusses the piece of equipment which may have failed, but it too has a section for 'none' and 'not known' whereas section 12 on the questionnaire tries to determine the reason for human error, listing four probable reasons with a 'none' and a 'not known'. Table III gives the cause, in terms of equipment, using the details extracted from section 11 of the questionnaire. Of these, 41% are not known. The largest item is 'leaking valve', which represents 24% of all the spills. On the other hand, Table IV gives the reason for the fault on the basis of the information in section 10, and under this heading 46% of the spills are put down to human error. This is actually 297 instances, which must be compared with the 265 instances in Table III where there was no mechanical fault. Table V examines the possible reasons for the human error. In this case 270 cases were put down, or 44% of the total reported. Of that total the largest percentage under the heading of 'inattention' is also, coincidentally, 44%. 34% of the spills are apparently due to improper procedure, 14% to improper supervision and 8% to failure of communication. In the light of the way the evidence is collected, it is difficult to say whether some of the mechanical failures may not be due to

TABLE II
Operation in Progress (Section 7 of Advice of Oil Spillage Form)

Operation	*No.*	%
Loading	155	22
Discharging	225	33
Bunkering	88	13
Ballasting/Deballasting	98	14
Other	124	18
Total	690	100

Note: The varying totals given in the tables indicate that there are missing figures in the returns fed to the computer. Percentages to one figure may not add exactly to 100%.

TABLE III
Cause/Equipment Material Element (Section 11 of Form)

Reason	*No.*	%
Defective pipeline	41	6·3
Hose failure	22	3·4
Loading arm failure	6	0·9
Open valve	12	1·8
Leaking valve	154	23·8
Manifold failure	9	1·4
Sea suction	21	3·2
None	265	41·0
Not known	86	13·2
Other	32	5·0
Total	648	100

TABLE IV
Reason for Spill (Section 10 of Form)

Reason for fault	*No.*	%
Equipment failure	222	34·4
Human error	297	46·0
Hull failure or defect	57	8·8
Offence denied	23	3·6
Shore fault	9	1·4
Other and not known	38	5·9
Total	646	100

TABLE V
Human Error (Section 12 of Form)

Cause	*No.*	%
Improper supervision	38	6·2
Improper procedure	92	15·0
Inattention	118	19·2
Lack of communication	22	3·6
Sub-total	270	44·0
None	96	15·7
Not known	246	40·1
Other	1	0·2
Total	613	100

human error either directly, in the way it has been operated, or in the failure to carry out proper maintenance. This factor may well apply particularly in the case of valve leakage. It is difficult to believe that valves are as unsatisfactory as the statistics suggest.

As was said earlier, it is the collisons and groundings, particularly of large tankers, which produce the greatest public disquiet. In the sample examined, as shown in Table VI, 24 of the 690 spills were the result of either stranding, grounding or collision. As explained earlier, an attempt is made to get details of the quantity of oil spilled; although these figures are not very reliable they have been included in the table, and it will be seen that the strandings and collisions result in general in larger spills than those from other causes. Thus 33% of that group are in the 50–5000 barrel range and 25% are over 5000 barrels, whereas for all spills 34% of the spills occur in the 0·5–5 barrel range and a total of 39% are in fact less than half a barrel.

TABLE VI
Number of Spills as a Result of Strandings/Groundings and Collisions, and the Relative Amounts of Oil Spilled

Size of spill	*No. in group*	%	*Total no. all incidents*	%
Trace	0	0	102	14·8
<0·5 barrel	1	4·2	169	24·5
0·5–5 barrels	3	12·5	238	34·5
5–50 barrels	5	20·8	78	11·3
50–5000 barrels	8	33·3	44	6·4
>5000 barrels	6	25·0	9	1·3
Unknown	1	4·2	50	7·2
Total	24	100	690	100

It will be noted that on the advice form, in section C, the costs of clean-up and third-party fines are also included. So far, owing to the relatively short time that the computer analysis has been carried out, sufficient data have not been received. It will be appreciated that details about the spill and the reasons for it are known almost at once, but clean-up can take a long time and it can take even longer before the costs have been collected and finally passed to the insurers who make the return to the Federation. In some cases the final details will not be received for up to two years after the date of the incident.

CONCLUSION

Although the computer program has only been running for a short time, and the number of returns entered has been far less than was hoped for, nevertheless it appears that the method of analysis will soon produce valuable results which can be used either to modify practices existing on tankers or even to enable equipment modifications to be made, all with a view to reducing the number of accidental spillages. The perhaps more serious accidents which occur while the tanker is in motion and which result from collisions or groundings can only be dealt with by more traditional maritime methods: improvements in crew training, in navigational equipment both carried by the ship and on shore, and in some areas by the enforcement of routing and traffic control at harbour entrances, etc. All these matters are, of course, being dealt with by the IMCO Maritime Safety Committee.

REFERENCES

Charter, D. B., Sutherland, R. A. and Porricelli, J. (1973). Quantitative estimate of petroleum in the ocean, in *Inputs, Fates and Effects of Petroleum in the Marine Environment*, Vol. I, Ocean Affairs Board, National Academy of Sciences, Washington, DC.

Jeffery, P. G. (1971). *Oil in the Marine Environment*, Report LR 156(PC), Warren Spring Laboratory, Stevenage, Herts.

Templeton, W. L. *et al.* (1975). Oil pollution studies on Lake Maracaibo, Venezuela, *Conference of Prevention and Control of Oil Pollution*, San Francisco, Calif., p. 489.

Wardley Smith, J. (1973). Occurrence, cause and avoidance of oil spills on tankers, *Joint Convention on the Prevention and Control of Oil Spills*, Washington, DC.

Wilson, R. D. (1973). Estimates of the annual input from natural marine seepage, in *Inputs, Fates and Effects of Petroleum in the Marine Environment*, Vol. I, Ocean Affairs Board, National Academy of Sciences, Washington, DC.

20
A Review of World Oil Spillages 1960–1975

SHEILA VAN GELDER-OTTWAY
(*formerly with Oil Pollution Research Unit*)
and MARION KNIGHT
(*Orielton Field Centre, Pembroke, Wales*)

INTRODUCTION

During the course of examining the reports of oil spillages in recent years, it was found to be useful to compile a list of the major spillages, together with basic details of these incidents and any subsequent biological effects. This information is presented in a tabulated form on the following pages. The apparent preponderance of oil pollution incidents in more recent years is partly a reflection of both an increasing total annual quantity of oil handled, and also the ever-increasing size of oil tankers. A more significant reason for the apparent preponderance of oil pollution incidents since 1967, however, is the increasing awareness of potential biological damage of these spillages since the *Torrey Canyon* disaster which occurred in the spring of that year. Subsequent oil spillages received far more attention from biologists, oceanographers and also the general public, with the result of far greater coverage of these incidents in scientific journals, information-relaying organisations and all news media. A great number of important oil pollution incidents almost certainly occurred in the early and mid-1960s, but reports of these are infrequent in the literature, thus accounting for the relatively small number of spillages during this period described in the following tables.

	Date	Name of tanker/ installation	Location	Size of tanker	Amount of oil spilt	Type of oil spilt	Sprayed with emulsifier	Biological Damage	Reference
1	Mar.–Apr. 1960	*	Lower Detroit River, U.S.A.	—	*	*	*	*c.* 12 000 ducks killed	Clark, 1968
2	May 1960	*Manchester Spinner*	Peckford Reef, Newfoundland, Canada.	*	360 tons	Bunker	*	No oil on shores apparently, but thousands of ducks reported killed. Also a few other types of birds oiled	Gillespie, 1968
3	9 July 1960	*Esso Portsmouth*	Milford Haven, S. Wales, U.K.	*	*c.* 300 tons	Crude	*	No investigations made	Dudley, 1971
4	Sept. 1960	Marine oil Terminal	Esso Refinery, Fawley, Southampton, U.K.	—	200 000 kg	Fuel	*	Extensive pollution of mud flats at mouth of R. Hamble. No specific details of biological damage	George, 1970a
5	Jan. 1961	Land installation	Holton Heath, nr. Poole Harbour, Dorset, U.K.	—	25 tons	*	*	Nearby salt marsh polluted, with localised damage	Ranwell and Hewett, 1964
6	25 Jan. 1961	Un-named tanker	Poole Harbour, Dorset, U.K.	*	270 tons	*	*	Salt marsh extensively polluted, with much oil trapped on *Spartina* leaves. *c.* 300 birds contaminated, of which *c.* 150 died	Ranwell and Hewett, 1964
7	20 Mar. 1962	*Benjamin Coates*	Just outside Milford Haven, S. Wales, U.K.	*	Over 100 tons	Crude	Yes	High mortalities among many inter-tidal invertebrates. Algae apparently unharmed	Dudley, 1971; Nelson-Smith 1968

8	16 July 1962	*Argea Prima*	S. shore of Puerto Rico	*	10 000 tons	Crude	*	Extensive damage. High mortalities among many shallow water and shore-dwelling organisms, including a wide variety of invertebrates. Also extensive damage to intertidal and sublittoral algae, and to the mangrove swamp habitat	Diaz-Piferrer, 1962
9	7 Jan. 1964	Un-named freighter	Few miles from Dry Tortugas, Gulf of Mexico	*	138 000 gals	Fuel	No	Oil treated with dredged ocean fill to prevent contamination of nearby tern colonies. Some terns oiled.	Clarke *et al.*, 1965
10	20 Feb. 1966	*Anne Mildred Brøvig*	Elbe estuary, W. Germany	40 000 tons	Several thousand tons	Crude	*	Oil disappeared, presumably sunk. Some reports of large duck mortality	Drost, 1966; Nelson-Smith 1970; Vermeer and Vermeer, 1974
11	Mar. 1966	*South America*	S.W. coast of Netherlands	*	*	*	*	Very high bird mortalities, especially eider and redthroated diver	Tanis and Mörzer Bruijns, 1968
12	Aug. 1966	*British Crown*	Persian Gulf	*	*	Qater crude	*	Extensive oil pollution prevented by burning much of cargo. No mention of biological damage	Beynon, 1971
13	18 Sept. 1966	*Seestern*	Medway Estuary, S.E. Britain	*	1700 tons	Nigerian light crude	*	8000 acres of saltings heavily polluted. Estimated 5000 birds killed. Some algae and salt marsh plants killed, also many intertidal invertebrates and several fish	Harrison and Buck, 1968
14	13 Jan. 1967	*Chryssi P. Goulandris*	Outside Milford Haven, S. Wales, U.K.	*	Over 250 tons	Crude	Yes	Most damage to intertidal organisms. Gastropod molluscs badly affected, also barnacles and sea anemones, on a number of shores. No apparent damage to algae	Dudley, 1971; Nelson-Smith 1968

	Date	Name of tanker/installation	Location	Size of tanker	Amount of oil spilt	Type of oil spilt	Sprayed with emulsifier	Biological Damage	Reference
15	18 Mar. 1967	*Torrey Canyon*	15 miles west of Land's End, Cornwall, U.K.	Cargo 117 000 tons oil	Most of cargo ($\frac{1}{3}$–$\frac{1}{5}$) burnt	Kuwait crude	Yes	Shores of S.W. England and Brittany extensively polluted. Very high mortalities of intertidal shore life, mostly due to use of toxic emulsifiers. Many invertebrates and algae killed on shores. Fisheries and plankton apparently unaffected. 10 000 sea bird bodies recovered, estimated kill 30 000+ (Bourne). A few seals observed in trouble (Spooner)	Smith, 1968; Spooner, 1967; Bourne, 1968
16	6 Sept. 1967	*R.C. Stoner*	Wake Island, mid-Pacific Ocean	18 000 tons	6 million gallons	Gasoline, jet fuel, turbine fuel, diesel oil, bunker C	No	S. coast of island polluted. Many dead fish stranded on shores. Also abundant dead molluscs, sea-urchins and crabs. 1360 kg dead fish collected; estimated 2500 kg fish washed up. Either gasoline or black oil thought to be lethal to fish. No birds reported oiled	Gooding, 1971
17	27 Oct. 1967	*Fina Norvege*	Off coast of Cornwall, U.K.	*	9·50 tons	Bunker C	Yes	Beaches from Bovisand to Renney Pt. polluted. Dead limpets and discoloured rock pool algae found on sprayed areas of shore	Holme and Spooner, 1968
18	'Fall' of 1967/1968	*The Sea Transport*	Harbour Grace, Newfoundland, Canada	*	30 000 gallons	*	*	No reports of seabird mortalities. No other damage mentioned	Cowell, 1969

19	3 Feb. 1968	*Sivella*	600 yds off Green Pt., Cape Town, S. Africa	80 596 dwt	Unknown	Crude	*	High seabird mortalities. No other damage mentioned	Stander and Venter, 1968; Anon., 1969a
20	29 Feb. 1968	*Tank Duchess*	Tay Estuary, Scotland, U.K.	13 229 tons	87 tons	Topped Venezuelan crude	Yes	Extensive shore pollution in estuary. Oil alone, smothering mussel beds and algae had no apparent toxic effect, although effects of emulsified oil are not mentioned. Over 1300 birds known to have died from pollution, mostly slow-breeding species	Greenwood and Keddie, 1968; McManus, 1968; Technical Advisory Committee for oil pollution in the Tay, 1968
21	3 Mar. 1968	*Ocean Eagle*	Entrance to San Juan harbour, Puerto Rico	*	*	*	Yes	Harbour, nearby beaches and lagoons polluted. Many intertidal organisms killed or damaged by oil or oil and emulsifier, including molluscs, crustaceans and algae, although subsequent recovery good. 10 species of fish found dead or in state of stress	Cerame-Vias, 1968
22	7 Mar. 1968	*General Colocotronis*	Eleuthera, Bahamas	* Cargo 18 000 tons of oil	*	Bunker	Yes	Localised area of shore heavily oiled, 4 types of emulsifier, of varying toxicity, applied. Few details of biological damage caused by spill; chitons and crabs unharmed by oil	Spooner and Spooner, 1968
23	29 Apr. 1968	*Esso Essen*	2½ miles off Cape Peninsula coast, S. Africa	48 535 dwt	3000–4000 tons	Arabian heavy crude	Yes	15 miles of shoreline polluted. High mortalities of sandhoppers (amphipods) but otherwise little damage on shores. High bird mortalities included 500 gannets and 750 out of 1700 oiled penguins	Stander, 1968a; Anon. 1969a; Westphal and Rowan, 1971

	Date	Name of tanker/installation	Location	Size of tanker	Amount of oil spilt	Type of oil spilt	Sprayed with emulsifier	Biological Damage	Reference
24	Apr. 1968	*Andron*	10 miles offshore from Durban, South Africa	16 222 dwt	Unknown	Crude	Yes	Oil attributed to *Andron* spill came ashore near Cape Agulhas. High bird mortalities. Localised damage to shore, with mortalities of molluscs and sea bamboo	Rowan, 1968; Stander and Venter 1968; Anon., 1969a
25	June 1968	*World Glory*	65 miles ENE of Durban, S. Africa	48 823 dwt	Probably most of cargo (45 572) tons of oil)	Crude	Yes	No oil reached shore. Very low mortalities of birds and fish. Near-inshore slicks sprayed with low toxicity emulsifiers	Stander, 1968b
26	29 Oct. 1968	Refinery storage lagoon	Allegheny River, Pennsylvania, U.S.A.	—	*c.* 3000 gallons	Industrial oil waste	No	Foamy mass of oil moved slowly downstream. Some fish killed, mostly minnows	S.I. Event No. 55–68
27	1 Nov. 1968	Gulf Refinery	Milford Haven, S. Wales, U.K.	—	*c.* 100 tons	Crude	Yes	Many small stream fish killed, also high mortalities of gastropods on shores cleaned with emulsifiers	Dudley 1971; Crapp, 1971
28	13 Nov. 1968	*Hess Hestler* (barge)	Rehoboth Beach, Delaware, U.S.A.	* Cargo: 1 m. gals oil	*	No. 6 fuel	No	1 mile of beach polluted. No mention of biological damage	S.I. Event No. 60–68

29	13 Dec. 1968	*Witwater*	Outside Atlantic entrance to Panama Canal	3400 tons tons	*c.* 20 000 barrels	Diesel oil and Bunker C	No	Much oil came ashore, especially on Galeta Island. On rocky shores, extensive mortality of supralittoral vegetation and tide pool life. On sandy beaches, great population decreases among meiofauna, especially crustaceans. Many young mangroves killed in swamp areas, also algae and many invertebrates. Coral reefs apparently unharmed	Rützler and Sterrer 1969; S.I. Event No. 64–68
30	28 Jan. 1969	*London Harmony*	Near Vaxholm on E. coast of Sweden	* Cargo 20 000 tons	200–900 m^3	No. 4 fuel	Yes	Shoreline polluted. Emulsifier used toxic to many organisms	Hasselhuhn and Englund, 1970; Ganning, 1970
31	28 Jan. 1969	Santa Barbara Channel oil well blow-out	Oil well platform in S. Barbara Channel, 5 miles offshore from coast of Calif., U.S.A.	—	Unknown Estimated 11 000–112 000 MT during first 100 days	Crude	*	Extensive research programme initiated after spill. *c.* 100 miles shoreline oiled. High mortalities of intertidal organisms covered with oil. High seabird mortalities: estimated 3600 killed. No apparent effects on fish or plankton. No directly attributable damaging effects of oil on large marine mammals (whales, seals and sea lions) or on benthic fauna, although oil layers found in sediments	Straughan, 1971a
32	Feb. 1969	Unknown	N. sea coast of Holland	—	150 tons +	Residual fuel	No	35 000–41 000 birds of 42 species killed in the Waddenzee	Swennen and Spaans, 1970

	Date	Name of tanker/ installation	Location	Size of tanker	Amount of oil spilt	Type of oil spilt	Sprayed with emulsifier	Biological Damage	Reference
33	4 Mar. 1969	*Yukon*	Cook Inlet, Alaska	*	5000 barrels	*	*	No mentioned of biological effects	S.I. Event No. 24–69 and correspondence
34	16 Mar. 1969	Oil rig	Off Louisiana coast, U.S.A.	—	Unknown	Crude	*	No evidence of damage to marine life	S.I. Event No. 25–69
35	Mar. 1969	Naval Storage tank	Suva Harbour, Fiji	—	*	*	Yes	No mention of biological effects	Anon., 1969b
36	Mar. 1969	Unknown	English Channel off Kent coast	—	* Heavy slick	*	No	50 km of Kent coast polluted. Many dead birds found, but no immediate significant damage to intertidal life	Tittley, 1970
37	30 Apr. 1969	*Hamilton Trader*	Liverpool Bay, U.K.	12 708 tons	*c.* 700 tons	Heavy fuel	Yes	Large slick at sea for several days, coalescing into lumps. Oil came ashore in May on Cumbs. and N. Lancs. shores. Over 4400 birds known to be killed, mostly guillemots, and another 1500–6000 more estimated killed at sea. No effects on fisheries or marine life observed	Rees, 1969 O'Sullivan, 1969; Hope Jones *et al.* 1970; S.I. Event No. 49–69
38	12 May 1969	*Hamsley I*	Fox's Cove, nr. St Merryn, Cornwall, U.K.	1100 tons	—	Heavy fuel	Yes	Locally severe damage on shores caused by emulsifiers, killing limpets, anemones, crabs and fish. No bird casualties reported	Gomm and Spooner, 1969

39	22 May 1969	*	Piscataqus River, New Hampshire, U.S.A.	—	Over 200 000 gals	Fuel	*	Intertidal zone hit by oil, resulting in deaths of clams oysters, clam-worms, shrimps and algae	S.I. Event No. 54–69
40	31 May 1969	*Benedict*	Few miles s. of Trelleborg, Sweden in S. Baltic	71 000 tons	*c.* 2000 tons	Bunker and crude	*	No damage to seabirds reported and no other biological damage reported	Anon., 1969c
41	End May 1969	Unknown	Maidencombe Devon, U.K.	—	Unknown	* In small lumps	Yes	Only 1 small beach polluted	Troake, 1969
42	End May 1969	Unknown	Torbay, Devon, U.K.	—	Unknown	*	Yes	14 out of 26 beaches in Torbay polluted. Oil probably of same origin as above	Hawkins, 1969
43	May 1969	Unknown	Beaches at Southend, Essex, U.K.	—	*c.* 60–100 tons	Bunker fuel	Yes	No ecological investigations made. *c.* 5 miles beach polluted. Subsequent nearby fish and shrimp landings normal	Simpson, 1969
44	5 May 1969	*Palva*	Archipelago of Åland, Finland	*	150 tons Slick 5 × 5 N. miles	Russian crude	Yes	Some of cargo burnt. Shore life apparently badly affected by oil. High mortality of breeding eider ducks	Dybern, 1969; Soikkali and Virtanen, 1972
45	7 Jun 1969	*	Weymouth, Massachusetts, U.S.A.	—	100 000 gals	No. 2 home heating fuel	*	Crabs, clams and other intertidal molluscs reported killed	S.I. Event No. 60–69.
46	16 Jun 1969	*	Alma, Wisconsin, U.S.A.	—	40 000 gals	No. 2 home heating fuel	*	Extensive pollution of (inland) shoreline. Some vegetation killed, also fish and insects	S.I. Event No. 68–69
47	23 June 1969	Un-named Libyan tanker	Cook Inlet, Alaska	*	* Small	Fuel	*	No significant damage to marine life	S.I. Event No. 74–69

	Date	Name of tanker/installation	Location	Size of tanker	Amount of oil spilt	Type of oil spilt	Sprayed with emulsifier	Biological Damage	Reference
48	3 July 1969	German industrial factory	Leine River, W. Germany	—	*	Oil, grease and organic solvents	*	Explosion and fire in factory. Large numbers of fish killed in river	S.I. Event No. 83–69
49	17 July 1969	*Onward Venture*	Lowca Pt., 3 miles N. of Whitehaven, Cumbs., U.K.	400 tons	Over 200 tons	Light diesel (gas) oil	*	No adverse effect on marine life found in area of wreck	Gaukroger, 1969; Anon., 1969d
50	July 1969	*Silja*	S. of Iles d'Hyeres, nr. Toulon, S. France	*	* Slick 1–2 × 3–4 miles, *c.* 1 mm thick	*	*	No mention of biological effects	Peres, 1969
51	July 1969	Unknown	Barmouth and Fairbourne beaches, N. Wales, U.K.	—	Unknown	* In lumps	No	2½ miles sandy beach polluted. No sea bird casualties reported. Otherwise no biological damage mentioned	Fish, 1969
52	8 Aug. 1969	Unknown	4 miles off Kent coast, U.K. (slick)	—	Slick 15 miles long 50–100 yds wide	*	No	Some shore pollution. No biological effects mentioned, except very few oiled sea birds	Stuttard, 1969

53	19 Aug. 1969	*Gironde*	Near Bay of St Brieuc, Brittany, France	* Full cargo 1500 tons	*c.* 1000 tons	Fuel	No	4 km of shore polluted. No reports of substantial damage to marine life	S.I. Event No.103–69
54	27 Aug. 1969	Refinery jetty	S. shore of Milford Haven, S. Wales, U.K.	—	* Spill 'moderately severe'	Crude	Yes	Most of oil not dispersed came ashore on salt marsh at Martinshaven. Oil trapped on vegetation: shoots later died but rhizomes probably survived	Baker, 1969
55	16 Sept. 1969	*Florida* (barge)	Buzzards Bay, W. Falmouth, Mass., U.S.A.	200 ft long (cargo 14 000 barrels) oil)	650–700 tons	No. 2 fuel (diesel fuel)	Yes	4 miles shoreline affected. Severe pollution of sublittoral zone, with 95% kill of all fauna, including many fish, worms, molluscs, crabs, lobsters and other crustaceans and invertebrates. Local shellfish industry severely affected	Blumer *et al.*, 1970; Allen, 1969
56	15 Oct. 1969	Paint Co. storage tank	R. Severn at Worcester, U.K.	—	1300 gals	Heavy fuel	No	25 swans contaminated; several died later. No reports of fish mortalities	Anon., 1969e
57	19 Oct. 1969	Storage tank (leak)	Loch Indaal, I. of Islay, Scotland	—	23 000–30 000 gals	Heavy fuel	Yes	No reports of fish mortalities. Local lobster batches normal. 450 + birds killed	Anon., 1969f; Ogilvie and Booth, 1970
58	5 Nov. 1969	*Keo*	*c.* 120 miles S.E. of Nantucket I., Mass., U.S.A.	15 797 tons gross	Presumed whole cargo	Fuel	No	None mentioned. Oil spilt possibly responsible for Martha's Vineyard bird kill (see below, Feb. 1970)	Anon, 1969g
59	21 Nov. 1969	Railway goods train (derailed)	S. of Ravenglass, Cumbs., U.K.	—	24 000 gals	Crude and fuel	No	Esk estuary slight polluted. Pollution confined to marsh fringe. Minimal contamination of mussel beds and *Corophium* spp. *c.* 20–30 gulls oiled	Anon., 1969h; Perkins, 1970

	Date	*Name of tanker/ installation*	*Location*	*Size of tanker*	*Amount of oil spilt*	*Type of oil spilt*	*Sprayed with emulsifier*	*Biological Damage*	*Reference*
60	Nov. 1969	Pipeline (fracture)	Trans-Arabian pipeline (sabotaged)	—	Over 175 000 gals	Crude	—	None mentioned	Nelson-Smith, 1970
61	2 Dec. 1969	Unknown	Channel between Islands of Emäsalo and Kalwö, Finland	—	* Small slick (3–4 km × 200 m)	*	*	No significant shore pollution	Haahtela, 1970
62	9 Dec. 1969	*Eira* (grounded)	In Ajax shallows, 17 km S.E. of Hanko at entrance to Gulf of Finland	5860 dwt	*c.* 15 000 litres	Thought to be diesel	*	Oil seriously affected shores of Island of Jussarö. Only a few sea bird mortalities recorded	Haahtela, 1970
63	15 Dec. 1969	*Raphael*	W. of Emäsalo, Finland	50 000 dwt	Over 60 tons	Crude	No	Shores and bays of nearby islands polluted. No seabird mortalities reported	Haahtela, 1970
64	15 Dec. 1969	*Marpessa*	100 miles NNW of Dakar, W. Africa	206 700 tons	Unknown	Possibly just ballast water	*	None mentioned	Anon., 1970a; and correspondence
65	18 Dec. 1969	Santa Barbara Channel oil pipeline leak	Oil well platform in Santa Barbara Channel off coast of California, U.S.A.	—	*c.* 400 barrels	Crude	*	2 beaches polluted. Slight damage, as for first Santa Barbara oil spill, 28 Jan. 1969 (see above)	Straughan, 1971a, 1971b

66	Dec. 1969–Feb. 1970	Unknown	N.E. coast of Britain	—	Unknown	Two types of weathered heavy fuel oil	*	Over 12 000 birds found dead, at least 50 000 thought to have been killed	Greenwood *et al.*, 1971; S.I. Event No. 12–70
67	4 Feb. 1970	*Arrow*	Chedabucto Bay, Nova Scotia, Canada	18 000 tons	1·5 million gals	Venezuelan Bunker C	Yes	12 miles shoreline polluted. Localised damage to intertidal life, where most mortalities were of crabs, limpets and algae, probably killed by smothering. Local fish catches normal. High sea bird mortalities	Boyd 1970,; Anon., 1970b; S.I. Event No. 15–70, Brown *et al.*, 1973
68	13 Feb. 1970	*Delian Apollon*	Tampa Bay, Florida, U.S.A.	*	10 000 gals	*	Yes	More than 1000 birds killed	I.C.B.P., 1971
69	Feb. 1970	Unknown	Martha's Vineyard, Mass., U.S.A.	—	Unknown	*	*	640 sea birds found dead. Oil possibly originated from *Keo* disaster (see above, 5 Nov., 1969).	S.I. Event No. 13–70
70	3 Mar. 1970	*Oceanic Grandeur*	Torres Strait, N. Australia	58 000 tons	*c.* 1100 tons	Crude	Yes	Heavy mortalities of pearl oysters near Thursday Is., Torres Strait, reported	S.I. Event 24–70
71	10 Mar. 1970	Oil rig	Breton Sound, off Louisiana, U.S.A.	—	65 000 barrels	Crude	Yes	No significant damage to benthos demonstrated	McAuliffe *et al.*, 1975
72	20 Mar. 1970	*Othello*	Trälhavet Bay, E. of Vaxholm, Sweden	*	60 000–100 000 MT	Bunker C	No	No indications of damage to wildlife	S.I. Event No. 42–70; Anon., 1970c
73	Feb.–Mar. 1970	Unknown	Kodiak Is., Alaska, U.S.A.	—	Unknown	Possibly dumped ballast	*	1000 miles of coastline polluted. Estimate of at least 100 000 birds killed (but see event 76). Also some oiled seals and sea-lions	S.I. Event No. 26–70; Bartonek *et al.*, 1971

	Date	Name of tanker/ installation	Location	Size of tanker	Amount of oil spilt	Type of oil spilt	Sprayed with emulsifier	Biological Damage	Reference
74	8 Apr. 1970	*Efthycosta II*	Off Lavernock Pt., nr. Cardiff, U.K.	15 895 tons	700–800 tons	Heavy fuel	Yes	16 km coastline polluted. About half oil spilt sucked back into tanker. Survey 36 h after spill revealed 206 contaminated birds and 7 dead. No other effects mentioned	Anon., 1970d
75	20 Apr. 1970	Pipeline (fracture)	NW shore of Tarut Bay, Saudi Arabia	—	100 000 barrels	Arabian light crude	Yes	Moderate damage to some shore and marsh life. Mortalities suffered by some fish, crabs, bivalves, and gastropods. Marshplants damaged but not killed. Subsequent recovery of all habitats good	Spooner, 1970; Anon., 1970c
76	25 Apr. 1970	Attributed to 2 Japanese vessels sunk previous week	SW coast of Alaska	*	*	Light diesel	*	100 000 + dead birds, but these were not oiled and probably died through bad weather and malnutrition	S.I. Event No. 36–70; Bailey and Davenport, 1972
77	5 May	*Poly-commander*	Mouth of R. Vigo, N.W. coast of Spain	* Cargo 500 000 MT oil	10 000–50 000 MT	Light crude	Yes	No biological damage apparent. 1 week after spill fish, birds, mussels and algae apparently normal	S.I. Event No. 46–70, Anon., 1970c
78	1 June 1970	*Ennerdale*	Off Port Victoria Mahe, Seychelles	50 000 MT dw	Unknown Up to 41 000 MT	Fuel	Yes	Few sea bird mortalities. No other biological damage mentioned	Anon., 1970e

79	27 July 1970	Pipeline (fracture)	Thames estuary, U.K.	—	400 000 kg	Arabian light crude	Yes	Some oil came ashore at Southend. Emulsifier used damaged some algae and killed many barnacles, periwinkles and crabs. Intertidal mud flats badly affected by emulsifier/fresh-water run-off: many dead polychaetes, molluscs and crabs	George, 1970b; Anon., 1970f
80	7 Sept. 1970	*Irving Whale*	Gulf of St. Lawrence, Canada	* Cargo: 1 million gals oil	Unknown Extensive slicks	Bunker C	*	Some oil came ashore on Magdalen Islands. No mention of biological effects	Loucks and Lawrence, 1971
81	16 Oct. 1970	*Kasamatsu Maru*	12 km off C. Irozaki, Izo Peninsula, Japan	800 tons	375 000 gals	Gasoline	Yes	None mentioned	S.I. Event No. 93–70
82	23 Oct. 1970	*Pacific Glory*	Off coast of Isle of Wight, U.K.	42 000 tons	Several thousand tons lost but partly burnt	Crude and some fuel	Yes	Major biological damage averted by pumping oil into other vessels, some pollution of Sussex beaches	S.I. Event No. 96–70; Shell Int. Mar. 1971
83	23 Oct. 1970	*Kazimah*	Robben Island, Table Bay, S. Africa	29 051 tons	*	Crude oil sludge	*	Jackass (blackfooted) penguins affected. 35 found dead, and 523 contaminated	Cooper, 1971
84	13 Nov. 1970	Oil storage tank	Schuylkill River, Douglasville, Penn., U.S.A.	—	3 million gals	slop (crank case) oil	No	50 mile stretch of Schuylkill River affected, also part of Delaware River. Minimal damage to fish. 200 wild geese contaminated by spill	S.I. Event No. 101–70; Anon., 1971a
85	Nov. 1970	*Marlena*	Near Syracuse, off coast of Sicily, Mediterranean	Over 16 000 MT	14 000 MT	Middle East crude	Yes	No mention of biological effects	Anon., 1970g

	Date	Name of tanker/installation	Location	Size of tanker	Amount of oil spilt	Type of oil spilt	Sprayed with emulsifier	Biological Damage	Reference
86	1 Dec. 1970 onwards	Oil rig (fire)	10 miles off Louisiana coast, Gulf of Mexico	—	* (mostly burnt)	Crude	No	None mentioned. (Fire lasted for *c.* 1 month)	S.I. Event No. 105–70
87	1 Dec. 1970	Unknown	Slick nr. Pennekamp coral reef, Florida Keys, U.S.A.	—	Unknown. Slick 75 miles long, ½ mile wide	Bunker C	No	No oil came ashore. No apparent effects on wildlife	S.I. Event No. 106–70
88	1 Dec. 1970	2 official naval oil dumping barges	50 miles offshore from Jacksonville, Florida, U.S.A.	—	500 000–700 000 gals released	Residual light type of diesel	*	No mention of oil coming ashore. No biological damage mentioned, except no threat anticipated	S.I. Event No. 107–70
89	27 Dec. 1970	Power Station storage tank	Amer River, nr. Jertindenberg, Netherlands	—	8000–16 000 tons	Heavy fuel (or crude?)	No	Estimated 10 000 birds affected. Waterfowl sanctuary, the Biesbos, extensively polluted	S.I. Event No. 2–71; Gregory, 1971; Vaas, 1971
90	18 Jan. 1971	*Oregon Standard*	Under Golden Gate Bridge, San Francisco, U.S.A.	*	1·5–1·9 million gals	Bunker C	No	Some damage to shore life, but no details given. Over 7000 birds affected	S.I. Event No. 6–71; Smail *et al.*, 1972
91	23 Jan. 1971	*Esso Gettysburg*	New Haven Harbour, Long Island, Conn., U.S.A.	23 665 tons	386 000 gals	No. 2 fuel	*	Some shores polluted. Possible damage to shellfish. Small number of bird casualties reported	S.I. Event No. 9–71

92	27 Feb. 1971	*Wafra*	5 miles S. of Cape Agulhas, Cape Province, S. Africa	68 000 tons dwt	64 000 tons	Stripped Arabian crude	Yes	*c.* 30 miles beach polluted, but little damage to intertidal life (N.B. emulsifier only used at sea). 1135 black footed penguins found oiled	S.I. Event No. 24–71; Stander, 1971
93	15 Mar. 1971	*Thuntank 6*	Milford Haven, S. Wales, U.K.	5025 dwt	159 tons	Light fuel	Yes	Several small beaches polluted, but little overall biological damage. Some limpets and small crustaceans killed. Some cockles killed, probably by emulsifier. No reports of oiled birds	S.I. Event No. 30–71
94	21 Mar. 1971	Pipeline (fracture)	N.W. Kazakhstan, U.S.S.R.	—	*	Russian crude	*	Damage to sturgeon fishing grounds and fertile agricultural land feared, although no definite reports	Anon., 1971b
95	27 Mar. 1971	*Texaco Oklahoma*	*c.* 120 miles N.E. of Cape Hatteras, N. Carolina, U.S.A.	* Cargo 220 000 barrels oil	Presumed most of cargo	Heavy sulphur fuel	*	No report of damage to wildlife. Slick remained far out at sea	S.I. Event No. 29–71
96	30 Mar. 1971	*Panther*	Goodwin Sands, 5 miles off Kent coast, U.K.	52 400 MT	Over 11 000 litres	Crude	*	36 m of Kent coast severely polluted. Over 200 sea birds (gulls) seen oiled but apparently unharmed. No other details of biological effects	Dixon and Dixon, 1971
97	26 Apr. 1971	Unnamed barge	Texaco Refinery, Anacortes, Wash., U.S.A.	*	5000 barrels	No. 2 diesel oil	No	Some oil on shores, damaging shellfish, limpets, crabs, clams and oysters. *c.* 1000 birds estimated contaminated	S.I. Event No. 49–71; Chia, 1971
98	26 Apr. 1971 onwards	Unknown	Off Sule Skerry Islet, W. of Orkney, Scotland, U.K.	—	Unknown	*	No	*c.* 60 oiled puffins seen near colony	S.I. Event No. 56–71; Bourne and Johnstone, 1972

	Date	Name of tanker/installation	Location	Size of tanker	Amount of oil spilt	Type of oil spilt	Sprayed with emulsifier	Biological Damage	Reference
99	5 May 1971	Pipeline (fracture)	Mouth of York River, Yorktown, Virginia, U.S.A.	—	10 000–70 000 gals	Bunker C	Yes	Northern shore oiled. Some damage to marsh vegetation, especially swamp grass and also to some bottom dwelling organisms	S.I. Event No. 45–71
100	30 May 1971	Unknown	S.E. shores of Shetland, also at sea between Sumburgh and Lerwick, Scotland, U.K.	—	Unknown	Old fuel oil	No	1200 bird bodies found. Mortality may have reached 10 000	S.I. Event No. 57–71; Bourne and Johnstone, 1972
101	14 July 1971	*Towle*	Bayonne, New Jersey, U.S.A.	*	38 000 gals	Bunker	No	*c*. 20 miles of polluted shoreline. No mention of biological effects	S.I. Event No. 70–71
102	20 Aug. 1971	*Manatee*	San Clemente, Cal., U.S.A.	*	1200 gals	Heavy fuel	*	*c*. 20 miles of sandy beaches polluted. No significant damage to wildlife	S.I. Event No. 82–71
103	16 Oct. 1971	Oil rig (fire)	*c*. 100 miles S. of New Orleans, Louis., U.S.A.	—	* several small slicks	Crude	*	None mentioned	S.I. Event No. 98–71
104	30 Nov. 1971	*Juliana*	Near port of Niigata, Japan	11 700 tons	4000–5000 tons	Crude	*	Extensive coastline pollution. Fears of destruction of fisheries and cultured sea weed (food supply)	S.I. Event No. 94–71; also press reports
105	2 Dec. 1971	Oil Well	80 mls S.W. Laban, Persian Gulf	*	100 000 barrels	*	Yes	No mention of biological effects	S.I. Event No. 98–71

106	24 Dec. 1971	*MV Solar Trader*	West Fayu, Caroline Is.	70 000 tons	520 tons tons	Fuel and lubricating	*	Survey taken during July/Aug. 72 find numerous dead and inedible lobsters and clams. Large algal growth on coral in the area. It was unidentified and no samples were taken	S.I. Event No. 1–72
107	28 Dec. 1971	*Elizabeth Knudsen*	Hook of Holland	*	*	*	Yes	Prompt treatment prevented damage	Anon., 1972a
108	Feb. 1972	*Meigen Maru*	Kisarazu City, Tokyo Bay	*	2·5 tons	Bunker C Heavy	*	Laver crop ruined in the bay	Anon., 1972b
109	4–9 Feb. 1972	Unknown	Firth of Forth, Seafield, Scotland	*	Small slick, probably came down sewer	Light	*	Although small, this slick occurred where 30 000 wintering scaup and numerous sea duck gather in a dense mass to feed. Hundreds of birds oiled mainly scaup	Bourne, 1972a; S.I. Event No. 10–72
110	20 Mar. 1972	*Dewdale*	Cromarty Firth, Scotland	*	30 tons	Fuel	Yes	The oil caught the local flock of 2000–3000 pink footed geese, roosting at Nigg Bay. At least 1000 were badly oiled. Hundreds of gulls, ducks and waders were also affected	Bourne, 1972b; S.I. Event No. 21–72
111	21 Mar. 1972	*F.L. Hayes*	Bartlett Reef, New London, Connecticut, U.S.A.	*	80 000 gals	No. 2 fuel	*	Some birds seen coated with oil but otherwise no damage reported	S.I. Event No. 19–72; Anon., 1972c
112	11 May 1972	*Tien Chee*	River Plate, Montevideo	*	*	Mixture Nequen/ Rio Negr crudes	*	No information regarding damage to birds or marine life	S.I. Event No. 31–72
113	30 June 1972	Pipeline rupture	Biribili, Ecuador, S.A.	*	50 000 barrels (est.)	*	*	Hopes that effects on flora and fauna will be limited as the leak was detected early	S.I. Event No. 42–72; Anon., 1972d

	Date	Name of tanker/installation	Location	Size of tanker	Amount of oil spilt	Type of oil spilt	Sprayed with emulsifier	Biological Damage	Reference
114	22 July 1972	*Tamano*	Portland Harbour, Maine, U.S.A.	*	100 000–200 000 gals	No. 6 fuel	*	Shores 40 miles away were polluted. Marine damage not known but clam, lobster and other fisheries were affected. Many reports of oiled herring gulls, eider duck, scoters, tern and loons	S.I. Event No. 46–72; Anon., 1972e
115	21 Aug. 1972	*Oswego Guardian*	S. coast of Africa	48 320 tons	*	Crude	*	Little danger of major pollution but 450 oil soaked penguins were collected and many others may have died at sea	S.I. Event No. 58–72; Anon., 1972f
116	14 Sept. 1972	*Republica De Colombia*	Coast of Cape Hatteras, N. Carolina, U.S.A.	*	24 000 gals	Banker	*	No reports of damage to birds or marine life	S.I. Event No. 56–72
117	27 Sept. 1972	Oil Tank	Salem Harbour, Salem, Mass., U.S.A.	*	30 000 gals	No. 2 & 5 fuel	*	Between 4 and 5 miles of coastline affected, but no oiled birds or damage to marine life were reported	S.I. Event No. 64–72
118	1 Oct. 1972	*Genimar*	English Channel	3535 tons gross	*	Bunker	Yes	Pollution was minimal. No evidence that seabirds were affected	S.I. Event No. 72–72
119	10 Oct. 1972	Pipeline rupture	N.W. of Shiprock, Arizona, U.S.A.	*	Thousands of gallons	Crude	*	Some oil entered the San Juan river but no reports of damage to plants or animals	S.I. Event No. 70–72

120	24 Oct. 1972	Tanker truck	Cimarron Canyon, New Mexico, U.S.A.	*	7000 gals	Diesel	*	Approximately 19 100 brown and rainbow trout were killed and all life in the stream was destroyed for a distance of 10 miles. Prediction of 3 year period for invertebrate recovery, and the recovery of the fish will be determined by this	S.I. Event No. 86–72
121	25 Oct. 1972	*Ocean 80* Barge	GATX Pier, Carteret, N.J., U.S.A.	*	*	Gasoline, Diesel	*	No bird or marine life damage reported	S.I. Event 73–72
122	27 Nov. 1972	*	North Atlantic	*	Slick approx. 100 sq ml	*	*	Source unknown	S.I. Event No. 85–72
123	28 Nov. 1972	Unknown	Arthur Kill, New Jersey, U.S.A.	*	5000–10 000 gals	Heavy bunker	*	The oil formed into globules and touched the shore but no damage to bird or marine life reported	S.I. Event No. 84–72
124	Early Dec. 1972	Unknown	N. Sea coast off Denmark	*	*	*	*	Damage to birds in Danish Waddensea due to this spill. 20 000 eiders and 5000 scoters found with oil soaked feathers, 5000 were killed. Off the SW coast of Jutland approximately 30 000 birds killed	S.I. Event No. 93–72
125	3 Dec. 1972	Well out of control	Southwest Passage, Mouth of Mississippi, Gulf of Mexico	*	10 000 gals	Diesel	*	No damage reported	S.I. Event No. 90–72
126	5 Dec. 1972	*Vita*	East River off Welfare Is., New York, U.S.A.	20 889 tons	350 000 gals	No. 6 and Bunker C	*	The spill stretched for 16–18 miles down river. 26 oiled ducks found, 20 died	S.I. Event No. 91–72

	Date	Name of tanker/ installation	Location	Size of tanker	Amount of oil spilt	Type of oil spilt	Sprayed with emulsifier	Biological Damage	Reference
127	25 Dec. 1972	*Bouchard No. 40* Barge	Reynolds Chan., Long Island, New York, U.S.A.	240 ft long	11 000 gals	Diesel	*	No bird or marine life damage reported	S.I. Event No. 97–72
128	26 Dec. 1972	*Connecticut* Barge	Old Saybrook breakwater, Connecticut, U.S.A.	*	12 000 gals	No. 6 fuel	*	No reports of damage to wildlife	S.I. Event No. 99–72
129	29 Dec. 1972	*Teesfield*	Ijmuiden, Netherlands	12 146 tons	30 000 litres	Gasoline	*	No reports of damage to wildlife	S.I. Event No. 1–73
130	5 Jan. 1973	IBS 26, 27, 28 & 30 Barges	Mississippi River, Helena, Arkansas, U.S.A.	*	450 000 gals	No. 2 fuel	*	Slick 100 miles long but no reports of damage to wildlife	S.I. Event No. 2–73
131	8–9 Jan. 1973	*Ternsjoe* and *Gruziadz*	Kattegat, Sweden	*	2300 tons	Heavy fuel	*	Heaviest recorded oil spill in or near Danish waters. Few birds, if any, approached the area due to fog and noxious oil odours	S.I. Event No. 5–73
132	10 Jan. 1973	Pipeline rupture	Nr. Bellingham, Washington, U.S.A.	*	500 000 gals	Canadian crude	*	No damage to wildlife reported	S.I. Event No. 7–73
133	19 Jan. 1973	Valves on holding tank	Oakland, California, U.S.A.	*	175 000 gals	Waste	*	5 miles of estuary affected. 292 birds treated. 146 died including coots, grebes, mallard and scaup	S.I. Event No. 9–73

134	24 Jan. 1973	*Irish Stardust*	Albert Bay, Vancouver Is., Brit. Columbia	19 500 tons	120 000 gals	Bunker Crude	*	Small islands and the shoreline were affected. Some birds were oiled	S.I. Event No. 18–73
135	Feb. 1973	*	Ardmore, Firth of Clyde, Scotland	*	Several hundred gallons	Crude	*	Ardmore nature reserve was threatened by this spill but only 9 birds were affected	Anon., 1973a
136	15 Feb. 1973	Unknown	The Caribbean	*	Slick 3 ml × 880 yds	Diesel	*	No damage reported	S.I. Event No. 22–73
137	Mid-Feb. 1973	Coil fracture in storage tank	Rouge River, Allen Park, Michigan	*	6000–10 000 gals	Vegetable	—	No biological damage reports as the river was iced over	S.I. Event No. 27–73
138	26 Feb. 1973	*George T Tilton.* Barge	Hudson River, New York	384 ft	201 000 gals	Premium gasoline	*	No damage to wildlife reported	S.I. Event No. 21–73
139	18 Mar. 1973	Pipeline rupture	Cambridge, Wisconsin	*	189 000 gals	Canadian crude	—	$1\frac{1}{2}$ sq. miles of marshland was affected. No bird mortality reported	S.I. Event No. 36–73
140	9 Apr. 1973	*Pennant*	Rhode Island, U.S.A.	644 ft	90 000 gals	No. 6 fuel	*	Oil spread approximately 8 miles both sides of the river, in globules. Damage to wildlife considered minimal	S.I. Event No.56–73; Anon., 1973b
141	29 Apr. 1973	Pipeline rupture	Casper (Soda Lake), Wyoming	*	5970 litres	Crude	Yes	Blizzard forced postponement of clean-up for 3 days. Oil had spread to all shores and 5000–10 000 birds died. An entire colony of grebes wiped out	S.I. Event No. 57–73
142	2 May 1973	Pipeline rupture	Murry, Idaho, U.S.A.	*	126 000–200 000 gals	Diesel No. 2 fuel	*	No biological damage reported	S.I. Event No. 62–73
143	3 May 1973	*Conoco Italia*	Ionian Sea, Mediterranean	*	10 miles × 200 yds slick	*	*	Vessel apparently pumping bilges and cleaning deck. Photographs taken	S.I. Event No. 66–73

	Date	Name of tanker/ installation	Location	Size of tanker	Amount of oil spilt	Type of oil spilt	Sprayed with emulsifier	Biological Damage	Reference
144	8 May 1973	Open valve on railroad car	Perinton, New York, U.S.A.	*	30 000 gals	No. 6 fuel	*	Vandals opened bottom valve. Oil confined to a small area. No serious effects on local ecology	S.I. Event No. 60–73
145	14 May 1973	Unknown	Windward Pass, S.E. Guantanamo, Cuba	*	8 naut. mls × 60 yds	Petroleum	*	No action taken	S.I. Event No. 65–73
146	2 June 1973	*Esso Brussels*	New York City harbour, U.S.A.	*	*	Viscous tar-like	*	No bird or marine life damage reported	S.I. Event No. 77–73
147	4 June 1973	*	Coal Oil Point, California, U.S.A.	*	5 mls × 50–70 yds	*	*	Natural seepage increase. Smelt, anchovies and seals seen swimming normally. No damage reported	S.I. Event No. 79–73
148	15 June 1973	*Napier*	Isle of Guamblin, Chonas Archipelago, Chile	35 000 tons	5 mls × 50–70 yds	Light Bolivian crude	*	Tanker was fired after grounding as it was a potential danger to mussel culture station 35 km away	Anon., 1973c
149	24 June 1973	*Conoco Brittania*	Humber River, England	57 084 metric tons	98 metric tons	Liberian crude	*	A slick 5 ft wide settled along the beaches. Bird sanctuaries threatened, and a number of birds found oiled	S.I. Event No. 89–73; Anon., 1973d
150	27 June 1973	Buckeye Pipeline Co.	Ottawa River, Ohio, U.S.A.	*	150 000 gals	Jet aviation fuel & naphtha	*	The spill destroyed most of the plant and animal life in an already contaminated part of the river	S.I. Event No. 98–73

151	5 Aug. 1973	*Dona Marika*	Lindsway Bay, Milford Haven, S. Wales, U.K.	11 000 tons	3000 tons	Petroleum	—	Numerous molluscs were killed. These included limpets, mussels, topshells, periwinkles and dog whelks	Blackmen *et al.*, 1973; S.I. Event No. 114–73
152	10 Sept. 1973	M/T *Splendid Arrow*	Shell oil docks, Houston, Texas, U.S.A.	*	126 000 gals	Qatar Marine crude	*	No damage to birds or marine life reported	S.I. Event No. 127–73
153	25 Sept. 1973	*Erawan*	English Bay, Vancouver, B.C.	10 000 tons	80 000 gals	Bunker	*	Two beaches and shoreline affected. Dead fish were found and oiled birds were seen	S.I. Event No. 120–73
154	17 Oct. 1973	*Seaboard 31* Barge	Exxon Terminal, S. Albany, New York, U.S.A.	*	20 000 gals plus	No. 6 fuel	*	1 mile of shore-line heavily coated. No damage reported	S.I. Event No. 137–73
155	9 Nov. 1973	Unknown	Bodrog River, Felsöberecki, NE Hungary	*	Tons	Heavy	*	Oil appeared to be passing down river from Czechoslovakia. Fauna and flora were threatened	S.I. Event No. 145–73
156	15 Nov. 1973	*British Mallard*	Laksfjorden, Norway	15 866 tons	2000 tons	Light fuel	Yes	1000 sea-birds reported killed. 90% eider duck, remainder long-tailed duck and razorbill. No marine or intertidal damage recorded	S.I. Event No. 148–73; Anon., 1974a
157	30 Nov. 1973	Oil well expl. and spill	61 ml N.W. of Casper, Wyoming, U.S.A.	*	200–400 barrels per hour	*	*	Oil covered an area of $\frac{1}{2}$ sq. ml. No ecological damage reported	S.I. Event No. 158–73
158	21 Dec. 1973	*Jawacta*	Western Baltic Sea, off Swedish Coast	*	20 000 tons	Heavy fuel	*	No damage reported	S.I. Event No. 1–74
159	21 Dec. 1973	*Lalibella*	Cape Cod, Massachusetts, U.S.A.	*	220 000 gals	Heavy	*	No damage reported	S.I. Event No. 16–74

	Date	Name of tanker/ installation	Location	Size of tanker	Amount of oil spilt	Type of oil spilt	Sprayed with emulsifier	Biological Damage	Reference
160	22 Dec. 1973	*Restless.* Tug towing oil barge	Atchafalaya River, W. Baton Rouge, Louisiana, U.S.A.	*	7–8000 barrels	Crude	*	No damage reported although 50 000 ducks nest in marshes along the river	S.I. Event No. 14–74
161	29 Dec. 1973	*Private Joseph Merrell*	N. Pacific coast off Monterey, California, U.S.A.	*	16 000 gals	Diesel	*	200 gallons washed up on local beach. 6 oiled gulls found. No other damage reported	S.I. Event No. 2–74
162	3 Jan. 1974	Oil tank rupture	Duck Island, nr. Trenton, New Jersey, U.S.A.	*	600 000 gals	No. 2 fuel	*	10 000 gals entered a swamp, 20 000 gals went into Delaware River, 5 miles affected. Extent of biological damage unknown	S.I. Event No. 5–74; Anon., 1974b
163	7 Jan. 1974	Unknown	Danube River, S. Hungary	*	2800 tons	Fuel oil residue	—	No ecological damage reported	S.I. Event No. 12–74
164	15–19 Jan. 1974	Pipeline rupture	Norco, Mississippi River, Louisiana, U.S.A.	*	Several thousand barrels	Crude	*	The spill extended 90 miles down river and numerous oil covered fowl and were found. No other damage reported	S.I. Event No. 13–74; Anon., 1974c
165	20 Jan. 1974	Release plug in pump	Prudhoe Bay, N. Alaska, U.S.A.	*	7000 gals	Diesel	*	4000 gals recovered the rest was burnt. No damage reported	S.I. Event No. 15–74
166	20 Jan. 1974	Pipeline rupture	Arkansas River, E. Belle Plain, Kansas, U.S.A.	*	37 800 gals	Crude	*	No damage reported	S.I. Event No. 17–74

167	Early Feb. 1974	Overfilled oil tank	Cromarty Firth, East Ross, Scotland	*	300 gals plus	Fuel	*	Birds oiled were 200 mute swan, whooper swan, goldeneye, tufted duck and scaup. 40 mute swans died. 13th incident since 1966. Salt marsh at nearby Nigg Bay worst affected area, already the site of Dewdale spill, see 110	Bourne, 1974; Currie, 1974; S.I. Event No.30–74
168	15 Feb. 1974	Pipeline rupture	Trinity River, Livingston, Texas, U.S.A.	*	168 000 gals	Crude	*	No visible damage to fish or wildlife	S.I. Event No. 37–74
169	20 Feb. 1974	*Athos*	Delaware, New Jersey, U.S.A.	*	285 000 gals	Heavy bunker	*	The Tinicum Wildlife Preserve was affected. 200–300 ducks were killed	S.I. Event No.26–74; Anon., 1974d
170	4 Mar. 1974	Barge	Mississippi, nr. Helena, Arkansas, U.S.A.	*	21 000 gals	Gasoline	*	No damage reported	S.I. Event No. 44–74
171	8 Mar. 1974	Boiler exploded	Greenville, S. Carolina, U.S.A.	*	16 000 gals	No. 6 fuel	*	Vegetation along river edge coated but no oiled birds found	S.I. Event No. 45–74
172	8 Mar. 1974	Pipeline rupture	Osage County, Skiatook, Oklahoma, U.S.A.	*	126 000 gals	No. 2 fuel	*	Three creeks affected. Due to floods no surveys made although carp in distress and oiled duck were seen	S.I. Event No. 38–74
173	11 Mar. 1974	Pipeline rupture	Venchoner Creek, Wise City, Texas, U.S.A.	*	78 750 gals	Crude	*	3–4 miles of creek affected. Oiled birds seen, oak trees and scrub covered in oil	S.I. Event No. 39–74
174	11 Mar. 1974	Faulty valve on storage tank	Nr. Poyner, Henderson City, Texas, U.S.A.	*	15 750 gals	Crude	*	1·5 miles of creek affected. 50–60 dead sun-fish and bull-head catfish found. Vegetation damaged due to oil burn-off	S.I. Event No. 40–74

	Date	Name of tanker/ installation	Location	Size of tanker	Amount of oil spilt	Type of oil spilt	Sprayed with emulsifier	Biological Damage	Reference
175	16 Mar. 1974	Pipeline rupture	Village Creek, nr. Kountze, Texas, U.S.A.	*	9450 gals	Absorption oil (kerosene-like)	*	There was a small fish kill associated with this spill	S.I. Event No. 41–74
176	19 Mar. 1974	Pipeline rupture	Okmulgee County, Oklahoma, U.S.A.	*	Large amount	Crude	*	No damage reported	S.I. Event No. 42–74
177	9 Apr. 1974	*Elias*	Delaware River, Philadelphia, P.A. and N.J., U.S.A.	31 000 tons	1000 gals	Venezuelan crude	*	500 ducks were found oiled	S.I. Event No. 64–74
178	22 Apr. 1974	Leaking valve in distillery	Cromarty Firth, East Ross, Scotland	*	Substantial quantity	*	Yes	2 mute swans died. 10 were oiled. Considerable damage to habitat due to numerous spills. See 163 and 110.	Currie, 1974
179	Apr. 1974	Indian tanker	Fumicino, Port of Rome, Italy	*	20 tons	*	*	30 km of coastline affected. No damage to bird or marine life reported	Anon., 1974e
180	3 May 1974	Valve on storage tank	Price, Estill Cnty., Kentucky, U.S.A.	*	25 000 gals	Crude	*	No damage reported. Vandals were responsible	S.I. Event No. 61–74
181	Late May 1974	*Sygna*	Stockton Bight, East coast of Australia	53 000 tonnes	400 tonnes	Heavy fuel	*	13 km of beaches affected. No damage to bird or marine life reported	Hughes, 1974

182	14 June 1974	Cannaport Buoy	Bay of Funday, St John, New Brunswick	*	33 075 gals	Crude	*	No damage reported	S.I. Event No. 92–74; Anon., 1974f
183	24 June 1974	Barge	Mississippi River, New Orleans, U.S.A.	*	5000 barrels	Crude	*	No damage reported	S.I. Event No. 90–74
184	28 June–15 July 1974	*Dongan Hills* Ferry	Arthur Kill, Staten Island, New York, U.S.A.	*	10 000 gals	Bunker	*	No reports of damage to wildlife in this already severely polluted area	S.I. Event No. 94–74
185	June 1974	*	Getty wharves, Mena Saud, Saudi Arabia	*	35 000 barrels	Crude	—	No mention of biological effects	Anon., 1974g
186	July 1974	*	Getty wharves, Khafji, Kuwait, Saudi Arabia	*	Several thousand barrels	*	Yes	No mention of biological effects	Anon., 1974g
187	31 July 1974	Storage tank rupture	Albany, New York, U.S.A.	*	800 000–840 000 gals	No. 2 fuel	*	A small number of ducks, muskrats and a few other animals were found dead	S.I. Event No. 98–74
188	2 Aug. 1974	*Jos Simard*	Saglek, Newfoundland, Canada	*	500 000 gals	Diesel	*	No damage reported	Anon., 1974h
189	9 Aug. 1974	*Metula*	Straits of Magellan, Chile	206 000 dwt	52 000 tons	Light Arabian crude and bunker oil	No	Mousse landed on many miles of coastline. About 1200 oiled birds found, most dead	Baker, *et al.*, 1975 (this volume)

	Date	Name of tanker/installation	Location	Size of tanker	Amount of oil spilt	Type of oil spilt	Sprayed with emulsifier	Biological Damage	Reference
190	29 Aug. 1974	*Garden State*	Tramandal, Rio Grande, Brazil	*	Undetermined	*	*	P69 compound was used to sink the oil but a slick 5 km × 150 m reached the local beach. No damage to bird or marine life was reported	S.I. Event No. 112–74
191	26 Sept. 1974	*Trans Huron*	Kiltan Is., India	*	5000 tons	Furnace fuel	*	The entire coast of the island was affected. Initially there was mortality to fishes and other marine life in and around the area	S.I. Event No. IPAN 16–74
192	30 Sept. 1974	*Golden Robin*	Dalhousie, New Brunswick, Canada	28 000 dwt	238 455–317 940 litres	No. 6 residual	*	Approximately 16·1 km of beach affected and 600 birds killed. The impact on the local ecology is being studied	S.I. Event No. 150–74
193	2 Oct. 1974	Natural seepage	Montecito-Ventura, California, U.S.A.	*	*	Light	*	Soft tar-like globules. No damage reported	S.I. Event No. 126–74
194	6 Oct. 1974	*Messiniaki Bergen*	New Haven Hbr., Connecticut, U.S.A.	*	379 000 litres	Venezuelan No. 4 and 6 crude	*	No immediate damage to wildlife. 2 oiled seabirds found. The effects of the oil on the bottom life are being monitored	S.I. Event No. 129–74
195	9 Oct. 1974	Unknown	Corpus Christi, Texas, U.S.A.	*	1 160 481 litres	Arabian crude	*	Little or no wildlife present in the harbour	S.I. Event No. 153–74

196	17 Oct. 1974	3 Barges	Ohio River, nr. Sciotoville, Ohio, U.S.A.	*	22 710 litres	Crude toluene, benzene and xylene	*	The mixture rapidly evaporated. No immediate adverse effects on wildlife was observed	S.I. Event No. 138–74
197	17 Oct. 1974	Pipeline fracture	Alta Loma, Texas, U.S.A.	*	79 485 litres 4769·6 litres	Gasoline Butane	*	Approximately 90% of gasoline recovered. Remainder and butane evaporated. No damage reported	S.I. Event No. 151–74
198	21 Oct. 1974	*Ercole*	Mississippi River, Donaldsville, Louisiana, U.S.A.	*	349 600 litres	Crude	*	Pollution occurred between Mile 110 and Mile 176 down river. No damage reported	S.I. Event No. 148–74
199	22 Oct. 1974	*Universe Leader*	Bantry Bay, Ireland	85 500 tons	2 460 510 litres	Kuwait crude	Yes	Periwinkles, sea urchins and mussels affected. Small number of swans oiled. Possible effects on herring eggs—herrings were spawning at the time of spill	O'Sullivan, 1974; S.I. Event No. 152–74
200	26 Oct. 1974	Storage tank	Taunton River, Somerset, Massachusetts, U.S.A.	*	79 493 litres	No. 6 C bunker	*	Ecological damage not apparent	S.I.Event No. IPAN 3–74
201	27 Oct. 1974	Storage tank	Tonowanda, New York, U.S.A.	*	276 334 litres	Leaded gasoline	*	Although entering a swamp and the Niagara river, only 2 muskrats and 2 teal were killed	S.I. Event No. 154–74
202	4–5 Nov. 1974	Barge ATC135	Cape Fear River, Wilmington, N. Carolina, U.S.A.	*	94 635 litres	No. 6 fuel	*	Minimal environmental damage	S.I. Event No. IPAN 5–74

	Date	Name of tanker/installation	Location	Size of tanker	Amount of oil spilt	Type of oil spilt	Sprayed with emulsifier	Biological Damage	Reference
203	9 Nov. 1974	Pipeline rupture	Feyodi Creek, nr. Jennings, Oklahoma, U.S.A.	*	79 493 litres	Crude	*	Some vegetation oiled but no fish or wildlife mortality reported	S.I. Event No. IPAN 7–74
204	9 Nov. 1974	2 train cars derailed	Lake Messalonskee, Maine, U.S.A.	*	37 854–52 995 litres	Bunker C	*	So far no damage to wildlife has been reported	S.I. Event No. IPAN 4–74
205	13 Nov. 1974	Barge No. 75	Hudson River, New Hamburg, U.S.A.	*	70 137–75 708 litres	No. 6	*	No reports of damage to wildlife	S.I. Event No. IPAN 6–74
206	16 Nov. 1974	Pipeline rupture	Cornie Creek, nr. Magnolia, Arkansas, U.S.A.	*	55 645 litres	Crude	*	8 km of creek affected but no damage to wildlife reported	S.I. Event No. IPAN 10–74
207	21 Nov. 1974	*Roy A. Jodrey*	Alexandria Bay, New York, U.S.A.	194·7 metres	Unknown amount from total cargo of: 9842 litres 189 270 litres	Lubricating Diesel	*	Severe weather conditions prevented estimate of spill	S.I. Event No. IPAN 11–74
208	23 Nov. 1974	Pipeline rupture	Terlton, Oklahoma, U.S.A.	*	39 746·7 litres	Crude	*	Approximately 0·4 km of a stream was affected. A minor fish kill did take place involving individuals of the genus *Lepomis*	S.I. Event No. IPAN 14–74

209	30 Nov. 1974	2 railroad tank cars	S. Marlow, Texas, U.S.A.	*	189 270 litres	Vegetable	*	No reports of damage to wildlife	S.I. Event No. IPAN 19–74
210	2 Dec. 1974	Pipeline rupture	Corpus Christi, Texas, U.S.A.	*	135 138 litres	Light crude	*	The area in which the spill occurred supports little wildlife. Dead fish were sighted however, but it was not known whether this was due to the spill or cold weather	S.I. Event No. IPAN 15–74
211	6 Dec. 1974	Pipeline rupture	North Channel, Gloucester, Massachusetts, U.S.A.	*	52 995 litres	No. 2 home heating fuel	*	Although the shore-line was affected, no immediate damage to wildlife was observed	S.I. Event No. IPAN 17–74
212	18 Dec. 1974	Oil tanks	Kurashiki, Japan	*	42 000 kilolitres	Heavy	*	The oil affected waters of four prefectures (Okayama, Kagawa, Tokushima and Hyogo), an area rich in seaweed. No damage to wildlife was reported	S.I. Event No. IPAN 20–74
213	3 Jan. 1975	Unknown	Hampton Roads, Virginia, U.S.A.	*	Approx. 114 000 litres	*	*	No reports of damage to wildlife	S.I. Event No. IPAN 1–75
214	4 Jan. 1975	Pipeline rupture	Texas City, Texas, U.S.A.	*	1 915 000 litres	Crude	*	About 638 000 litres entered the Texas City Industrial Canal, which was already polluted before the spill	S.I. Event No. IPAN 2–75
215	6 Jan. 1975	*Showa Maru*	Malacca Strait, S. Singapore	237 000 tons	844 000 gals	Crude	*	As yet no estimate of damage available	Press reports, 1975
216	10 Jan. 1975	*Afton Zodiac*	Glengarriff, N. Bantry Bay, Ireland	210 000 tons	500 tons	Bunker C	Yes	Oil 1 cm to several inches thick heavily affected $2\frac{1}{2}$ miles of beach	Press reports, 1975

FURTHER REFERENCES TO ORNITHOLOGICAL ASPECTS

Bird kills until the spring of 1968 are listed in 'A chronological list of ornithological oil pollution incidents' by W. R. P. Bourne (*Seabird Bulletin*, **7**, 1969, 3–8); many important bird kills in Danish waters from 1935 to mid-1971 in 'Oil pollution and seabirds in Denmark 1935–1968' and 'Studies on oil pollution and seabirds in Denmark 1968–1971' by A. H. Joensen (*Danish Review of Game Biology*, **6**(8–9), 1972, pp. 24 and 32); and a chronological list of abstracts and alphabetical bibliography of ornithological oil pollution literature is supplied by R. and K. Vermeer, 'Oil pollution of birds: an abstracted bibliography' in Canadian Wildlife Service, Pesticide Section Manuscript Reports No. 29 (1974) 68 pp. Most notable bird kills are listed in the annual report on the north-west European beach surveys in *Birds*.

ACKNOWLEDGEMENTS

We are grateful to Dr W. R. P. Bourne, Mr E. W. Mertens, Dr R. B. Schwendinger and Mr J. Wardley Smith for helpful comments during the updating of this review.

BIBLIOGRAPHY

Allen, J. A. (1969). West Falmouth oil spill, Massachusetts, September, 1969, *Mar. Poll. Bull.* (16), 16–17.

Anon. (1969a). Oil Pollution in South Africa, January–April 1968, *Mar. Poll. Bull*, (14), 2–7.

Anon. (1969b). Oil Pollution Incident in Suva Harbour, Fiji, *Mar. Poll. Bull.* (15), 22.

Anon. (1969c). *Benedict* Oil Pollution Incident, May–June 1969, *Mar. Poll. Bull.* (13), 17–18.

Anon. (1969d). Oil Pollution Incident, Cumberland Coast, July 1969, *Mar. Poll. Bull.* (14), 20.

Anon. (1969e). River Severn Oil Pollution Incident, October 1969, *Mar. Poll. Bull.* (17), 15–16.

Anon. (1969f). Oil Pollution Incident of the Island of Islay, W. Scotland, October 1969, *Mar. Poll. Bull.* (17), 16.

Anon. (1969g). *Keo* Disaster, November 1969, *Mar. Poll. Bull.* (18), 22–3.

Anon. (1969h). Cumberland Oil Spill, November 1969, *Mar. Poll. Bull.* (18), 23.

Anon. (1970a). Sinking of Super Tanker, December 1969, *Mar. Poll. Bull.*, **1** (1), 2.

Anon. (1970b). Another Torrey Canyon? February 1970, *Mar. Poll. Bull.*, **1** (3), 34–5.

Anon. (1970c). Oil Spills Continue, March 1970, *Mar. Poll. Bull.*, **1** (6), 83.

Anon. (1970d). Oiled Beaches Successfully Cleaned, April 1970, *Mar. Poll. Bull.*, **1**(5), 77.

Anon. (1970e). Seychelles Oil Scare, June 1970. *Mar. Poll. Bull.*, **1**(7), 98–9.

Anon. (1970f). Burst Pipeline in Essex, July 1970. *Mar. Poll. Bull.*, **1**(8), 117.

Anon. (1970g). Sicilian Oil Spill, November 1970. *Mar. Poll. Bull.*, **1**(12), 181.

Anon. (1971a). Schuylkill Oil Deluge, November 1970, *Mar. Poll. Bull.*, **2**(1), 6.

Anon. (1971b). A Foretaste of TAPS?, March 1971, *Mar. Poll. Bull.*, **2**(5), 66.

Anon. (1972a). Dutch Oil Spill, December, 1971, *Mar. Poll. Bull.*, **3**(2), 24.
Anon. (1972b). Oil Ruins Laver Crop, February 1972, *Mar. Poll. Bull.*, **3**(2), 23.
Anon. (1972c). Light Oil Spillage, February 1972, *Mar. Poll. Bull.*, **3**(5), 69.
Anon. (1972d). Transandean Pipeline Leak, June 1972, *Mar. Poll. Bull.*, **3**(9), 134.
Anon. (1972e). Tamano Oil Spill, July 1972, *Mar. Poll. Bull.*, **3**(9), 133.
Anon. (1972f). More on Tanker Collisions, 1969–1972, *Mar. Poll. Bull.*, **3**(11), 163.
Anon. (1973a). Ardmore Threatened by Oil on the Clyde, February 1973, *Mar. Poll. Bull.*, **4**(3), 40.
Anon. (1973b). Oil Spill in Rhode Island, April 1973, *Mar. Poll. Bull.*, **4**(6), 84.
Anon. (1973c). Firing Chilean Oil, June 1973, *Mar. Poll. Bull.*, **4**(11), 163–4.
Anon. (1973d). Yorkshire Coast Oiled, June 1973, *Mar. Poll. Bull.*, **4**(7), 101.
Anon. (1974a). Eiders Oiled in Norway, November 1973, *Mar. Poll. Bull.*, **5**(1), 4.
Anon. (1974b). Duck Soup, January 1974, *Mar. Poll. Bull.*, **5**(2), 19.
Anon. (1974c). Mop Up, January 1974, *Mar. Poll. Bull.*, **5**(2), 19.
Anon. (1974d). Delware Hit Again, February 1974, *Mar. Poll. Bull.*, **5**(4), 53.
Anon. (1974e). Spill off Port of Rome, April 1974, *Mar. Poll. Bull.*, **5**(4), 52.
Anon. (1974f). Docking System Fails, June 1974, *Mar. Poll. Bull.*, **5**(9), 133.
Anon. (1974g). Oil Escapes at Getty Wharves, June–July 1974, *Mar. Poll. Bull.*, **5**(9), 132.
Anon. (1974h). Saglek Slip-up, August 1974, *Mar. Poll. Bull.*, **5**(11), 165.
Bailey, E. P. and Davenport, G. H. (1972). Die-off of Common Murres on the Alaska Peninsula and Unimak Island, *Condor*, **74**, 215–19.
Baker, J. M. (1969). Oil Pollution at Milford Haven, August 1969, *Mar. Poll. Bull.* (16), 18.
Baker, J. M. *et. al.* (1975). An oil spill in the Straits of Magellan (this volume).
Bartonek, J. C., King, J. G. and Nelson, H. K. (1971). Problems confronting migratory birds in Alaska, *Trans. N. Amer. Wildlife and Natural Resources Conf.*, **36**, 345–61.
Beynon, L. R. (1971). Dealing with Major Oil Spills at Sea, Water Pollution by Oil; proc. symp. ed. Hepple, Inst. of Pet., London, 187–193.
Blackman, R. A. A., Baker, J. M., Jelly, J. and Reynard, S. (1973). *Dona Marika* Oil Spill, August 1973, *Mar. Poll. Bull.*, **4**(12), 181–2.
Blumer, M., Sass, J., Souza, G., Sanders, H. L., Grassle, J. F. and Hampson, G. R. (1970). The West Falmouth Oil Spill, Technical Report, Woods Hole Oceanographic Institution, Ref. No. 70-44 (32 pp.).
Bourne, W. R. P. (1968). Oil Pollution and Bird Populations, *Fld. Stud.*, **2**, (suppl), 99–121.
Bourne, W. R. P. (1972a). Ducks Die in Forth, February 1972, *Mar. Poll. Bull.*, **3**(4), 53.
Bourne, W. R. P. (1972b). *Dewdale* Oil Spill, March 1972, *Mar. Poll. Bull.*, **3**(5), 66–7.
Bourne, W. R. P. (1974). Scotched Birds, February 1974, *Mar. Poll. Bull.*, **5**(4), 52.
Bourne, W. R. P. and Johnstone, L. (1972). The threat of oil pollution to N. Scottish seabird colonies, *Mar. Poll. Bull.*, **2**(8), 117–20.
Boyd, H. (1970). Oil Poses Urgent Problems in Canada, *Mar. Poll. Bull.*, **1**(5), 69–71.
Brown, R. G. B., Gillespie, D. I., Locke, A. R., Pearce, P. A. and Watson, G. H. (1973). Bird mortality from oil slicks off eastern Canada, Feb.–Apr. 1970, *Canadian Field Naturalist*, **87**, 225–34.
Cerame-Vivas, M. J. (1968). The Wreck of the *Ocean Eagle*, *Sea From Sea Frontiers*, **15**, 22–231.

Chia, Fu-Shiang (1971). Diesel Oil Spill at Anacortes, April 1971, *Mar. Poll. Bull.*, **2**(7), 105–6.

Clark, R. B. (1968). Oil Pollution and Conservation of Sea Birds, *Proc. Int. Conf. Oil Pollution Sea*, Rome, 76–112.

Clarke, C. H. D., Gabrielson, N., Kessel, B., Robertson, W. B., Wallace, G. J. and and Cahalane, V. H. (1965). Report of the Committee on Bird Protection, 1964, *Auk* (1st series), **82**, 477–91.

Cooper, J. (1971). *Mar. Poll. Bull.*, **2**(4), 52.

Cowell, E. B. (1969). Oil Pollution on the Newfoundland Coast, *Mar. Poll. Bull.* (14), 8–16.

Crapp, G. B. (1971). The Biological Consequences of Emulsifier Cleansing, proc. symp., ed. Cowell, Inst. of Pet., London, 150–68.

Currie, A. (1974). Oil Pollution in the Cromarty Firth, *Mar. Poll. Bull.*, **5**(8), 118–19.

Diaz-Piferrer, M. (1962). The effects of Oil on the Shore of Guanica, Puerto Rico, Association Island Marine Laboratories, 4th meeting, Curacao, 12–13.

Dixon, T. and Dixon, T. (1971). The *Panther* Affair, March 1971, *Mar. Poll. Bull.*, **2**(7), 107–8.

Drost, R. (1966). Über den Seeunfall des Tankers *Anne Mildred Brövig*: Folgen und Folgerungen, Internat Rat. f. Vogelschutz, Dt. Sekt. Ber. Nr. 6, 52–5.

Dudley, G. (1971). Oil Pollution in a Major Oil Port; the Incidence, Behaviour and Treatment of Oil Spills, The ecological effects of oil pollution on littoral communities, symp. proc., ed. Cowell, Inst. of Pet., London, 5–12.

Dybern, B. I. (1969). Aland Archipelago Oil Pollution Incident, May 1969, *Mar. Poll. Bull.* (13), 18.

Fish, J. D. (1969). Oil Pollution on Barmouth and Fairbourne Beaches, July 1969, *Mar. Poll. Bull.* (15), 26.

Ganning, B. (1970). Östersjön-oljan-saneringen (The Baltic—the oil—the clean-up), *Forsk. Framsteg.* (8), 25–27.

Gaukroger, P. (1969). *Onward Venture*, Cumberland coast, July 1969, *Mar. Poll. Bull.* (16), 19–20.

George, J. D. (1970a). Sublethal Effects on Living Organisms, *Mar. Poll. Bull.*, **1**(7), 107–9.

George, J. D. (1970b). Mortality at Southend, July 1970, *Mar. Poll. Bull.*, **1**(9), 131.

Gomm, R. and Spooner, M. (1969). *Hemsley 1* Incident, May 1969: Biological aspects, *Mar. Poll. Bull.* (15), 24.

Gooding, R. M. (1971). Oil Pollution on Wake Island from the Tanker *R.C. Stoner*, U.S. Dept. of Commerce; Fisheries report No. 636 (12 pp.).

Greenwood, J. J. D. and Keddie, J. P. F. (1968). Birds killed by oil in the Tay Estuary, *Scott. Birds*, **5**(4), 189–96.

Greenwood, J. J. D., Donally, R. J., Feare, C. J., Gordon, N. J. and Waterston, G. (1971). A massive wreck of oiled birds: NE Britain, Winter 1970, *Scott. Birds*, **6**, 235–50.

Gregory, K. G. (1971). Oiled birds in Holland, December 1970, *Mar. Poll. Bull.*, **2**(2), 23.

Haahtela, I. (1970). Oil Spills off Finoand, December 1969, *Mar. Poll. Bull.*, **1**(2), 19–20.

Harrison, J. G. and Buck, W. F. A. (1968). The second winter survey following the Medway oil pollution of 1966, Wildfowlers' Association of Great Britain and Ireland, *Annual Report and Handbook* 1967–1968, 68–71.

Hasselhuhn, B. and Englund, H. (1970). *London Harmony*—ett oljeutsläpp, *Forsk. Framsteg.*, **8**, 28–29.

Hawkins, M. R. (1969). Paignton Beach, May 1969, *Mar. Poll. Bull.* (13), 17.

Holme, N. A. and Spooner, G. M. (1968). Oil Pollution at Bovisand—an Interim Report, *J. Devon. Trust Nat. Conserv.*, 665–7.

Hope-Jones, P., Howells, G., Rees, E. I. S. and Wilson, J. (1970). Effect of *Hamilton Trader* Oil on Birds in the Irish Sea in May, 1969, *British Birds*, **63**, No. 3.

Hughes, D. (1974). *Sygna* slick, May 1974, *Mar. Poll. Bull.*, **5**(7), 99.

International Council for Bird Preservation (1971). *Annual Report for* 1970.

Loucks, R. H. and Lawrence, D. J. (1971). Reconnaissance of an Oil Spill, Sept. 1970, *Mar. Poll. Bull.*, **2**(6), 92–94.

McAuliffe, C. D. *et. al.* (1975). Chevron Main Pass Block 41 Oil Spill, API/EPA/USCG Conference, 555–66.

McManus, J. (1968). Problems of Oil Pollution in the Tay Estuary, February 1968, *Mar. Poll. Bull.* (6), 2–5.

Nelson-Smith, A. (1968a). Biological Consequences of Oil Pollution and Shore Cleansing, *Fld. Stud.*, **2** (suppl.) 73–80.

Nelson-Smith, A. (1970). The Problem of Oil Pollution of the Sea, *Adv. mar. Biol.*, **8**, 215–306.

Ogilvie, M. A. and Booth, C. G. (1970). An oil spillage on Islay in October 1969, *Scott. Birds*, **6**, 149–53.

O'Sullivan, J. (1969). *Hamilton Trader* Oil Pollution Incident, April–May 1969, some Preliminary Observations, *Mar. Poll. Bull.* (13), 19–26.

O'Sullivan, J. (1974). Massive Oil Spillage in Bantry Bay, October 1974, MSS.

Peres, J. M. (1969). *Silja* Incident in Mediterranean, S.E. of Toulon, July 1969, *Mar. Poll. Bull.* (15), 21.

Perkins, E. J. (1970). Oil Pollution at Ravenglass, November 1969, *Mar. Poll. Bull.*, **1**(2), 18.

Ranwell, D. S. and Hewitt, D. (1964). Oil Pollution in Poole Harbour and its Effects on Birds, *Bird Notes*, **31**, 192–7.

Rees, E. I. S. (1969). *Hamilton Trader* Oil Pollution Incident, April–May 1969, *Mar. Poll. Bull.* (12), 10–13.

Rowan, M. K. (1968). Oiling of marine birds in South Africa, *Proc. Int. Conf. Oil Pollut. Sea, Rome*, 121–4.

Rutzler, K. and Sterrer, W. (1969). Bioscience, **20**(4).

Shell International Marine Ltd. (1971). Salvaging the *Pacific Glory*, November 1970, *Mar. Poll. Bull.*, **2**(2), 27.

S. I. = Smithsonian Institution, Centre for short-lived phenomena.

Simpson, A. C. (1969). Oil Pollution Incident, Southend, May 1969, *Mar. Poll. Bull.* (13), 15–16.

Smail, J., Ainley, D. G. and Strong, H. (1972). Notes on birds killed in the 1971 San Francisco oil spill, *California Birds*, **3**(2), 25–32.

Smith, J. E. (ed.). *Torrey Canyon* Pollution and Marine Life, Cambridge University Press.

Soikkali, M. and Virtanen, J. (1972). The *Palva* oil disaster in the Finnish south-western archipelago: II. Effects of oil pollution on the eider *Somateria Mollissima* population in the archipelagos of Kökar and Föglo, SW Finland, *Aqua Fennica*, 122–8.

Spooner, M. F. (1967). Biological Effects of the *Torrey Canyon* Disaster, *J. Devon Trust Nat. Conserv.* (suppl.), 12–19.

Spooner, M. F. and Spooner, G. M. (1968). *The Problem of Oil Spills at Sea*, Marine Biology Association, Plymouth.

Spooner, M. F. (1970). Oil Spill in Tarut Bay, Saudi Arabia, April 1970. *Mar. Poll. Bull.*, **1**(11), 166–7.

Stander, G. H. (1968a). The *Esso Essen* Incident, *S. Afr. Fishing News & Fish. Ind. Rev.*, Aug. 1968, **33**(8), 41–4.

Stander, G. H. (1968b). Oil Pollution off the South African Coast—the *World Glory* Disaster, *S. Afr. Fishing News & Fish. Ind. Rev.*, Nov. 1968, **23**(11), 47–51.

Stander, G. H. (1971). Stilibaii Oil Spill Fumble, February, 1971, *Mar. Poll. Bull.*, **2**(6), 83–4.

Stander, G. H. and Venter, J. A. V. (1968). Oil Pollution in South Africa, *Proc. Int. Conf. Oil Pollut. Sea, Rome*, 251–9.

Straughan, D. (ed.). Biological and Oceanographical Survey of the Santa Barbara Channel Oil Spill, 2 Vols, Allan Hancock Foundation, University of S. California.

Straughan, D. (1971b). What Has Been the Effect of the Spill on the Ecology in the Santa Barbara Channel? Biological and Oceanographical Survey of the Santa Barbara Channel Oil Spill (ed. Straughan), Allan Hancock Foundation, Univ. of S. California, Ch. 18, pp. 401–19.

Stuttard, P. (1969). Oil pollution on the coast of Kent, August 1969, *Mar. Poll. Bull.* (15), 26.

Swennen, D. C. and Spaans, A. L. (1970). De sterfte van zeevogels door Olie in Februari 1969 in het Waddengebied, *Het Vogeljaar*, **18**(2), 233–45.

Tanis, J. J. C. and Mörzer Bruijns, M. F. (1968). The Impact of Oil Pollution on Sea Birds in Europe, *Proc. Int. Conf. Oil Pollut. Sea, Rome*, 67–74.

Technical Advisory Committee for Oil Pollution in the Tay (1968). Oil Pollution in the Tay Estuary Following the *Tank Duchess* Incident, Corporation of the City of Dundee publ. dept., Dundee, Scotland, 30 pp.

Tittley, I. (1970). Kent Coast in 1969, *Mar. Poll. Bull.*, **1**(6), 83.

Troake, R. P. (1969). Oil Pollution Incident at Maidencombe, May–June 1969, *Mar. Poll. Bull.* (13), 16.

Vass, K. F. (1971). Oil Ravages the Biesbos, December 1970, *Mar. Poll. Bull.*, **2**(4), 51–52.

Vermeer, R. and Vermeer, K. (1974). Oil pollution of birds: An abstracted bibliography. Canadian Wildlife Service, Pesticide Section Manuscript Reports No. 29, 68 pp.

Westphal, A. and Rowan, M. K. (1970). Some observations on the effect of oil pollution on the Jackass Penguin, *Ostrich* (suppl.) **8**, 521–6.

Discussion

Offshore Biological Monitoring

Dr A. Myers (University College, Cork) said that since bacterial degradation and food chain transfers are slow processes, the effects of any oil spill could be expected to be very widespread. Were any data available to suggest that there could be very localised effects on the benthos around offshore oil installations? **Dr Dicks** replied that owing to dispersion of pollutants in the offshore situation, effects may be very widespread if concentrations are high. However, the widespread dispersion is likely to dilute pollutants to the point where effects are difficult to detect. In this sense the controls in the Ekofisk survey will be exposed to much smaller amounts of pollutant than those sites close to source and will act as controls of the closer sites. He took the point about dispersion and the possibility of widespread effect and in this sense all sample sites, including controls, are monitoring the overall condition of the sample area. It was important to point out, however, that the Ekofisk survey was not just concerned with detecting effects of offshore spillages but also with a low content ballast water output. The sample sites were arranged in radiating transect lines from the oil field centre in order to detect any gradient of effect likely to occur from the pollutant source. There were no data available at present to show that localised detrimental effects can occur around offshore installations. Information is available on the establishment of installations as offshore reef systems with the consequent attraction of many marine species.

Dr A. G. Bourne (Hunting Technical Services Ltd) asked Dr Dicks if he had carried out any physical oceanographic measurements, such as vertical and horizontal movements in the water column and movements of the uncompacted sediments. The sampling period was exceptionally calm, for the North Sea. Dr Bourne suggested that the thermocline would be destroyed during most of the year owing to mixing by wave action. **Dr Dicks** replied that physical oceanographic measurements were not carried out other than water temperature profiles and he had no data on sediment movements. However, echo soundings of the sea bed in Ekofisk do not show any evidence of large 'sand waves' and the uniformity and diversity of the community from site to site over the whole oil field suggests that sediment movement, if any, is not extensive. The water thermocline is very likely to break down under wave action normally encountered in this area.

Dr R. G. J. Shelton (Department of Agriculture and Fisheries for Scotland) thought that perhaps the most important ecological effect of offshore structures in the northern North Sea was to provide fixed substrates for the attachment of epifauna. This may well be a clearer ecological effect of offshore structures than any changes in the benthos of soft sediments. The likelihood that toxic effects on the planktonic larvae of benthic animals will affect subsequent recruitment to the adult population depends on the shape of the stock/recruitment curves for the various species. His own opinion was that offshore oil pollution is unlikely to affect benthic recruitment in any measurable way.

Mr A. D. McIntyre (Department of Agriculture and Fisheries for Scotland) said that there had been extensive studies of benthic communities in the North Sea over the past 60 years or so, and on the basis of these he felt any major changes would be detectable. In view of this, and because offshore benthic surveys are expensive and labour-intensive, he had some reservations on the relevance of monitoring new operations of this kind. These reservations applied in particular to the immediate area around rigs where one might be willing to accept that some damage to benthic communities must occur as part of the price to be paid for the oil. **Dr Dicks** agreed that any major population changes in the North Sea will probably be detected by the past and present extensive studies of benthic communities. However, as it was not known at present what ecological effects, if any, occur around North Sea oil installations, or their extent, it would seem important to quantify these changes—to find out the biological price being paid for the oil. Occasional intensive surveys such as the Ekofisk survey also provided information valuable in the research sense on the benthos of a previously unsampled area.

Mr A. J. O'Sullivan (Atkins Research and Development) commented that following the emplacement of a large structure such as the Ekofisk tank on a soft substrate, disturbance effects may be observed similar to those which have occurred during and after the emplacement of long sea outfalls. Did Dr Dicks see problems in distinguishing these changes from those that may be caused by the discharge of oily ballast water or other wastes from the oil production activities? Mr O'Sullivan added that he was glad that the benthic programme was being extended to the Celtic Sea—perhaps one of the most important areas around the coasts of north-west Europe from the biogeographical point of view, yet very little known. **Dr Dicks** replied that it was important to bear in mind any factors which may cause variation in benthic communities, particularly when trying to attribute any changes to one specific factor. In this case he thought disturbance caused by tank installation had not seriously affected any of the 24 sample sites, the sampling being carried out after installation but before operation started. The 24 sites were very uniform in composition with little evidence that sites close to the tank were different from sites further away. Physical disturbance of the benthos may well be important nearer to the tank than the nearest sample which was 200 m away. There may well be problems of distinguishing effects of pollution from physical presence of installations. It was necessary to wait and see what future surveys and correlated hydrocarbon measurements produced.

Mr P. H. Monaghan (Exxon Production Research Company) commented briefly on a two-year study of the offshore producing area in the Gulf of Mexico that had just been completed by a group of people at the Gulf University Research Consortium. The plan of attack was similar to the Ekofisk approach of taking transects away from producing platforms, and there were two or three platforms chosen. In addition, there were areas about 20 or 30 miles from these sites which could be used as controls. Sampling was carried out six or eight times over the two-year study and there were great variations in both the numbers and kinds of organisms that were found at each site, but the changes correlated with the natural variations in the area, e.g. the seasonal flood of the Mississippi river. The conclusion so far was that there are no observable effects from the producing activities there. It was not possible to say that there were no effects, rather that any effects are non-observable because of the big natural variations that take place.

Mr D. W. Mackay (Clyde River Purification Board) stated that for surveys of the type described by Dr Dicks, aimed at detecting the effects of pollutants on the marine environment, there was little point in measuring in great detail variations in the structure of the benthic community if insufficient data were gathered on the nature and strengths of the pollutants suspected of producing the changes. The biological investigation should be paralleled by physical and chemical measurements designed to identify the pollutants and quantify the concentrations achieved adjacent to the fauna.

An Oil Spill in the Straits of Magellan

M. C. R. Gatellier (Institut Français du Pétrole) asked about the natural fate of the oil, bearing in mind that no dispersant had been used to clean the shore. Would some analytical data be available, especially about the amount of hydrocarbons remaining in sand or mud? **Dr Baker** replied that it seemed likely that buried mousse deposits would remain for a long time with unknown, probably slow, degradation rates. Some analytical data should be available from America, but she did not know what plans existed for continuing this work in the future.

Dr W. R. P. Bourne (Aberdeen University) largely agreed with the comments about the impact of *Metula* oil on birds frequenting the Magellan Strait. These are largely local populations of numerous and widespread species which were lucky to get off lightly, although it should be pointed out that birds killed at the time of the disaster may also have included migrants from the south, so that damage would not necessarily be evident among the local breeding populations. The main matter for concern appeared to be the amount of oil which may have gone out to sea in an area frequented by many pelagic birds some of which, such as the great albatrosses, have comparatively small world populations. Owing to uncertainties about the amount of water in the mousse it seemed difficult to make out how much oil remained unaccounted for. When the *Torrey Canyon* oil finally went out to sea off Cornwall people forgot about it too, until when it arrived in Brittany it proved to be the larger part of the total. When it reached the main French seabird colony on the Sept Îles it wiped out some five-sixths of the entire French Puffin *Fratercula arctica* population, which has not recovered (*Ibis*, **112**, 120–5), and much greater havoc could have been caused if the *Metula* oil had reached the Falkland Islands, which hold millions of birds. As it was, it seemed likely that any birds killed at sea would have drifted on into the open Southern Ocean and would have escaped observation entirely. It might be questioned whether this may also have occurred in another area where the winds and currents tend to set offshore and where there was surprisingly little observed bird mortality, namely the Santa Barbara incident in California. Dr Bourne had also noticed a marked difference in the impact of oil pollution on birds at different temperatures. In colder areas such as those of the Magellan Strait the oil remains liquid or forms a mousse, and the impact on birds is severe. In warmer seas it is soon reduced to solid residues which form tar-balls, which apparently have less effect on birds (*Marine Pollution Bulletin*, **6**, 77–80).

Dr D. Straughan (University of Southern California) described her intertidal work conducted in the Straits of Magellan during January 1975, to determine effects of the *Metula* oil spill. Using Dr Baker's preliminary report, she had selected specific areas for study chosen for graded damage of oil. The area is one in which it is impossible to match control and experimental sites because areas outside those contaminated by the oil have, in general, a different physical structure than those in the area of the oil spill. Hence the sampling programme in the intertidal zone was designed to measure all possible natural physical parameters that influence the distribution and abundance of species, e.g. intertidal height and sediment size. As far as visibly possible, control and experimental quadrats were matched. Quadrats were selected at each oiled site to account for natural physical variables in the area as well as visible variations in oil exposure:

1. Two sites (H, I) still heavily oiled near the ferry at Punta Espora (including part of the marsh area).

2. One site (G) at Punta Espora more exposed to wind and waves than H or I, that had been heavily oiled and now only had a small amount of oil remaining.
3. One site (E) between Punta Baja and Punta Remo with heavy oil contamination in the upper 50 ft of the intertidal zone only.
4. One site (C) with evidence of only light oil contamination in the upper intertidal zone only (including part of the marsh area).
5. One site (A) at Porvenir with no visible signs of oil contamination in quadrats.
6. A site (X) on the northern shore which was not visibly oil-contaminated.

A total of 30 quadrats were examined. Samples were collected for analysis for oil contamination of sediments. In addition, samples were collected from mussel communities for tissue analysis. Samples were being analysed for oil content and species were being identified.

World Oil Spillages

Captain G. Dudley (Milford Haven Conservancy Board) asked if the figures for human error as a cause of spills were for all spills however small. Were figures available for human error as a cause of spills over say 50 tons? His impression was that spills of over 1000–5000 tons said to be the result of human error are better classed as errors of navigation. **Mr Wardley Smith** replied that all spills reported, regardless of size, are included in the tables, but it would be possible to segregate them for any particular size of spill. He agreed that spills over 1000 tons are almost always the result of collision or grounding of the tanker and so could be classed as errors of navigation.

Dr M. Spooner (Marine Biological Association of the UK) reported a spill of *c*. 4000 tons of a marine diesel oil of about 40% aromatic content, from a tank farm in Hong Kong in November 1973. This spill had a disastrous effect on floating fish farms. So far as she was aware, this caused the highest compensation ever paid out for damage to a fishery resource; there was no damage to the main fisheries. The toxicity of the oil and the hydrography of the area were clearly very important in the effects seen in this spill. During the follow-up research, studies were made on depuration of fish. Despite the presence of a chronic source of somewhat weathered oil in an adjacent beach, fish farming was progressing satisfactorily about six months after the spill, to her considerable surprise and relief, because it had been impracticable to clean this beach at all thoroughly. An account was to be published.

Dr W. R. P. Bourne pointed out that west Iberian beaches were in a bad state before the *Jakob Maersk* incident. He visited those in south-west Portugal early in 1973 and they all had varying amounts of tar-balls on them. The worst-oiled beach he had ever seen was at Canet, north of Cape St Vincent, which was so covered with tar-balls up to 50 cm in diameter that it was impossible to set foot on it without becoming oiled. The larger balls were spherical and appeared to be composed of a series of fragments of different consistency, as if they had picked up smaller fragments while rolling about in the surf. He had some specimens tested and they were composed of waxy tanker residues of the sort which have also been causing trouble in such places as Bermuda. **Mr Wardley Smith** said the comments on tar-balls were of great interest and it would appear that any windward ocean coast has to suffer this chronic pollution.

Closing Summary

M. W. HOLDGATE

(Institute of Terrestrial Ecology, Natural Environment Research Council)

We are concerned about pollutants because of their effects. The extent to which oil pollution has a deleterious effect on the ecology of the sea has been argued in some circles with more emotion than science. In this Conference we have recognised from the outset that unless we can quantify the problem and so define the scale of the effects, the cost of the damage, the scale and cost of the preventive or remedial measures needed, and the amount of monitoring we require to tell us whether we are succeeding or failing in our goals, we are liable to waste resources we can ill afford.

This was the keynote of both Mr Dewhurst's and Mr Jagger's introductory papers. Both pointed out that there need be no conflict between industry, concerned to win and market oil, and the custodians of marine life, concerned to preserve the functions of the marine ecosystem, its yields of fish and shell-fish, and the amenities of the shores on which so many humans bask. Indeed, within many oilmen there is a conservationist trying to get out, or a sunbather trying to get brown. The industry has done a great deal in planning and in developing new technology to prevent pollution or to clean up any that occurs. The issue for us has been to define with greater precision just how much oil pollution is likely in certain circumstances, and how much it matters, and to define the needs for such things as environmental impact assessment before development, a certain level of standard and control during development and operation, and monitoring of those operations afterwards. And we were reminded of two goals: the controls must be adequate to prevent unacceptable damage and they must be cost-effective.

Perhaps it would have been helpful to have reminded ourselves in the beginning of where oil comes from and how it threatens the sea. Mr Wardley Smith has dealt only with sources of oil at the end of the meeting. His figures differ somewhat from others I have seen quoted, but emphasise two things: that shipping and oil industry operations only contribute a minor part of the oil at sea, and that human error still accounts for a disturbing amount of this pollution. As for the effects of this oil, the main general points have emerged during the meeting (Fig. 1). While all crude oils contain substances toxic to marine life, there is no evidence that pollution of the oceans, at today's level, poses any threat to the stability of the oceanic ecosystems. Only some of these toxins are water-soluble. Concentrations in the sea are low. Allegations of the accumulations of carcinogens by organisms that could pass them along food

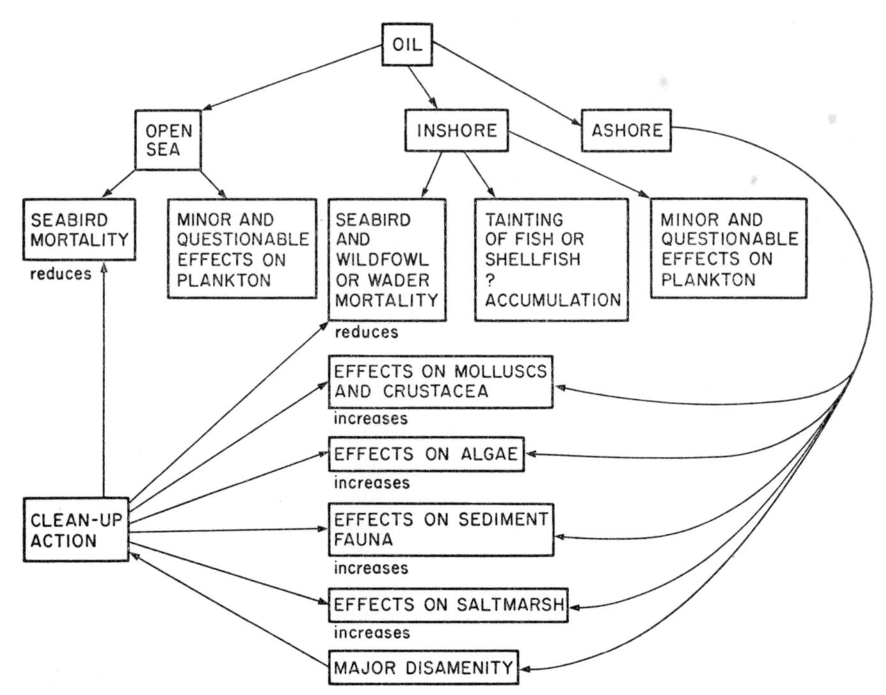

FIG. 1. Schematic diagram of oil pollution and its effects.

chains to man are unsubstantiated. Tainting of food and shellfish is a local problem in inshore waters. At sea the main proven hazard is to seabirds and is a physical effect. Oiling has depleted some bird populations, but only locally to danger point. This is socially repugnant rather than ecologically hazardous. We cannot expect to prevent such oiling altogether (and it remains the most likely catastrophic biological consequence of an accident); we can aim to keep deaths from this cause below the recruitment potential of the populations so that total numbers are sustained. For this, we shall need constant effort to prevent carelessness. On shore, where effects are both chemical and physical, the proven effects of oil, accentuated by clean-up practice using dispersants, are to cause short-term and local ecological changes. In our meeting we have heard many details especially of the effects of chronic pollution by repeated small spills and by effluents. Unless there is repeated contamination at the same spot, there is clearly unlikely to be a major permanent or long-term ecological change. As at sea the most significant impact may well be on birds feeding on mud flats and in sheltered inlets. It can be argued that, ecologically speaking, oil pollution today is a non-problem, except locally—and will remain so, so long as we sustain our vigilance and preventive efforts.

Charles Sinker reminded us that in setting our goals we could not evade judgement about the quality of the environment we wanted and, equally, must avoid distorting our science to support the preferred hypothesis of a sectional interest. We have to live with some pollution: the difficult question is 'how much?' It is to this question that we must address ourselves as scientists, by examining the scale and significance of the problem. This brings me to my last general point from the first session. Pollution control is an applied science. We need to ensure that the data we gather are relevant to policy—that they can be used to guide decisions about the nature and adequacy of controls and standards. We have to ensure that the right research is done, that the information is evaluated and related to the vast amount of existing knowledge, and that it is communicated to policy-makers and industrial operators so that they can make sensible judgements in that dialogue which, as Henry Jagger reminded us, is vital if we are to get the right solutions. And we operate (as Fig. 2 may remind us) within a system of public debate and concern, not all of it well informed. We must ensure that our results and evaluations help shape public thinking and goal setting.

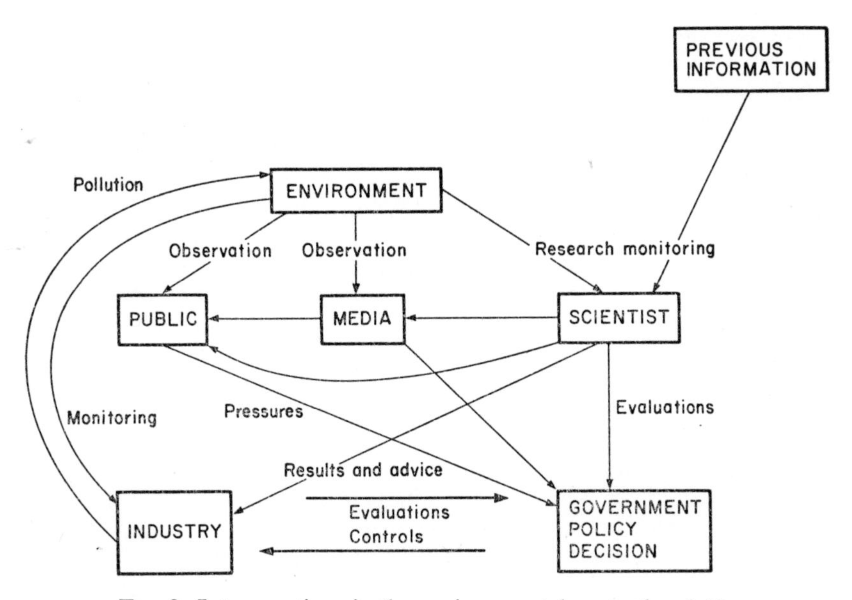

FIG. 2. Inter-reactions in the environmental protection field.

From this background we turned to Milford Haven as one of the best available examples of oil pollution and its control in a potentially vulnerable situation—a major terminal, land-locked, with a prevailing wind blowing into the port and so tending to hold oil there. Captain Dudley made three points that particularly struck me. First, it was possible to operate such a port so that pollution is kept to an acceptable level—but the Harbour Authority has to be professionally knowledgeable and tough (and as was said in discussion, recent events in Bantry Bay suggest that you cannot leave it to industry to police itself). Second, even with good management, some pollution incidents will occur. Third, damaged vessels must be permitted to enter those ports especially equipped to handle their cargoes. The pollution is likely to be less if this is done than if they are left to ooze outside. Perhaps we should learn from this that if, as we may expect, the Law of the Sea Conference (incidentally, I was amazed that this was not mentioned once in our meeting) gives coastal states pollution control rights in a 200 mile zone offshore, we must ensure there are enough facilities around Britain to handle damaged ships—and enough controls to prevent 'cowboy' tankers or other vessels with inadequate standards of design and seamanship barging unchecked through our congested shipping lanes. The price of cleanliness is clearly unceasing vigilance.

Breaking the sequence in which the papers were delivered, Jenifer Baker's account of changes in the Haven since it became an oil port seems to me to provide the other side of Captain Dudley's coin, in documenting the kind of thing that will happen in a well-run port. Clearly there have been changes. Green algae (*Enteromorpha*) have increased, and this increase is at least in part unrelated to changes in limpet populations. Barnacle and top shell populations have also altered. Her paper highlighted the basic point that it is much easier to detect change than to interpret it: we need more work if we are to elucidate cause and so move to prediction. Her saltmarsh observations underlined the same point. However, the *Dona Marika* incident did give a clearer pattern of successive changes in molluscs and algae, and yet may provide a model we can use predictively—though not yet in quantitative terms.

On our second day, after a beautifully illustrated general introduction to the benthic invertebrates, John Addy came to grips (or to grabs) with the sublittoral fauna of Milford. His paper stressed the daunting challenge such a fauna presents in sampling: the number of stations clearly needs to be enough to demonstrate the range of variation and point to some environmental correlations; enough grab samples are needed to catch at least the majority of the species present; the grab and sieve dimensions can affect the apparent balance of the composition of the fauna; and finally, three weeks' sampling can lead to several years of sorting and identifying and counting material. We need such work, but it is not to be undertaken lightly. Brian Dicks' account of his North Sea benthic survey equally demonstrated the mass of data such work can generate.

The next major section of the meeting, and one of its most valuable parts, was concerned with the effects of oil pollution. We seek to control pollution so as to prevent harmful or unacceptable consequences. To do this, we need to be able to quantify the relationship between effect and exposure. And we are increasingly aware that the first effects, most useful as indicators and warnings, are subtle, sublethal, often behavioural and undetected by standard LD_{50} acute toxicity tests.

Mr Levell showed that both Kuwait crude and BP 1100X reduce lugworm numbers on sand flats, especially when there is repeated application. It seems that recruitment of new young worms is the key to the re-establishment of a population. The use of dispersants in an attempt to clean oil from sandy shores—except perhaps on a rising tide with a choppy sea—can make effects worse by driving oil into the sediments. I was struck by the incidental point that sediment stability may also be affected if *Arenicola* numbers are altered. One of my own colleagues (Dr S. M. Coles), working in the Wash, has demonstrated that living diatoms play an important role in stabilising certain sediments. Massive and repeated oil pollution and clean-up attempts might have a dramatic effect on erosion and accretion in sandy or muddy estuaries.

We should investigate this: indeed we need to know a good deal more about how the benthic faunas of sediments between and below tide marks respond to pollution.

Mr Roberts reviewed for us another aspect of oil pollution: refinery effluents, their sources and the costs of attaining various standards. Although he himself said it was a crude over-simplification, many of us will have noted the relative figures implicit in the stated £300 000 for an API separator to bring an effluent from 220 ppm to a normal 10–15, with 30 ppm as a high oil content, and the further £1–1·5 million needed to get that level down to 1 ppm. These are capital costs: we were reminded in discussion that the operating costs of the more sophisticated equipment are not negligible. His paper and the discussion further emphasised the need to have an agreed standard method for sampling and analysing effluent for oil—recognising, incidentally, that what we call 'oil' is a complex mixture of substances of varying toxicity. We also recognised the need to plan a works from the outset not to exceed the capacity of receiving waters, and to set emission standards that work, via the complex pathways of environmental dispersion (summarised in Fig. 3), to protect the targets we want to safeguard, adequately but not wastefully. As Dr Spooner said, we need to know the toxicity of particular oils and the hydrography of the receiving waters.

Refinery effluents, like oiling incidents, clearly do have local effects on living things. Jenifer Baker reviewed how they could alter the density and population structure of limpets and other molluscs near points of release, especially in confined waters. As when grazing limpets are reduced by oil, the herbivore/plant balance can be dramatically disturbed. Her overall summary was most valuable:

1. Offshore release 2. Shore release with good dispersion	Small effects only
3. Shore release with intermediate dispersion	Reduced species and populations in the immediate vicinity
4. Shore with poor dispersion	Damage to flora and impoverishment of fauna

Several points emerged and recurred later. We learned that reedswamp of *Phragmites* could tolerate effluents with low oil levels, act as a polishing agent and maybe make a final treatment lagoon an amenity or even a useful harvest field for thatchers! We were reminded how hard it is to be sure what component in an effluent actually causes observed change. And the paper led directly to Dr Dicks' valuable detailed chronicle of the destruction of saltmarsh at Fawley by chronic pollution from a very large effluent volume, and the recent recovery process. It was interesting that this latter seemed to be leading to a different type of marsh. One does not necessarily find ecological processes readily reversible.

Mrs van Gelder-Ottway's absence may have caused us to do less than justice to her papers. I was interested to note her documentation of another type of effect in the reduction of photosynthesis, oxygen levels and animal

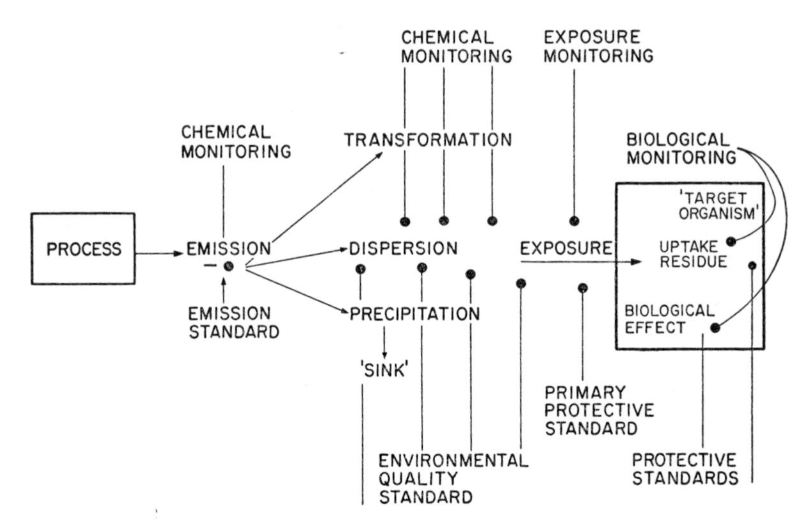

FIG. 3. The pathways of a pollutant from source to target, and the points at which standards may be imposed and monitoring occur.

activity in pools covered by an oil film, especially where they are high up a beach and cut off for some hours.

These studies are all very well. It is valuable to observe and quantify effects of this kind, but they do not submit readily to explanation unless there is also experiment, both in the field (as Mr Levell described) and in the laboratory. Dr Dicks' paper on behavioural patterns in toxicity testing and prediction was valuable in this connection. He demonstrated how subtle sublethal effects can be—and how complex the balance of toxicity due to pollution and simpler influences like that of salinity. Barnacle settlement is clearly deterred near refinery outfalls—but the low salinity of the effluent is clearly the major factor swamping the toxicity component. The responses of limpets and winkles in effluent are equally clearly multivariate in cause, and a fruitful field for study. Experiments of this kind are clearly needed to explain and quantify gross correlations between pollution and abundance and performance observed in the field, and also to develop methods of using the responses of organisms we have calibrated in this way as a monitoring tool. We are a long way from the conclusion of this research: indeed so far we do not know just what makes effluent toxic.

Many points arose in the discussion of these papers on effects. We noted the need, in setting standards for discharge, for models of the dispersion pattern so that we could predict and regulate effects. Bioassays and field surveys clearly complement one another: we can use organisms as an aid to chemical monitoring because they accumulate substances, as indicators because their responses integrate the influence of many contaminants, known and unknown, and as measuring devices, where we have been able to calibrate a specific response to a specific situation. We noted that in this second area synergism could be important, and we talked of possible effects of detergents, herbicides and traces of persistent organochlorine pesticides in mixed effluent. There was debate over the value of LD_{50} tests—especially the 96 h test—

and a general conclusion that each test has its own value, the important thing being to know what each was really revealing and hence when to use each. And to strike two cautionary notes, we were told that the properties of refinery effluent changed with time so that storage of samples for toxicity testing was a problem, while we did not touch at all on the possibility that the toxicity of a refinery effluent or ballast water would vary widely according to the nature of the crude oil from which toxic components came.

Inevitably we were much concerned with Milford Haven this week. Brian Dicks took us away from this port, asking the wider question 'Are all oil ports capable of being equally pollution-free?' The answer, rather evidently, was 'Of course not'. His paper emphasised the dangers of uncritical extrapolation from one port to another. It also emphasised, to me anyway, the need for a model (mathematical rather than physical) of such ports, to explore such things as tidal exchange, surface drift, probable patterns of floating and dissolved pollutant dispersal, and desirable sampling density for monitoring. Such models must go outside the actual port; on a point of detail, for example, I query anyone considering Sullom Voe without also looking at the fierce tide races of Yell Sound to which the Voe opens. And if we are concerned with birds, you have to look at the whole area where oil could spread. Murphy's two laws remain valid in this whole connection. (The first states 'If it is possible for something to go wrong, sooner or later it will', and the second, a less statistical variant, 'It is impossible to make any system foolproof because fools are so ingenious.') Jenifer Baker showed that they certainly applied in the Straits of Magellan—though as one who has sailed that beautiful waterway I would comment that the industry (and the environment) was lucky. The *Metula* stranded in the accessible, fair-weather end of the Strait.

Finally, monitoring. This was a recurrent theme. We became tangled in semantics: what is monitoring? We were at variance over what it is for. We noted alternative methods without always examining their applicability, accuracy and inter-comparability. I should like to end on this theme and maybe throw out some points to structure later discussion.

First, going back to a point raised in Dr Baker's first paper, 'What do we mean by monitoring?' May I suggest:

> *Monitoring* is the repeated measurement of a variable so that a trend is defined and can be related to a goal or standard, whether or not it is mandatory.
>
> *Surveillance* is the general process of repeated measurement of variables to define trends, often in order to establish correlations that can be explored further by research.

In this meeting we have tended to use 'monitoring' in a way that did not discriminate, and lost precision in consequence.

The first 'monitoring' paper usefully defined chemical and biological monitoring and certain categories of indicator organism. It was also clear that any monitoring or surveillance must be calibrated—that is, the accuracy of the method must be determined, and inter-comparison with other methods facilitated. It was also evident (and this came out in discussion) that there were many methods and choice must be related to the variables being examined, the environmental circumstances, and the precision of discrimination required in space and time. Brian Dicks put this admirably on the final morning, when he stressed that the accuracy of monitoring and surveillance was related to the adequacy of quantification, that crude methods were adequate only if we were content to detect gross change, and that if we need to detect fine-order change we need more sophisticated methods. Hence the need for selection of sampling density and frequency, both being also related to the variability of the system under study. Two other points emerge from Brian Dicks' final paper. First, that before choosing what to monitor we need initial surveys, and an initial analysis of the ecosystem, so that we select the most sensitive component, be it plankton, benthos or fish, and those components whose responses we can interpret. Dr Shelton reminded us in this context that year-class variations and variations in recruitment could obscure changes due to pollution. Second, that these initial surveys may well need to be more detailed than the later monitoring and surveillance: they define the field of variation, and later sampling can then be related to that variation and predictions of wider effects made and tested.

We talked a good deal about standardisation in monitoring and surveillance. We were reminded that while there are dangers in over-standardisation, all serious monitoring programmes must be designed to give a meaningful sample, quantified accuracy, and inter-comparability. We were also reminded that biological and physico-chemical monitoring must proceed together and that there is a need for feedback from monitoring and surveillance to research to seek to establish the causes of observed correlations and hence improve the selection of variables to monitor. Here the monitoring papers and those on effects came together.

I suggest that in our discussion today we come back to these points, sticking in this meeting to oil and related oil industry pollutants, and to marine ecosystems, and recalling, as Figs. 3 and 4 indicate, that monitoring

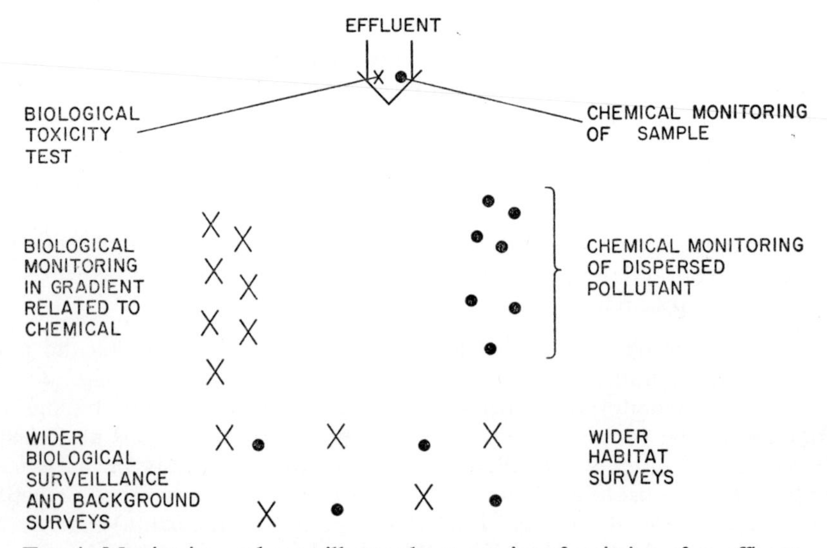

FIG. 4. Monitoring and surveillance about a point of emission of an effluent.

tends to be done at points along complex pathways from sources to targets. Can we proceed by:

1. Agreeing on the definitions, but avoiding semantic debate?
2. Asking ourselves more precisely what we are monitoring for, e.g.
   (a) to record the chemical composition of an effluent and compare with a standard;
   (b) to record the biological changes in indicator organisms or communities associated with the discharge of that effluent with levels of the contamination it produces in the environment;
   (c) using 'sentinel' organisms with calibrated response, exposed to that effluent, to give greater precision or relate to a biological standard;
   (d) in a wider area about a source of pollution, to survey the changes in flora and fauna in case there are general effects needing study because they may result from the pollution or associated disturbance;
   (e) to record specifically the condition of sites of especial interest (such as Nature Reserves);
   (f) to record the scale of effect of accidental pollution and recovery after treatment;
   (g) to record the efficacy of policies designed to improve the environment, e.g. by land-use planning?
3. Considering the resolution needed in each case, methods available and the quantitative accuracy needed and obtained for each? Are there difficulties in chemical monitoring? What are the limitations to the OPRU quick transect method on rocky shores? How reliable is it? How reliable are simple behaviour effects like limpet drop-off, barnacle settlement or seaweed growth?

How should we design wider surveillance schemes? How far do we need to monitor accident and recovery?

In each case, how do we decide on the frequency of sampling in space and time—and avoid vast taxonomic complications?

Finally, let us remember the need to be cost-effective. Just as there is a danger in environmental managers pursuing excessively costly purity, scientists tend to pursue excessive perfectionism in information gathering, which can lead to inaction pending the evaluation of excessive data, or the ignoring of simple warning signals. Eric Cowell reminded me this morning of A. A. Milne's snail, James, who 'gave the cry of a snail in danger. And nobody heard him at all.' Simple observation of endangered snails—or limpets—does not need electronic instrumentation and may give us quick, cheap warning systems with adequate sensitivity.

Last of all, I have two things to say, one practical and one philosophical. This meeting has been built around the work of the Orielton Unit. I am sure you would wish me to congratulate its members on the quality of their work and the lucidity of their presentations. Please do not think me unappreciative if I suggest to the organisers, however, that if the Institute of Petroleum holds another conference on this subject, it might be better to take a narrower theme and pursue it more widely, drawing speakers from the many other groups working on marine pollution by oil and the petroleum industry. OPRU has grown up. Its work can stand such wider comparison.

Finally, my philosophical point. Charles Sinker reminded us that what we are really concerned with is man's relationship with the natural world. Only in recent centuries—or even decades—has the admiration of natural beauty, shown by the ancients and well illustrated in the quotations he gave us, been followed by the concept of stewardship. Industry a century ago showed poor stewardship of the environment in which its people lived. Riches were won at a price of squalor and degradation. These three days have shown a new and welcome change. We can run a great industry and also have a clean environment, if we want it enough. But we all make mistakes, and we shall continue to have accidents. We can be thankful that nature is resilient. Charles Sinker quoted Gerard Manley Hopkin's lines to a girl mourning for the falling leaves. But Hopkins also wrote:

> 'Yet, for all this, nature is never spent
> There breathes the dearest freshness deep down things,
> And though the last lights off the black west went,
> Lo! morning at the brown brink eastward springs.'

Let us make sure that morning still lights a world that is fair to look at and good to live on. That is our collective aim, in which all narrow boundaries between industrialist and conservationist, scientist and citizen, become irrelevant.

Final Discussion on Monitoring

Edited by M. W. HOLDGATE

(An abridged record of a discussion on the monitoring of oil pollution and its effects, held on the final afternoon of the meeting under the chairmanship of Dr M. W. Holdgate, Director of the Institute of Terrestrial Ecology.)

INTRODUCTION

In introducing the session, the Chairman proposed that the discussion be organised under the following headings:

1. *The monitoring of emissions:*
 (a) Analysis of effluents. This type of monitoring is undertaken in order to determine whether controls set on discharges to the environment are being met. It was suggested that the discussion should concentrate particularly upon the methods available for the measurement of the quantities of oil present in dilute effluents.
 (b) Bioassay of the effects of samples of effluents on selected organisms in the laboratory. This type of study is undertaken in order to test the toxicity of effluents to species of known sensitivity and provides some basis for predicting the consequences of emitting the effluent to the environment.

2. *Monitoring of the environment around points of emission:*
 (a) Monitoring of effects on 'sentinel' organisms of known sensitivity to components of the emission.
 (b) Observation of changes in communities of plants and animals at various distances from points of emission.

3. *Background surveillance of flora and fauna:*
 This heading covers the general observation of marine life, including plankton, and of seabird populations as general indicators of the ecological well-being of broad areas of ocean and as a means of detecting widespread change.

4. *Methods of intercalibration and data storage and retrieval:*
 It is inevitable that programmes of monitoring will be localised and undertaken by many different individuals and laboratories and strict uniformity of methods may not be desirable, but it is important that there is a machinery for cross-checking on the accuracy of different sets of results and for assembling information for evaluation so as to give a general picture of environmental effects.

Dr Holdgate said that it was important to bear in mind that monitoring was a practical exercise undertaken for a specific purpose and the meeting should be similarly practical in its approach.

MONITORING OF EMISSIONS

Measurement of Oil and Other Components of Effluents

Mr P. J. Osbaldeston (North West Water Authority) in this section of the discussion referred to problems encountered on Merseyside where there were a number of oil refineries, petrochemical works and other industries dependent upon oil and producing a wide range of by-products including bitumen, tetraethyl lead and fuel additives. Samples of effluent commonly contain no visible oil or grease but when subjected to the standard oil analysis test involving extraction with petroleum ether have been recorded as containing 100–200 ppm. This called in question the meaning of various types of test.

Mr T. Rielly (BP Chemicals Ltd, Grangemouth) pointed out that the standard method of extracting an effluent with petroleum ether would only reveal the amount of heavy residual oil. Material with a boiling point much below 100°C would be lost by evaporation. If a complete spectrum of what is present in the sample is required, gas chromatography should be employed. **Mr D. W. Mackay** (Clyde River Purification Board) said that this technique was used by his authority, who could often identify the particular oils involved since they had a 'library' of the oil types likely to be encountered.

Dr R. H. Cook (Environmental Protection Service, Canada) said that with respect to the use of chemical parameters to control effluent, GC, IR and fluorescent spectrophotometry can give very specific answers, but this intensity of analysis is not generally needed for effluent control purposes on a routine basis. His laboratory in Halifax used these three methods, singly or in combination, to identify responsibility for oil spills. Using one method alone did not generally provide sufficient information. For court cases, GC data were supported by infra-red 'fingerprinting' and with fluorescent spectrophotometry. In discussion with the Chairman, Dr Cook agreed that these techniques should be used for an initial detailed analysis of the composition of a particular effluent so as to ascertain whether its effects were likely to be so serious that highly sophisticated monitoring would be needed under operational circumstances.

Mr H. Jagger (Esso Petroleum Co. Ltd) pointed out that there was a considerable discrepancy between the analytical methods used by authorities and individual companies. Figures purporting to show levels of oil in effluent were consequently very hard to interpret. Some were for total oil and others only for persistent heavy oils. Traditional methods only recorded heavy oils, yet the more volatile and more soluble components of an effluent were often the most toxic. The 1973 IMCO Convention required that ships must have equipment to measure oil concentrations, but no completely satisfactory instrument was available commercially. Some, like the Bailey meter, gave a measure for a given oil provided they were correctly calibrated. For low traces of oil, extraction followed by infra-red photometry was used. If ambient oil in the oceans as opposed to levels of oil in an effluent was being looked for, the methods must separate naturally occurring hydrocarbons from those derived from released petroleum products, and this added yet more complications.

Dr C. Johnston (Heriot-Watt University) stressed the dangers in treating 'oil' as if it was a single substance. People talked about the need for standards for the maximum level of oil in discharges from the shore and the need to relate such a standard to the dispersion capacity of the receiving water, but it was important also to look at the precise composition of the effluent. While 25 ppm of oil in effluent might be a safe normal standard, should the oil happen to be nearly all aromatic materials of high toxicity then 10 or 1 ppm could give rise to serious problems. It was therefore necessary to determine the range of hydrocarbons in the particular 'oil', particularly from new sources. Much of the information available so far related to Kuwait or other Middle Eastern oils; the special properties of North Sea oil were not well documented. **Mr P. D. Holmes** (British Petroleum Co. Ltd) agreed that individual hydrocarbons needed to be determined if any valid predictions about the toxicities of an emission were to be made. Undoubtedly, aromatics are the most toxic components of crude oils. Benzene is the most toxic of these, but also the lightest, boiling at 80°C and becoming rapidly volatilised on exposure to wind and waves; it is also soluble and disperses into the water. Toluene and the xylenes and the subsequent more complex members of the aromatic series become decreasingly soluble in water and also less toxic. Gas chromatography should be used to determine the frequency of these substances in an effluent, and allowance must be made for variation in the composition of crude oil even between individual wells within a small field.

Deterioration of Samples in Handling

Dr Holdgate referred to a comment made earlier in the meeting about changes in the composition of samples of effluent during storage and suggested that this would pose yet another series of complicating variables. **Mr E. B. Cowell** (British Petroleum Co. Ltd) said that another factor commonly ignored by people working with dilute samples of oily effluent was that the type of container could affect the results: for example, hydrocarbons could be lost through polythene or polypropylene. There were also problems of biodegradation. Inhibitors could be added to prevent this, but these could further complicate gas–liquid chromatography. The real need was to conduct the analysis very soon after the samples had been obtained or at least to extract the sample ready for analysis. **Dr W. R. P. Bourne** (Aberdeen University) commented that oil hydrocarbons were far from unique in these problems and that many biological samples were more difficult to handle: it was a matter of being aware of the problem and choosing proper methods. **Mr A. J. O'Sullivan** (Atkins Research and Development) said that sodium azide used as an inhibitor did not interfere with the GLC process.

Assay of Toxicity

Dr Cook commented that a degree of standardisation was needed in bioassay just as in chemical analysis. If samples were tested by one laboratory against molluscs and others used limpets, *Daphnia* or rainbow trout it was going to be extremely difficult to make a general statement about relative toxicities. It was clearly appropriate to use a range of organisms but equally important that the same organisms were widely used. The laboratory tests for toxicity, moreover, ought to be capable of extension into field studies, and should therefore include organisms living in the area of environment to which the effluent was released.

Mr. K. Hiscock (Coastal Surveillance Unit, Menai Bridge) made the point that bioassay in the laboratory might be difficult to extend into the field because of fluctuations in the nature of effluent emitted from an installation. His observations on Anglesey had revealed that occasional accidental releases of extremely concentrated effluents for very short periods of time could lead to biological changes in the field, but since these events were abnormal they would not be likely to be predicted from routine toxicity testing on the standard effluent. If there was to be extrapolation from the laboratory to the field it would be necessary to link bioassay to continual monitoring of the effluent being released. **Mr Cowell** commented that while there might be standardised biological techniques for toxicity testing, since it was questionable whether refinery effluents could be sampled in a representative way the interpretation of the bioassay test would inevitably be extremely difficult. **Mr O'Sullivan** commented that electronic or infra-red means of detecting small oil spills or accidental discharges at night would be valuable: without these, releases might be occurring and biological damage caused without anyone being aware that the oil had actually been spilled.

Conclusion

Summing up this section of the discussion, **Dr Holdgate** commented that the problem appeared to be that it was impossible to rely on simple visual inspection to tell us whether there was 'oil' in an effluent. Sophisticated techniques including gas chromatography were needed to determine the spectrum of substances in an effluent, and particular attention needed to be paid to the lightest and most volatile aromatic components. Moreover, continuous monitoring to detect short periods of highly toxic emission was really required if the environmental effects were to be explained and traced to source. On the other hand, in practical terms it would be extremely costly to maintain highly sophisticated continuous analysis indefinitely. Perhaps the best course was to conduct very thorough initial characterisation of an effluent and then to relate the level of continuing monitoring to the likely hazards. The meeting appeared to agree that while laboratory bioassay testing for toxicity of effluents against carefully chosen organisms was valuable, there were also difficulties in extrapolating from such tests into the highly variable field situation. It was extremely important not to obscure the variability and complexity of the situation in the quest for standardisation of controls on effluents, whether on a national, European Community or worldwide basis. It was the responsibility of the scientist to make certain that if standard test methods or standards for effluent composition were laid down, they had scientific validity across the whole spectrum of actual situations encountered.

THE EFFECTS OF OIL DISCHARGES IN THE ENVIRONMENT

Detection of Oil Slicks at Sea

Dr A. G. Bourne (Hunting Technical Services Ltd) said that work had been going on for some time on finding a suitable all-weather method of detecting oil slicks. The method currently being pursued is that of using side-looking radar, and although it is too early to predict the success or otherwise of this method, it might be useful to describe some of its advantages and disadvantages. The advantages are that side-looking radar can detect an oil slick in all weather conditions, its imagery is dependent on its own power source and therefore it can be used night or day and through thick cloud cover or fog. Unfortunately its disadvantages are that it is expensive, requiring surveillance by aircraft, and it cannot detect a slick in dead calm sea conditions. It is the difference between the 'smoothness' of the slick and the broken sea that provides the contrasting imagery. The image is converted into a light image and fixed on a photographic plate. Discussions are in progress that might result in side-looking radar satellite surveillance, which, in the long term, would provide a cheaper and continuous surveillance technique, but so far the problem of sea-state has not been overcome. However, work would continue, and combinations of side-looking radar and new imaging techniques using other parts of the electromagnetic spectrum are being tried. Some success has been achieved in combining side-looking radar and infra-red imaging methods. This with laser measuring techniques may provide a method of detecting and accurately measuring the dimensions of an oil slick and even of providing a means of 'fingerprinting' the discharge.

Mr Cowell commented that remote sensing from satellites offered the first chance of keeping watch over the oil contamination of those two-thirds of the earth's surface that are covered by ocean. Although resolution might not be adequate to detect small oil slicks, the evidence was that these were not too damaging biologically. Aircraft flights with sophisticated equipment policing enormous bodies of water would be prohibitively costly: surely it would be sensible to concentrate on satellite technology to look for the really large areas of pollution. In reply, **Dr Bourne**

said that satellite techniques had already been demonstrated as of potential value for oceanic surveillance, but his group had been concentrating on areas like the North Sea or the Gulf where accidents were particularly likely. The cost of aircraft surveillance in these limited areas was really quite low.

MONITORING AROUND POINTS OF DISCHARGE

Mr N. C. Morgan (Nature Conservancy Council) said that the basic criterion that must be used was that when a change was detected its cause must be capable of identification. At a public inquiry, for example, there was no point in stating that the biology of a shoreline had altered unless this alteration could be clearly related to the particular activity being investigated. One of the most important tasks for research at the present was to discover which organisms should be monitored to meet specific needs.

In discussion the need for integrated biological and chemical monitoring was stressed. Several speakers believed that it would rarely if ever be possible, except in an experimental situation, to prove conclusively that a pollutant was the cause of an observed effect. There were too many environmental variables. A high order of probability could, however, be demonstrated if the scientific programme was well designed. Biological monitoring represented a useful intermediate stage from which chemical investigation and laboratory experiment could then follow to explore further the probability that a pollutant could have caused an observed change. A programme of investigation of the effects of a discharge to the environment needed to begin with both chemical and biological baseline studies. Thereafter, chemical monitoring, recording of changes in the abundance and distribution of organisms and the collection of organisms that accumulated pollutants and provided an aid to assay should all proceed, possibly for many years. The chemical investigation of levels of pollutants in selected organisms could provide a valuable indication of whether the accumulation of contaminants was indeed likely to have been the cause of the observed effects. **Dr D. Scarratt** (Department of the Environment, Canada) stressed particularly the vital importance of the time at which observations were made. Chemical monitoring after an event was no substitute for observations at the actual time. For example, if an oil spillage in Bantry Bay coincided with herring spawning and a year-class failed, it was the recording of the coincidence of the events that was important. No subsequent analysis of a non-existent group of fish for non-existent oil would prove anything! Conversely, surveillance needed to extend sufficiently in time to cover the full period of likely effect: after the *Arrow* accident in Canada a great deal of effort was expended on an immediate study of the consequences, but it was the longer-term investigations that demonstrated that marsh plants which had not seemed significantly affected in the first season showed very significant changes in the second one.

Dr I. C. White (Ministry of Agriculture, Fisheries and Food) commented that while he entirely agreed that biological monitoring should not be divorced from physical or chemical monitoring, the difficulty was that this statement implied a need to measure everything all the time. This was obviously impossible: a critical path analysis was therefore needed to determine what should be included. Dose/response information was obviously vital together with an idea of the acceptable levels of harm to target organisms. Sometimes man would be the target; at other times we would be concerned about effects on an edible species and subsequently on man as a consumer. The model should record the substances going into the sea, their fate and their pathways to sensitive or accumulator species, especially those of concern to man. By discovering and then monitoring a highly sensitive organism and establishing a standard so that it does not undergo unacceptable change, we can keep an adequate watch over and give adequate protection to the entire system with reasonable economy of effort. This view was supported by **Mr A. D. McIntyre** (Department of Agriculture and Fisheries for Scotland). He stressed that any analysis led back to the problem of making value judgements about the amount of change that was acceptable and, once changes occurred, the difficulties of deciding the point at which action had to be taken to arrest further change.

Mr Hiscock stressed the need to record the natural changes occurring in the marine organisms which one was hoping to use to interpret pollutant effects. Many organisms had annual cycles of abundance and also showed variations in abundance from one year to the next. Because of these fluctuations, it would be extremely difficult to extrapolate from a survey of an area of environment carried out in July to evaluate the effects of some catastrophic event like an oil spillage in the winter. After that event the reduction in numbers of some organisms might be interpreted as due to the spillage but might equally be regarded as part of the normal winter decline in numbers. A continual watch on seasonal changes in abundance at least of key organisms was therefore required. His group was examining these changes in all species found on rocky shores, but concentrating especially on widely dispersed key species known to be tolerant or intolerant of pollution. **Dr Holdgate** said that ideally one would like to extend from this type of research into a predictive approach. Were there adequate knowledge of the ecological system it would be possible to state on learning that an oil spill had occurred that certain changes in the abundance of named organisms should occur and the investigation then be set up in order to test this hypothesis. It would be a much more rigorous scientific procedure.

Miss S. Hainsworth (Oil Pollution Research Unit) commented that if dose/response relationships were to be established in the laboratory there was a considerable constraint imposed by the kind of animal that could be used. The animal had to show a specific response, it had to be common and easily collected and it had to survive well under experimental conditions. This narrowed the range to only a few species. **Dr Scarratt** added that in laboratory testing physical as well as chemical responses needed to be borne in mind: the response of organisms was not necessarily to toxic components in the oil but to its physical properties. **Mr Cowell** reminded the meeting that Brian Dicks had shown diurnal fluctuations in sensitivity and Geoffrey Crapp had likewise demonstrated variations in the responses of organisms with the seasons. Laboratory studies needed to look for such changes, or generalisations about the effects of a substance on a species would clearly be invalidated. **Dr Holdgate** commented that it would be strange to find an organism that did not have something both of a circadian and a circannual rhythm and also variations in its response with age and nutritional state, and this stressed the need to use adequate samples adequately spread over time. **Dr J. M. Baker** (Oil Pollution Research Unit) accepted that both seasonal and diurnal rhythms could be important but stressed that it was necessary to draw the line at some point and commence tentative predictions. Diurnal rhythms could readily be modified. Limpets in Milford Haven are most active between midnight and 3 a.m. and therefore her Unit does drop-off tests at this time. The tests are done on limpets carried back to the laboratory attached to stones. Control studies have been done on the shore at night and these demonstrated the very considerable influence of weather under field conditions. Although the limpets' rhythm makes them normally much more active at the period of testing, in a hailstorm on the beach they clamped down very tightly and the drop-off rate went back to nearly zero. These kinds of additional variables make judgement very difficult: factors of this kind must be taken into account but a complete analysis is surely extremely difficult. **Miss Hainsworth** commented that this was the whole basis for control experiments: near-simultaneous studies were needed both between field and laboratory and between areas of environment affected by an effluent discharge or accident and matched areas not so affected. **Dr W. R. P. Bourne** re-emphasised the importance of physical properties. Temperature has an enormous effect on solubilities, volatilities and biological reactions and hence the persistence of oil and other variables. The physical state of the oil has a big influence on the character of the slicks. Slicks vary enormously in thickness and even very thin layers can be extremely conspicuous from aircraft, even if they do not have a very large effect on the roughness of the sea.

This increasing stress on the complexities of the situation made it difficult, it was suggested, for those concerned with the practical management of estuaries to predict the success or failure of expensive programmes of pollution control which might involve the spending of £100–200 million of public money. Yet it was a fact of life that substantial improvements had been made to grossly polluted estuaries (such as the Thames), even though many potentially toxic industrial discharges continued to be made to them. There was a danger that the increasingly detailed level of scientific investigation might provide diminishing practical returns.

Conclusions

Summing up this section of the discussion, **Dr Holdgate** said that where an effluent was being discharged into the environment it was natural that the fate and the effects of the substances discharged should be subject to some programme of monitoring. The discussion had brought out the highly multivariate nature of ecological systems. It would not be possible to monitor every potential target. In selecting variables to monitor, some kind of model was required, and to construct this, baseline survey and background research were needed in order to give understanding about the composition and functioning of the ecosystem. Dose/response relationships were needed in order to interpret the likely effects of particular levels of emission and to select species which were of especial sensitivity and therefore valuable as 'sentinels'. Monitoring would naturally concentrate on targets of known sensitivity whose performance has been calibrated or on targets important to man. The selection of variables to monitor would also be affected by the uses to be made of the particular area of the environment. Despite the difficulties in analysing complex systems and predicting their responses to pollution, it was a fact that useful chemical, physical and biological monitoring was being done around points of effluent discharge and these programmes would undoubtedly continue and be improved progressively. In biology as in many other sciences one moved forward by progressive approximation. If a model could be constructed that broadly described the system, the wisest course was then to formulate some predictions and test them: inevitably imperfections in the model would be revealed and this would lead to its improvement. It would be unwise to seek a high order of perfection in the model before any prediction and useful application was attempted. This section of the discussion had at least revealed that where known emissions of known materials were entering the environment, monitoring programmes could be designed and there was knowledge of at least some organisms (not necessarily the most important or the most sensitive) whose responses could be interpreted as a result of combined field observation and laboratory experiment.

BACKGROUND SURVEILLANCE

In opening this section of the discussion **Dr Holdgate** contrasted the specific monitoring of effects around identified emissions with the broader surveillance of wide areas of environment to detect change which might result from the influence of pollution at low level and might be the first indication that some substance was causing unforeseen problems. In our present state of ignorance it was essential that the specific monitoring of known emissions should be supplemented by this wider type of observation. The value of such studies had been demonstrated by the Irish Sea seabird wreck in 1969 which drew attention to the problems posed by polychlorinated biphenyls in the marine environment. It had also demonstrated another feature of this kind of monitoring: the detection of change may be relatively straightforward and sometimes result from some catastrophic event, but the detection of cause needed a substantial follow-up programme of laboratory study.

Seabirds have been particularly used as the subject of monitoring programmes of this kind, and Dr Holdgate pointed out that two participants in the meeting had come specifically to cover ornithological matters. The discussion would commence by hearing from them.

Monitoring of Seabird Populations

Miss C. Lloyd (Royal Society for the Protection of Birds) reported that a census of breeding seabirds is organised annually by the Seabird Group and the Royal Society for the Protection of Birds. The aims of this are twofold: it enables a short-term monitoring of immediate effects of increased mortality, for example due to oil spills, and also provides a method of detecting long-term population trends. The seabirds in which these organisations are interested are all cliff-nesters: the three commoner species of auk (Razorbill, Guillemot and Puffin) which are most frequently affected by oil, and the Kittiwake and Fulmar. A widespread sample of colonies is necessary for this survey and the RSPB and the Seabird Group try to cover sites throughout Britain and Ireland. All counters are volunteers and vary in experience; the only requirement for taking part in the survey is the ability to visit a colony year after year and to count the birds according to instructions.

Counting units are extremely important. Kittiwakes build conspicuous nests and Fulmars occupy nest sites which are also fairly obvious. A census of these species is quite straightforward, as apparently occupied nest sites can be counted with relative ease. Censuses of the Razorbill and Guillemot, which build no nests, involve counts of individual birds. These can cause trouble as the birds often nest at very high densities and the interpretation of the head counts obtained is most important. The third species of auk, the Puffin, is a burrow-nester. In this species an accurate census can be obtained only by counting occupied burrows. This can be done by sampling a small area within the larger occupied area of the colony.

The main problem with Razorbill and Guillemot counts is that large variations in colony attendance occur during the breeding season. A number of the study colonies are manned throughout the summer by wardens, though many of the other colonies are visited only once or twice each year. Valuable information has been collected by RSPB and bird observatory wardens who were able to count a study colony daily throughout the season. On the basis of the data collected for several years, patterns of attendance have been drawn up and these play a vital part in monitoring seabird populations.

For example, Guillemot numbers on the ledges in a breeding colony vary widely from day to day at the beginning of the season. The variation in attendance becomes progressively smaller as the season advances. Numbers are least variable and, hence, counts are most accurate during the brief nestling period, when the young birds are on the ledges. This is therefore the optimum time for census counts. The gradual build-up in overall numbers with the season is due (as colour ringing studies have shown) to the return of young immature non-breeding birds to the colony. The most stable period of attendance for these and other non-breeding birds also coincides with the nestling period.

By looking in detail at the nestling period counts, the expected error of different numbers of census counts can be estimated (see table). Quite clearly, very high errors are to be expected if censuses are based on counts carried out on a single visit to the colony. This is a problem peculiar to the two cliff-nesting species for which

Estimated Percentage Error (95% Confidence Limits) of Means of Different Numbers of Counts in 1973 at Handa, Sutherland

	Razorbill			*Guillemot*		
Timing of count	*1 count*	*5 counts*	*10 counts*	*1 count*	*5 counts*	*10 counts*
Nestling period	43·9	19·5	13·7	19·2	8·6	6·0
Month of June	40	18·1	12·7	20·2	9·0	6·3

head counts are necessary (Razorbill and Guillemot). For example, a single count of Razorbills can give a figure up to 44% away from the true value of the number of birds occupying the colony (as indicated by the average daily count). An acceptable error of about 10% in Guillemot counts can usually be obtained by carrying out five daily counts at the colony. Fortunately errors for counts conducted any time during the month of June indicate levels of accuracy similar to those confined to the nestling period. This may prove more convenient for the amateur counters contributing nationally to a scheme of this kind.

Attendance also varies throughout the day. Once again census counts must be timed to coincide with the period of the day when numbers are least variable. This is especially important for species where individual birds rather than nest sites are used as a counting unit. It is therefore strongly recommended that census counts take place well before midday, preferably between about 8 a.m. and 10 a.m.

The Seabird Group is also responsible for a monitoring programme at tern colonies. Information on numbers of birds breeding in the main colonies throughout Britain and Ireland is collected annually, though no species census technique is necessary. For the rarer species, the Roseate, Sandwich and Little Terns, coverage is almost complete. This survey is of special importance in relation to onshore oil developments as construction sites tend to turn up in the same sort of habitat as breeding terns. Loss of breeding habitat and disturbance are probably the major factors influencing the distribution and numbers of terns.

Britain and Ireland's tern population is of considerable international importance. The Roseate Tern breeds at only twelve colonies, all of which are on reserves or wardened areas, and numbers about 1400 pairs. 60% of the population in Britain and Ireland breed on a single colony in south-east Ireland. Elsewhere in northern Europe there are only about 500 pairs of Roseate Terns breeding in Brittany.

Since the drastic decline in Sandwich Terns in the Netherlands (from 40 000 pairs in 1940–57 to 650 pairs in 1967) following pesticide poisoning, the British and Irish population of some 12 500 pairs has also been of considerable international importance. These birds are especially vulnerable to disturbance, notably early in the season. Colour marking studies have shown large-scale movements between breeding colonies usually associated with disturbance.

The rarest species is the Little Tern which breeds in small, widely scattered colonies, most vulnerable to shore-based oil developments. Thus, protection not only from casual disturbance and egg-collecting, but also from loss of nesting habitat, is essential if numbers are to be maintained.

Miss Lloyd went on to say that participants in the beached bird survey are asked to score the amount of oil they find on the beaches as well as the number of dead birds. For this, a four-point scale is used which is 'no oil', two degrees of 'slight oiling', and 'heavy oiling', mainly according to the coverage of the oil on the beach.

Dr W. R. P. Bourne said that there are two approaches by which you can assess an animal: its breeding population and success and its mortality rate and causes.

Bird mortality is about the most conspicuously obvious effect of oil pollution. It was the first one which attracted general public attention and it is the one which invariably attracts public concern whenever there is a major spill.

The interpretation of ornithological damage is not easy, because if an oiled bird is still alive it is liable to swim ahead of the oil and either come ashore because it is getting cold or, alternatively, go where it originally intended. If it is killed it drifts at 2% of the wind speed, whereas the oil drifts at 3%, so the bird arrives behind the oil and is found too late to give warning of the oil's movements or future threats.

For 50 years now, the Royal Society for the Protection of Birds has been collecting information on the more conspicuous ornithological disasters due to oil (*Marine Pollution Bulletin*, **6**, 77–80). Originally this was merely a question of collecting anecdotes and presenting them to Parliament until they passed legislation. In the mid-1960s it became obvious that the Society needed to have a clearer idea of precisely what the effect was, and the factors causing damage, so regular surveys were started five times a year during the winter months when the damage is worst. The late February count coincides with big counts organised by youth groups on the Continent of Europe, so that surveys from much of the west coast of Europe are received. In the British Isles, somewhere between 1000 and 2500 km are covered during the main weekend surveys by something over 500 observers, which gives a fairly good indication of the amount of oil about and whether many oiled birds are coming ashore.

If there are major incidents, these are then followed up individually. This usually results in a good deal of bad publicity, and it has been noticeable over the last five years or so that this has been followed by a perceptible reduction in the number of oiled birds found in the north-west of Europe.

The auks—guillemot, razorbill and puffin—are the main sufferers. Sea duck suffer very severely on the Continent. Large numbers of oiled gulls are recorded but fewer of them die. It is obviously the swimming species that get into trouble. Evidence for the impact of oil pollution is provided by the high proportion of bird bodies that are oiled, in particular when they are washed up in large numbers, and variations in the incidence of oiling from place to place. The groups of which the largest numbers of bodies are found are the ones which have the highest proportion oiled, and the areas where most bird bodies are found are the areas where there is most oil. This provides a running indication of the amount of damage that is caused. It has the great advantage that practically anybody can walk along a beach and make some sort of record of the number of oiled birds present, whereas it requires a certain amount of knowledge and skill to assess what happens to limpets.

A point which has become increasingly obvious is the seasonal variation in the ornithological damage. There is a very marked peak in the winter, especially the late winter, and a very marked trough in the summer, which is paralleled by the incidence of oiling on the bodies and on beaches. At all seasons oiling is very much more prominent in the north of Europe. When all the puffins went missing in the winter it was thought that they might possibly have gone to the south of Europe. Trial surveys round about the Straits of Gibraltar, on the Balearic Islands and Madeira were made. Only eight birds in 100 km were found, which is a very low level by north European standards. Not a single one of them was oiled, although there were very large quantities of tar-balls present.

A large proportion of the oiled birds found in the north occur along the Continental coasts, and there is also frequently an accumulation in the north-east Irish Sea. This can be attributed to the prevailing westerly winds which cause the bodies to drift east and land on the west-facing coasts close to the main shipping lanes. It is also noticeable in winters with a predominance of easterly winds that more are found on the east coast of Britain. By making surveys on both coasts of the North Sea, one gets some indication of what is happening in every year.

Dr Bourne went on to mention a few critical situations where oil pollution incidents would be particularly disastrous. At the moment, Seafield Sewer in the Edinburgh area discharges the untreated sewage from half a million people into the southern Firth of Forth. A minimum of about 15 000 duck winter off its outfall. At the present time there are intermittent incidents when oil comes down the sewer and kills birds along the south shore of the Firth. An oil terminal is about to be opened immediately upstream and it was imagined that any leak from that would drift straight along the south shore of the Firth, sweeping up the birds feeding off the sewer, until it reached the famous Creedy station on the Bass Rock. The Royal Society for the Protection of Birds is then going to be confronted with the problem of measuring the mortality with the prevailing westerly wind. The dead birds are likely to drift straight across the North Sea and the Society will have to calculate what proportion are washed up on the coast. It may be necessary to fall back on estimates of whether there is a decline in the duck flocks, which are hard to count. It is going to be extremely difficult to measure the mortality.

In Shetland there is a big accumulation of feeding birds off the east coast during most of the year. This is a rocky, indented shore and it is hard to estimate the birds' numbers. There is also a very big movement of feeding birds through the channels and past the mouth of the Yell Sound in the north, which will be vulnerable to any leaks from the oil terminals in Sullom Voe. In 1971 a small slick that was virtually undetectable on the sea passed Noss, which is halfway up the east coast of Shetland. On the comparatively small area of beach it was possible to investigate there were at least 1200 bodies found and it is not known how many more went on out to sea. The colonies are hard to count with an accuracy sufficient to detect the loss of 100 000 birds. If there was an east wind, they would drift straight to Norway. The Society has a very big problem in trying to estimate what the damage is going to be in this situation which involves a large proportion of the pelagic sea birds breeding in the North Atlantic.

Dr Holdgate pointed out that there was a further reason why seabirds were useful in monitoring. They accumulated certain residues, especially in their body fats. The surveillance of levels of polychlorinated biphenyls and persistent organochlorine pesticides in samples of seabird body fat and in seabird eggs had been continuing for many years and provided a valuable general indicator of levels of marine contamination by these substances.

Dr Holdgate asked for information about other programmes of general surveillance, for example of oceanic changes or changes in fish populations. **Dr A. G. Bourne** referred to the observations on seabirds. Tragic as they were, the losses to our seabird populations have provided a useful indicator. An oiled seabird may not in fact have been killed by the oiling, for there might be other causes which have gone undetected through being masked by the oil. The real cause should be looked for and the trophic level at which the causative agent is entering the ecosystem. Was the agent entering at the seabird level, the afflicted birds' immediate food species, or lower down in the webs or at more than one trophic level? The only reliable method of monitoring would be to detect depression in the energy transfer between trophic levels in the food web or impairment of an indicator organism's metabolism, and then to look for dangerous levels building up in that organism. It may turn out that a pollutant is entering a system at the level of the primary converters, in which case it could affect all the organisms in the system.

Dr I. C. White referred to research extending over 40 years by the then Oceanographic Laboratory at Edinburgh and now the Institute for Marine Environmental Research on the plankton of the North Atlantic. This has involved the towing of continuous plankton recorders and has yielded a great deal of information about the abundance of various species of planktonic organisms and their annual and longer-term changes. Quite substantial changes have been detected and the time of onset of the spring bloom has been found to differ quite markedly today from when the work began. The problem lies in establishing the cause of this change.

There are probably better statistics available for fish than for any other marine organisms. They have been collected for many species since the late 19th century. Again, the difficulty is that natural fluctuations have had an overriding influence. For example, the 1962–63 winter resulted in a very powerful year-class for some North Sea species, which has been the dominant influence for many subsequent years on the plaice fishery and has now carried right through the population. The other problem about fishery statistics, although they are regionally based, is that the fishing industry itself has a very great influence on the information. For many fisheries there is very good information about catch per unit of effort and about size distribution and these do point to some classic effects of overfishing in which the size of fishes in the fishery is decreasing even if the abundance is remaining more or less the same. The information is evidently useful and on the basis of it there appears to be no indication of a decline in fisheries that can be attributed to pollution. Dr White admitted that this was to some extent a contentious statement but might merely mean that any such effect was masked by the greater influence of natural factors and the impact of fisheries.

Surveys of Oil Pollution on Beaches

Mr O'Sullivan said that the Advisory Committee on Oil Pollution of the Sea has for several years been carrying out a survey of the amount and severity of oil on beaches throughout England and Wales and in 1974 the survey was extended to Scotland. Questionnaires are sent out to a range of authorities and organisations which are represented on the Advisory Committee on Oil Pollution; these include the British Ports Authorities, local government agencies, local authorities, hotels and resorts, HM Coastguard, sea fishery officers and regional water authorities. A map and a list of incidents is drawn up describing the state of beaches throughout the country in terms of type and severity of oil pollution. The results are compared with RSPB beached bird surveys in the ACOPS annual report, and are similar to those obtained by schoolchildren under the Advisory Centre for Education project. The schoolchildren's map was very similar in its outline of severe areas to the map that the Advisory Committee produces yearly.

Mr O'Sullivan pointed out that it is difficult to achieve some kind of quantitative scale suitable for use by people such as fishery officers and hotel owners.

In discussion, **Dr Baker** described how she organised primary schoolchildren to measure oil on beaches. A simple four-point abundance scale (rare, occasional, frequent, abundant) was used and photographs were supplied to the children so that they could obtain a visual idea of the meaning of these terms. The children enjoyed the work and the costs were minimal: she wondered whether it was necessary to expend manpower and money on the kind of survey the Advisory Committee had undertaken.

In reply **Mr O'Sullivan** said that the Advisory Committee acted by co-ordinating information which would probably be gathered anyway by fishery officers and coastguards. It received the information without charge. He did feel that the survey could be extended by more precise determination of the kinds of oil being washed ashore. This would demand the submission of samples for analysis.

Mr Jagger recalled an earlier survey conducted by the Institute of Petroleum in 1962 or 1963 in collaboration with the then Ministry of Housing and Local Government. It was conducted through local authorities who were supplied with questionnaires. It led to agreement by the Ministry that they should make a 50% grant to local authorities to clean up contaminated beaches. On a smaller scale, a complete record is kept of oil spills in places such as Milford Haven and Fawley to which all the major oil companies and port authorities contribute. This is a valuable indicator of the frequency and scale of spillages. The general point is that to be useful, surveys of this kind must have a purpose. The information needs to be pooled together by somebody and used in a way that has a feedback to policy.

Surveillance in Advance of Development

Dr Holdgate commented that in addition to the specific monitoring of the environment about points of emission and the general background monitoring of organisms over very wide areas, there was an intermediate situation where a major development was likely to affect a considerable area and a monitoring system was designed both to record the impact of its specific effluents and any associated changes in the general environment. He invited Mr N. C. Morgan of the Nature Conservancy Council to comment on the situation in Orkney and Shetland, which illustrated this kind of situation rather well.

Mr Morgan said that the Nature Conservancy Council had been concerned first with collecting baseline information both for Shetland and Orkney. In Shetland a detailed study had been carried out by the Institute of Terrestrial Ecology, partly under an NCC contract. This had included a general survey of the land, freshwater and marine environments and a more detailed survey of certain areas which were of outstanding importance for nature conservation. There had been a re-survey of the major seabird breeding colonies for which Shetland was of international importance. In addition, a symposium was held in Edinburgh early in 1974 bringing together specialists who had worked on various aspects of the Shetland environment in the past.

In Orkney the County Council, on the advice of their planning consultants and of the Nature Conservancy Council, commissioned the University of Dundee to undertake an environmental assessment of Scapa Flow at an early stage in the planning for oil-associated developments. This work was on the marine environment and included some hydrology. It was supplemented by information on the bird fauna from the Royal Society for the Protection of Birds. Again, the Nature Conservancy Council organised a symposium at the end of 1974 bringing together a range of specialists.

Having drawn together the baseline information, the next step had been to appraise the impact on the environment of the likely developments. By definition an impact study requires a detailed knowledge of the kind of industrial works proposed. In Orkney a pipeline will bring oil to a terminal on the island of Flotta from which it will be carried by tankers. The developers themselves have an obvious interest in appraising the likely impact of their developments, both to gain information to support their planning applications and subsequently to check on the effectiveness of their own environmental safeguards. The Occidental Oil Company accordingly placed contracts with consultants to study the impact of development on the landscape, land and marine ecology, hydrology and siting of ballast water pipeline outfalls on Flotta. Dundee University have also undertaken impact studies and are now commencing to select sites for monitoring, while the ornithological work has also been expanded. To obtain co-ordination and avoid unnecessary duplication in this complex field of activity the Nature Conservancy Council has exercised its advisory role on behalf of the County Council and has arranged liaison meetings between the many operators and agencies that are concerned.

In Shetland a different procedure has been followed for the appraisal of environmental impact. Plans for the running of the Sullom Voe port and terminal are being prepared by an organisation called the Sullom Voe Association composed of the Shetland Islands Council and the oil companies. In parallel a Sullom Voe Environmental Advisory Group has been established on which the oil companies, the Shetland Islands Council, the Nature Conservancy Council, the Countryside Commission, the Natural Environment Research Council and independent expert interests are represented. This group has been closely examining all the major developments related to the growth of the oil terminal and has suggested ways of ameliorating any harmful effects; it is also designing a monitoring programme.

In Mr Morgan's opinion, the most important lessons to be learned from both the Orkney and Shetland situation are first that very early consultation is needed between the developers, the local authorities and specialist environmental organisations before too many options are closed. Second, there needs to be some form of co-ordination between all the agencies involved in the environmental impact assessment and monitoring exercises.

Dr Cook in commenting on Mr Morgan's presentation, drew a distinction between ecological and environmental impact assessments and baseline surveys. He placed the former in a political and the latter in a scientific arena. In Canada decisions tended to be made about where a development would be located and there was a subsequent rationalisation of the decision in scientific terms. Had environmental scientists been consulted earlier an alternative site might well have been suggested, but problems of land acquisition and land value prevented this type of open discussion. The decision also frequently reflected political factors such as the distribution of employment. Dr Cook considered that the most important work for scientists lay not in rationalising decisions already made but in conducting thorough baseline surveys. Such surveys should cover all trophic levels in the marine environment, combined with physical and chemical measurements. Sampling should be conducted with adequate frequency and should be as quantitative as possible. Studies concerning the distribution and strength of currents were vital in predicting the dispersion of effluent and its concentration in the environment.

Conclusion

In concluding the discussion on general programmes of surveillance, **Dr Holdgate** suggested that the first conclusion was once again that when establishing such a programme, selection was required either of the range of organisms or of the regions to be studied. Sometimes the region might be selected politically by a decision to develop a particular area. Once the choice has been made, the second step quite evidently was to conduct a baseline survey. This should examine all the trophic levels in the system and must have adequate physical and chemical measurements associated with it. Ideally the baseline survey should be carried on for long enough to reveal the natural fluctuations within the system: at least the annual fluctuations should be documented. Both in extensive programmes of general surveillance and in the monitoring of areas around points of development it was important to establish a hypothesis of cause and to have some hope of testing this. Around areas of development this would, among other things, demand the establishment of control sites carefully matched to those nearer the development itself. It was important so to design monitoring programmes that they did not themselves cause significant change.

CO-ORDINATION AND INTERCALIBRATION

The final section of the discussion was concerned with the co-ordination of monitoring programmes and the intercalibration of their methods and results.

Intercalibration

Mr Cowell expressed worry about the tendency to demand standard monitoring processes and techniques. He feared that as soon as such techniques were established they became fossilised in legislation and incapable of improvement by research.

He suggested it was preferable to adopt a code of practice in which people were supplied with guidelines concerning replication and sampling errors.

Dr Holdgate suggested that a basic core of synoptic observations common to most programmes might be adopted but extended by wider observations peculiar to each individual programme. The need to link the monitoring results from different areas of the world did demand at least knowledge of the margins of accuracy to be attached to the observations made in each programme. This had posed difficulties in the past: Dr Holden of DAFS, Pitlochry, had chaired a working party sponsored by the Organisation for Economic Co-operation and Development which had interchanged samples of material between a number of laboratories analysing for organochlorine residues. The results obtained varied by as much as 30% although the material supplied was identical. This kind of check on the accuracy of laboratories and stimulus to improvement was clearly important.

Dr W. R. P. Bourne said that one of the problems with organochlorine results was that some laboratories presented their analyses as lipid levels while some were on a wet weight and some a dry weight basis, and unless the percentage of lipid in the sample was quoted it was completely impossible to compare the series. It was essential that the results published were comparable the one with the other. The other need was for co-ordination so that there would not be a duplication of programmes wasting scarce resources.

Mr O'Sullivan felt that intercalibration should not be too difficult for chemical monitoring, but the system of distributing standard or split samples was clearly impossible in the biological field. There the only possible way of ensuring intercalibration would be for laboratories to exchange workers so that each learned the techniques of the others.

Dr Scarratt considered that one collaborative exercise might be in the field of development of predictive models and the construction of monitoring experiments to test them.

Organisation

Dr White drew attention to the Marine Pollution Monitoring Management Group chaired by Dr A. Preston of MAFF. This was concerned with the co-ordination of the various schemes for monitoring the marine environment around the United Kingdom. It would not confine itself to data collected by MAFF, although that Ministry would clearly be a major contributor as it carried out regular, often twice-yearly, samples of fish and shellfish which were analysed for metals and pesticides. Regional Water Authorities and the River Purification Boards in Scotland were supplying information about the input of these and other materials from land-based sources. The whole point of the Marine Pollution Monitoring Management Group was to bring together all the available information from whatever source.

Dr Holdgate explained that this group was only one of a series established following the acceptance of a recent Government report on the Monitoring of the Environment in the United Kingdom. Other groups were concerned with land pollution, air pollution, freshwater pollution, effects of pollution on human health and biological monitoring in its widest sense. The aim of the various Management Groups which were co-ordinated by the Central Unit on Environmental Pollution of the Department of the Environment was to promote a harmonised and economical way of determining the changes occurring in the British environment, assessing the adequacy of control measures and deciding where new controls were needed. It was necessary to co-ordinate national activities especially because the European Communities and the United Nations were both moving towards linking national monitoring networks into regional or global schemes.

Mr Hiscock asked what had happened in this process of rationalisation to the

Working Party on Marine Biological Surveillance established by the Natural Environment Research Council.

Dr Holdgate said that the three NERC Working Parties reviewing biological surveillance in the terrestrial, freshwater and marine environments were all actively preparing their reports. These reports were likely to be considered by the NERC Council towards the end of 1975 and the conclusions would probably be published.

CONCLUSION OF THE SESSION

In concluding the session **Dr Holdgate** stressed the way in which it had demonstrated that monitoring and surveillance were inescapably linked to experimental science. It had become apparent that monitoring and surveillance schemes needed to be selective in what they measured and this selection depended upon an understanding of environmental systems. It was evident that the methods used needed to be capable of relationship one to another so that the data could be built into more widely applicable models. It was also apparent that monitoring schemes would often lead to hypotheses that needed to be tested by research. The meeting had not led to the construction of guidelines for any ideal scheme for monitoring the impact of oil and oil-related products on the environment, but it had not been designed to do so. Nonetheless, the Institute of Petroleum and the Field Studies Council might consider, in planning future meetings, the case for a specialist conference on surveillance and monitoring and the intercalibration and interpretation of the results. It was clear that there was going to be a rapid increase in demand for monitoring and surveillance schemes as offshore resources were brought into use and as coastal state controls over contiguous waters increased. The management of the world's oceans needed to be on a sound scientific footing, and to establish this a dialogue between scientists was essential.

In closing the meeting Dr Holdgate thanked all those who had participated in the discussion and paid special tribute to Dr Baker and her colleagues from the Oil Pollution Research Unit. The session ended with an expression of thanks to the Chairman from Mr Jagger.

Index